P9-CRP-176

Sustainable Construction

Sustainable Construction

Green Building Design and Delivery

Second Edition

Charles J. Kibert

JOHN WILEY & SONS, INC.

This book is printed on acid-free paper. ♾

Published by John Wiley & Sons, Inc., Hoboken, New Jersey
Published simultaneously in Canada

Anniversary Logo Design: Richard Pacifico

Library of Congress Cataloging-in-Publication Data:

Kibert, Charles J.
Sustainable construction : green building design and delivery / Charles J. Kibert.—2nd ed.
p. cm.
Includes bibliographical references and index.
ISBN 978-0-470-11421-6 (cloth)
1. Sustainable buildings—Design and construction. 2. Sustainable architecture. 3. Construction industry—Management. I. Title.
TH880.K53 2007
690—dc22 2007029990

Printed in the United States of America

10 9 8 7 6 5 4

For Charles, Nicole, and Alina,
and in memory of Howard T. Odum

Contents

Chapter 3

Green Building Assessment 55

Chapter 4

The Green Building Process 79

Chapter 5

Ecological Design 99

Part II

Green Building Systems 129

Chapter 6

Sustainable Sites and Landscaping 133

Foreword

According to the United Nations, there are 6.7 billion of us on Earth, a number that will rise, perhaps, to more than 9 billion by midcentury. Last year, the level of carbon in the atmosphere took its largest jump (three parts per million) since measurements began, and the evidence of climate change is now everywhere. The World Meteorological Organization estimates that 150,000 people died last year due to climate change–driven weather extremes—15,000 in France alone. *Nature* magazine, the British counterpart to *Science,* reports that 90 percent of large predatory fishes are now gone. We have good reason to believe that the health of the oceans is in a sharp downward spiral due to pollution, overfishing, and loss of estuaries. Forests are disappearing at record rates, along with many species of life. We are at or near the midpoint of oil extraction while world demand for oil is rising sharply. Worldwide, wealth continues to concentrate in the very upper echelons of a new class of superwealthy, while 2 billion live in conditions of absolute poverty. And presently there is no system of governance—national or global—adequate to the challenge of reversing these and related trends in a timely, orderly, and humane manner.

The challenge of the twenty-first century requires that we make a transition to a new order of things that can be sustained within the limits of natural systems. This will require rethinking old and threadbare assumptions about economics, governance, ethics, and security as parts of a revolution in how humankind is provisioned with food, energy, water, materials, and shelter. But the transition to something far better than what is now in prospect is not a technological problem as much as it is a matter of politics and leadership.

In this sense, perhaps the most promising developments in the transition to sustainability are now under way in the latter category, evident in the emergence of a robust movement in the design and building industry. With the leadership of many people, including Bob Berkebile, Bill Browning, Amory Lovins, Susan Maxman, Bill McDonough, Steven Strong, Adrian Tuluca, Sim van der Ryn, Steven Winter, and, not the least, the author of this book, the art and science of green building is rapidly gaining ground. These are practical visionaries, aware of the promise and perils that lie ahead. The results are evident in the growing number of buildings certified by the U.S. Green Building Council and the larger number now being designed to LEED standards. But this is far from a U.S. monopoly. High-performance buildings are becoming the norm throughout the world as the design professions are being transformed by a sense of ecological emergency, the awareness of new design possibilities based on the science of ecology, a revolution in materials and technology, and a more acute sense of the full costs of buildings, both to the owner and to the larger society.

Green, or high-performance, buildings are not, however, more clever adaptations of existing practice or merely the application of better intentions. They are rather like the transition from typewriters to notebook computers, both of which produce paper inscribed with symbols. One, however, requires a periodic ribbon change; the other, software and networking to do many other things as well. As described in this book, the process of green design requires "front-loading" the design process to regard buildings, surrounding landscapes, and entire communities as systems embedded in ecologies and cultures. This further requires that designers and clients see buildings and landscapes as systems, not as unrelated parts, and the design process as an integrated conversation involving programming intentions, locality, energy systems, controls, materials, water, and form. The transition from linear and piecemeal planning and design to systems design in a microcosm is indicative of the larger paradigmatic changes under way worldwide.

The barriers to green or systems design are often said to be economic—a matter of cost. But the evidence cited here and elsewhere indicates that high-performance buildings can be made to be cost-effective. For example, some of the costs associated with higher energy efficiency (tighter shell, more insulation, better glazing, daylighting, smarter landscaping, etc.) should be offset by lower costs to buy, operate, and eventually replace larger HVAC equipment. Second, better buildings are known to increase the comfort, the happiness, and, therefore, the productivity of those living and working in them, which is directly translated into increased profit and value. Third, there are collateral benefits that arise from better design that are no less real for being unpredictable and difficult to quantify. In other words, the real barriers to green design are not so much economic as they are those of imagination and design competence.

On both counts, Charles Kibert stands out as a leader in the field and a reliably competent guide through the complexities of delivering high-performance green buildings. As director of the Powell Center for Construction and Environment at the University of Florida, he is both a practitioner and a theorist. Few, if any, have done more to transform our sense of construction and buildings as ecological processes and practical good sense. There is a revolution in building design quickly gathering steam—a revolution that is manifest in buildings that function as ecologies existing in larger ecologies. The question is no longer whether we build with ecological intelligence, but how.

David W. Orr

Preface to the Second Edition

Several major changes to the first edition of this book are the result of providing more foundational material regarding the rationale for green building. A range of ethical arguments supporting sustainability and, by extension, sustainable construction is provided in Chapter 2, "Background." These are essential to any discussion of sustainability because at the end of the day, sustainability distills to what is the right thing to do. In addition to the well-known Precautionary Principle, a wide variety of other supporting arguments are provided, for example, the Reversibility Principle, Distributional Equity, and Protecting the Vulnerable. A much wider discussion of ecological design is provided in Chapter 5, "Ecological Design," particularly about future ecological design and the thinking that is likely to influence it, for example, biomimicry, cradle-to-cradle design, adaptive management, natural capitalism, and Factor 4/Factor 10. Although the USGBC LEED rating system remains the main driving force behind green building in the United States, other approaches, such as Green Globes, have begun to emerge as competitors, providing alternative paths and sorely needed competition. The chapters of this book on site and landscaping, water, energy, materials, and indoor environmental quality provide information about the structure of both LEED for New Construction (LEED-NC) 2009 and Green Globes v.1.

In this edition there are also contributions from several experts on key green building issues. In Chapter 3, "Green Building Assessment," Alex Zimmerman provides some innovative thinking on how to develop the next generation of LEED rating tools by basing them as much as possible on the laws of physics. Chapter 4, "The Green Building Process," now contains an extended discussion of the integrated design process, an important facet of green building design, the goal of which is to develop synergistic working relationships between project team members. Chapter 5, "Ecological Design," summarizes the perspective of Bill Reed on how to shift from sustainable design, which represents today's best practices, to a state where regenerative design is the norm. Rather than simply straddle the breakeven point by doing minimal damage, which sustainable design represents, regenerative design suggests restoration as the norm and as a more holistic relationship between humans and nature. Chapter 7, "Energy and Atmosphere," now contains an extended case study of an exceptional building energy system design for River Campus One, a medical facility of the Oregon Health and Science University in Portland, Oregon. This case study was kindly provided by Interface Engineering, Inc. Similarly in Chapter 9, "Closing Materials Loops," David Hobbs, President of InterfaceFLOR Commercial, Inc., details the company's journey in pursuit of meeting the mandate of Ray Anderson to become the world's first truly sustainable corporation by describing the evolution of the design of their key product, carpet tiles. In the final chapter, Chapter 14, "The Cutting Edge and Beyond," there are two new case studies that describe the key aspects of the San Francisco Federal Building, designed by Morphosis Architects, and the Forensic Science Center in Philadelphia, designed by the Croxton Collaborative. Chapter 14 also covers a new cutting-edge topic associated with green buildings, passive survivability.

In addition to the acknowledgments for the first edition, I would like to express my deepest appreciation to Randy Croxton (again) for allowing me to use the Forensic Science Center as a case study. Similarly, Tim Christ at Morphosis Architects provided enormous help in obtaining materials about the San Francisco Federal Building. Thanks to Interface Engineering for the contribution of the case study of River Campus One. Bill Reed and Alex Zimmerman made thoughtful and thought-provoking contributions about future aspects of green building. And thanks once again to Paul Drougas and the other Wiley editors for their assistance in guiding this update. For this edition, Donna Conte was the production editor and Helen Greenberg was the copy editor. I am grateful to Donna and Helen for their care and attention in developing a greatly improved revision. Finally, I would like to thank the reviewers of the Second Edition: Margot McDonald, Cal Poly, San Luis Obispo; Rives Taylor, University of Houston; and Carol Diggelman, Milwaukee School of Engineering.

Charles J. Kibert
Gainesville, Florida

Preface to the First Edition

The legend of how the chief executive officer of Interface, Inc., Ray Anderson, embraced sustainability is part of the lore of the U.S. green building movement. In a cross-country flight from Los Angeles to Atlanta in 1994, he read a book by Paul Hawken, *The Ecology of Commerce,* which opened his eyes to a new array of ethical responsibilities to people and the planet. I had a similar experience after attending the first conference of a new organization in March 1994, the U.S. Green Building Council (USGBC), at which Paul Hawken spoke. After hearing his presentation, I immediately bought and read *The Ecology of Commerce,* then invited Paul to speak at an international meeting I was organizing for November 1994, the First International Conference on Sustainable Construction in Tampa, Florida, at which he deepened the convictions of an international audience from 35 countries. Many others were similarly converted by Hawken's writings and because of an emerging array of greening activities across the United States and around the world, all occurring at remarkably the same time, in the early 1990s. In this same era, the American Institute of Architects formed the Committee on the Environment (COTE), the U.S. Green Building Council was founded, the first international green building conferences were held in England and Florida, and work began in earnest to develop the first tools for measuring the environmental and resource impacts of buildings. The American Society for Testing and Materials (ASTM) initiated work on standards to address green building design and construction, an effort that later evolved into the development of the USGBC's Leadership in Energy and Environmental Design (LEED) building assessment standard.

Green building is a broad subject, and for this book I decided to focus on those areas of greening that I had become more familiar with as a result of my professional work and research: green building as it applies to larger commercial, institutional buildings. I have attempted to lay a solid foundation of basic principles that the reader can use to test any of the myriad decisions that have to be made in designing and constructing a green building, from materials selection to considering the use of natural systems for wastewater processing. I have kept the number of case studies to an essential few because in my view the recitation of green building components and numerous pictures of green buildings are not ultimately of great use to the owners, designers, or builders. It is my reckoning that gaining an understanding of the problem and the best general approaches is of the greatest value.

This book is meant to enhance the collaboration espoused by the green building movement by providing a reasonably detailed overview of the entire process so that the participants will be able to learn about the role of all the actors in the project. It also is meant to serve as a reference for building owners and buyers of construction services to learn about a new emerging process that makes good economic sense and that also addresses the ethical dilemma of how to both modify the Earth and protect the environment. Finally, it is designed to give a detailed insight into the USGBC LEED building assessment standard.

As a participant in some of the early activities of the USGBC and the international effort to define sustainable construction, I began teaching classes on these concepts in 1995 at the University of Florida, a course with the title Principles of Sustainable Construction. At that time there was a definite lack of reference materials, textbooks, film, and other resources to support a course purporting to address environmentally responsible building. One of the purposes of this book is to serve as a textbook on green building that can be used by professors in colleges that have built environment departments. It is meant to be broad enough to cover the needs of faculty and students in architecture, engineering, landscape architecture, interior design, and construction management. A website (www.wiley.com/go/sustainableconstruction) is designed to accompany this book, and it is hoped that this site will evolve into a location where faculty post lectures, notes, syllabi, student projects, and project and other information to bring the spirit of collaboration so characteristic of the green building movement into academia, into the teaching of subject matter related to green building.

In writing this volume, I acknowledge the dominant position of the USGBC's LEED building assessment standards in defining contemporary green building in the United States. The USGBC and LEED have had arguably unprecedented success in developing and promoting a tool for both rating buildings as to their "greenness" as well as guiding building design. Since 1998, this movement has doubled in size each year by several measures, such as membership and floor area of buildings certified under LEED. If one were to project this growth rate into the future, by the end of the decade the majority of all commercial and institutional buildings would be created and built using LEED as a design document. But for all its virtues, LEED also has many flaws, and I have tried to point those out as well. I consider LEED to be, in essence, an addendum to the building code, and as with a building code, where if you were to build it any cheaper it would be illegal, if a project does not meet the minimum level of LEED certification, it should probably be considered unethical and immoral from the viewpoint of sustainability. The enormous success of LEED has also brought with it a number of problems. The USGBC is struggling to keep up with the large number of building registrations and consequent paperwork and documentation needed to maintain its integrity. From a single LEED product in 1998, LEED is now a suite of nine products, all of which must be synchronized to adhere to a core philosophy and a relatively standard approach. The original LEED standard has been through several evolutions and is now known as LEED 2009 for New Construction, or LEED-NC 2009. The jury is still out on whether this proliferation of products helps or hinders the green building movement. The good news is that there is a strong sense of collaboration among the members of the USGBC and the hundreds of volunteers participating in the development of new standards and the redesign of existing ones.

Other issues are the lack of science in many of the components of LEED, the inflexible weighting system for major areas such as energy efficiency versus water conservation, and overreliance on anecdotal information for justifying indoor environmental quality measures. Although the goal of LEED and other green building efforts is to minimize environmental impacts and resource consumption, it is very difficult to connect the measures that provide points to their environmental and resource benefits. The credits in LEED do not carry equal weight, and certain of the credits that may have great environmental benefits may be especially difficult to achieve. As a result, a certain level of gamesmanship is often used to attain a desired level of certification without consideration of environmental impacts or benefits. The good news is that LEED, in spite of its deficiencies, does in fact work surprisingly well, and its great success is evidence of its broad acceptance, relative ease of use, and acceptable costs.

As a final comment, I have tried to balance the theoretical and the practical in this book, and have clearly had just partial success. But if it does nothing else but force the reader toward deeper thinking on high-performance green buildings, rather than following the same worn path, then it will have achieved most of its intent.

Acknowledgments

In acknowledging the people who helped bring this book to reality, I would like to first recognize those who provided the inspiration and its philosophical foundation. Paul Hawken, who spoke at the first USGBC conference in 1994, and who continues to write about and inspire the sustainability movement, started the shift in my thinking. William McDonough spoke at the same conference, provided another nudge, and continues to inspire and push for deeper thinking in this arena. The late H.T. Odum, founder of systems ecology, and for many years a colleague of mine at the University of Florida, provided clarity about sustainability and what it really meant to humankind. David Orr, author and professor, articulated a vision for the movement and motivated the creation of the Lewis Environmental Center at Oberlin College as an outstanding physical example of the potential for green buildings. His writings have given me and many others the impetus to sustain the effort to facilitate change in how we design, build, and operate our human-built environment.

Several individuals provided support and inspiration during 10 years of trying to implement sustainability at the University of Florida. Gisela Bosch offered tremendous support for many years in organizing Greening the University of Florida, organized several conferences and workshops, and kept the whole effort moving forward. She also was the inspiration behind a course, Challenges of Sustainable Development in Poland, that she, Jan Sendzimir, and several Polish colleagues have taught each summer since 1998. Nicole Kibert, my daughter and now a successful attorney with Carlton Fields in Tampa, Florida, assisted many of the greening efforts, participated in research projects, and then introduced sustainability to the University of Florida Law School, where a new round of collaborations was established.

Brad Guy, whom I met as a graduate student at the University of Florida, was a deep source of ideas and inspiration for almost 10 years; he later took on a national and international leadership role in deconstruction, an effort that I believe is the key to success in the building materials arena. A number of other graduate students provided assistance and enthusiasm over time: Johnny Peng, Domenic Scorpio, James Baker, Jitendra Jain, Kyle Abney, Hal Knowles, Paul Shahriari, Jennifer Languell, and Donna Isaacs, to name just a few of many. More recently, James Sullivan, John Dryden, Luke Nicholson, and Bilge Celik, all doctoral students in the Rinker School, helped to further develop new concepts that are sorely needed to create more robust green buildings.

I must also acknowledge my many colleagues at the University of Florida who were instrumental in the evolution of many of the thoughts that appear in this volume: Abdol Chini, Kevin Grosskopf, Kwaku Tenah, Kim Tanzer, Tina Gurucharri, Helena Moussatche, Mark Brown, Martin Gold, Bob Buschbacher, and Glenn Acomb. I was also fortunate to have a number of forward-thinking colleagues on the operational side of the University of Florida who helped implement green buildings as standard practice, among them J.T. McCaffrey, Carol Walker, Bahar Armaghani, and Howie Ferguson.

Dave Newport played a critical role in promoting sustainability at the University of Florida and worked with me for many years, keeping greening efforts a priority and fostering the establishment of a Sustainability Task Force. I appreciate all his energy, enthusiasm, and friendship over the last few years.

A special thanks is owed to Michele Moretti for her invaluable efforts in proofreading this book and for creating and editing many of the photographs used as illustrations. Her many suggestions and fresh point of view helped to dramatically improve the delivery of its content.

I must also thank the publisher of *Environmental Building News,* Dan Woodbury, and its executive editor, Alex Wilson, the former for allowing me a great deal of latitude in using their publication as a source of materials for this book and the latter for many years of fruitful exchanges about green building.

Thanks to the reviewers of the draft manuscript of this book: Bruce Haglund of the University of Idaho, Mary Ellen Nobe of Colorado State University, and Walter Grondzik of Florida State University, for their input and suggestions. The following individuals reviewed the proposal for this book, and I am grateful for their general recommendation for its publication: Wayne Trusty, James A. Wise, Ph.D., Bion D. Howard, Robert Ries, Donald Fournier, and Tim Richard, AIA. Thanks to the editors at John Wiley & Sons, Inc., who assisted me in keeping this effort on track: the acquisitions editor, Paul Drougas, and his editorial assistant, Lauren LaFrance; the production editor, Shannon Egan; and the copy editor, Janice Borzendowski.

And last but certainly not least, thanks to my wife, Pat, and my younger daughter, Alina, for their months of patience during my long hours of focus on green building issues. Thanks also to my daughter Nicole and my son Charles for their encouragement and patience.

Charles J. Kibert
Gainesville, Florida

Chapter 1

Introduction and Overview

The high-performance green building movement is said to be the most successful environmental movement in the United States, certainly the fastest-growing and highly successful at creating partnerships with a broad cross section of manufacturers, builders, and others who are not often allies with environmentalists. In addition to having enormous success, the green building movement provides a model for other sectors of economic endeavor about how to create a consensus-based, market-driven program that has rapid uptake, not to mention broad impact. With respect to buildings, unprecedented forces are reshaping the building construction industry, forcing professionals engaged in all phases of building construction, design, operation, financing, insurance, and public policy to fundamentally rethink their roles in the building delivery process. The main impetus is the *sustainable development* movement, which is changing not only physical structures but also the workings of the companies and organizations that populate the built environment, as well as the hearts and minds of individuals who inhabit it.[1] Fueled by examples of personal and corporate irresponsibility and negative publicity resulting from events such as the Enron debacle and the Wall Street mutual fund and stock trading scandals, *accountability* and *transparency* are becoming the watchwords of today's corporate world. Heightened corporate consciousness has embraced comprehensive sustainability reporting as the new standard for corporate transparency. *Corporate transparency* refers to complete openness of companies about all financial transactions and all decisions that affect their employees and the communities in which they operate. Major companies such as DuPont, the Ford Motor Company, and the Hewlett-Packard Company now employ *triple bottom-line reporting,*[2] which refers to a corporate refocus from mere financial results to a more comprehensive standard that also includes environmental and social impacts. By including these cornerstone principles of sustainability in their annual reporting, corporations acknowledge their environmental and social impacts and ensure improvement in all arenas.

Still, other major forces such as *climate change* and the rapid depletion of the world's oil reserves threaten national economies and the quality of life in developed countries. Both are connected to our dependence on fossil fuels, especially oil. Climate change, caused at least in part by increasing concentrations of human-generated carbon dioxide, methane, and other gases in the Earth's atmosphere, is believed by many authoritative scientific institutions and Nobel laureates to profoundly affect our future temperature regimes and weather patterns.[3] Much of today's built environment will still exist during the coming era of rising temperatures and sea levels; however, little consideration has been given to how human activity and building construction should adapt to potentially significant climate alterations. Global temperature increases must now be considered when forming assumptions about passive design, the building envelope, materials selection, and the types of equipment required to cope with higher atmospheric energy levels.

The *oil rollover point* describes the time when peak worldwide production of oil will occur and when approximately 50 percent of the world's oil supply will have been depleted (see Figure 1.1).[4] At the rollover point, the energy value of oil (the amount of energy into which the oil can be converted) will be less than the energy needed to extract it. Experts predict that between 2010 and 2020, oil prices will skyrocket as production falls and demand begins to exceed supply, sending shock waves through a world economy predicated on growth subsidized by cheap energy. The Chinese economy officially grew 9.3 percent in 2005 with some estimates that it will continue at this rate and others stating that it will grow 9.5 percent in 2007. China produced about 2 million automobiles in 2000, tripling to about 6 million in 2005. China's burgeoning industries are in heavy competition with the United States and other major economies for oil and other key resources such as steel and cement. The combination of increasingly scarce supplies of oil, rapid economic growth in China and India, and concerns over the contribution of fossil fuel consumption to climate change will inevitably force the price of gasoline and other fossil-fuel-derived energy sources to increase rapidly in the coming decades. At present, there are no foreseeable technological substitutes for the world's rapidly depleting oil supplies. Alternatives such as hydrogen or fuels derived from coal and tar sands threaten to be prohibitively expensive. The expense of operating buildings that are heated and cooled using fuel oil and natural gas will likely increase, along with the cost of fossil-fuel-dependent industrial, commercial, and personal transportation. A shift toward hyperefficient buildings and transportation cannot begin soon enough.

A unique vocabulary is emerging to describe concepts related to sustainability and global environmental changes. Terms such as *Factor 4* and *Factor 10, ecological footprint, ecological rucksack, biomimicry, Natural Step, eco-efficiency, ecological economics, biophilia,* and the *Precautionary Principle* describe the overarching philosophical and scientific concepts that apply to a paradigm shift toward sustainability. Complementary terms such as *green building, building assessment, ecological design, life-cycle assessment, life-cycle costing, high-performance building,* and *charrette* articulate specific techniques in the assessment and application of principles of sustainability to the built environment.

The sustainable development movement has been evolving worldwide for almost two decades, causing significant changes in building delivery systems in a relatively short period of time. A subset of sustainable development, *sustainable construction,* addresses the role of the built environment in contributing to the overarching vision of sustainability. In the United States, the founding of the U.S. Green Building Council (USGBC) in 1993 heralded government and industry's newfound

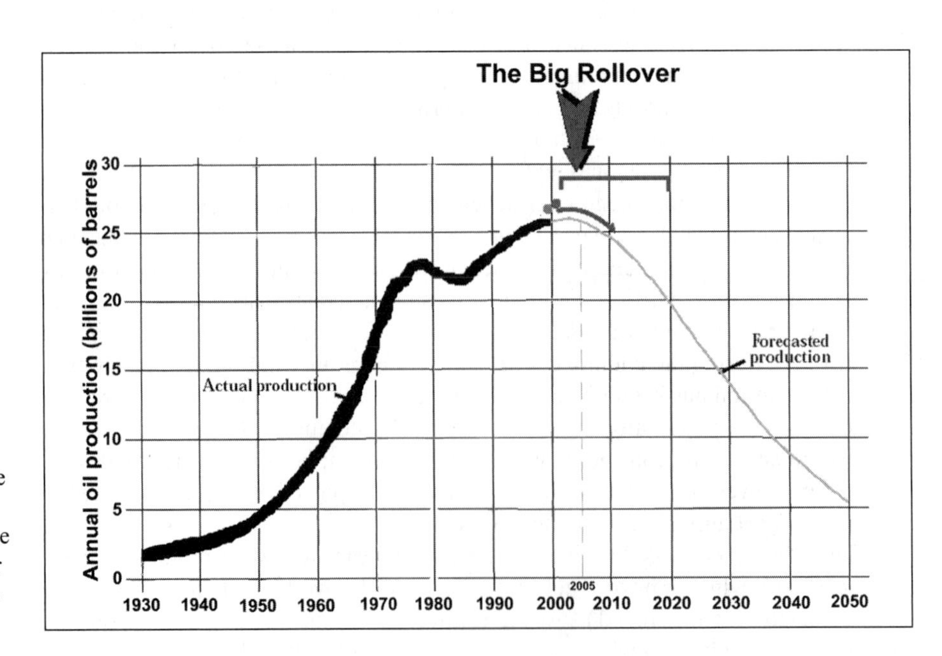

Figure 1.1 The oil rollover point is the year in which the worldwide production rate of oil peaks. Although there are varying points of view as to when this will occur, the probability is that it has already occurred or that it will occur in just a few years. (Drawing by Bilge Çelik.)

commitment to high-performance green building practices. From 1993 to 1998, a USGBC task force diligently developed a rating system to evaluate a building's resource efficiency and environmental impacts. This rating system, Leadership in Energy and Environmental Design (LEED), was the watershed event that precipitated an exponential shift from conventional to sustainable building delivery systems from 1998 on. The LEED rating system removed ambiguity in the loosely interpreted concepts associated with sustainability and green building. LEED's newly articulated, cohesive rating system rapidly gained wide acceptance in both the private and public sectors and has significantly impacted the construction industry in the most energy- and materials-intensive economy in the world. In recent years, the USGBC has developed other LEED rating systems including LEED for core and shell (LEED-CS), LEED for Commercial Interiors (LEED-CI), LEED for Schools, LEED for Homes (LEED-H), LEED for Retail, LEED for Neighborhood Development (LEED-ND), LEED for Existing Buildings—Operations and Maintenance (LEED-EB), and LEED for Healthcare. By the end of 2008, over 2100 buildings had been certified under one of these rating systems. LEED projects are expected to constitute about 10 percent of new construction in the United States by 2010.

An example of an exceptional LEED-certified building is illustrated in Figure 1.2A–C. The Robert Redford Building, the Santa Monica, California, office of the Natural Resources Defense Council, which opened in November 2003, received a Platinum certification from the USGBC, the highest of the four levels of certification, and was one of just a handful to receive this rating. The building is cooled largely by ocean breezes, uses about one-third of the energy of a typical office building in Santa Monica, obtains 100 percent of its energy from carbon-free renewable resources, and has a 7.5-kW solar array on the roof. An existing building on the site was deconstructed, and 98 percent of its materials were recycled into the new building. The building uses rainwater and graywater (recycled water from sinks) for flushing toilets and landscape irrigation. Waterless urinals in the building each save about 40,000 (151,000 liters) gallons of water a year by not requiring flushing, and the toilets have a dual-flush option, light or heavy, depending on the need. The building also boasts an exceptionally high level of indoor environmental quality due to its use of emissions-free materials and exceptional daylighting for its occupants.

Although other building assessment standards had been developed and implemented, LEED now predominates in the United States and has been wholly or partially adopted in several other countries. Spain, Canada, and China are all considering LEED-based approaches to green building. LEED's wide acceptance has likely resulted from its authors' focus on fashioning LEED as a consensus-based rating system and on creating buildings that would have higher market value. Studies show that green buildings command up to 6% higher rents and have higher occupancy levels than non-green buildings. One study by CoStar Group in early 2008 found that LEED buildings sell for $171 more per square foot than their counterparts, a trend that could signal greater attention from investors.

Another organization that has recently emerged as part of the growing green building movement is the Green Building Initiative (GBI). In cooperation with the National Association of Home Builders (NAHB), GBI has helped initiate more than 15 city and state-level green home building programs based on the NAHB's *Model Green Home Guidelines.* GBI's entrance into the market has had several other impacts on this movement. In 2005 GBI became the first organization to earn accreditation as a Standard Developer under the American National Standards Institute (ANSI). Following the trend in corporate America, GBI has led the green building movement toward a higher-profile commitment to consensus and transparency. Following GBI's lead, the USGBC also earned ANSI accreditation in 2006. In the same year, NAHB committed to take their green home guidelines through an ANSI consensus process, and organizations including the American Society for Heating,

(A)

(B)

(C)

Figure 1.2 The Robert Redford Building, offices of the Natural Resources Defense Council (NRDC) in Santa Monica, California, is one of the first Platinum-certified buildings under the USGBC's LEED building assessment standard. (A) The front elevation shows its urban setting and its close connection to the street and the adjoining farmer's market; (B) rear elevation; (C) interior second-floor lightwell. (Photographs courtesy of the NRDC.)

Refrigeration and Air Conditioning Engineers (ASHRAE), National Institutes of Building Sciences (NIBS), and the American Society for Testing and Materials (ASTM) have all announced their intention to promulgate high performance or sustainable building standards. GBI's Green Globes system, originally developed in Canada and one of the newcomers to contribute a commercial building rating systems to the market, is undergoing a technical review by GBI's ANSI technical committee. GBI expects to promulgate Green Globes as an ANSI standard in 2008.

Organization

This book describes the *high-performance green building delivery system,* a rapidly emerging building delivery system that satisfies the owner while addressing sustainability considerations of economic, environmental, and social impact, from design through the end of the building's life cycle. A *building delivery system* is the process used by building owners to ensure that a facility meeting their specific needs is designed, built, and handed over for operation in a cost-effective manner. This book will examine the design and construction of state-of-the art green buildings in the United States, considering the nation's unique design and building traditions, products, services, building codes, and other characteristics. Best practices, technologies, and approaches of other countries will be used to illustrate alternative techniques. Although intended primarily for a U.S. audience, the general approaches described could apply broadly to green building efforts worldwide.

Much more so than in conventional construction delivery systems, the high-performance green building delivery system requires close collaboration among building owners, developers, architects, engineers, constructors, facility managers, building code officials, bankers, and real estate professionals. New certification systems with unique requirements must be considered. This book will focus largely on practical solutions to the regulatory and logistical challenges posed in implementing sustainable construction principles, delving into background and theory as needed. The USGBC's green building certification program will be covered in detail. Other complementary or alternative standards such as the Green Building Initiative's Green Globes building assessment system, the U.S. government's Energy Star Program, and the United Kingdom's BREEAM building certification program will be discussed. Economic analysis and the application of life-cycle costing, which provides a more comprehensive assessment of the economic benefits of green construction, will also be considered.

Following this introduction, this book is organized in three parts, each of which describes an aspect of this emerging building delivery system. Part I, "Green Building Foundations," covers the background and history of green buildings, the most significant rating and assessment systems, and green building design. Part II, "Green Building Systems," more closely examines several important subsystems of green buildings: siting and landscaping, energy and atmosphere, the building hydrologic cycle, materials selection, and indoor environmental quality. In Part III, "Green Building Implementation," the subjects of construction operations, building commissioning, economic issues, and future directions of sustainable construction are addressed. Additionally, five appendixes containing supplemental information on key concepts are provided. To support the readers, a website, www.wiley.com/go/sustainableconstruction, contains hyperlinks to relevant organizations, references, and resources. This website also references supplemental materials, lectures, and other information suitable for use in university courses on sustainable construction.

Rationale for High-Performance Green Buildings

High-performance green buildings marry the best features of conventional construction methods with emerging high-performance approaches. Green buildings are achieving rapid penetration in the U.S. construction market for three primary reasons:

1. Sustainable construction techniques provide an ethical and practical response to issues of environmental impact and resource consumption. Sustainability assumptions encompass the entire life cycle of the building and its constituent components, from resource extraction through disposal at the end of the materials' useful life. Conditions and processes in factories are considered, along with the actual performance of their manufactured products in the completed building. High-performance green building design relies on renewable resources for energy systems; recycling and reuse of water and materials; integration of native and adapted

species for landscaping; passive heating, cooling, and ventilation; and other approaches that minimize environmental impact and resource consumption.

2. Green buildings virtually always make economic sense on a life-cycle cost (LCC) basis, though they may be more expensive on a capital, or first-cost, basis. Sophisticated energy-conserving lighting and air-conditioning systems with an exceptional response to interior and exterior climates will cost more than their conventional, code-compliant counterparts. Rainwater harvesting systems that collect and store rainwater for nonpotable uses will require additional piping, pumps, controls, storage tanks, and filtration components. However, most key green building systems will recoup their original investment within a relatively short time. As energy and water prices rise due to increasing demand and diminishing supply, the payback period will decrease. LCC provides a consistent framework for determining the true economic advantage of these alternative systems by evaluating their performance over the course of a building's useful life.[5]
3. Sustainable design acknowledges the potential effect of the building, including its operation, on the health of its human occupants. A 1984 World Health Organization report suggested that as many as 30 percent of new and remodeled buildings worldwide may generate excessive complaints related to indoor air quality.[6] Estimates peg the direct and indirect costs of building-related illnesses, including lost worker productivity, as exceeding $150 billion per year.[7] Conventional construction methods have traditionally paid little attention to sick building syndrome (SBS), building-related illness (BRI), and multiple chemical sensitivity (MCS) until prompted by lawsuits. In contrast, green buildings are designed to promote occupant health, including measures such as protecting ductwork during installation to avoid contamination during construction; specifying finishes with low to zero volatile organic components to prevent potentially hazardous chemical off-gassing; more precise sizing of heating and cooling components to promote dehumidification, thereby reducing mold; and the use of ultraviolet radiation to kill mold and bacteria in ventilation systems.[8]

Defining Sustainable Construction

The terms *high performance, green,* and *sustainable construction* are often used interchangeably; however, the term *sustainable construction* most comprehensively addresses the ecological, social, and economic issues of a building in the context of its community. In 1994, the Conseil International du Batiment (CIB), an international construction research networking organization, defined the goal of sustainable construction as ". . . creating and operating a healthy built environment based on resource efficiency and ecological design."[9] The CIB articulated seven Principles of Sustainable Construction, which would ideally inform decision making during each phase of the design and construction process, continuing throughout the building's entire life cycle (see Table 1.1).[10] These factors also apply when evaluating the components and other resources needed for construction (see Figure 1.3). The Principles of Sustainable Construction apply across the entire life cycle of construction, from planning to disposal (here referred to as *deconstruction* rather than *demolition*). Furthermore, the principles apply to the resources needed to create and operate the built environment during its entire life cycle: land, materials, water, energy, and ecosystems.

TABLE 1.1

The Principles of Sustainable Construction

1. Reduce resource consumption (reduce).
2. Reuse resources (reuse).
3. Use recyclable resources (recycle).
4. Protect nature (nature).
5. Eliminate toxics (toxics).
6. Apply life-cycle costing (economics).
7. Focus on quality (quality).

RESOURCE-CONSCIOUS DESIGN

The issue of resource-conscious design is central to sustainable construction, which ultimately aims to minimize natural resource consumption and the resulting impact

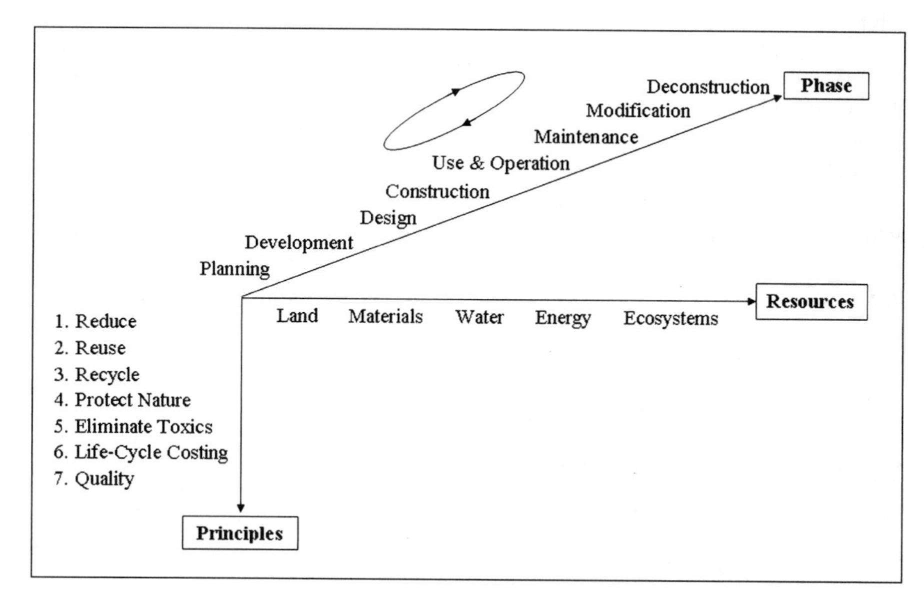

Figure 1.3 Framework for sustainable construction developed in 1994 by Task Group 16 (Sustainable Construction) of the CIB for the purpose of articulating the potential contribution of the built environment to the attainment of sustainable development. (Drawing by Bilge Çelik.)

on ecological systems. Sustainable construction considers the role and potential interface of ecosystems to provide services in a synergistic fashion. With respect to materials selection, closing materials loops and eliminating solid, liquid, and gaseous emissions are key sustainability objectives. *Closed loop* describes a process of keeping materials in productive use by reuse and recycling rather than disposing of them as waste at the end of the product or building life cycle. Products in closed loops are easily disassembled, and the constituent materials are capable and worthy of recycling. Because recycling is not entirely thermodynamically efficient, dissipation of residue into the biosphere is inevitable. Thus, the recycled materials must be inherently nontoxic to biological systems. Most common construction materials are not completely recyclable, but rather *downcyclable,* for lower-value reuse such as for fill or road subbase. Fortunately, aggregates, concrete, fill dirt, block, brick, mortar, tiles, terrazzo, and similar low-technology materials are composed of inert substances with low ecological toxicity. In the United States, the 140 million tons (127 million metric tons) of construction and demolition waste produced annually comprise about one-third of the total solid waste stream, consuming scarce landfill space, threatening water supplies, and driving up the costs of construction. As part of the green building delivery system, manufactured products are evaluated for their life-cycle impacts, to include energy consumption and emissions during resource extraction, transportation, product manufacturing, installation during construction, operational impacts, and the effects of disposal.

LAND RESOURCES

Sustainable land use is based upon the principle that land, particularly undeveloped, natural, or agricultural land (*greenfields*), is a precious finite resource, and its development should be minimized. Effective planning is essential to creating efficient urban forms and minimizing urban sprawl, which leads to overdependence on automobiles for transportation, excessive fossil fuel consumption, and higher pollution levels. Like other resources, land is recyclable and should be restored to productive use whenever possible. Recycling disturbed land such as former industrial zones (*brownfields*) and blighted urban areas (*grayfields*) back to productive use facilitates land conservation and promotes economic and social revitalization in distressed areas.

ENERGY AND ATMOSPHERE

Energy conservation is best addressed through effective building design, which integrates three general approaches: (1) designing a building envelope that is highly resistant to conductive, convective, and radiative heat transfer; (2) employing renewable energy resources; and (3) fully implementing passive design. *Passive design* employs the building's geometry, orientation, and mass to condition the structure using natural and climatological features such as the site's solar insolation,[11] thermal chimney effects, prevailing winds, local topography, microclimate, and landscaping. Since 30 percent of domestic primary energy[12] is consumed by buildings in the United States, increased energy efficiency and a shift to renewable energy sources can appreciably reduce carbon dioxide emissions and mitigate climate change.

WATER ISSUES

The availability of potable water is the limiting factor for development and construction in many areas of the world. In the high-growth Sun Belt and western regions of the United States, the demand for water threatens to rapidly outstrip the natural supply, even in normal, nondrought conditions.[13] Climate alterations and erratic weather patterns precipitated by global warming threaten to further limit the availability of this most precious resource. Since only a small portion of the Earth's hydrological cycle yields potable water, protection of existing ground and surface water supplies is increasingly critical. Once water is contaminated, it is extremely difficult, if not impossible, to reverse the damage. Water conservation techniques include the use of low-flow plumbing fixtures, water recycling, rainwater harvesting, and *xeriscaping,* a landscaping method that utilizes drought-resistant plants and resource-conserving techniques.[14] Innovative approaches to wastewater processing and stormwater management are also necessary to address the full scope of the building hydrologic cycle.

Figure 1.4 The Lewis Environmental Center at Oberlin College was designed by a team of top designers, led by William McDonough, a leading green building architect, and including John Todd, developer of the Living Machine. In addition to the superb design of the building's hydrologic system, the extensive photovoltaic system makes it a net exporter of energy. (Photograph courtesy of Oberlin College.)

ECOSYSTEMS: THE FORGOTTEN RESOURCE

Sustainable construction considers the role and potential interface of ecosystems in providing services in a synergistic fashion. Integration of ecosystems with the built environment can play an important role in resource-conscious design. Such integration can supplant conventional manufactured systems and complex technologies in controlling external building loads, processing waste, absorbing stormwater, growing food, and providing natural beauty, sometimes referred to as *environmental amenity*. For example, the Lewis Environmental Center at Oberlin College in Oberlin, Ohio, uses a built-in natural system, referred to as a "Living Machine," to break down waste from the building's occupants; the effluent then flows into a reconstructed wetland (see Figure 1.4). The wetland also functions as a stormwater retention system, allowing pulses of stormwater to be stored, reducing the burden on stormwater infrastructure. The restored wetland also provides environmental amenity in the form of native Ohio plants and wildlife.[15]

DEFINING *GREEN BUILDING*

The term *green building* refers to the quality and characteristics of the actual structure created using the principles and methodologies of sustainable construction. Green buildings can be defined as "healthy facilities designed and built in a resource-efficient manner, using ecologically based principles." Similarly, *ecological design, ecologically sustainable design,* and *green design* are terms that describe the application of sustainability principles to building design. Despite the prevalent use of these terms, truly sustainable green commercial buildings with renewable energy systems, closed materials loops, and full integration into the landscape are rare to nonexistent. Most existing green buildings feature incremental improvement over, rather than radical departure from, traditional construction methods. Nonetheless, this process of trial and error, along with the gradual incorporation of sustainability principles, continues to advance the industry's evolution toward the ultimate goal of achieving complete sustainability throughout all phases of the built environment's life cycle.

HIGH-PERFORMANCE BUILDINGS, WHOLE BUILDING DESIGN, AND SYSTEMS THINKING

The term *high-performance building* has recently become popular as a synonym for green building in the United States. According to the U.S. Office of Energy Efficiency and Renewable Energy (EERE), a high-performance commercial building ". . . uses *whole-building design* to achieve energy, economic, and environmental performance that is substantially better than standard practice."[16] This requires that the design team fully collaborate from the project's inception in a process often referred to as *integrated design*.

Whole building, or integrated, design considers site, energy, materials, indoor air quality, acoustics, and natural resources, as well as their interrelation with one another. In this process, a collaborative team of architects, engineers, building occupants, owners, and specialists in indoor air quality, materials, and energy and water efficiency utilizes *systems thinking* to consider the building structure and systems holistically, examining how they best work together to save energy and reduce the environmental impact. A common example of systems thinking is *advanced daylighting strategy,* which reduces the use of lighting fixtures during daylight, thereby reducing daytime peak cooling loads and justifying a reduction in the size of the mechanical cooling system. This, in turn, results in reduced capital outlay and lower energy costs over the building's life cycle.

According to the Rocky Mountain Institute (RMI), a well-respected nonprofit organization specializing in energy and building issues, *whole-systems thinking* is a process through which the interconnections between systems are actively considered and solutions are sought that address multiple problems. Whole-systems thinking is often promoted as a cost-saving technique that allows additional capital to be invested in new building technology or systems. RMI cites developer Michael Corbett, who applied just such a concept in his 240-unit Village Homes subdivision in Davis, California, completed in 1981. Village Homes was one of the first modern-era developments to successfully create an environmentally sensitive, human-scale residential community. The result of designing narrower streets was reduced stormwater runoff. Simple infiltration swales and on-site detention basins handled stormwater without the need for conventional stormwater infrastructure. The resulting $200,000 in savings was used to construct public parks, walkways, gardens, and other amenities that improved the quality of the community. A more recent example of systems thinking is Solaire, a 27-story luxury residential tower in New York City's Battery Park (see Figure 1.5). The façade of Solaire contains photovoltaic cells that convert sunlight directly into electricity, and the building itself uses 35 percent less energy than a comparable residential building. It provides its residents with abundant natural light and excellent indoor air quality. The building collects rainwater in a basement tank for watering roof gardens. Wastewater is processed for reuse in the air-conditioning system's cooling towers or for flushing toilets. The roof gardens not only provide a beautiful urban landscape, but also assist in insulating the building to reduce heating and cooling loads. This interconnection of many of the green building measures in Solaire indicates that the project team carefully selected approaches that would have multiple layers of benefit, the core of systems thinking.[17]

STATE AND LOCAL GUIDELINES FOR HIGH-PERFORMANCE CONSTRUCTION

Several states have taken the initiative in articulating guidelines aimed at facilitating high-performance construction. The Pennsylvania Governor's Green Government Council (GGGC) uses mixed but very appropriate terminology in its "Guidelines for Creating High-Performance Green Buildings: A Document for Decision Makers" (1999). The lengthy but instructive definition of high-performance green building (Table 1.2) focuses as much on the collaborative involvement of the stakeholders as it does on the physical specifications of the structure itself.[18]

Similar guidance is provided by the City of New York Department of Design and Construction in its "High Performance Building Guidelines," in which the end product, the building, is hardly mentioned and the emphasis is on the strong collaboration of the participants (see Table 1.3).[19]

The "High Performance Guidelines: Triangle Region Public Facilities," published by the Triangle J Council of Governments in North Carolina (1999), focuses on three principles:

- *Sustainability,* which is a long-term view that balances economics, equity, and environmental impacts
- *An integrated approach,* which engages a multidisciplinary team at the outset of a project to work collaboratively throughout
- *Feedback and data collection,* which quantifies both the finished facility and the process that created it and serves to generate improvements in future projects

Like the other state guidelines, North Carolina's "High Performance Guidelines" emphasize collaboration and process, rather than merely the physical characteristics of the completed building. Historically, building owners have assumed that they were

Figure 1.5 Solaire, a 27-story residential tower on the Hudson River in New York City, which opened in July 2003, is the first high-rise residential building in the United States specifically designed to be environmentally responsible. (Photograph courtesy of the Albanese Development Corporation.)

TABLE 1.2

High-Performance Green Building as Defined by GGGC

- A project created via cooperation among building owners, facility managers, users, designers and construction professionals through a collaborative team approach.
- A project that engages the local and regional communities in all stages of the process, including design, construction, and occupancy.
- A project that conceptualizes a number of systems that, when integrated, can bring efficiencies to mechanical operation and human performance.
- A project that considers the true costs of a building's impact on the local and regional environment.
- A project that considers the life-cycle costs of a product or system. These are costs associated with its manufacture, operation, maintenance, and disposal.
- A building that creates opportunities for interaction with the natural environment and defers to contextual issues such as climate, orientation, and other influences.
- A building that uses resources efficiently and maximizes use of local building materials.
- A project that minimizes demolition and construction wastes and uses products that minimize waste in their production or disposal.
- A building that is energy- and resource-efficient.
- A building that can be easily reconfigured and reused.
- A building with healthy indoor environments.
- A project that uses appropriate technologies, including natural and low-tech products and systems, before applying complex or resource-intensive solutions.
- A building that includes an environmentally sound operations and maintenance regimen.
- A project that educates building occupants and users to the philosophies, strategies, and controls included in the design, construction, and maintenance of the project.

benefiting from this integrated approach as a matter of course. In practice, however, the actual lack of coordination among design professionals and their consultants often resulted in facilities that were problematic to build. Now the green building movement has begun to emphasize that strong coordination and collaboration is the true foundation of a high-quality building. This philosophy promises to influence the

TABLE 1.3

Goals for High-Performance Buildings According to the City of New York Department of Design and Construction

- Raise expectations for the facility's performance among the various participants.
- Ensure that capital budgeting design and construction practices result in investments that make economic and environmental sense.
- Mainstream these improved practices through (1) comprehensive pilot high-performance building efforts and (2) incremental use of individual high-performance strategies on projects of limited scope.
- Create partnerships in the design and construction process around environmental and economic performance goals.
- Save taxpayers money through reduced energy and material expenditures, waste disposal costs, and utility bills.
- Improve the comfort, health and well-being of building occupants and public visitors.
- Design buildings with improved performance, which can be operated and maintained within the limits of existing resources.
- Stimulate markets for sustainable technologies and products.

entire building industry and, ultimately, to enhance confidence in the design and construction professions.

Green Building Progress and Obstacles

Although still considered a fringe movement, in the early twenty-first century the green building concept has won industry acceptance, and it continues to impact building design, construction, operation, real estate development, and sales markets. Detailed knowledge of the options and procedures involved in "building green" is invaluable for any organization providing or procuring design or construction services. The number of buildings registered with the USGBC for the LEED-NC building assessment system grew from a few in 1999 to almost 11,000 registered and certified in late 2008. The area of LEED-certified buildings increased from a few thousand square feet in 1999 to over 4.2 billion square feet (390 million square meters) in late 2008. Federal and state governments, many cities, several universities, and a growing number of private sector construction owners have declared sustainable or green materials and methods as their standard for procurement.

Despite the success of LEED and the U.S. green building movement in general, challenges abound when implementing sustainability principles within the well-entrenched traditional construction industry. Although proponents of green buildings have argued that whole-systems thinking must underlie the design phase of this new class of buildings, conventional building design and procurement processes are very difficult to change on a large scale. Additional impediments may also apply. For example, most jurisdictions do not yet permit the elimination of stormwater infrastructure in favor of using natural systems for stormwater control. Daylighting systems do not eliminate the need for a full lighting system, since buildings generally must operate at night. Special low-e window glazing, skylights, light shelves, and other devices increase project cost. Controls that adjust lighting to compensate for varying amounts of available daylight, and occupancy sensors that turn lights on and off depending on occupancy, add additional expense and complexity. Rainwater harvesting systems require dedicated piping, a storage tank or cistern, controls, pumps, and valves, all of which add cost and complexity.

Green building materials often cost substantially more than the materials they replace. Compressed wheatboard, a green substitute for plywood, currently costs as much as 10 times more than the plywood it replaces. The additional costs, and those associated with green building compliance and certification, often require owners to add a separate line item to the project budget. The danger is that during the course of construction management, when costs must be brought under control, the sustainability line item is one of the first to be "value-engineered" out of the project. To avoid this result, it is essential that the project team and the building owner clearly understand that sustainability goals and principles are paramount, and that LCC should be the applicable standard when evaluating a system's true cost. Yet, even LCC does not guarantee that certain measures will be cost-effective in the short or long term. Where water is artificially cheap, systems that use rainwater or graywater are difficult to justify financially, even under the most favorable assumptions. Finally, more expensive environmentally friendly materials may never pay for themselves in an LCC sense.

A summary of trends in, and barriers to, green building is presented in Table 1.4. These trends are an outcome of the Green Building Roundtable, a forum held by the USGBC for members of the U.S. Senate Committee on Environment and Public Works in April 2002.[20]

TABLE 1.4

Trends and Barriers to Green Building in the United States

Trends

1. Rapid penetration of the LEED green building rating system and growth of USGBC membership
2. Strong federal leadership
3. Public and private incentives
4. Expansion of state and local green building programs
5. Industry professionals taking action to educate members and integrate best practices
6. Corporate America capitalizing on green building benefits
7. Advances in green building technology

Barriers

1. Financial disincentives
   a. Lack of LCC analysis and use
   b. Real and perceived higher first costs
   c. Budget separation between capital and operating costs
   d. Security and sustainability perceived as trade-offs
   e. Inadequate funding for public school facilities
2. Insufficient research
   a. Inadequate research funding
   b. Insufficient research on indoor environments, productivity, and health
   c. Multiple research jurisdictions
3. Lack of awareness
   a. Prevalence of conventional thinking
   b. Aversion to perceived risk

Emerging Directions

Measures to cope with problems of constructability, cost, coordination of drawings, and attention to client requirements have been woven into the high-performance green building delivery system. Three powerful approaches coexist to ensure the creation of a truly high-performance building: performance-based fees, the charrette, and building commissioning. Although none of these concepts is new, each has rapidly gained acceptance among procurers and providers of green construction.

PERFORMANCE-BASED FEES

Ensuring collaboration and cooperation among the building team members during the design and construction phases is a challenge inherent in any building project. The use of performance-based fees (PBFs) has been suggested as an effective and ethical incentive for cooperation in which the savings derived from highly efficient design increase the designers' compensation. Since PBFs are dependent on the building systems' performance and efficiency rather than on initial cost, the greater the savings in electricity, natural gas, liquid fuels, and other resources, the higher the fees earned by the architects and engineers. For example, PBFs were used for the Clackamas Senior High School in Clackamas, Oregon, a 265,000-square-foot (25,000 square meters) facility that opened in April 2002, with a projected energy consumption level 44 percent below Oregon Building Code specifications. Optimizing the building's design meant reducing the mechanical plant size by investing in a high-performance building envelope and low-emissivity glass. Unlike a conventional project, mechanical plant size reduction actually resulted in *increased* fees due to projected energy savings.[21]

Establishing building efficiency goals and objectives at a project's inception is essential for the effective use of PBFs. Presently, practical goals are limited to energy consumption. It is also necessary to define and establish specific methods for quantifying system performance, which are typically expressed in terms of energy use or energy cost per unit of building area and may be impacted by other variables. For example, differences in heating and cooling degree-days, compared to the year in which the building's energy uses were modeled by computer simulation, can make enormous differences in the results.

THE CHARRETTE

According to the National Charrette Institute, "[t]he 'charrette' is often used to describe the final, intense work effort expended by art and architecture students to meet a project deadline."[22] The term originates from the École des Beaux Arts in Paris during the nineteenth century, where proctors circulated a cart, or *charrette,* to collect final drawings while students frantically put finishing touches on their work. Today's charrette brings a wide range of stakeholders together to facilitate a dynamic exchange of ideas, with the benefit of immediate feedback to all participants. An ideal charrette would include the owner, design team, builder, facility managers, members of the community, nonprofit organizations, children—literally anyone affected by the building. By incorporating the building into the fabric of the community, local opposition is lessened; the approval process is expedited; and community concerns, along with owner and builder needs, are addressed in a holistic process. Although not required by the USGBC's LEED standard, the charrette has been an enormously successful feature of the green building delivery process. The role of the charrette in the green building process is covered in more detail in Chapter 4.

BUILDING COMMISSIONING

Building commissioning has also become a standard, critical component of the green building delivery process. According to the Oregon Office of Energy, building commissioning is the process of ensuring that building systems are designed, installed, functionally tested, and capable of being operated and maintained according to the owner's operational needs.[23] Although building commissioning does not specifically address issues of sustainability, it demonstrates how the high-performance building process is improving the overall building delivery industry in the United States.

The building commissioning process has evolved from the mere testing and balancing of heating, ventilation, and air-conditioning (HVAC) systems at the project's completion to include an array of services. Ideally, a commissioning organization becomes involved in the conceptual stage of the project, providing expertise during the design phase. Prior to the completed building's delivery to its owner, the commissioning agent performs a thorough review of all systems, including but not limited to roofing, interior finishes, power, lighting, HVAC, plumbing, fire protection, telecommunications, and elevators. The building commissioning process is covered in depth in Chapter 12.

Summary and Conclusions

The rapidly evolving and exponentially growing green building movement is arguably the most successful environmental movement in the United States today. In contrast to many other areas of environmentalism that are stagnating, sustainable building has proven to yield substantial beneficial environmental *and* economic advantages. The USGBC's market and consensus-based LEED standards have catapulted sustainable construction into wide adoption. Despite this progress, however, there remain significant obstacles, erected by the inertia of the building professions and the construction industry and compounded by the difficulty of changing building codes. Industry professionals, in both the design and construction disciplines, are generally slow to change and tend to be risk-averse. Likewise, building codes are inherently difficult to change, and fears of liability and litigation over the performance of new products and systems pose appreciable challenges. Furthermore, the environmental or economic benefit of some green building approaches has not been scientifically quantified, despite their often intuitive and anecdotal benefits. Finally, lack of a collective vision and guidance for future green buildings, including design, components, systems, and materials, may affect the present rapid progress of this arena.

Despite these difficulties, the robust U.S. green building movement continues to gain momentum, and thousands of construction and design professionals have made it the mainstay of their practices. Numerous innovative products and tools are marketed each year, and in general, this movement benefits from an enormous air of energy and creativity. Like other processes, sustainable construction may one day become so common that its unique distinguishing terminology may be unnecessary. At that point, the green building movement will have accomplished its purpose: to transform fundamental human assumptions that create waste and inefficiency into a new paradigm of responsible behavior that supports both present and future generations.

Notes

1. Sustainability was first defined in 1981 by Lester Brown, a well-known American environmentalist and for many years the head of the Worldwatch Institute. In "Building a Sustainable Society," he defined a sustainable society as ". . . one that is able to satisfy its

needs without diminishing the chance of future generations." In 1987, the Bruntland Commission, headed by then Prime Minister of Norway, Gro Bruntland, adapted Brown's definition, referring to sustainable development as ". . . meeting the needs of the present without compromising the ability of future generations to meet their needs." Sustainable development, or sustainability, strongly suggests a call for intergenerational justice and the realization that today's population is merely borrowing resources and environmental conditions from future generations. In 1987, the Bruntland Commission's report was published as a book, *Our Common Future,* by the UN World Commission on Environment and Development.

2. The World Business Council on Sustainable Development (WBCSD) promotes sustainable development reporting by its 170 member international companies. The WBCSD is committed to sustainable development via the three pillars of sustainability: economic growth, ecological balance, and social progress. Its website is www.wbcsd.org.
3. In November 1992, more than 1,700 of the world's leading scientists, including the majority of the Nobel laureates in the sciences, issued the "World Scientists' Warning to Humanity." The preamble of this warning stated: "Human beings and the world are on a collision course. Human activities inflict harsh and often irreversible damage on the environment and critical resources. If not checked, many of our current practices put at serious risk the future that we wish for human society and the plant and animal kingdoms, and may so alter the living world that it may be unable to sustain life in the manner we know. Fundamental changes are urgent if we are to avoid the collision our present course will bring about." The remainder of this warning addresses specific issues, global warming among them, and calls for dramatic changes, especially on the part of the high-consuming developed countries, particularly the United States.
4. See, for example, Campell and Laherrere (1998).
5. A recent report, "The Cost and Benefits of Green Buildings," made to California's Sustainable Buildings Task Force, describes in detail the financial and economic benefits of green buildings. The principal author of this report is Greg Kats of Capital E. Several other reports on this theme by the same author are available online. See the References for more information
6. From World Health Organization (1983).
7. The losses are estimated productivity losses as stated by Mary Beth Smuts, a toxicologist with the U.S. Environmental Protection Agency, in Marsha Zabarsky (2002).
8. From "Ultra-violet Radiation Could Reduce Office Sickness" (2004).
9. At the First International Conference on Sustainable Construction held in Tampa, Florida, in November 1994, Task Group 16 (Sustainable Construction) of CIB formally defined the concept of sustainable construction and articulated six principles of sustainable construction, later amended to seven principles.
10. Sustainable construction and the model are described in Kibert (1994).
11. *Insolation* is an acronym for **in**coming **sol**ar radia**tion.**
12. Primary energy accounts for energy in its raw state. The energy value of the coal or fuel oil being input to the power plant is primary energy, while the electricity being generated, which has lower energy value due to the inefficiency of the generation system, is simply the output energy. Consequently, primary energy accounts for the losses in energy conversion, generation, and transmission.
13. A description of severe water resource problems beginning to emerge even in water-rich Florida can be found in the May/June 2003 issue of *Coastal Services,* an online publication of the NOAA Coastal Services Center, available at www.csc.noaa.gov/magazine/2003/03/florida.html. A similar overview of water problems in the western United States can be found in Young (2004).
14. An overview of xeriscaping and the seven basic principles of xeriscaping can be found at http://aggie-horticulture.tamu.edu/extension/xeriscape/xeriscape.html.
15. The Adam Joseph Lewis Center at Oberlin College was designed by a highly respected team of architects, engineers, and consultants and is a cutting-edge example of green buildings in the United States. An informative website, www.oberlin.edu/envs/ajlc, shows real-time performance of the building and its photovoltaic system.
16. *The Whole Building Design Guide* can be found at www.wbdg.org.
17. The design approach used in creating Solaire in Battery Park, New York City, plus updates on construction progress can be found at www.batteryparkcity.org. Another website with detailed information and illustrations is www.thesolaire.com.

18. See *Guidelines for Creating High-Performance Green Buildings* (1999).
19. Excerpted from *High Performance Building Guidelines* (1999).
20. The outcomes of the Green Building Roundtable can be found in *Building Momentum* (2003).
21. A detailed description of the application of PBFs to the Clackamas School is available at the RMI website www.rmi.org/sitepages/pid715.php.
22. Information on the recommended approach for conducting a charrette can be found at the website of the National Charrette Institute, www.charretteinstitute.org.
23. Extensive information about building commissioning can be found at the website of the Oregon Department of Energy, www.energy.state.or.us/bus/comm/bldgcx.htm.

References

Brown, Lester. 1981. *Building a Sustainable Society.* New York: Norton.

Building Momentum: National Trends and Prospects for High-Performance Green Buildings. February 2003. U.S. Green Building Council. Available at www.usgbc.org/Docs/Resources/043003_hpgb_whitepaper.pdf.

Campbell, C.J., and J.H. Laherrere. March 1998. "The End of Cheap Oil," *Scientific American,* 273(3), pp. 78–83.

Creating High-Performance Green Buildings: A Guide for Decision Makers. 1999. Pennsylvania Department of Environmental Protection. Available at www.gggc.state.pa.us.

Guidelines for Creating High-Performance Green Buildings. 1999. Pennsylvania Governor's Green Government Council (GGGC). Available at www.gggc.state.pa.us/publictn/gbguides.html.

High Performance Building Guidelines. April 1999. City of New York Department of Design and Construction. Available at www.ci.nyc.ny.us/nyclink/html/ddc/home.html.

High Performance Guidelines: Triangle Region Public Facilities, Version 2.0. September 2001. Available at www.tjcog.dst.nc.us/hpgtrpf.htm.

Kats, Gregory H. October 2006. "Greening America's Schools," available at www.cap-e.com.

Kats, Gregory H., and Jeff Perlman. December 2005. "National Review of Green Schools," a report for the Massachusetts Technology Collaborative, available at www.cap-e.com.

Kats, Gregory H. October 2003. "The Costs and Financial Benefits of Green Buildings," a report developed for California's Sustainable Building Task Force. Available at the Capital E website, www.cap-e.com.

Kibert, Charles J. November 1994. "Principles and a Model of Sustainable Construction," in *Proceedings of the First International Conference on Sustainable Construction,* 6–9 November 1994, Tampa, Florida, pp. 1–9.

Our Common Future. 1987. UN World Commission on Environment and Development, Oxford, England: Oxford University Press. *Our Common Future* is also referred to as the Brundtland Report, because Gro Bruntland, Prime Minister of Norway at the time, chaired the committee writing the report.

"Ultra-violet Radiation Could Reduce Office Sickness." January 4, 2004. Available at the website of the Lung Association of Saskatchewan, www.sk.lung.ca/content.cfm/xtra0129.

World Health Organization. (1983). "Indoor Air Pollutants: Exposure and Health Effect," *EURO Report and Studies,* No. 78.

Young, Samantha. March 10, 2004. *Las Vegas Review Journal.* Available at www.reviewjournal.com/lvrj_home/2004/Mar-10-Wed-2004/news/23401764.html.

Zabarsky, Marsha. August 16, 2002. "Sick Building Syndrome Gains a Growing Level of National Awareness." *Boston Business Journal.* Available at www.bizjournals.com/boston/stories/2002/08/19/focus9.html.

Part I

Green Building Foundations

This book is intended to guide construction and design professionals through the process of developing commercial and institutional high-performance green buildings in the United States. A green building is a healthy facility that is designed, built, operated, and disposed of in a *resource-efficient* manner using *ecologically sound* approaches. The nonprofit USGBC has successfully defined the parameters of a nonresidential green building in the United States.[1] The organization's LEED standard or its variants have provided design guidance for the vast majority of U.S. buildings currently described as green and have been implemented in several other countries.[2] From 1998 to 2006, the number of LEED-certified buildings has almost doubled each year in both number and area. The value of green building construction is projected to increase to $60 billion by 2010, according to McGraw-Hill Construction. At this rate of growth, high-performance green buildings could eventually exceed the annual value of conventionally constructed buildings. More recently, an alternative to LEED know as Green Globes has been competing with LEED as a tool for assessing and certifying high-performance green buildings in the United States.[3]

This book addresses the application of building assessment systems such as LEED and Green Globes in the United States, as well as several noteworthy building assessment systems used in other countries. Part I addresses the background and history of the sustainable construction movement, various green building rating systems, the concept of life-cycle assessment, and green building design strategies. It is intended to provide the working professional with sufficient information to implement the techniques necessary to create high-performance green buildings. This part contains the following chapters:

Chapter 2: Background

Chapter 3: Green Building Assessment

Chapter 4: The Green Building Process

Chapter 5: Ecological Design

Chapter 2, "Background," describes the emergence of the green building movement, its rapid evolution and growth over the past decade, and current major influences. This chapter also addresses the unusual scale of resource extraction, waste, and energy consumption associated with construction, and it examines the resource and environmental impacts of the built environment. And although this book focuses on the United States, the context, organizations, and approaches of other countries are also mentioned.

Measuring the performance of buildings to determine their relative environmental impacts, resource efficiency, and potential effects on human health is necessary to determine if key green building performance objectives are being met. Methods for measuring building performance are covered in Chapter 3, "Green Building Assessment." Rating systems that indicate a building's greenness, or environmental friendliness, are now being used in many countries. The term *building assessment system* is often used synonymously with the term *building rating system*. The first successful method for measuring green building performance was the Building Research Establishment Environmental Assessment Method (BREEAM), widely used in England and adopted for use in Hong Kong and Canada. LEED, the USGBC's building rating system, is rapidly transforming the commercial construction marketplace in the United States. Green Globes provides an alternative system for assessing building performance and is emerging in the United States as a potential competitor to LEED. Two other systems covered are CASBEE, a Japanese rating system in its initial stages of implementation, and the Australian Green Star building rating system.

Chapter 4, "The Green Building Process," addresses the high-performance green building delivery system as a distinctly identifiable construction delivery system, analogous to individually recognized design-build systems. A hallmark of the high-performance green building delivery system is the high level of coordination and integration required of the design and construction team members. New services, such as building commissioning and the charrette, are necessary to participate fully in this new delivery system. Performance-based design contracts provide financial incentives to implement certain sustainable design features, such as relying on nature for some building services, thus enabling a downsizing of mechanical and electrical systems to reduce energy consumption and cost. Documenting the green building process and gathering system performance data are necessary to demonstrate that the building has met all certification requirements.

General design strategies for green building are covered in Chapter 5, "Ecological Design." Fundamentally, green design is based on an ecological model or metaphor commonly referred to as *ecological design*. The recent works of Sim van der Ryn and Stuart Cowan, Ken Yeang, and David Orr, along with earlier works of R. Buckminster Fuller, Frank Lloyd Wright, Ian McHarg, Lewis Mumford, John Lyle, and Richard Neutra, are reviewed in this chapter. Bill Reed provides thoughtful closure by suggesting that sustainable design is simply a starting point in a path to what he calls *regenerative design*.

In spite of the impulse to apply the highest ecological ideals to the built environment, a vast majority of contemporary designers lack an adequate understanding of ecology. Claims of a building's "ecological design" are often tenuous in fact, and greater participation by ecologists and industrial ecologists is necessary to reduce the gap between the ideal of ecological design and its expression in reality. To that end, the LEED standard is probably the first step in a long process of achieving truly ecological design. The products, systems, techniques, and services needed to create buildings in harmony and synergy with nature are rare. Buildings are often assembled from components manufactured by a variety of manufacturers that have paid little or no attention to the environmental impacts of their activities. Installation is performed by a workforce largely unaware of the impacts of the built environment and often results in enormous waste. Conventional buildings are designed by architects and engineers who have had little training in sustainable construction. In spite of these obstacles, LEED-certified buildings are usually superior to conventional projects in terms of energy and water efficiency, materials selection, building health, waste generation, and site utilization. The USGBC has created an ambitious training and publicity program to disseminate LEED concepts. Innovative products for sustainable construction have become more prevalent, greatly easing the process of materials selection. Of equal importance, the green building process has necessitated a deeper integration of the client, the designer, and the general public. New projects are generally initiated via the charrette, which includes construction and design professionals as well as community members, who together brainstorm the project's initial design.

Exceeding the requirements of the contemporary assessment standards such as LEED and Green Globes is the next rung on the ladder of truly sustainable construction. The following are some of the features of future sustainable construction:

- The built environment would fully adopt closed-loop materials practices, and the entire structure, envelope, systems, and interior would be composed of products easily disassembled to permit ready recycling. Waste material throughout the structure's life cycle would be capable of biological (composting) or technological recycling. The building itself would be deconstructable; in other words, it would be possible to disassemble it economically for reuse and recycling. Only materials with future value, either to human or to biological systems, would be incorporated into buildings.
- Buildings would have a synergistic relationship with their natural environment and blend with the surrounding environment. Materials exchanges across the building-nature interface would benefit both sides of the boundary. Building and occupant waste would be processed to provide nutrients to the surrounding biotic systems. Toxic or harmful emissions of air, water, and solid substances would be eliminated.
- The built environment would incorporate natural systems at various scales, ranging from individual buildings to bioregions. The underexplored integration of natural systems with the built environment has staggering potential to produce superior human habitats at lower cost. Landscaping would provide shade, food, amenities, and stormwater uptake for the built infrastructure. Wetlands would process wastewater and stormwater and often eliminate the need for enormous and expensive infrastructure. The integration of nature, which is barely addressed in LEED or other assessment systems, is currently considered under the comprehensive category of *design innovation.* Ideally, the integration of human and natural systems would be standard practice rather than being considered an innovation.
- Energy use by buildings would be reduced by a Factor 10 or more below that of conventional buildings.[4] Rather than the typical 100,000 or more BTU per square foot (292 kWh per square meter) consumed by today's commercial and institutional structures, truly green buildings would be relatively deenergized, using no more than 10,000 BTUs per square foot (29 kWh per square meter). The source of this energy would be the sun or other solar-derived sources such as wind power or biomass. Alternatively, geothermal and tidal power, both nonsolar energy sources, would also be employed as renewable forms of energy derived from natural sources.

In summary, the green building movement has come a long way in a short time. Its exponential growth promises its longevity, and numerous public and private organizations support its agenda. It is exciting to contemplate the possibility of extending the boundaries of ecological design and construction as global environmental problems become exigent and as solutions, if not survival itself, demand a radical departure from conventional thinking. The evolution of products, tools, services, and, ultimately, Factor 10 buildings cannot occur soon enough. Only then may we alter the trajectory of the human quality of life from one of certain disaster to one that finally exists within the carrying capacity of nature. Although humanity is halfway through the race, the ultimate question remains unanswered: Can we change the built environment rapidly enough to save both nature and ourselves?

Notes

1. The USGBC (www.usgbc.org) is now the de facto U.S. leader in promoting commercial and institutional green buildings. The greening of single-family home residential construction and land development is far more decentralized and varies from state to state. An example of an organization leading change at the state level in the residential and land development sectors is the Florida Green Building Coalition (FGBC) (www.floridagreenbuilding.org). The Florida Green Residential Standard and the Florida Green Development Standard can be downloaded from the FGBC website.
2. Although intended for the greening of U.S. buildings, LEED is being adopted by other countries, such as Canada, Spain, and Korea. The World Green Building Council (www.worldgbc.org), organized by Michael Gottfried, one of the founders of the USGBC, is promoting the international use of LEED. Several versions of LEED are being adopted in Canada: LEED-BC for British Columbia; LEED-Canada; and LEED-prairie version, which is under development. Organization of the Canadian Green Building Council (www.cagbc.org), which will coordinate and implement LEED in Canada, is under way.
3. The genesis of Green Globes was BREEAM, which was developed in the United Kingdom in the early 1990s, brought to Canada in 1996, and eventually developed as an online assessment and rating tool. In 2004, the Green Building Initiative (GBI) acquired the rights to distribute Green Globes in the United States. In 2005 GBI became the first green building organization to be accredited as a standards developer by the American National Standards Institute (ANSI) and began the process of establishing Green Globes as an official ANSI standard. The GBI ANSI technical committee was formed in early 2006.
4. Factor 10, a concept developed by the Wuppertal Institute in Wuppertal, Germany (www.wupperinst.org), suggests that long-term sustainable development can be achieved only by reducing resource consumption (energy, water, and materials) to 10 percent of its present levels. Another concept, Factor 4, suggests that technology presently exists to reduce resource consumption immediately by 75 percent. The book *Factor 4: Doubling Wealth and Halving Resource Consumption,* by Ernst von Weizsäcker, Amory Lovins, and L. Hunter Lovins (London: Earthscan, 1998), popularized this concept.

Chapter 2

Background

The implementation of the high-performance green building delivery system is a complex process propelled by three major forces. First, there is growing evidence of accelerated destruction of planetary ecosystems, alteration of global biogeochemical cycles, and enormous increases in population and consumption. The threat of global warming, depletion of major fisheries, deforestation, and desertification are among likely outcomes that some environmentalists have labeled the *Sixth Extinction,* referring to the human species' massive destruction of life and biodiversity on the planet.[1]

Second, increasing demand for natural resources is pressuring developed and developing countries such as China and India, resulting in shortages and higher prices for materials and agricultural products. Illustratively, China adds about 11 million people each year to its population of 1.3 billion, and its economy is expanding at a rate of about 10 percent annually. China produced over 35 percent of the world's steel in June 2006. Chinese steel production has increased at a prodigious rate, from approximately 12 million tons (11 million metric tons) per month in 2001 to 37 million tons (34 million metric tons) in June 2006, an annual rate of 440 million tons (399 million metric tons) and rising rapidly. In comparison, the U.S. steel production rate was 100 million tons (91 million metric tons) per year. In early 2004 steel prices rose sharply due to Chinese domestic demand, which then led to a 20 percent rise in steel costs to the U.S. industry over a 6-month period. Since that time, through June 2006, steel prices have remained at about the same relatively high level. Likewise, Chinese demand for fossil fuel is growing at a rate of 30 percent per year, just as world oil production is beginning to peak. Similarly, the nation's growing economy and improving quality of life have increased the demand and, consequently, the prices for meat and grain. The negative consequences of rapid urban expansion in China have included water shortages and increasing desertification, leading to the growth of the Gobi desert by 4,000 square miles per year.

Third, the green building movement is coinciding with similar transformations in manufacturing, tourism, agriculture, medicine, and the public sector, which have adopted various approaches toward greening their activities. From redesigning entire processes to implementing administrative efforts such as adopting green procurement policies, new concepts and approaches are emerging that deem the environment, ecological systems, and human welfare to be of equal importance to economic performance. For example, the Xerox Corporation has announced the strategic environmental goal of creating "waste-free products and waste-free facilities for waste-free workplaces." A recently introduced Xerox product, the DocuColor iGen3 Digital Press, uses nontoxic dry inks and has a transfer efficiency of almost 100 percent. Up to 97 percent of the machine's parts and 80 percent of its generated waste can be reused or recycled. Furthermore, by reclaiming copy machines at the end of their useful life, recovering components for reuse and recycling, and instituting sophisticated remanufacturing processes, Xerox conserves materials and energy, dramatically reduces waste, and reduces potential liability by eliminating hazardous materials.[2]

Figure 2.1 The structural system for Rinker Hall, a LEED-NC 2.1 Gold-certified building at the University of Florida, is steel. Steel is an excellent material due to its high recycled content, almost 100 percent for some building components, and is readily recyclable. High demand for scrap steel due to rapidly growing Chinese consumption resulted in shortages and high prices in the United States in mid-2004. Although some would consider metals such as steel to be green building materials, their embodied energy—that is, the energy required for resource extraction, manufacturing, and transport—is fairly high and results in the consumption of nonrenewable fossil fuels and the generation of global warming gases and air pollution. Of all the challenges in creating high-performance green buildings, finding or creating truly environmentally friendly building materials and products is the most difficult task facing construction industry professionals.

In the automotive industry, the European "End-of-Life Vehicle (ELV)" directive has been in effect since the year 2000. This legislation requires manufacturers to accept the return of vehicles at the end of their useful life, with no charge to the consumer. The measure requires extensive recycling of the returned vehicles and minimizes the use of hazardous materials in automobile production. Spurred by European efforts, Ford Motor Company is utilizing European engineering expertise in its Aachen, Germany, research center to develop recycling technologies that will raise the recovery yield of recycled materials above their current 80 to 85 percent level (see Figure 2.1).

This chapter describes the effect of these three forces on the green building movement and their influence on defining new directions for design and construction of the built environment. It lays out the ethical arguments supporting sustainability and, by extension, sustainable construction. It explores the relatively new vocabulary associated with various efforts that attempt to reduce human environmental impact, increase resource efficiency, and ethically confront the dilemmas of population growth and resource consumption. Finally, it covers the history of the green building movement in the United States, acknowledging that an understanding of its roots is necessary to appreciate its evolution and current status.

Ethics and Sustainability

In the context of sustainable development and sustainable construction, ethics must be broadened to address a wide range of concerns that are not usually a basis for consideration. Ethics addresses relationships between people by providing rules of conduct that are generally agreed to govern the good behavior of contemporaries. Sustainable development requires a more extensive set of ethical principles to guide behavior because it addresses relationships between generations, calling for what is sometimes referred to as *intergenerational justice.* The classic definition of sustainable development is ". . . meeting the needs of the present without compromising the ability of future generations to meet their needs."[3] It is clear that intertemporal considerations, the responsibility of one generation to future generations as well as the rights of future generations vis-à-vis a contemporary population, are fundamental concepts of sustainable development. The result of intertemporal or intergenerational considerations with respect to morality and justice must be an expanded concept of ethics that extends not only to future generations, but also to the nonhuman living world and arguably to the nonliving world because the alteration or destruction of nonhuman living and nonliving systems affects the quality of life of future generations by reducing their choices. The result of destroying biodiversity today, for instance, is the removal of important information for future populations that could have been the basis for biomedicines, not to mention the removal of at least some portion of environmental amenity. It is clear that the choices of a given population in time will directly affect the quantity and quality of resources remaining for future inhabitants of Earth, impact the environmental quality they will experience, and alter their experience of the physical world. With this in mind, the purpose of this chapter is to expand on the foundations of classical ethics to provide a robust set of principles that are able to address questions of intergenerational equity.

THE ETHICAL CHALLENGES

Humans are unique among all species with respect to control over their destiny. Gary Peterson (2002), an ecologist, articulated this very well when he stated:[4]

> Humans, individually or in groups, can anticipate and prepare for the future to a much greater degree than ecological systems. People use mental models of varying complexity and completeness to construct views of the future. People have developed elaborate ways of exchanging, influencing, and updating these models. This creates complicated dynamics based upon access to information, ability to organize, and power. In contrast, the organization of ecological systems is a product of the mutual reinforcement of many interacting structures and processes that have emerged over long periods of time. Similarly, the behavior of plants and animals is the product of successful evolutionary experimentation that has occurred in the past. Consequently, the arrangement and behavior of natural systems are based upon what has happened in the past, rather than looking in anticipation toward the future. The difference between forward-thinking human systems and backwards-looking natural systems is fundamental. It means that understanding the role of people in ecological systems requires not only understanding how people have acted in the past, but also how they think about the future.

Following this chain of thinking, humans are certain to create materials and develop processes that have not evolved in a natural sense, that have no precedent in nature. The question then becomes: what constraints should society place on the development of new materials, products, and processes? The ongoing debates about genetically modified organisms (GMOs) and cloning are indicative of the uncertainty about the outcomes of human tinkering with the blueprints of life, not to mention the creation of materials that have uncertain long-term impacts. Other major developments such as biotechnology, genetic engineering, nanotechnology, robotics, and nuclear energy, to name but a few, present fundamental challenges to human society relative to decisions about implementing technologies with no precedent in nature and with potentially unprecedented negative and irreversible impacts. Decisions about how to move forward must be based on (1) an ethical framework that represents society's general moral attitudes toward life and future generations; (2) an understanding of and willingness to accept risk; and (3) the economic costs of implementation and resulting impacts.

INTERGENERATIONAL JUSTICE AND THE CHAIN OF OBLIGATION

The choices of today's generations will directly affect the quality and quantity of resources remaining for future inhabitants of Earth and environmental quality. This concept of obligation that crosses temporal boundaries is referred to as *intergenerational justice*. Furthermore, the concept of intergenerational justice implies a chain of obligation between generations that extends from today into the distant future. Richard Howarth (1992) expresses this obligation by stating, ". . . unless we ensure conditions favorable to the welfare of future generations, we wrong existing children in the sense that they will be unable to fulfill their obligation to their children while enjoying a favourable way of life themselves."[5] Howarth also suggests that the actions and decisions of the present generation affect not only the welfare but also the composition of future generations. He argues that when we create conditions that change resource availability or that alter the environment, future populations will be compositionally different than if the resource base and environmental conditions had been passed on, from one generation to future generations, unchanged. For instance, one can envision that mutations caused by excessive ultraviolet radiation through an ozone layer depleted by human activities, or by synthetic toxic chemicals used without adequate safeguards, will certainly result in different people and conditions. Consequently, the chain of obligation that underpins the key sustainability concept of intergenerational justice includes parents' responsibility for enabling their offspring to meet their moral obligations to their children and beyond. Clearly, this would include educating the offspring about these obligations and the basis for them.

DISTRIBUTIONAL EQUITY

There is an obligation to ensure the fair distribution of resources among present people so that the life prospects of all people are addressed. This obligation can be referred to as *distributional equity* or *distributive justice* and refers to the right of all people to an equal share of resources, including goods and services, such as materials, land, energy, water, and high environmental quality. Distributional equity is based on principles of justice, and the reasonable assumption that all individuals in a given generation are equal and that a uniform distribution of resources must be a consequence of *intragenerational equity*. The principle of distributional equity can be extended to relationships between generations because a given generation has moral responsibility for providing for their offspring, that is, intergenerational equity. Thus, distributional equity also underpins the chain of obligation concept. Distributional equity is a complex concept, and a number of principles underpin and are related to it: (1) the Difference Principle. (2) Resource-Based Principles, (3) Welfare-Based Principles, (4) Desert-Based Principles, (5) Libertarian Principles, and (6) Feminist Principles.

THE PRECAUTIONARY PRINCIPLE

The *Precautionary Principle* requires the exercise of caution when making decisions that may adversely affect nature, natural ecosystems, and global biogeochemical cycles. According to the Center for Community Action and Environmental Justice (CCAEJ), the Precautionary Principle states that "When an activity raises threats of harm to human health or the environment, precautionary measures should be taken even if some cause and effect relationships are not fully established scientifically." Global climate change is an excellent example of the need to act with caution. Notwithstanding debate about the effects of man-made carbon emissions on future planetary temperature regimes, the potentially catastrophic outcome should motivate humankind to behave cautiously and attempt to limit the emission of carbon-containing gases such as methane and carbon dioxide. The CCAEJ lists the four tenets of the Precautionary Principle:[6]

1. People have a duty to take anticipatory action to prevent harm.
2. The burden of the proof of harmlessness of a new technology, process, activity or chemical lies with the proponents, not the general public.
3. Before using a new technology, process, or chemical or starting a new activity, people have an obligation to examine a full range of alternatives including the alternative of not doing it.
4. Decisions applying the Precautionary Principle must be open, informed, and democratic and must include the affected parties.

With respect to the Precautionary Principle, a hypothetical danger of nanotechnology is the creation of so-called *gray goo.* Nanotechnology is an approach to building machines at the submicron level, that is, at the atomic scale. K. Eric Drexler (1987) suggested that one of the hallmarks of nanotechnology will be the ability of these invisible machines to self-replicate, with enormous potential benefits to humanity, but with the attendant danger that the replication will bring an out-of-control conversion of matter into machines. Drexler warned that "We cannot afford certain kinds of accidents with replicating assemblers," which could be restated, "We cannot afford the irresponsible use of powerful technologies."[7] Thermodynamics and energy requirements will limit the effects of the gray goo conversion process, but significant harm may still be the consequence. Similar concerns exist with regard to genetic engineering and nuclear engineering: that they will put future generations at risk. Clearly, the Precautionary Principle should be applied to each of these scenarios to eliminate as much as possible risks to future populations, both human and nonhuman, from the consequences of technologies that are not fully understood.

Despite the wisdom of exercising caution when addressing complex issues that may have unknown, far-reaching effects, the Precautionary Principle is controversial and is sometimes perceived as a threat to progress, since it fails to consider the negative consequences of its application. For example, refusing to utilize new drugs because society has not fully established their effects on nature and people may foreclose options for advancing human health. Nonetheless, the consequences of not applying the Precautionary Principle are becoming apparent in several areas. Most notably, the widespread use of estrogen-mimicking chemicals is believed to damage the reproductive systems of animal species and probably that of humans. With these concerns in mind, in 1999 the National Science Foundation developed the *Biocomplexity in the Environment Initiative* to address the interaction of human activities with the environment and on climate change and biodiversity.[8] At least the debate surrounding application of the Precautionary Principle has focused greater attention on the environmental impacts of technology, and has pressured technologists to acknowledge the potential consequences of their efforts on humans and nature.

THE REVERSIBILITY PRINCIPLE

Making decisions that can be undone by future generations is the foundation of the *Reversibility Principle.* Renowned science fiction author Arthur C. Clarke suggested a rule that well describes this principle: "Do not commit the irrevocable."[9] At its core, this principle calls for a wider range of options to be considered in decision making.

Addressing the issue of energy choices is an excellent example because a rapidly growing global economy is faced with looming energy shortages, exacerbated by depletion of finite oil supplies. In the United States, a shift is under way to reconsider nuclear plants as a major source of energy because they can probably generate electricity at an acceptable cost and also be a source of thermal energy for producing hydrogen from water for use in fuel cells. The Reversibility Principle would force today's society to confront the issue of whether or not the choice of nuclear energy as an option is reversible by a future society. Two questions would immediately emerge from this consideration. First, is the technology safe enough for widespread use? The nuclear industry suggests that over the past two decades of a national hiatus from building new plants, the technology has advanced to the point where a Chernobyl or Three Mile Island incident has been eliminated. The second question is: how would a future society cope with the nuclear waste from these plants? Converting the waste to harmless materials via a new technology is highly unlikely, and the power plants built today would force future generations to store and be put at risk by the radionuclides in the spent fuel rods. A subset of questions on this same subject would result as a consequence of assuming that if storage of the radioactive waste for periods of time in the 10,000-year range is feasible, what are the storage options? In addressing this question, Gene I. Rochlin suggests that there are two options.[10] One is to deposit the waste deep in a stable rock formation where it could be recovered, for example, if leaks in the storage containers were detected by future generations. A second option is to deposit it in inaccessible locations, for example, by allowing the waste to melt through the polar ice or to place it deep in the ocean, where sliding continental plates would gradually cover it. The former solution allows future generations access to the waste to take corrective action, while the latter forgoes the option.

The Reversibility Principle is related to the Precautionary Principle because it lays out criteria that must be observed prior to the adoption of a new technology. It is less stringent than the Precautionary Principle in some respects because it suggests reversibility as the primary criterion for making a decision to employ the technology, whereas the Precautionary Principle requires that a technology not be implemented if the effects of it are not fully understood and the risks are unacceptable.

THE POLLUTER PAYS PRINCIPLE AND PRODUCER RESPONSIBILITY

The fundamental premise of the Precautionary and Reversibility Principles is that those who are responsible for implementing technologies must be prepared to address the consequences of their implementation. The Precautionary Principle suggests that technologists should demonstrate the efficacy of their products and processes prior to allowing them to impact the biosphere. The Reversibility Principle permits implementation in the face of some level of risk as long as any negative effects can be undone. The Polluter Pays Principle addresses existing technologies that have not been subject to these other principles and places the onus for mitigating damage and consequences on the individuals causing the impacts. The Polluter Pays Principle originated with the Organization for Economic Cooperation and Development (OECD) in 1973 and is based on the premise that polluters should pay the costs of dealing with pollution for which they are responsible. Historically, the Polluter Pays Principle has focused on retrospective liability for pollution; for example, an industry causing pollution would have to pay for the cleanup costs arising from it. More recently, the focus of the Polluter Pays Principle has shifted toward avoiding pollution and addressing wider environmental impacts through Producer Responsibility. Producer Responsibility is an example of the extended version of the Polluter Pays Principle, as it applies to waste and resource management, placing responsibility for the environmental impact associated with a product on the producers of that product. Producer Responsibility is intended to address the whole life-cycle environmental problems of the production process, from initial minimization of resource use, through extended product life span, to recovery and recycling of products once they have been disposed of as waste. Producer Responsibility is increasingly used throughout the world as a means of addressing environmental impacts of certain products. The European Union (EU) has applied Producer Responsibility through directives on packaging and packaging waste, waste electronic and electrical equipment, and end-of-life vehicles.

PROTECTING THE VULNERABLE

There are populations, including those of the animal world, that are vulnerable to the actions of portions of the human species, due to the destruction of ecosystems under the guise of development, introduction of technology (including toxic substances, endocrine disruptors, and genetically modified organisms), and general patterns of conduct (war, deforestation, soil erosion, eutrophication, desertification, and acid rain, to name a few). People who are essentially powerless due to governing and economic structures are vulnerable to the decisions of those who are powerful because of their wealth or influence. This asymmetrical power arrangement is governed by moral obligation. Those in power have a special obligation to protect the vulnerable, those dependent on them. In a family, children's dependence on their parents gives them rights against their parents. Future generations are also vulnerable because they are subject to the effects of decisions we make today. In a technological society, many portions of the human population and certainly the animal world can be exposed to harm by the actions of individuals or companies performing medical research or because the government that is charged with protecting them fails in its responsibilities when it comes to pollution, the use of toxic substances, and a wide variety of other poorly controlled actions. Breaches of ethics are not uncommon when it comes to vulnerable populations such as prisoners, mentally disabled people, women, and people in developing countries. And as noted above, today's actions have consequences for future generations that have only recently been considered. Future people are certainly vulnerable to our actions, and both their existence and their quality of life are potentially compromised by short-term thinking and decisions based solely on the comfort and wealth of past populations. The ethical principle of Protecting the Vulnerable places an enormous responsibility on Earth's present population, one made even more difficult due to rampant global poverty.

PROTECTING THE RIGHTS OF THE NONHUMAN WORLD

The nonhuman world refers to plants and animals and could be extended to bacteria, viruses, mold, and other living organisms. The principle of protecting this world is an extension of the principle of Protecting the Vulnerable, particularly to animals but also to plants that are in danger of extinction. Animal rights fall under this principle. The nonliving portion of the Earth is essential to supporting life, and a set of sustainability principles should address the requirements for protecting this key element of the life support system. Some would argue that ethics should require the character of beautiful places such as the Grand Canyon to be protected in perpetuity. This principle is an important one because humans have become disconnected from both the living and nonliving human worlds when in fact we are utterly dependent on them for our survival. Indeed, the Biophilia Hypothesis, described in a subsequent section of this chapter, states that humans crave a connection with nature and that our health, at least in part, is dependent on being able to connect on a routine basis with nature. Human ingenuity in the form of technology is having quite the opposite effect. As noted by Andrew J. Angyal, ". . . this destructive myth of a technological wonderland in which nature is bent to every human whim is turning the Earth into a wasteland and threatening human survival. Western spiritual traditions have not been able to impede these lethal tendencies, but have encouraged them as part of God's plan for human domination of the Earth, and these traditions have understood human destiny as primarily involving a heavenly spiritual redemption. With their preoccupation with redemption and their neglect of creation, modern religious traditions are unable to offer a spirituality adequate to experience the divine in ordinary life or in the natural world."[11] Thomas Berry describes 10 precepts based on nature deriving its rights from universal law, and not human law, that provide an ethical framework for the rights of the nonhuman world:[12]

1. Rights originate where existence originates. That which determines existence determines rights.
2. Since it has no further context of existence in the phenomenal order, the universe is self-referent in its being and self-normative in its activities. It is also the primary referent in the being and activities of all derivative modes of being.
3. The universe is a communion of subjects, not a collection of objects. As subjects, the component members of the universe are capable of having rights.
4. The natural world on the planet Earth gets its rights from the same source that humans get their rights, from the universe that brought them into being.
5. Every component of the Earth community has three rights: the right to be, the right to habitat, and the right to fulfill its role in the ever-renewing processes of the Earth community.
6. All rights are species specific and limited. Rivers have river rights. Birds have bird rights. Insects have insect rights. Difference in rights is qualitative, not quantitative. The rights of an insect would be of no value to a tree or a fish.
7. Human rights do not cancel out the rights of other modes of being to exist in their natural state. Human property rights are not absolute. Property rights are simply a special relationship between a particular human "owner" and a particular piece of "property" so that both might fulfill their roles in the great community of existence.
8. Since species exist only in the form of individuals, rights refer to individuals and to their natural groupings of individuals into flocks, herds, packs, not simply in a general way to species.
9. These rights as presented here are based upon the intrinsic relations that the various components of Earth have to each other. The planet Earth is a single community bound together with interdependent relationships. No living being nourishes itself. Each component of the Earth community is immediately or mediately dependent on every other member of the community for the nourishment and assistance it needs for its own survival. This mutual nourishment, which includes the predator-prey relationships, is integral with the role that each component of the Earth has within the comprehensive community of existence.
10. In a special manner humans have not only a need for but a right of access to the natural world to provide not only the physical need of humans but also the wonder needed by human intelligence, the beauty needed by human imagination, and the intimacy needed by human emotions for fulfillment.

Clearly, putting nature on an equal footing with humans is a difficult leap for many people, but vigorously protecting nature is in the best interests of humanity. Indeed, simply protecting nature does not quite meet the imperatives of this principle. Rather, humans should consider restoring nature in all activities, righting the wrongs of the past, and in the process restoring the badly needed link between humans and nature.

RESPECT FOR NATURE AND THE LAND ETHIC

Respect for nature follows from acknowledging the rights of the nonhuman world described in the previous sections. An ethics of respect for nature is based on the fundamental concepts that (1) humans are members of the Earth's community of life, (2) all species are interconnected in a web of life, (3) each species is a teleological center of life pursuing good in its own way, and (4) human beings are not superior to other species. This last concept is based on the other three and shifts the focus from anthropocentrism to a biocentric outlook.[13]

Humans are part of precisely the same evolutionary process as all other species. All other species that exist today faced the same survival challenges as humans. The same biological laws that govern other species—for example, the laws of genetics, natural selection, and adaptation—apply to all living creatures. Earth does not depend on humans for its existence. On the contrary, humans are the only species that has ever threatened the existence of Earth itself. As relative latecomers, humans appeared on a planet that had contained life for 600 million years, and not only have to share Earth with other species, but are totally dependent on them for survival. Human beings threaten the soundness and health of the Earth's ecosystems by their behavior. Technology results in the release of toxic chemicals, radioactive materials, and endocrine disruptors. Forestry and agriculture destroy biologically dense and diverse forests. Emissions pollute land, water, and air. Unlike natural extinctions of the past from which the Earth recovered, the present human-induced extinction is causing disruption, destruction, and alteration at such a high rate that, even with the self-extinction of the human species, the planet may never recover. An ethics based on biocentrism would result in humans realizing that the integrity of the entire biosphere would benefit all communities of life, including nonhumans. It is debatable whether this concept is merely an ethical one because it is also a biological fact that humans cannot survive without the ecosystems upon which they depend. However, human beings have the capability to act and change behavior based on knowledge, in this case being aware of the causal relationship of behavior to the survival of other species. An ethics of respect for nature consists not only of realizing this causal relationship, but also of adopting behaviors that respect the rights of nonhuman species to both exist and thrive.

In addition to respecting the rights to survival of other species, as a consequence of careful observation and the application of scientific principles and the scientific method, humans understand the unique qualities and aspects of other organisms. These observations allow us to see these organisms as unique teleological centers of life, each struggling to survive and realize its good in its own way. This does not mean that organisms need to have the characteristic of consciousness, that is, self-awareness, to be "good" because each is oriented toward the same ends: self-preservation and well-being. The ethical concept here is that because each species is a teleological center of life, its universe or world can be viewed from the perspective of its life. Consequently, good (finding food), bad (being injured or killed), and indifferent (swimming in the ocean) events can be said to occur in each species' life, as is the case for the human species. Having respect for nature means that humans can view life events for nonhuman species in much the same fashion as they would for other humans.

Aldo Leopold (1949) suggests that there should be an ethical relationship to the land and that this relationship should and must be based on love, respect, and admiration for the land.[14] Furthermore, this ethical relationship should exist not just because of economic value but also should be based on value in the philosophical sense. The land ethic makes sense because of the close relationship and interdependence of humans with land that provides food and amenities and contributes to good air and water quality. Humans have tended to become disconnected from the land because of technological developments that give apparent but not actual independence from the land. Substitutes for natural material (for example, polyester) instead of cotton, further the notion that land is not essential for survival and that technology can provide suitable substitutes. Farm mechanization has also tended to separate the farmer from the land, the result being less care and attention for a critical resource.

Basic Concepts and Vocabulary

Although probably the greatest success story of the contemporary American environmental movement, sustainable construction is only one part of a larger transformation taking place via a wide range of activities throughout numerous economic sectors. Progressive ideas articulated with new vocabulary serve as the intellectual foundation for this evolution. The most notable and important include the concepts of sustainable development, industrial ecology, construction ecology, biomimicry, design for the environment, ecological economics, carrying capacity, ecological footprint, ecological rucksack, embodied energy, the Biophilia Hypothesis, eco-efficiency, the Natural Step, life-cycle assessment, life-cycle costing, the Precautionary Principle, Factor 4, and Factor 10. These concepts are briefly described in the following sections.

SUSTAINABLE DEVELOPMENT

Sustainable development, or *sustainability,* is the foundational principle underlying various efforts to ensure a decent quality of life for future generations. The Bruntland Report, more properly known as "Our Common Future" (1987), defines sustainable development as ". . . meeting the needs of the present without compromising the ability of future generations to meet their needs." This classic definition implies that the environment and the quality of human life are as important as economic performance and suggests that human, natural, and economic systems are interdependent. It also implies intergenerational justice, highlights the responsibility of the present population for the welfare of millions yet unborn, and implies that we are borrowing the planet, its resources, and its environmental function and quality from future generations. Intergenerational justice raises the question of how far into the future we should consider the impacts of our actions. Although no clear answer to this important question is readily apparent, the Native American philosophy of thinking seven generations, or 200 years, into the future is instructive. If in two centuries few contemporary buildings will be standing, we must ask whether our present stock of materials will provide recyclable resources for future generations or saddle them with enormous and difficult waste disposal problems. It is this question, originating in the philosophy of sustainability, that marks the fork in the road of our current industrial processes. Those on the path of "business as usual" will view the environment as an infinite source of materials and energy and a repository for waste. In contrast, those on the more ethical "road less traveled" will regard the quality of life of our descendants and question whether we are permanently stealing, versus temporarily borrowing, the environmental capital of future generations. At the philosophical core of the green building movement is the decision to embark on the latter path.

Figure 2.2 The publication of *Our Common Future* in 1987 is generally accepted as marking the initiation of the contemporary sustainable development movement.

INDUSTRIAL ECOLOGY

The science of *industrial ecology,* which emerged in the late 1980s,[15] refers to the study of the physical, chemical, and biological interactions and interrelationships both within and among industrial and ecological systems.[16] Applications of industrial ecology involve identifying and implementing strategies for industrial systems to emulate more closely harmonious and sustainable ecological ecosystems. The first major effort of industrial ecology was to reduce the massive quantities of waste generated by traditional manufacturing processes, from which only an estimated 6 percent of extracted resources end up as final products.[17] The first well-known example of the resulting process, known as *industrial symbiosis,* occurred at the industrial complex in Kalundborg, Denmark, where excess heat energy, waste, and water were shared among the five major partner companies: (1) the Asnæs Power Station, Denmark's largest power station, coal-fired with a 1,500-

megawatt capacity; (2) the Statoil Refinery, Denmark's largest, with a present capacity of 4.8 million tons per year; (3) Gyproc, a plasterboard factory producing 14 million square meters of gypsum wallboard annually (roughly enough to build all the houses in six towns the size of Kalundborg); (4) Novo Nordisk, an international biotechnological company, with annual sales of over $2 billion, and producers of industrial enzymes and pharmaceuticals, including 40 percent of the world's supply of insulin; and (5) the City of Kalundborg district heating system, which supplies heating to 20,000 residents and water to its homes and industries. The Kalundborg complex (diagrammed in Figure 2.3) was the world's first *eco-industrial park;* since its inception, similar waste exchange complexes have been created around the world.[18] Since the early 1990s, the concept of industrial ecology has expanded to encompass issues of Design for the Environment, product design, closing materials loops, recycling, and other environmentally conscious practices. Industrial ecology can be considered a comprehensive approach to implementing sustainable industrial behavior.

CONSTRUCTION ECOLOGY

Construction ecology is a subcategory of industrial ecology that applies specifically to the built environment. Construction ecology employs principles of industrial ecology combined with ecological theory that differentiates buildings from other industrial products such as automobiles, refrigerators, and copying machines. Construction ecology also supports the design and construction of a built environment that (1) has a closed-loop materials system integrated with eco-industrial and natural systems; (2) depends solely on renewable energy sources; and (3) fosters the preservation of natural system functions. Application of these principles should result in buildings that (1) are readily deconstructable at the end of their useful lives; (2) have components that are decoupled from the building for easy replacement; (3) are composed of products designed for recycling; (4) are built using recyclable, bulk structural materials; (5) have slow "metabolisms" due to their durability and adaptability; and (6) promote the health of their human occupants.[19]

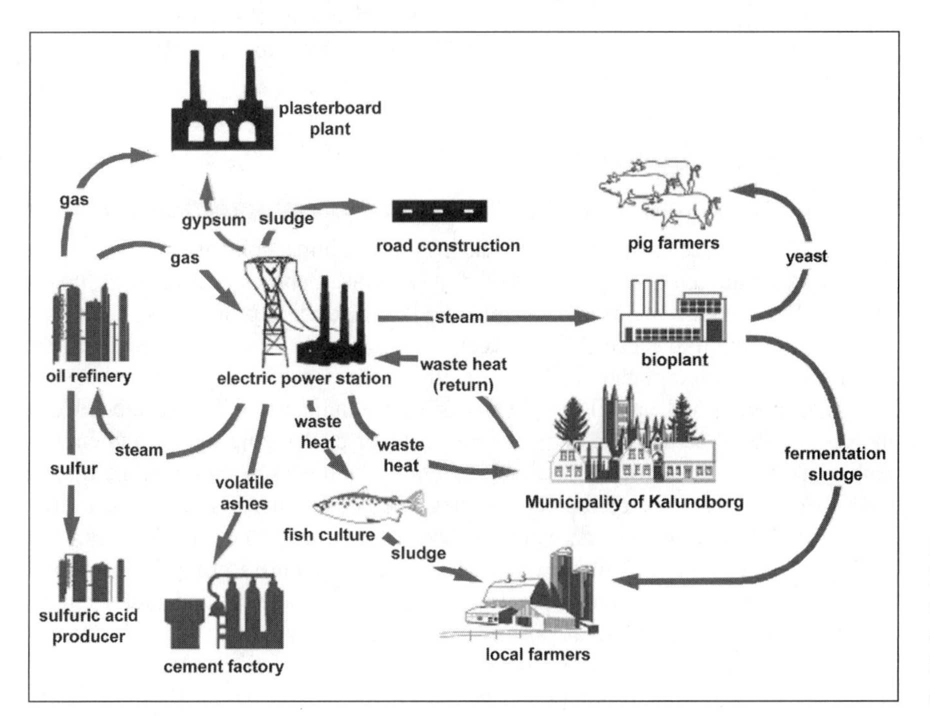

Figure 2.3 The industrial complex in Kalundborg, Denmark, exchanges energy, water, and materials among its member companies and organizations, demonstrating industrial symbiosis, one of the basic concepts of industrial ecology.

BIOMIMICRY

The term *biomimicry* was popularized by Janine Benyus in her book, *Biomimicry: Innovation Inspired by Nature,* and has since received widespread attention as a concept that demonstrates the direct application of ecological concepts to the production of industrial objects.[20] Biomimicry, or the "conscious imitation of nature's genius," suggests that most of what we need to know about energy and materials use has been developed by natural systems over almost 4 billion years of trial and error. Biomimicry advocates the possibility of creating strong, tough, and intelligent materials from naturally occurring materials, at ambient temperatures, with no waste, and using current solar "income" (sunlight) to power the manufacturing process. For example, nature produces strong, elegant, functional, and beautiful ceramic seashells from local materials in seawater at ambient temperatures. At the end of their useful lives as aquatic habitat, they degrade and provide future resources in a waste-free manner. In contrast, production of ceramic clay tiles requires high-fire temperatures of at least 2700°F and the extraction and transport of clay and energy resources, and results in emissions and waste. Unlike their natural counterparts, clay tiles do not degrade into useful products and are likely to be disposed of in landfills at the end of their useful lives.

DESIGN FOR THE ENVIRONMENT

Design for the Environment (DfE), sometimes referred to as *green design,* is a practice that integrates environmental considerations into product and process engineering procedures and considers the entire product life cycle.[21] The related concept of *front-loaded design* advocates the investment of greater effort during the design phase to ensure the recovery, reuse, and/or recycling of the product's components. Although DfE typically describes the process of designing products that can be disassembled and recycled, depending on the context, DfE may encompass design for disassembly, design for recycling, design for reuse, design for remanufacturing, and other applications. DfE's application to building design implies that to be considered green, significant effort was made in product design to enable the reuse and recycling of the product's components. Furthermore, the materials must possess and maintain value in order to motivate the industrial system to keep them in productive use. As applied to the built environment, DfE implies that entire buildings should be designed to be taken apart, or *deconstructed,* to recover components for further disassembly, reuse, and recycling.

ECOLOGICAL ECONOMICS

Contemporary, or neoclassical, economics fails to consider or adequately address the problems of resource limitations or the environmental impact of waste and toxic substances on productive ecological systems. In contrast, *ecological economics* posits that healthy, natural systems and the free goods and services provided by nature are essential to economic success. Ecological economics is a fundamental requirement of sustainable development that specifically addresses the relationship between human economies and natural ecosystems. Since the human economy is embedded in the larger natural ecosystem, and depends on it for exchanging matter and energy, both systems must coevolve. Ecological economic philosophy counters the human propensity to ignorantly or deliberately degrade ecosystems by extracting useful, high-quality matter and energy, which are ultimately transformed into useless, low-quality waste and heat. Ecological economics values nature's provision of goods, energy, services, and amenities, as well as humanity's cultural and moral contributions.[22] Valuing nature—that is, assigning a monetary worth to its goods and services—although antithetical to some, is essential to appreciate and understand the worth of natural system resources and services in the human economy.

Unfortunately, obstacles exist to replacing the shortsighted approach of contemporary neoclassical economics with the ecological economics consideration of the contributions and limitations of natural systems. Our present limited understanding of complex nonlinear natural systems, and the difficulty of accurately representing these systems in relevant economic models, present challenges. Nonetheless, ecological economics illuminates the dismal science of traditional economics and provides a more comprehensive framework for applying economic principles in the evolving, transformative era of sustainable development.

CARRYING CAPACITY

The term *carrying capacity* attempts to define the limits of a specific land's capability to support people and their activities. According to the Carrying Capacity Network,

> Carrying capacity is the number of people who can be supported in a given area within natural resource limits, and without degrading the natural social, cultural, and economic environment for present and future generations. The carrying capacity for any given area is not fixed. It can be altered by improved technology, but mostly it is changed for the worse by pressures which accompany a population increase. As the environment is degraded, carrying capacity actually shrinks, leaving the environment no longer able to support even the number of people who could formerly have lived in the area on a sustainable basis. No population can live beyond the environment's carrying capacity for very long.[23]

Carrying capacity focuses on the relationship between land area and human population growth and suggests the point at which the system may break down. Much debate surrounds the carrying capacity of the planet in general and the United States in particular. Although the United States may be able to carry 1 billion people with adequate resources, it is doubtful that a population of this magnitude is desirable. The concept of carrying capacity is also linked to the Precautionary Principle, discussed earlier in the chapter.

ECOLOGICAL FOOTPRINT

Mathis Wackernagel and William Rees suggested that an *ecological footprint,* referring to the land area required to support a certain population or activity, could serve as a surrogate measure for total resource consumption, thus allowing a simple comparison of the resource consumption of various lifestyles.[24] The ecological footprint is the inverse of carrying capacity and represents the amount of land needed to support a given population. An ecological footprint calculation indicates that, for example, the Dutch need a land area 15 times larger than that of The Netherlands to support their population. The population of London requires a land area 125 times greater than its physical footprint. If everyone on Earth enjoyed a North American lifestyle, it would take up to five planet Earths, owing to the increasingly consumptive U.S. lifestyle and the burgeoning world population, which exceeds 6 billion at the time of this writing. The ultimate problem that must be solved, especially in the context of sustainable development, is how all people can have a decent quality of life without destroying the planetary systems that support life itself. A partial solution requires developed countries to dramatically reduce consumption and to ensure that developing countries receive resources sufficient for more than mere survival. Such resource sharing lies at the heart of the original formulation of sustainable development, which values the goal of moving the developing world from mere survival to the ability to sustain a reasonably good quality of life. As William Rees notes in the preface to the book *Our Ecological Footprint,* coauthored with Mathis Wackernagel,

"On a finite planet, at human carrying capacity, a society driven mainly by selfish individualism has all the potential for sustainability of a collection of angry scorpions in a bottle."

ECOLOGICAL RUCKSACK AND MIPS

The term *ecological rucksack,* coined by Friedrich Schmidt-Bleek, formerly of the Wuppertal Institute in Wuppertal, Germany, attempts to quantify the mass of materials that must be moved in order to extract a specific resource. The concept of the ecological rucksack was developed to demonstrate that prosperity attributable to certain human activities has been achieved only by the destruction of natural resources through excavation, mining, channeling rivers and lakes, and processing gigatons[25] of materials to extract dilute resources. Schmidt-Bleek suggested that since these activities are responsible for significant environmental damage, extracted materials could be said to carry a "rucksack," or extraction burden. For example, the 10 grams of gold contained in a typical gold wedding band are extracted and concentrated from 300 tons of raw material.

The European Environmental Agency (EEA) defines ecological rucksack as the material input of a product or service minus the weight of the product itself.[26] The material input is defined as the life-cycle-wide total quantity (in pounds or kilograms) of natural material physically displaced in order to generate a particular product (see Table 2.1).[27] The environmental stress caused by an activity is proportional to the quantity of materials moved. The greater the mass moved, the higher the environmental impact. The concept of ecological rucksack focuses on these large displacements of earth and rock rather than on minute quantities of toxic materials. It has been the large land transformations occasioned by increasing material demands, coupled with depleted deposits of rich materials, that have been historically neglected by environmentalists and policymakers.

TABLE 2.1

Ecological Rucksack* of Some Well-Known Materials

Material	Ecological Rucksack
Rubber	5
Aluminum	85
Recycled aluminum	4
Steel	21
Recycled steel	5
Platinum	350,000
Gold	540,000
Diamond	53,000,000

*The rucksack indicates how many units of mass must be moved to produce one unit mass of the material. For example, 1 kilogram (2.2 pounds) of aluminum from bauxite requires displacing 85 kilograms (187 pounds) of materials, compared to moving only 4 kilograms (9 pounds) to produce 1 kilogram of recycled aluminum.

Materials Intensity per Unit Service (MIPS) is another concept originated by Friedrich Schmidt-Bleek to assist in understanding the efficiency with which materials are used. MIPS measures how much service a given product delivers. The higher or greater the service, the lower the MIPS value. MIPS is also an indicator of resource productivity, or eco-efficiency, and products with greater service are said to possess greater eco-efficiency and resource productivity.

THE BIOPHILIA HYPOTHESIS

E.O. Wilson, the eminent Harvard University entomologist, suggested that humans have a need and craving to be connected to nature and living things. He coined the term *Biophilia Hypothesis* to propose the concept that humans have an affinity for nature and that they "tend to focus on life and lifelike processes." The Biophilia Hypothesis asserts the existence of a fundamental genetically based human need and propensity to affiliate with life and lifelike processes. Various recent studies have shown that even minimal connection with nature, such as looking outdoors through a window, increases productivity and health in the workplace, promotes healing of patients in hospitals, and reduces the frequency of sickness in prisons. Prison inmates whose cells overlooked farmlands and forests needed fewer health-care services than inmates whose cells overlooked the prison yard.[28]

In their book *The Biophilia Hypothesis,* Wilson and Stephen Kellert, a professor in the Yale University School of Forestry and Environmental Studies, collected invited papers to both support and refute this hypothesis. Kellert suggests that for green buildings to eventually become truly successful, they must relate to natural processes and help humans achieve meaning and satisfaction. He suggests that there are nine values of biophilia, which offer a broad design template for sustainable building: (1) the *utilitarian value* emphasizes the material benefit that humans derive from exploiting nature to satisfy various needs and desires; (2) the *aesthetic value* emphasizes a primarily emotional response of intense pleasure at the physical beauty of nature; (3) the *scientific value* emphasizes the systematic study of the biophysical patterns, structures, and functions of nature; (4) the *symbolic value* emphasizes the tendency for humans to use nature for communication and thought; (5) the *naturalistic value* emphasizes the many satisfactions people obtain from the direct experience of nature and wildlife; (6) the *humanistic value* emphasizes the capacity for humans to care for and become intimate with animals; (7) the *dominionistic value* emphasizes the desire to subdue and control nature; (8) the *moralistic value* emphasizes right and wrong conduct toward the nonhuman world; and (9) the *negativistic value* emphasizes feelings of aversion, fear, and dislike that humans have for nature.[29]

Anecdotal evidence emerging about the effects of daylighting and views to the outside indicates that human health, productivity, and well-being are promoted by access to natural light and views of greenery. Hundreds of studies have demonstrated that stress reduction results from connecting humans to nature. Consequently, facilitating the ability of humans to interact with nature, even at a distance, from inside a building, is emerging as an issue for consideration in the creation of high-performance green buildings.

ECO-EFFICIENCY

Originated by the World Business Council on Sustainable Development (WBCSD) in 1992, the concept of *eco-efficiency* includes environmental impacts and costs as a factor in calculating business efficiency. The WBCSD considers the term *eco-efficiency* to describe the delivery of competitively priced goods and services that satisfy human needs and enhance the quality of life while progressively reducing ecological impacts and resource intensity throughout the products' life cycles to a level commensurate with the Earth's estimated carrying capacity. The WBCSD has articulated seven elements of eco-efficiency (see Table 2.2).[30]

TABLE 2.2

Seven Elements of Eco-Efficiency as Defined by the WBCSD

1. Reducing the material requirements of goods and services
2. Reducing the energy intensity of goods and services
3. Reducing toxic dispersion
4. Enhancing materials recyclability
5. Maximizing sustainable use of renewable resources
6. Extending product durability
7. Increasing the service intensity of goods and services

Furthermore, the WBCSD has identified four aspects of eco-efficiency that render it an indispensable strategic element in the contemporary knowledge-based economy:[31]

- *Dematerialization:* Companies are developing ways of substituting knowledge flows for material flows.
- *Closing production loops:* The biological designs of nature provide a role model for sustainability.
- *Service extension:* The world is moving from a supply-driven economy to a demand-driven economy.
- *Functional extension:* Companies are manufacturing smarter products with new and enhanced functionality and are selling services to enhance the products' functional value.

The WBCSD suggests that business can achieve eco-efficiency gains through:

- *Optimized processes:* Moving from costly end-of-pipe solutions to approaches that prevent pollution in the first place
- *Waste recycling:* Using the by-products and wastes of one industry as raw materials and resources for another, thus creating zero waste
- *Eco-innovation:* Manufacturing "smarter" by using new knowledge to make old products more resource-efficient to produce and use
- *New services:* For instance, leasing products rather than selling them, which changes companies' perceptions, spurring a shift to product durability and recycling
- *Networks and virtual organizations:* Sharing resources to increase the effective use of physical assets

As a concept, eco-efficiency describes most of the foundational principles underpinning the concept of sustainable development. Its promotion by the WBCSD, essentially an association of major corporations, is a positive sign that the business community is beginning to take sustainability seriously.

THE NATURAL STEP

Developed by Swedish oncologist Karl Henrik Robert in 1989, *The Natural Step* provides a framework for considering the effects of materials selection on human health. Robert suggested that many human health problems, particularly those of children, result from materials we use in our daily lives. The extraction of resources such as fossil fuels and metal ores from the planet's crust produces carcinogens and results in heavy metals entering the Earth's surface biosphere. The abundance of chemically produced synthetic substances that have no model in nature has similar deleterious effects on health. The Natural Step articulates the Four Systems Conditions, or basic principles, that should be followed to eliminate the effects of materials practices on our health. The Four Systems Conditions are listed here.[32] Their potential application to construction projects is described in greater detail in Chapter 9.

1. In order for a society to be sustainable, nature's functions and diversity are not systematically subject to increasing concentrations of substances extracted from the Earth's crust.
2. In order for a society to be sustainable, nature's functions and diversity are not systematically subject to increasing concentrations of substances produced by society.
3. In order for a society to be sustainable, nature's functions and diversity are not systematically impoverished by overharvesting or other forms of ecosystem manipulation.
4. In a sustainable society, resources are used fairly and efficiently in order to meet basic human needs globally.

LIFE-CYCLE ASSESSMENT

Life-cycle assessment (LCA) is a method for determining the environmental and resource impacts of a material, a product, or even a whole building over its entire life. All energy, water, and materials resources, as well as all emissions to air, water, and land, are tabulated over the entity's life cycle. The life cycle, or time period considered in this evaluation, can span the extraction of resources, the manufacturing process, installation in a building, and the item's ultimate disposal. The assessment also considers the resources needed to transport components from extraction through disposal. LCA is an important, comprehensive approach that examines *all* impacts of material selection decisions, rather than simply an item's performance in the building. LCA and the tools used to produce an LCA are described in greater detail in Chapter 9.

LIFE-CYCLE COSTING

The ability to model a building's financial performance over its life cycle is necessary to justify measures that may require greater initial capital investment but yield significantly lower operational costs over time. Using *life-cycle costing* (LCC), a cost/benefit analysis is performed for each year of the building's probable life. The present worth of each year's net benefits is determined using an appropriate discount rate. Net benefits for each year are tabulated to calculate the total present worth of a particular feature. For example, the financial return for installation of a photovoltaic system would be determined by amortizing the system's costs over its probable life;

the worth of the energy generated each year would then be calculated to determine the net annual benefit. Application of LCC may determine whether the payback for this system meets the owner's economic criteria. LCC analysis can also be combined with LCA results to weigh the combined financial and environmental impact of a particular system. LCC is covered in more detail in Chapter 13.

EMBODIED ENERGY

Embodied energy refers to the total energy consumed in the acquisition and processing of raw materials, including manufacturing, transportation, and final installation. Products with greater embodied energy usually have higher environmental impact due to the emissions and greenhouse gases associated with energy consumption. However, another calculation, which divides the embodied energy by the product's time in use, yields a truer indicator of the environmental impact. More durable products will have a lower embodied energy per time in use. For example, a product with high embodied energy such as aluminum could have a very low embodied energy per time in use because of its extremely high durability. Additionally, certain products have relatively low embodied energy when recycled. Recycled aluminum has just 10 percent of the embodied energy of aluminum made from bauxite ore. Similarly, recycled steel has about 20 percent of the embodied energy of steel made from ores. A list of typical embodied energies for common construction materials is presented in Table 2.3.[33]

TABLE 2.3

Embodied Energy of Common Construction Materials

Material	Embodied Energy MJ/kg*	MJ/m³†
Aggregate	0.1	150
Concrete (30 Mpa)	1.3	3,180
Lumber	2.5	1,380
Brick	2.5	5,170
Cellulose insulation	3.3	112
Mineral wool insulation	14.6	139
Fiberglass insulation	30.3	970
Polystyrene insulation	117.0	3,770
Gypsum wallboard	6.1	5,890
Particleboard	8.0	4,400
Plywood	10.4	5,720
Aluminum	227.0	515,700
Aluminum (recycled)	8.1	21,870
Steel	32.0	251,200
Steel (recycled)	8.9	37,210
Zinc	51.0	371,280
Copper	70.6	631,164
PVC	70.0	93,620
Linoleum	116.0	150,930
Carpet (synthetic)	148.0	84,900
Paint	93.3	117,500
Asphalt shingles	9.0	4,930

*Megajoules per kilogram of material.
†Megajoules per cubic meter of material.

Figure 2.4 The Factor 4 concept originated in the book *Factor Four: Doubling Wealth, Halving Resource Use,* by Ernst von Weizsäcker, Amory Lovins, and L. Hunter Lovins (1997).

FACTOR 4 AND FACTOR 10

The concepts of *Factor 4* and *Factor 10* provide a set of guidelines for comparing design options and for evaluating the performance of buildings and their component systems. The notion of Factor 4 was first suggested in the book *Factor Four: Doubling Wealth, Halving Resource Use,* written in 1997 by Ernst von Weizsäcker, Amory Lovins, and L. Hunter Lovins.[34] Factor 4 suggests that for humanity to live sustainably today, we must rapidly reduce resource consumption to one-quarter of its current levels. Fortunately, the technology to accomplish Factor 4 reductions in resource consumption already exists, and requires only public policy prioritization and implementation. A parallel approach originated by Friedrich Schmidt-Bleek hypothesizes that, in order to achieve long-term sustainability, we must reduce resource consumption by a factor of 10.[35] An example of applying this principle to the built environment is provided by Lee Eng Lock, a Chinese engineer in Singapore. He has challenged many of the fundamental assumptions made by mechanical engineers in their systems design and layout. Rather than oversizing chillers, air handlers, pumps, and other equipment, he ensures that they are precisely the correct size for the job. This commonsense approach achieves the same cooling and comfort while using only 10 percent of the energy of conventional designs, thus accomplishing a Factor 10 reduction in energy.[36] The Factor 10 concept has had a significant effect internationally and is now being implemented by the EU.

Major Environmental and Resource Concerns

Concerns about environmental degradation, resource shortages, and human health impacts are promoting widespread acceptance of green building, the ultimate goal of which is to mitigate the enormous pressures on planetary ecosystems caused by human activities. The major environmental issues to be addressed by sustainable construction methods are shown in Table 2.4. Some of these are covered in more detail in the following sections.

TABLE 2.4

Major Environmental Issues Connected to Built Environment Design and Construction

Climate change
Ozone depletion
Soil erosion
Desertification
Deforestation
Eutrophication
Acidification
Loss of biodiversity
Land, water, and air pollution
Dispersion of toxic substances
Depletion of fisheries

CLIMATE CHANGE AND OZONE DEPLETION

As defined by the National Oceanographic and Atmospheric and Administration (NOAA), *climate change* consists of long-term fluctuations in temperature, precipitation, wind, and all other aspects of the Earth's climate. The United Nations Convention on Climate Change describes the phenomenon as a change of climate attributable directly or indirectly to human activity that alters the composition of the global atmosphere and that is, in addition to natural climate variability, observable over comparable time periods. The Intergovernmental Panel on Climate Change (IPCC) was established by the World Meteorological Organization (WMO) and the United Nations (UN) in 1988 to assess, on a comprehensive, objective, open, and transparent basis, the scientific, technical, and socioeconomic information relevant to understanding the scientific basis of the risk of human-induced climate change, its potential impacts, and options for adaptation and mitigation. The Third Assessment Report of the IPCC, published in 2001, concludes that the globally averaged surface temperatures increased by 1.2 ± 0.4°F (0.6 ± 0.2°C) over the twentieth century. For a range of scenarios, the globally averaged surface air temperature is projected by models to warm 2.5 to 10.4°F (1.4 to 5.8°C) by 2100 relative to 1990. Furthermore, the globally averaged sea level is projected by models to rise 0.3 to 2.9 feet (0.09 to 0.88 meter) by 2100. These projections indicate that the warming would vary by region and be accompanied by increases and decreases in precipitation.[37]

Moreover, there would be changes in climate variability, as well as in the frequency and intensity of some extreme climate phenomena. It is important to note that systems theory shows that the behavior of global systems such as climate is nonlinear. Each increase in carbon dioxide will not necessarily produce a proportional change in global temperature. However, the dynamic, chaotic character of the Earth's climate is such that climate can suddenly "flip" from one temperature regime to another in a relatively short time. Indeed, fossil records indicate that previous flips have occurred, with temperatures increasing or decreasing almost 10°F (5.6°C) in about a decade. The potential for climate change has profound implications for every aspect of human activity on the planet. Shifting temperatures, more violent storms, rising sea levels, melting glaciers, and other effects will displace people, affect food supplies, reduce biodiversity, and greatly reduce the average quality of life. The responsibility for creators of the built environment, which is a major energy consumer, is to dramatically reduce energy consumption, particularly reliance upon fossil fuels.

In addition to causing climate change, certain chemicals used in building construction and facility operations have been thinning the ozone layer, the protective sheath of the atmosphere consisting of three-molecule oxygen (O_3), which is located 10 to 25 miles (16 to 40 kilometers) above the Earth and serves to attenuate harmful ultraviolet radiation. In 1985, scientists discovered a vast hole the size of the continental United States in the ozone layer over Antarctica. By 1999, the size of the hole had doubled. Ozone depletion is caused by the interaction of halogens—chlorine- and bromine-containing gases such as chlorofluorocarbons (CFCs) used in refrigeration and foam blowing, and halons used for fire suppression. Table 2.5 provides a summary of the main contributors to the destruction of the ozone layer.[38] In one of the few successful examples of international environmental cooperation, the United Nations Montreal Protocol of 1987 produced an international agreement to eventually halt the production of ozone-depleting chemicals. Assuming that the Montreal Protocol is faithfully adhered to by the international community, the ozone layer is projected to be fully restored by the year 2050.

TABLE 2.5

Gases Used for Typical Building Functions (e.g., Refrigeration and Fire Suppression)

Halogen Gas*	Lifetime† (years)	Global Emissions (1000s of metric tons/year)	Ozone Depletion Potential (ODP)‡
Chlorine			
CFC-12	100	130–160	1
CFC-113	85	10–25	1
CFC-11	45	70–110	1
HCFCs	1–26	340–370	0.02–0.12
Bromine			
Halon 1301	65	~3	12
Halon 1211	16	~10	6

*The chlorine gases are used in refrigerants, the bromine gases in fire suppression systems.
†Lifetime refers to their duration in the atmosphere, and ODP is their ozone depletion impact.
‡The ODP of CFC-11 is defined as 1. With an ODP of 12, halon 1301 depletes ozone at a rate 12 times greater than CFC-11.

DEFORESTATION, DESERTIFICATION, AND SOIL EROSION

Natural forests are estimated to contain half of the world's total biological diversity, possessing the greatest level of biodiversity of any type of ecosystem. Sadly, worldwide deforestation is occurring at a rapid rate, with 2 acres (0.8 hectare) of rainforest disappearing every second[39] and temperate zone forests losing about 10 million acres (4 million hectares) per year. Although about one-third of the total land area is forested worldwide, about half of the Earth's forests have disappeared. In the United States, only 1 to 2 percent of the original forest cover still remains. This pattern of large-scale forest removal, known as *deforestation,* is linked to negative environmental consequences such as biodiversity loss, global warming, soil erosion, and desertification.

Deforestation defeats the capability of forests to "lock up," or *sequester,* the large quantities of carbon dioxide stored in tree mass; instead, it is released into the atmosphere as gaseous compounds, which contribute to accelerated climate change. Between 1850 and 1990, worldwide deforestation released 134 billion tons (122 billion metric tons) of carbon. Currently, deforestation releases about 1.8 billion tons (1.6 billion metric tons) of carbon per year, compared to the burning of fossil fuels such as oil, coal, and gas, which releases about 6.6 billion tons (6 billion metric tons) per year. And because trees and their root systems are necessary to prevent soil erosion, landslides, and avalanches, their removal contributes to soil loss and changes the rate at which water enters the watershed. Forest-sustained freshwater supplies are an important source of oxygen, which fosters biodiversity, especially in rainforests. Additionally, large-scale deforestation affects the *albedo,* or reflectivity, of the Earth, altering its surface temperature and energy, rate of surface water evaporation, and, ultimately, patterns and quantity of rainfall.

Deforestation also causes *soil erosion,* a key factor in land degradation. More than 2 billion tons (1.8 billion metric tons) of topsoil are lost annually due to human agricultural and forestry land development. More than 5 billion acres (2 billion hectares) of land, an area equal to the United States and Mexico combined, is now considered degraded.[40] In arid and semiarid regions, degradation results in *desertification,* or the destruction of natural vegetative cover, which prevents desert formation. The United Nations Convention to Combat Desertification, formed in 1996 and ratified by 179 countries, reports that over 250 million people are directly affected by

Figure 2.5 Deforestation, such as this clear-cut in north Florida, destroys animal habitat, causes soil erosion, and affects biodiversity. Green building standards call for the use of wood products from sustainably managed forests. (Photograph: M.R. Moretti.)

Figure 2.6 Desertification in South Niger, Africa, is consuming not only land but also local villages.

desertification.[41] Furthermore, drylands susceptible to desertification cover 40 percent of the Earth's surface, putting at risk a further 1.1 billion people in more than 100 countries dependent on these lands for survival. China, with a rapidly growing population and economy, loses about 300,000 acres (121,000 hectares) of land each year to drifting sand dunes.

EUTROPHICATION AND ACIDIFICATION

Two environmental conditions that frequently threaten water supplies are eutrophication and acidification. *Eutrophication* refers to the overenrichment of water bodies with nutrients from agricultural and landscape fertilizer, urban runoff, sewage discharge, and eroded stream banks. Nutrient oversupply fosters algae growth, or algae blooms, which block sunlight and cause underwater grasses to die. Decomposing algae further utilize dissolved oxygen necessary for the survival of aquatic species such as fish and crabs. Eventually, decomposition in a completely oxygenless, or *anoxic,* water body can release toxic hydrogen sulfide, poisoning organisms and making the lake or seabed lifeless. Eutrophication has led to the degradation of numerous waterways around the world. For example, in the Baltic Sea, huge algae blooms, now common after unusually warm summers, have decreased water visibility by 10 to 15 feet (3 to 4.6 meters) in depth.

Acidification is the process whereby air pollution in the form of ammonia, sulfur dioxide, and nitrogen oxides, mainly released into the atmosphere by burning fossil fuels, is converted into acids. The resulting *acid rain* is well known for its damage to forests and lakes. Less obvious is the damage caused by acid rain to freshwater and coastal ecosystems, soils, and even ancient historical monuments. The acidity of polluted rain leaches minerals from soil, causing the release of heavy metals that harm microorganisms and affect the food chain. Many species of animals, fish, and other aquatic animal and plant life are sensitive to water acidity. As a result of European directives that forced the installation of desulfurization systems and discouraged the use of coal as a fossil fuel, Europe experienced a significant decrease in acid rain in the 1990s. Nonetheless, a 1999 survey of forests in Europe found that about 25 percent of all trees had been damaged, largely due to the effects of acidification.[42]

Figure 2.7 Agricultural runoff, urban runoff, leaking septic systems, sewage discharges, eroded stream banks, and similar sources can increase the flow of nutrients and organic substances into aquatic systems, overstimulating algae growth, causing eutrophication and interfering with the recreational use of lakes and estuaries, and adversely affecting the health and diversity of indigenous fish, plant, and animal populations. (Photograph: M.R. Moretti.)

LOSS OF BIODIVERSITY

Biodiversity refers to the variety and variability of living organisms and the ecosystems in which they occur. The concept of biodiversity encompasses the number of different organisms, their relative frequencies, and their organization at many levels, ranging from complete ecosystems to the biochemical structures that form the molecular basis of heredity. Thus, biodiversity expresses the range of life on the planet, considering the relative abundances of ecosystems, species, and genes. Species biodiversity is the level of biodiversity most commonly discussed. An estimated 1.7 million species have been described out of a total estimated 5 to 100 million species. However, deforestation and climate change are causing such a rapid extinction of many species that some biologists are predicting the loss of 20 percent of existing species over the next 20 years.

Deforestation is particularly devastating, especially in rainforests, which comprise just 6 percent of the world's land but contain more than 500,000 of its species. Biodiversity preservation and protection is important to humanity since diverse ecosystems provide numerous services and resources, such as protection and formation of water and soil resources; nutrient storage and cycling; pollution breakdown and absorption; food; medicinal resources; wood products; aquatic habitat; and undoubtedly many undiscovered applications.[43] Once lost, species cannot be replaced by human technology, and potential sources of new foods, medicines, and other technologies may be forever forfeited.

Furthermore, destruction of ecosystems contributes to the emergence and spread of infectious diseases by interfering with natural control of disease vectors. For example, the fragmentation of North American forests has resulted in the elimination of the predators of the white-footed mouse, which is a major carrier of Lyme disease, now the leading vector-borne infectious illness in the United States. Finally, species extinction prevents discovery of potentially useful medicines such as aspirin, morphine, vincristine, taxol, digitalis, and most antibiotics, all of which have been derived from natural models.[44]

TOXIC SUBSTANCES AND ENDOCRINE DISRUPTORS

One dangerous by-product of the human propensity to invent has been the creation of an enormous number of chemical compounds that have no analogue in nature and often affect biological systems toxically. A *toxic substance* is a chemical that can cause death, disease, behavioral abnormalities, cancer, genetic mutations, physiological or reproductive malfunctions, or physical deformities in any organism or its offspring, or that can become poisonous after concentration in the food chain or in combination with any other substances.[45] Toxic substances can be carcinogenic or mutagenic, or affect developmental, reproductive, neurological, or respiratory systems. Ignitable or corrosive substances are also classified as toxic. As an aside, *toxins* are biological poisons that are the by-products of living organisms. A toxin may be obtained naturally, that is, from secretions of various organisms, or it may be synthesized.

The rate of synthetically produced chemicals in the United States has increased from 1 million tons (0.9 million metric tons) per year in 1940 to over 125 million tons (113 million metric tons) per year in 1987. And in spite of the fact that, in 1994, approximately 13.5 million chemicals were listed with the Chemical Abstract Service (CAS), the National Academy of Sciences stated that adequate information to assess public health hazards existed for only 2 percent of these chemicals. Each year, more than 6,000 new chemical compounds are developed; however, industry is required to report the environmental release of only 320 specific substances. Over 3 billion pounds (1.4 billion kilograms) of toxic chemicals enter the environment each year, with official hazardous waste production amounting to 1,400 trillion pounds (635 trillion kilograms) per year. Each year, U.S. industry produces about 12 pounds (5.4 kilograms) of toxic waste per capita.[46] Since 1987, industries have been required to report the release of certain chemicals to the government through the Toxic Release Inventory (TRI), but TRI does not cover all chemicals or all industries, and only the largest facilities are required to report. A recent report by the U.S. Public Interest Research Group (PIRG) Education Fund summarized that the following amounts of chemicals were released into the atmosphere in the year 2000:[47]

- *Cancer-causing chemicals:* 100 million pounds (45.4 million kilograms), with dichloromethane being the most frequent
- *Chemicals, such as toluene, linked to developmental problems:* 138 million pounds (63 million kilograms)
- *Chemicals, such as carbon disulfide, related to reproductive disorders:* 50 million pounds (23 million kilograms)
- *Respiratory toxicants:* 1.7 billion pounds (0.8 billion kilograms), most commonly acid aerosols of hydrochloric acid
- *Dioxins:* 7,000 grams (15.4 pounds)
- *Persistent toxic substances:* lead (275,000 pounds (125,000 kilograms)), lead compounds (1.3 million pounds (0.6 million kilograms)), mercury (30,000 pounds (13,600 kilograms)), and mercury compounds (136,000 pounds (62,000 kilograms))

During the past decade, it has become apparent that many chemicals damage animal and human hormonal systems. Endocrine-disrupting chemicals (EDCs) interfere with the hormones produced by the endocrine system, a complex network of glands and hormones that regulates the development and function of bodily organs, physical growth, development, and maturation. Some commonly known EDCs are dioxin, polychlorinated biphenyls (PCBs), DDT, and various pesticides and plasticizers. EDCs have been implicated in the occurrence of abnormally swollen thyroid glands in the eagles, terns, and gulls found in the fish-bird food chain of the Great Lakes. EDCs have contributed to the appearance of alligators with diminished reproductive organs and are blamed for the declining alligator populations in Lake Apopka, Florida. The most notorious example occurring in the human population was the use of diethylstilbestrol (DES), a synthetic estrogen prescribed until 1971 to prevent miscarriages in pregnant women. DES has since been linked to numerous health problems in offspring exposed to DES in the womb, including reproductive complications and infertility in DES daughters.[48] Although a "Better Life through Chemistry," the tagline of American industry of the 1950s, can still be claimed, the uncxpcctcdly high pricc tag is still bcing tallicd.

DEPLETION OF METAL STOCKS

The depletion of key resources needed to support the energy and materials requirements of today's technological, developed world societies is a threat to the high quality of life enjoyed by North Americans, Europeans, Japanese, and the other countries that make up modern industrialized societies. The subject of oil depletion is covered in Chapter 1 of this book, and evidence to date seems to indicate that we have maximized our ability to extract oil and that we are in an era of probably far higher prices for oil-based products, among them gasoline, diesel fuel, jet fuel, and oil-based polymers. A similar scenario is playing out with other key resources, most notably metals. A recent study of the supply and usage of copper, zinc, and other metals has determined that supplies of these resources—even if recycled—may fail to meet the needs of the global population.[49] Even the full extraction of metals from the Earth's crust and extensive recycling programs may not meet future demand if all countries try to attain the same standard of living enjoyed in developed nations. The researchers, Robert Gordon, Marlen Bertram, and Thomas Graedel, based their study on metal still in the Earth, in use by people and lost in landfills. Using copper stocks in North America as a starting point, they tracked the evolution of copper mining, use, and loss during the twentieth century. They then applied their findings and additional data to an estimate of the global demand for copper and other metals if all nations were fully developed and used modern technologies. The study found that all of the copper in ore, plus all of the copper currently in use, would be required to bring the world to the level of the developed nations for power transmission, construction, and other services and products that depend on copper. Globally, the researchers estimate that 26 percent of extractable copper in the Earth's crust is now lost in nonrecycled wastes, while lost zinc is estimated at 19 percent. Interestingly, the researchers said that current prices do not reflect those losses because supplies are still large enough to meet the demand, and new methods have helped mines produce material more efficiently. While copper and zinc are not at risk of depletion in the immediate future, the researchers believe that scarce metals, such as platinum, are at risk of depletion in this century because there is no suitable substitute for their use in devices such as catalytic converters and hydrogen fuel cells. And because the rate of use for metals continues to rise, even the more plentiful metals may face similar depletion risks in the not too distant future. The impact on metal prices due to a combination of demand and dwindling stocks has been dramatic. In a single year 2005–2006, zinc and copper experienced a 300 percent rise, and metals such as nickel, brass, and stainless steel rose by about 250 percent. In spite of the higher prices, the good news is that there is a renewed emphasis on recycling, using only the quantity of metals required and ensuring that all in-plant scrap is recovered during manufacturing.[50]

The Green Building Movement

More than any other human endeavor, the built environment has direct, complex, and long-lasting impacts on the biosphere. In the United States, the production and manufacture of building components, along with the construction process itself, involves the extraction and movement of 6 billion tons of basic materials annually. The construction industry, representing about 8 percent of the U.S. gross domestic product

(GDP), consumes 40 percent of extracted materials in the United States. Some estimates suggest that as much as 90 percent of all materials ever extracted reside in today's buildings and infrastructure. Construction waste is generated at a rate of about 0.5 ton (0.45 metric tons) per person each year in the United States, or about 5 to 10 pounds per square foot (24 to 49 kilograms per square meter) of new construction. Waste from renovation occurs at a level of 70 to 100 pounds per square foot (344 to 489 kilograms per square meter). The demolition process results in truly staggering quantities of waste, with little or no reuse or recycling occurring. Of the approximately 145 million tons (132 million metric tons) of construction and demolition waste generated each year in the United States, about 92 percent is demolition waste, with the remainder being waste from construction activities. In addition to the enormous quantities of waste resulting from built environment activities, questionable urban planning and development practices also have far-reaching consequences. Since transportation consumes about 40 percent of primary energy consumption in the United States, the distribution of the built environment and the consequent need to rely on automobiles for movement between work, home, school, and shopping results in disproportionate energy consumption, air pollution, and the generation of carbon dioxide, which contributes to global warming.

The green building movement is the response of the construction industry to the environmental and resource impacts of the built environment. As was noted in Chapter 1, the term *green building* refers to the quality and characteristics of the actual structure created using the principles and methodologies of sustainable construction. In the context of green buildings, resource efficiency means high levels of energy and water efficiency, appropriate use of land and landscaping, the use of environmentally friendly materials, and minimizing the life-cycle effects of the building's design and operation.

Figure 2.8 Annual construction and demolition waste in the United States is estimated to be about 145 million tons (132 million metric tons), or about one-half ton per capita. Buildings are not generally designed to be disassembled, and the result is that only a small percentage of demolition materials can be recycled. The partial demolition of the Law School Library at the University of Florida in mid-2004 illustrates the quantities of waste typically generated in renovation products, on the order of 70 pounds per square foot (344 kg per square meter). (Photograph: M.R. Moretti.)

GREEN BUILDING ORGANIZATIONS—UNITED STATES

Key American organizations promoting the implementation of sustainable construction practices include the U.S. Green Building Council, the Green Building Initiative, the U.S. Department of Energy, the U.S. Environmental Protection Agency, the National Association of Home Builders, the Department of Defense, and other public agencies and nonprofit companies. The private sector has been led by several manufacturers. Notably, Ray Anderson, Chairman of Interface, Inc., has guided the company's transition from a conventional carpet tile manufacturer to one with a corporate philosophy based on industrial ecology. Anderson's efforts to move Interface toward sustainability prompted competition among other manufacturers to produce "green" carpet tiles, among them Milliken and Collins Aikman. In the U.S. commercial building arena, the prime green building organization is the U.S. Green Building Council (USGBC), located in Washington, DC. A relatively new organization, the Green Building Initiative (GBI), which is headquartered in Oregon, acquired the rights to a Canadian building assessment standard known as Green Globes in 2004. The GBI has adapted Green Globes to the U.S. building market and is offering it as an alternative to the USGBC LEED building rating systems.

Homebuilding and residential development are represented by a proliferation of organizations, many of which preceded the USGBC and arose independently in homebuilding organizations and municipalities across the United States. The city of Boulder, Colorado, took an aggressive stance in 1998 with respect to green building by passing an ordinance requiring specific measures. Pennsylvania established the Governor's Green Government Council (GGGC) in part to address the implementation of green building principles in the state. The city of Austin, Texas, is perhaps best known for its efforts in green building, and was the recipient of an award at the first UN conference on sustainable development in Rio de Janeiro in 1992. Local residential green building movements have emerged in Denver, Colorado; Kitsap County, Washington; Clark County, Washington; Baltimore, Maryland, with the Suburban Builders Association; and, more recently, in Atlanta, Georgia, with the EarthCraft Houses Program.

The National Association of Homebuilders now provides guidance to its 800 state and local associations to assist in implementing green building programs in local homebuilding associations. Reliable and independent information and critical analysis is published by BuildGreen, Inc., in its monthly newsletter, *Environmental Building News*. BuildGreen, Inc., also publishes GreenSpec, a directory of products addressed to high-performance building needs, and provides the Green Building Advisor, computer software that facilitates green building design.

GREEN BUILDING ORGANIZATIONS—INTERNATIONAL

The international green building movement came of age in the early 1990s thanks to the activities of task groups within CIB, a construction research networking organization based in Rotterdam, and the International Union for Experts in Construction Materials, Systems, and Structures (RILEM), based in Bagneux, France. In 1992, CIB Task Group 8 on Building Assessment provided international impetus for the development and implementation of building assessment tools and standards. CIB Task Group 16 on Sustainable Construction helped consolidate international standards regarding the application of sustainability principles to the built environment. And the relatively new International Institute for a Sustainable Built Environment (iiSBE)[51] provides a clearinghouse for an extensive range of green building information. The iiSBE also organizes the biannual Green Building Challenge and Sustainable Building Conference and facilitates international sustainable building assessment with its main assessment method, the Green Building Tool (GBTool), which is used at the biannual conferences to assess or rate entrant exemplary buildings worldwide. For the Green Building Challenge held in Tokyo in 2005, the U.S. team, sponsored by the U.S. Department of Energy, submitted buildings as American Best Practices in five categories:

1. Retail: Big Horn Improvement Center, Silverthorne, Colorado
2. Office: Phillip Merrill Environmental Center, Chesapeake Bay Foundation, Annapolis, Maryland
3. School: Clearview Elementary School, Hanover, Pennsylvania
4. Multifamily residential: Twenty River Terrace, Battery Park City, New York
5. Office/laboratory: National Oceanic and Atmospheric Administration Building, Honolulu, Hawaii

HISTORY OF THE U.S. GREEN BUILDING MOVEMENT

In the United States, the green building movement has a long history, with its philosophical roots traceable to the late nineteenth century. Subsequently, it developed in tandem with the country's environmental movement, and since the 1990s it has been enjoying a renaissance. Notable dates include 1970, the year the first Earth Day was celebrated and the U.S. Environmental Protection Agency was created, both events marking a major philosophical shift. Other influential events include the publication of Rachel Carson's landmark book *Silent Spring* in 1962 and the efforts of early environmentalists such as Barry Commoner, Lester Brown, Dennis Hayes, and Donnella Meadows. Concern over resource availability, particularly reliance on fossil fuels, was magnified by the oil shocks of the early 1970s, which resulted from the Arab-Israeli conflict of the time. This further piqued public interest in energy efficiency, solar technologies, retrofitting homes and commercial buildings with insulation, and energy recovery systems. As a result, the federal government began to provide tax credits for investment in solar energy, and funded development and testing of innovative technologies ranging from solar air conditioning to eutectic salt energy storage batteries. By the late 1970s, many new efficiency standards were embodied in model energy codes adopted by the states. After this burst of activity, however, interest in energy conservation began to wane as energy prices began to decline.

The early 1990s saw a renewed interest in energy and resource conservation as humans began to seriously consider more complex global environmental issues such as ozone depletion, global climate change, and destruction of major fisheries. Three events in the late 1980s and early 1990s helped to focus attention on problems associated with global environmental impacts: the publication in 1987 of *Our Common Future,* commonly referred to as the Bruntland Report; the 1989 meeting of the American Institute of Architects (AIA), at which it established its Committee on the Environment (COTE); and the United Nations Conference on Sustainable Development in 1992, commonly known as the Rio Conference.

The recent American resurgence in sustainable construction was precipitated in 1993 by a joint meeting of the International Union of Architects (UIA) and the AIA, known as "Architecture at the Crossroads." The UIA/AIA World Congress of Architects promulgated the Declaration of Interdependence for a Sustainable Future, which articulated a code of principles and practices to facilitate sustainable development (see Figure 2.9).

Although many energy-efficient buildings emerged after the oil crises of the 1980s, the first U.S. buildings that considered a wider range of environmental and resource issues did not emerge until the 1980s. The earliest examples of green buildings were the result of major U.S. environmental organizations requiring holistic approaches to the design of their office buildings. In 1985, William McDonough was hired by the Environmental Defense Fund to design its New York offices. The design featured natural materials, daylighting, and excellent indoor air quality, all part of a

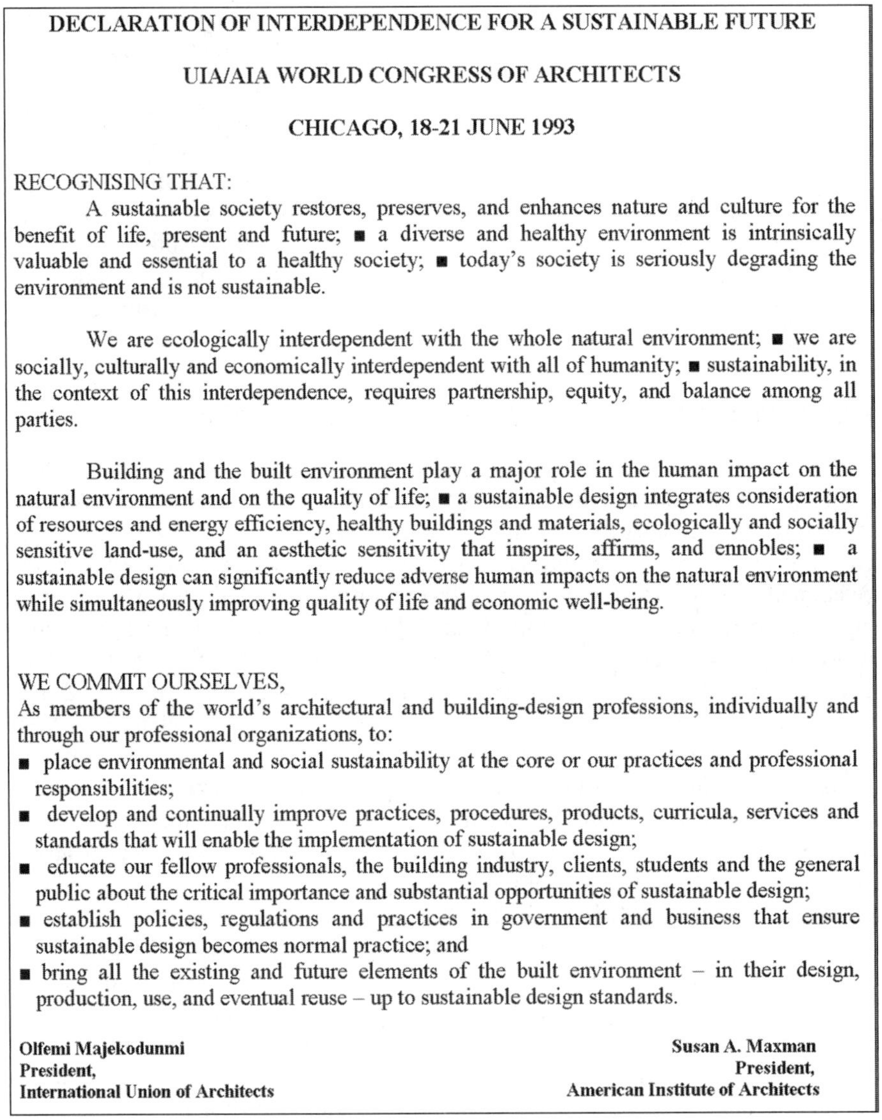

DECLARATION OF INTERDEPENDENCE FOR A SUSTAINABLE FUTURE

UIA/AIA WORLD CONGRESS OF ARCHITECTS

CHICAGO, 18-21 JUNE 1993

RECOGNISING THAT:

A sustainable society restores, preserves, and enhances nature and culture for the benefit of life, present and future; ■ a diverse and healthy environment is intrinsically valuable and essential to a healthy society; ■ today's society is seriously degrading the environment and is not sustainable.

We are ecologically interdependent with the whole natural environment; ■ we are socially, culturally and economically interdependent with all of humanity; ■ sustainability, in the context of this interdependence, requires partnership, equity, and balance among all parties.

Building and the built environment play a major role in the human impact on the natural environment and on the quality of life; ■ a sustainable design integrates consideration of resources and energy efficiency, healthy buildings and materials, ecologically and socially sensitive land-use, and an aesthetic sensitivity that inspires, affirms, and ennobles; ■ a sustainable design can significantly reduce adverse human impacts on the natural environment while simultaneously improving quality of life and economic well-being.

WE COMMIT OURSELVES,

As members of the world's architectural and building-design professions, individually and through our professional organizations, to:

- place environmental and social sustainability at the core or our practices and professional responsibilities;
- develop and continually improve practices, procedures, products, curricula, services and standards that will enable the implementation of sustainable design;
- educate our fellow professionals, the building industry, clients, students and the general public about the critical importance and substantial opportunities of sustainable design;
- establish policies, regulations and practices in government and business that ensure sustainable design becomes normal practice; and
- bring all the existing and future elements of the built environment – in their design, production, use, and eventual reuse – up to sustainable design standards.

Olfemi Majekodunmi
President,
International Union of Architects

Susan A. Maxman
President,
American Institute of Architects

Figure 2.9 The joint Declaration of Interdependence for a Sustainable Future by the International Union of Architects (UIA) and the American Institute of Architects (AIA) in Chicago in 1993 was an important event in the history of the high performance green building movement.

green solution for then endemic sick building problems. In 1989, the Croxton Collaborative, a design firm founded by Randy Croxton, designed the offices of the Natural Resources Defense Council in the Flatiron district of New York City. In this project, natural lighting and energy-conserving technologies were employed to reduce energy consumption by two-thirds compared to conventional buildings. The 1992 renovation of Audubon House, also in New York City, was a significant early effort in the contemporary green building movement. The organization sought to reflect its values as a leader of the environmental movement and directed architect Randy Croxton to design the building in the most environmentally friendly and energy-efficient manner possible. In the process of achieving that goal, the extensive collaboration required by the many building team members provided a model of cooperation that has now become a hallmark of the contemporary green building process in the United States.[52]

The first highly publicized green building project in the United States, the "Greening of the White House," was initiated in 1993 and included renovation of the Old Executive Office Building, a 600,000-square-foot (55,700 square meters) structure across from the White House. The participation in this project of a wide array of architects, engineers, government officials, and environmentalists drew national attention and resulted in dramatic energy cost savings (about $300,000 per year), emissions reductions (845 tons (767 metric tons) of carbon per year), and significant reductions in water and solid waste associated costs. The success of the White House project spurred the federal government's sustainability efforts and prompted the U.S. Post Office, the Pentagon, the Department of Energy, and the Government Services Administration to address sustainability concerns within their organizations. The National Park Service, too, opened green facilities at several national parks, including the Grand Canyon, Yellowstone, and Alaska's Denali. The Naval Facilities Engineering Command (NAVFAC), the U.S. Navy's construction arm, began a series of eight pilot projects to address sustainability and energy conservation concerns. The highly visible effort at its 156,000-square-foot (14,500 square meters), 150-year-old headquarters in the Washington Navy Yard reduced energy consumption by 35 percent and resulted in annual savings of $58,000.[53]

In addition, several important guides to green building or sustainable design appeared in the early to mid-1990s. The *Environmental Building News,* first published in 1992, remains an independent, dispassionate, and authoritative guide to sustainable construction.[54] In 1994, the AIA first published its "Environmental Resources Guide," followed by a more detailed version in 1996.[55] The "Guiding Principles for Sustainable Design," produced by the National Park Service in 1994, provides one of the first overviews of green building production.[56] Similarly, the "Sustainable Building Technical Manual" was developed and published jointly by the U.S. Department of Energy and Public Technology, Inc., in 1996.[57] The Rocky Mountain Institute's "A Primer on Sustainable Building," published in 1995, also contributed to the public understanding of sustainable construction.

Other international efforts and organizations interacted with and influenced the U.S. movement during this period. The British green building rating system, BREEAM, was developed in 1992. As noted previously, in 1992 CIB convened Task Group 8 (Building Assessment) and Task Group 16 (Sustainable Construction), which held influential international conferences in 1994 in the United Kingdom and Tampa, Florida. Also, as noted earlier, the USGBC, headquartered in Washington, DC, was formed in 1993 and held its first major meeting in March 1994.[58] Early articulations of the organization's LEED Standard appeared at this time, along with green building standards developed by the American Society for Testing and Materials (ASTM). The ASTM standards were eventually set aside in favor of the USGBC's LEED assessment standard.

Development of the USGBC's LEED building rating system took 4 years and culminated in a 1998 test version known as LEED Version 1.0. It was enormously successful, and the Federal Energy Management Program sponsored a pilot effort to test its assumptions. Eighteen projects comprising more than 1 million square feet (93,000 square meters) were evaluated in this beta testing. A greatly improved LEED 2.0 was launched in 2000, and provided for a maximum of 69 credits and four building certifications: Platinum, Gold, Silver, or Bronze. A further refined LEED 2.1 was published in 2003, changing the lowest, Bronze, level of certification to the designation "Certified." The formal name of LEED 2.1 was modified to include New Construction (NC), distinguishing it from LEED rating systems for other applications; currently it is known as LEED-NC 2009. Other LEED rating systems include LEED-EB for existing buildings, LEED-CI for commercial interiors, LEED-H for homes, and LEED-CS for core and shell.

New approaches, including the GBI's *Green Globes Design* and *Green Globes for Continual Improvement of Existing Buildings,* as well as the National Association of Home Builders *Model Green Home Guidelines,* are reinforcing the enormous growth in green building by providing a variety of approaches to rating green building and creating competition to improve green building rating systems.

(A)

(B)

Figure 2.10 (A) Audubon House in New York City was designed by the Croxton Collaborative as the headquarters of the Audubon Society. It is one of the projects marking the start of the contemporary U.S. green building movement. (B) Desk illumination from a skylight. (Photographs courtesy of the Croxton Collaborative Architects, P.C.)

Figure 2.11 The "Greening of the White House" project was the first widely publicized federal government green building project. (Illustration used by permission of View by View, Inc.)

Summary and Conclusions

Significant global environmental problems increasingly threaten food supplies, water and air quality, and the survival of ecosystems upon which humanity depends for a wide variety of goods and services. Because it uses enormous quantities of resources and replaces natural systems with human artifacts, the built environment sector of the economy has disproportionate environmental impacts on the planet. Consequently, the construction industry has a special obligation to behave proactively and shift rapidly from wasteful, harmful practices to a paradigm under which construction and nature work synergistically rather than antagonistically. This new model of sustainable construction is referred to as *high-performance green building.*

The green building movement is a relatively recent phenomenon, and in the United States it is presently growing at an exponential rate. The USGBC's LEED building assessment standard has emerged as the definitive guideline. It articulates the parameters for green buildings in the United States and several other countries. Parallel efforts in other economic sectors are occurring simultaneously as manufacturers attempt to design and produce goods with low environmental impact. The concepts of closed materials loops, efficient resource use, and the redesign of products and buildings to emulate natural systems are indispensable to preserve humanity's quality of life, along with the constant acknowledgment that nature is the source of that quality.

Notes

1. The five prior extinctions were the Orodovcian (440 million years ago), Devonian (365 million years ago), Permian (245 million years ago), Triassic (210 million years ago), and Cretaceous (66 million years ago). The as yet unnamed Sixth Extinction is not being caused by major geologic upheavals, as was the case for the previous five, but instead by the activities of just one of the millions of species inhabiting the planet: humans.

2. Xerox's activities to redesign its product line and incorporate sustainability in the company's philosophy are described by Maslennikova and Foley (2000).
3. This commonly accepted definition of sustainable development was first stated in *Our Common Future* (1987).
4. From Peterson (1999).
5. From Howarth (1992).
6. As stated on the website of the Center for Community Action and Environmental Justice, www.ccaej.org.
7. From Drexler (1987).
8. A description of the NSF's Biocomplexity in the Environment Program can be found at www.eng.nsf.gov/be.
9. From Goodin (1983).
10. From Rochlin (1978).
11. From Angyal (2003).
12. From Berry (2002).
13. From Taylor (1981).
14. From Leopold (1949).
15. In Frosch and Gallopoulos (1989), the term *industrial ecology* was used for the first time in the popular scientific press. This marked the beginning of the widespread use of this phrase to describe a wide variety of environmentally responsible approaches to industrial production.
16. This definition of *industrial ecology* is from Garner and Keoleian (1995).
17. Robert Ayres has written extensively on the subject of industrial materials flows. More detailed information on the problem of enormous waste can be found in Ayres (1989).
18. An excellent summary of industrial ecology in general and the Kalundborg plant specifically can be found at the website of Indigo Development, www.indigodev.com. Several excellent references and handbooks are also available from the website. Indigo Development, founded by Ernie Lowe, is devoted to the further development of industrial ecology, which he refers to as ". . . an interdisciplinary framework for designing and operating industrial systems as living systems that are interdependent with natural systems."
19. Construction ecology is defined in the context of industrial ecology and sustainable construction in "Defining an Ecology of Construction," by Kibert, Sendzimir, and Guy (2002).
20. In addition to Janine Benyus's book on biomimicry, a useful website providing an overview of this concept is www.biomimicry.net.
21. This definition of DfE is from Keoleian and Menerey (1994).
22. An excellent short overview of ecological economics by Stephen Farber of the Graduate School of Public and International Affairs, University of Pittsburgh, can be found at www.fs.fed.us/eco/s21pre.htm.
23. The definition of carrying capacity is from the Carrying Capacity Network at www.carryingcapacity.org.
24. A thorough description of the ecological footprint concept can be found in Wackernagel and Rees (1996).
25. A gigaton is 1 billion tons.
26. The EEA has an excellent online glossary of environmental terms at http://glossary.eea.eu.int/EEAGlossary.
27. The ecological rucksack quantities are derived from a number of online and published sources. A good description of the concept, along with a diagram showing relative ecological rucksacks for a variety of materials can be found in von Weizsäcker, Lovins, and Lovins (1997).
28. From Kahn (1997).
29. From Stephen R. Kellert, "Ecological Challenge, Values of Nature, and Sustainability," in Kibert (1999) and Kahn (1997).
30. From World Business Council for Sustainable Development (1996).
31. As stated on the WBCSD website, www.wbcsd.org.
32. The website of the U.S. branch of the Natural Step is www.naturalstep.org.
33. Excerpted from the website of *Canadian Architect,* www.cdnarchitect.com.
34. The book *Factor Four: Doubling Wealth, Halving Resource Use* was written as a report to the Club of Rome as a follow-up to the 1972 book *The Limits to Growth,* written by

Dennis Meadows, Donella Meadows, Jorgen Randers, and William Behrens III, which was the original report to the club. *Limits to Growth* stated that exponential growth in population and the world's industrial system would force growth on the planet to be halted within a century, a result of environmental impacts and resource shortages.

35. The Factor 10 concept continues to be fostered by the Factor 10 Club and the Factor 10 Institute, whose publications and activities can be found at www.factor10-institute.org.
36. According to the authors of *Factor Four,* Lee Eng Lock's supply fans use 0.061 kW/ton of cooling versus 0.60 kW/ton in conventional practice. Similarly, his chilled water pumps use 0.018 kW/ton versus 0.16 kW/ton; condenser water pumps use 0.018 kW/ton versus 0.14 kW/ton; and cooling towers use 0.012 versus 0.10 kW/ton.
37. The Third Assessment Report of the IPCC (2001) can be found at www.ipcc.ch.
38. Excerpted from "Twenty Questions and Answers about the Ozone Layer" (2002) at www.epa.gov/ozone/science/unepSciQandA.pdf.
39. Rate of rainforest destruction according to the Rainforest Action Alliance available at www.rainforest-alliance.org.
40. Data are from the *Global Environmental Outlook 2002 Report* (GEO-3) (2002) at www.epa.gov/ozone/science/unepSciQandA.pdf.
41. The website of the United Nations Convention to Combat Desertification is www.unccd.int.
42. A group of Swedish nongovernmental organizations maintains a website promoting knowledge about the effects of acid rain, www.acidrain.org.
43. See "Global Environmental Problems: Implications for U.S. Policy" (January 2003).
44. Excerpted from "The Loss of Biodiversity and Its Negative Effects on Human Health" (2004).
45. The definition of toxic substances is adapted from the definition provided on the Great Lakes website of Environment Canada, www.on.ec.gc.ca/water/raps.
46. Excerpted from reports on the website of the Center for Community Action and Environmental Justice, www.ccaej.org.
47. Excerpted from Dutzik, Bouamann, and Purvis (2003).
48. Information on endocrine disruptors can be found on the website of the National Resources Defense Council (NRDC), www.nrdc.org.
49. From Gordon, Bertram, and Graedel (2006).
50. From "Materials Prices Dictate Creative Engineering" (2006).
51. The iiSBE website is www.iisbe.org.
52. The story of the Audubon House design process is recounted in Croxton Collaborative and the National Audubon Society (1992).
53. An excellent detailed overview of the history of the U.S. green building movement can be found in the "White Paper on Sustainability" (2003). This publication also contains other important background information about the green building movement and suggests an action plan to help improve and ensure the quality and outcomes of green building design and construction.
54. BuildingGreen, Inc., publishes *Environmental Building News* and produces a range of other useful products, including the GreenSpec Directory. All of its publications are also available by subscription at www.buildinggreen.com.
55. The "Environmental Resource Guide" is a thorough guide to the environmental and resource implications of construction materials. The first version was published by the AIA in 1994; the second, expanded version was published by John Wiley & Sons in 1996.
56. The National Park Service's "Guiding Principles of Sustainable Design" is available at www.nps.gov/dsc/d_publications/d_1_gpsd.htm.
57. The "Sustainable Building Technical Manual" is available at www.sustainable.doe.gov/freshstart/articles/ptipub.htm.
58. The USGBC's earliest organizers were David Gottfried and Michael Italiano, and its first president was Rick Fedrizzi, who at the time was with Carrier Corporation. The first annual meeting of the USGBC was held in Washington, DC, in March 1994 and featured as its keynote speakers Paul Hawken, who had just completed the groundbreaking book *Ecology of Commerce,* and William McDonough, recognized as one of the major architectural figures in the U.S. green building movement and the author of *The Hannover Principles.*

References

Angyal, Thomas J. 2003. "Thomas Berry's Earth Spirituality and the 'Great Work,'" *The Ecozoic Reader,* 3, pp. 35–44.

Ausubel, J.H., and H.E. Sladovich, eds. 1989. *Technology and the Environment.* Washington, DC: National Academy Press.

Ayres, Robert U. 1989. "Industrial Metabolism," in *Technology and the Environment.* (J.H. Ausubel, and H.E. Sladovich, eds.) Washington, DC: National Academy Press.

Benyus, Janine. 1997. *Biomimicry: Innovation Inspired by Nature.* New York: HarperCollins.

Berry, Thomas. 2002. "Rights of the Earth: Earth Democracy," *Resurgence,* 214, pp. 28–29.

Croxton Collaborative and the National Audubon Society. 1992. *Audubon House: Building the Environmentally Responsible, Energy Efficient Office.* New York: John Wiley & Sons.

Demkin, Joseph, ed. 1996. *Environmental Resource Guide.* New York: John Wiley & Sons.

Drexler, K. Eric 1987. *Engines of Creation.* New York: Anchor Books.

Dutzik, Tony, Jeremiah Bouamann, and Mehgan Purvis. January 2003. "Toxic Releases and Health: A Review of Pollution Data and Current Knowledge on the Health Effects of Toxic Chemicals." Written for the U.S. PIRG Education Fund. Available at www.uspirg.org/reports/toxics03/toxicreleases1_03report.pdf.

"Eco-Efficient Leadership for Improved Economic and Environmental Performance." 1996. World Business Council for Sustainable Development. Available at www.wbcsd.org.

Frosch, Robert, and Nicholas Gallopoulos. September 1989. "Strategies for Manufacturing," *Scientific American,* pp. 144–152.

Garner, Andy, and Gregory Keoleian. 1995. *Industrial Ecology: An Introduction,* National Pollution Prevention Center in Higher Education, University of Michigan, Ann Arbor, Michigan.

Global Environmental Outlook 2002 Report (GEO-3). 2002. Published by the United Nations Environmental Program. Available at www.unep.org/GEO/geo3.

"Global Environmental Problems: Implications for U.S. Policy." January 2003. Watson Institute for International Studies, Brown University. Available at www.choices.edu.

Goodin, Robert E. 1983. "Ethical Principles for Environmental Protection," in *Environmental Philosophy,* R. Elliot and A. Gare, eds. London: Open University Press.

Gordon, R.B., M. Bertram, and T.E. Graedel. 2006. "Metal Stocks and Sustainability," *Proceedings of the National Academy of Sciences,* 103(5), pp. 1209–1214.

Howarth, Richard B. 1992. "Intergenerational Justice and the Chain of Obligation," *Environmental Values,* 1, Isle of Harris, U.K.: White Horse Press.

Kahn, Peter H., Jr. 1997. "Developmental Psychology and the Biophilia Hypothesis: Children's Affiliations with Nature," *Developmental Review,* 17, pp. 1–61.

Kats, Greg. 2003. *Green Building Costs and Financial Benefits.* Addresses the economics of green buildings for the State of Massachusetts. Available at www.cap-e.com/publications/default.cfm.

Kats, Greg. October 2003. "The Cost and Benefits of Green Buildings," a report to California's Sustainable Buildings Task Force. Available at www.cap-e.com/publications/default.cfm.

Kellert, Stephen R., and E.O. Wilson, eds. 1993. *The Biophilia Hypothesis.* Washington, DC: Island Press.

Keoleian, Gregory, and D. Menerey. May 1994. "Sustainable Development by Design," *Air & Waste,* 44, pp. 645–668.

Kibert, Charles J., ed. 1999. *Reshaping the Built Environment: Ecology, Ethics, and Economics.* Washington, DC: Island Press.

Kibert, Charles J., Jan Sendzimir, and G. Bradley Guy, eds. 2002. *Construction Ecology: Nature as the Basis for Green Buildings.* London: Spon Press.

Leopold, Aldo. 1949. *A Sand County Almanac.* New York: Oxford University Press.

Lopez Barnett, Dianna, and William D. Browning. 1999. *A Primer on Sustainable Building.* Snowmass, CO: Rocky Mountain Institute.

Maslennikova, Irina, and David Foley. May–June 2000. "Xerox's Approach to Sustainability," *Interfaces,* 30(3), pp. 226–233.

"Materials Prices Dictate Creative Engineering." May 26, 2006. *Engineeringtalk,* available at www.engineeringtalk.com/news/lag/lag102.html.

Meadows, Donella H., Dennis I. Meadows, Jorgen Randers, and William W. Behrens III. 1972. *The Limits to Growth.* New York: Universe Books.

Our Common Future, 1987. World Commission on Environment and Development. Oxford: Oxford University Press.

Peterson, Gary. 1999. "Ecology of Construction," in *Construction Ecology: Ecology as the Basis for Green Buildings,* Charles J. Kibert, Jan Sendzimir, and Bradley Guy, eds. London: Spon Press.

Rochlin, Gene I. 1978. "Nuclear Waste Disposal: Two Social Criteria," *Science,* 195, pp. 23–31.

Sustainable Building Technical Manual. 1996. U.S. Department of Energy and the Public Technology Initiative, Inc. Available at www.sustainable.doe.gov/freshstart/articles/ptipub.htm.

Taylor, Paul W. 1981. "The Ethics of Respect for Nature," *Environmental Ethics,* 3, pp. 206–218.

"The Loss of Biodiversity and Its Negative Effects on Human Health." 2004. Available at the website of Students for Environmental Awareness in Medicine, seamglobal.com/lossofbiodiversity.html.

von Weizsäcker, Ernst, Amory Lovins, and L. Hunter Lovins. 1997. *Factor Four: Doubling Wealth, Halving Resource Use.* London: Earthscan Publications.

Wackernagel, Mathis, and William Rees. 1996. *Our Ecological Footprint.* Gabriola Island, British Columbia: New Society Publishers.

"White Paper on Sustainability: A Report on the Green Building Movement." 2003. *Building Design and Construction.* Available at www.bdcnetwork.com.

World Business Council for Sustainable Development. (1996). "Eco-Efficient Leadership for Improved Economic and Environmental Performance." Available at www.wbcsd.org.

Chapter 3

Green Building Assessment

During the pre-1998 era of sustainable construction in the United States, environmentally friendly buildings were conceptualized by teams of architects and engineers who relied on their collective interpretation of what constituted green building. Beyond the understanding that green buildings should be resource-efficient and environmentally friendly, no specific criteria existed to evaluate and compare the merits of green building design. In 1998, however, the USGBC dramatically changed the landscape with the launch of its LEED building assessment system for new construction, which identified criteria that specified not only whether a building was green, but what specific shade of green it was.

Now referred to as LEED for New Construction (LEED-NC), it employs a point system to award a Platinum, Gold, Silver, or Certified rating based on how many specific predetermined criteria in several categories the building successfully addresses. The generic term for LEED and similar systems used in other countries is *building assessment system*. As mentioned in Chapter 2, the primary building assessment system used in the United Kingdom is BREEAM, which also was the first widely adopted rating system in the world.[1] The Comprehensive Assessment System for Building Environmental Efficiency (CASBEE) is a relatively new building assessment approach created for Japanese construction; it is under development by the Japan Sustainable Building Consortium.[2] In Australia, Green Star is the building assessment system advocated by the Australian Green Building Council; it is fully implemented for commercial office design and construction.[3]

Building assessment systems score or rate the effects of a building's design, construction, and operation, among them environmental impacts, resource consumption, and occupant health. This can be a complicated determination, as each aspect has different units of measurement and applies at different physical scales. Environmental effects can be evaluated at local, regional, national, and global scales. Resource impacts are measured in terms of mass, energy, volume, parts per million (ppm), density, and area. Building health can be inferred by the presence or absence of chemical and biological substances within circulating air, as well as the relative health and well-being of the occupants. Comparing arrays of data for various building features presents further complications.

Why consider a building assessment standard or rating at all? In general, building assessment systems are created for the purpose of promoting high-performance buildings; and some, like LEED, are specifically designed to increase market demand for sustainable construction. Building assessment systems generally offer a label or plaque indicating a building's rating and displaying a public statement of the building's performance. A superior building assessment rating should create higher market value due to the building's lower operating costs and healthy indoor environment. Competition among owners and developers to achieve high building assessment ratings will ultimately create a high-quality, high-performance building stock. Parallel effects of successful building assessment systems could also help facilitate otherwise difficult political goals, for example, national requirements related to the Kyoto Protocols on climate change, which in effect call on the United States to significantly reduce fossil fuel consumption.[4]

Developers are faced with two major choices when designing a building assessment system: either to use a single number to describe the building's overall performance or to provide an array of numbers for the same purpose. A single number representing a score for the building has the virtue of being easy to understand. But if a single number is used to assess or rate a building, the system must somehow convert the many different units describing the building's resource and environmental impacts (energy usage, water consumption, land area footprint, materials, and waste quantities) and conditions resulting from the building design (building health, built-in recycling systems, deconstructability, percentage of products coming from within the local area) into a series of numbers that can be added together to produce a single overall score. This is a difficult and arbitrary method at best. Paradoxically, however, both the advantage and the disadvantage of the single-number assessment is its simplicity. The LEED standard provides a single number that determines the building's assessment or rating based on an accumulation of points in various impact categories, which are then totaled to obtain a final score.

Alternatively, a building assessment system can utilize an array of numbers or graphs that depict the building's performance in major areas, such as environmental loadings or energy and water consumption compared to conventional construction. Although this approach yields more detailed information, its complexity makes it difficult to compare buildings, depending on the range of factors considered. Green Building Tool (GBTool), a system used in the Green Building Challenge conferences to compare building performance in several countries, is an example of an assessment methodology that uses a relatively large quantity of information to assess the merits of a building's design.[5]

This chapter describes the two major U.S. building assessment standards, LEED and Green Globes. It also provides information about other major building assessment standards or systems used around the world, including the BREEAM system, the CASBEE system, the Green Star approach, and GBTool.

The USGBC LEED Building Assessment Standard

LEED is the predominant building assessment tool in the United States and, arguably, in the world. The success of LEED is the result of a long, careful development process that occurred between 1994 and 1998.[6] The earliest attempts at formulating a rating system, dating from 1994, were conducted under the aegis of the standards structure of the ASTM. This first iteration proved largely unsuccessful, and the effort that eventually produced the LEED rating system moved directly under the auspices of the USGBC. The most important decision of the USGBC members developing LEED was that green buildings should be market-driven rather than being required by regulation, meaning that the building owners would be the ultimate arbiters of the program's success. For commercial green buildings, this probably meant that they would have to distinguish themselves in the market by having higher resale value than comparable buildings.

A second significant decision in the development of LEED was to create a broad consensus-based process during its formulation. Typical building assessment standards are produced by national building research organizations such as the Building Research Establishment (BRE) in the United Kingdom. The standard is then "sold" to the respective building development market as a tool developed by a reputable institution that will help meet the public demand for more environmentally responsible behavior on the part of the building industry. In contrast, the USGBC was, and remains, a nongovernmental organization comprising many collaborators from industry, academia, and government. LEED was produced by a cross section of the USGBC's membership during a long, slow, and laborious process that sought to produce a green building rating system that would meet the needs of the wide range of participants in the building industry. The engagement of so many collaborators ensured acceptance when the standard was completed. In addition, government provided crucial financial support to the fledgling organization with an idea whose time had come: the U.S. Department of Energy offered critical funding in the form of grants to support LEED's development. Nonetheless, the USGBC was, and continues to be, a nonprofit, nongovernmental organization whose membership is drawn from diverse public and private stakeholders. Various LEED products continue to enjoy a high degree of success, largely as a result of the collaborative, consensus-based approach that marks both its products and the contemporary U.S. green building delivery system.

STRUCTURE OF LEED SUITE OF STANDARDS

Though referred to in the singular, LEED is not a single rating system but a *suite* of building rating systems. The first LEED product is now known as LEED-New Construction (NC) 2009, and it has evolved into a highly accepted measure of green building in the United States. In addition to LEED-NC 2009 there are several other LEED rating systems:

- LEED-EB: Existing Building-Operations and Maintenance
- LEED-CI: Commercial Interiors Projects
- LEED-CS: Core and Shell Projects
- LEED-H: Homes
- LEED-ND: Neighborhood Development
- LEED for Schools
- LEED for Retail
- LEED for Healthcare

LEED-NC was originally developed for office buildings and is now being used for almost every type of building except single-family homes. Therefore, the USGBC is developing a series of application guides for various building types. Guides are being created for healthcare facilities, lodging, volume building programs, multifamily residences, campuses, retail stores, and laboratories.[7]

BRIEF HISTORY OF LEED FOR NEW CONSTRUCTION

As noted previously, LEED, later renamed LEED-NC for commercial buildings, was developed by the USGBC during a 4-year process from 1994 to 1998 (see Table 3.1). The first version, known as LEED 1.0, was issued in 1998 as a beta version. Twenty buildings were certified using LEED 1.0 to obtain a rating that originally was either Platinum, Gold, Silver, or Bronze. LEED 2.0 was issued in 2000 as a dramatically changed version of the original LEED standard. Subsequent versions were marked as to their application and, in the case of the version for new construction, the descriptor NC was appended to the title. LEED-NC 2.1, issued in 2002, was virtually identical to LEED-NC 2.0, except that it had greatly simplified documentation requirements. LEED-NC 2.2, issued in 2005, added the USGBC LEED-online website. LEED-NC 2009 is the latest version of the LEED assessment standard for new construction.

As noted above, LEED-NC 2009 is the most recent USGBC standard for new commercial/institutional buildings and major renovations. It is structured with eight prerequisites and a maximum of 110 points divided into six major categories (see Table 3.2). Prerequisites are conditions that must all be successfully addressed for a building to be eligible for consideration for a LEED rating.

The number of points available in each category was established by the develop-

TABLE 3.1

LEED-NC Versions

LEED-NC Version Certified	Year Issued	Maximum Points	Buildings
1.0	1998		8
2.0	2000	69	282
2.1	2002	69	882
2.2	2005	69	177
2009	2009	110	N/A

TABLE 3.2

LEED-NC 2009

Category	Maximum Points
1. Sustainable Sites	26
2. Water Efficiency	10
3. Energy and Atmosphere	35
4. Materials and Resources	14
5. Indoor Environmental Quality	15
6. Innovation and Design Process	6
7. Regional Priority Credits	4
Total Possible Points	**110**

TABLE 3.3

Points Required for LEED-NC 2009 Ratings

LEED-NC 2009 Rating	Points Required
Platinum	80+
Gold	60–79
Silver	50–59
Certified	40–49
No rating	39 or less

ers of LEED-NC to indicate the weight they place on the various major issues addressed by this standard. Thus, the allocation of points to each category is arbitrary, based solely on the judgment of the developers. Consequently, it is arguable, for example, that Energy and Atmosphere (35 points maximum) is more important than Sustainable Sites (26 points maximum) and over three times as important as Water Efficiency (10 points maximum). This situation indicates some of the pitfalls inherent in a building assessment system, which attempts to reduce complex factors to a single number. Still, it does provide a logical and rational, if somewhat arbitrary, approach to producing numerical scores in each category. It is important to keep in mind that LEED was developed in an extensive collaborative process over a total of 6 years; hence, the outcome of this group thought process probably is fairly on target with respect to weighting the categories. Thus, in spite of its relative simplicity, it does an excellent job overall of taking complex information and converting it into a single number.

The total score from LEED-NC, computed by adding up the points earned in each category, results in a building rating (see Table 3.3). Current experience using the LEED-NC 2009 is that the Platinum and Gold ratings are fairly difficult to achieve and, unlike the typical American approach, in which the highest score possible is the only one worth achieving, a Silver rating is actually a very good assessment and a noteworthy accomplishment.

An expanded outline version of LEED-NC categories, credits, and prerequisites is shown in Appendix A. Outlines of the other major active LEED standards are shown in Appendix B (LEED-EB), Appendix C (LEED-CS), and Appendix D (LEED-CI).

The LEED Certification Process

The final stage of a successful LEED process is the award of a plaque designating the building as being *Certified* and indicating the LEED rating of the building: Platinum, Gold, Silver, or Certified. Prior to certification, the building is referred to by the USGBC as a LEED *Registered* project. Achieving the LEED Certified designation requires significant care and attention during the entire LEED certification process to successfully complete all steps in the documentation process. These steps include (1) ensuring that the building is eligible for certification; (2) registering the project with the USGBC; (3) ensuring and documenting that the project meets the prerequisites of the applicable LEED rating system; (4) documenting that the project attains at least the minimum number of points to achieve at least the minimum rating, the Certified level; (5) submitting online to the USGBC the required documentation demonstrating that both the prerequisites and points have been achieved; (6) if necessary, appealing points denied by the USGBC; and (7) receiving final notification from the USGBC that the project has been certified. If all these steps are completed, along with payment of registration and certification fees, the project receives the plaque showing the rating for display on the building (see Figure 3.1). This is a significant achievement because, as of late 2008, fewer than 1400 buildings have been certified under LEED-NC, with over 5,000 having been registered.

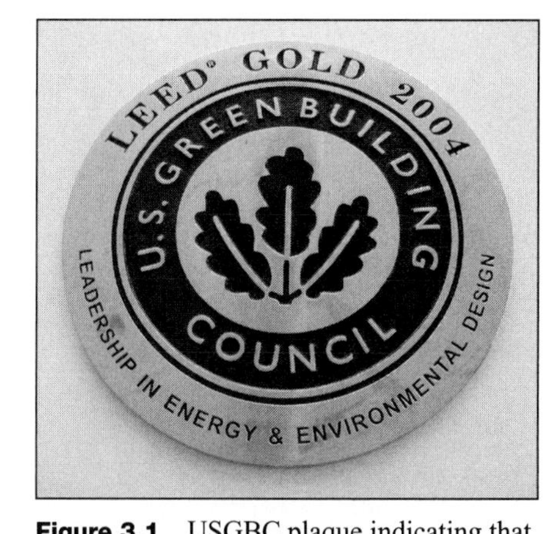

Figure 3.1 USGBC plaque indicating that Rinker Hall, an academic building at the University of Florida, is a Gold-certified green building.

LEED-ONLINE

The LEED building rating system has shifted from requiring certification hard-copy documentation to an Internet-based system know as LEED-Online. Project teams can submit 100 percent of their documentation online in an easy-to-use format. LEED-Online stores all LEED information, resources, and support in one centralized location. It enables team members to upload credit templates, track Credit Interpretation Requests (CIRs; documented responses to questions posed by previous project teams), manage key project details, contact customer service departments, and communicate with reviewers throughout the design and construction reviews. In addition to submitting documentation online, project teams have the option to submit documentation in two separate phases: first for the design phase and then for the construction phase.

ELIGIBILITY

Commercial buildings are eligible for certification using LEED-NC, LEED-CS, LEED-CI, and LEED-EB. The USGBC relies on standard building codes to designate what is a commercial building and include, but are not limited to, offices, retail and service establishments, institutional buildings (e.g., libraries, schools, museums, churches), hotels, and residential buildings of four or more habitable stories. If LEED is being applied to a unique building type that may be questionable as a commercial building, the USGBC encourages the project team to tally a potential point total using the LEED Rating System Checklist that is provided with each LEED Rating System. Documents known as LEED Reference Guides describe in detail each of the four commercial occupancy rating systems and are available as free downloads on the USGBC website. If a project can meet all prerequisites and achieve the minimum number of points to earn the basic Certified level of LEED project certification, the project is viable for LEED certification.

REGISTRATION

The first step in earning LEED certification is project registration. Projects are registered by visiting the LEED Registration page of the USGBC website, where information about the project is input and a registration fee is paid. Early registration is encouraged because starting the process as early as possible maximizes the potential for achieving certification. Registration establishes contact with the USGBC and provides access to essential information, software tools, and communications. Appointment of a Project Administrator occurs when the project is first registered at LEED-Online. The Project Administrator invites members of the project team to register with the project and then assigns roles to the individual project team members. Typical roles include architect, landscape architect, civil engineer, owner, and developer, to name just a few. The system also allows the Project Administrator to create new roles that are unique to the project if needed. The Project Administrator develops a project description, assigns responsibility for LEED credits to the project team members, and then monitors the submission of documentation to support the LEED credits. The Project Administrator should be a LEED Accredited Professional (LEED-AP) and be the project team member assigned to steer the project through the certification process. Once a project is registered and responsibilities are assigned, the project team begins to prepare documentation and calculations to satisfy the prerequisite and credit submittal requirements.

CREDIT INTERPRETATIONS

If a project team encounters difficulties applying a LEED prerequisite or credit to a specific project, the USGBC encourages the team to sort out the issue themselves first and contact the USGBC only as a last resort. To address the issue of questions that may arise about a LEED prerequisite or credit, CIRs are available in a database at the USGBC website. The CIR system ensures that rulings are consistent and available to other projects.

If a question about a prerequisite or credit arises, project teams should take the following actions, in order of priority:

1. Consult the appropriate LEED Reference Guide regarding credit intent, requirements, and calculations. View additional guidance for Innovation Credits.
2. Review the LEED CIR page for previously logged CIRs on relevant credits.
3. If a similar credit interpretation has not been logged or does not answer the question sufficiently, submit a new CIR via LEED-Online. The inquiry should be succinct and based on information found in the Reference Guide, with emphasis on the intent of the prerequisite or credit. Each CIR costs $220.

DOCUMENTATION AND CERTIFICATION

To earn LEED certification, the project must satisfy all of the prerequisites and attain the minimum number of points to attain the LEED Certified rating level. The certification review process includes the following:

1. Application Documentation Submittal: Documentation is submitted via LEED-Online, which is paperless; Letter Templates, additional documentation, and online certification payment are all submitted via the USGBC website.
2. LEED Technical Reviews.
   a. LEED-Online Review: LEED-Online is currently available for LEED-NC 2.1, LEED-NC 2.2, LEED-CI 2.0, and LEED-EB 2.0. All Letter Templates are on-line Adobe Acrobat pdf files, allowing for paperless submissions. The entire submission and review process is paperless using LEED-Online.
   b. LEED-EB Version 2.0 Review: The LEED Letter Templates and additional submittals for each prerequisite and credit are reviewed for compliance. Within 30 days of administrative approval, the USGBC issues a Preliminary LEED Review document noting the credit achievement anticipated, pending, and denied. Upon receipt of the Preliminary Review, the project team is given the opportunity to provide corrections and/or additional supporting documents (e.g., calculations and other backup) as a supplementary submittal to the application. The USGBC conducts a Final LEED Review of the application within 30 days of receiving the supplementary submittal and notifies the project contact of certification status.
   c. LEED-CI Version 2.0 Review: The LEED Letter Templates and additional submittals for each prerequisite and credit are reviewed for compliance. Within 30 days of administrative approval, the USGBC issues a Preliminary LEED Review document noting credit achievement anticipated,

pending for cause or pending for audit, and rejected. In addition, up to five prerequisites and/or credits are selected for audit. The project team has 30 days from the receipt of the Preliminary Review to provide corrections and/or additional supporting documents (e.g., calculations, cut sheets, and other backup) as a supplementary submittal to the application. The USGBC conducts a Final LEED Review of the application within 3 weeks of receiving the resubmittal and notifies the project contact of the certification status. If two or more audited credits are denied, additional credits may be selected for a second audit and may prompt a Second Preliminary LEED Review prior to a Final LEED Review.

AWARD OF CERTIFICATION

Upon notification of the LEED certification, the project team has 30 days to accept or appeal the awarded certification. Upon the project's acceptance, or if it has not appealed the rating within 30 days, the LEED certification is final. The project may then be referred to as a *LEED-certified building.* The USGBC presents the project team with an award letter, certificate, and metal LEED plaque indicating the certification level.

APPEAL

If the project team feels that sufficient grounds exist to appeal a credit denied in the Final LEED Review, it has the option to appeal. The appeal fee is $500 per credit appealed. A review of the documentation for the appealed credits will occur within 30 days, at which time an Appeal LEED Review will be issued to the applicant.
All appeals are submitted via LEED-Online. If an appeal is pursued, a different review team will assess the appeal documentation, which should include:

1. LEED registration information, including project contact, project type, project size, number of occupants, date of construction completion, and so on.
2. An overall project narrative including at least three project highlights.
3. The LEED Project Checklist/Scorecard indicating projected prerequisites and credits and the total score for the project.
4. Drawings and photos illustrative of the project, including:
   a. Site plan
   b. Typical floor plan
   c. Typical building section
   d. Typical or primary elevation
   e. Photo or rendering of project
5. Original, resubmittal, and appeal submittal documentation for only those credits being appealed. It is recommended that a narrative for each appealed credit describing how the documents address the reviewers' comments and concerns be included with the appeal documentation.

LEED REGISTRATION AND CERTIFICATION FEES

A registration fee must be paid for all LEED-NC, LEED-CI, LEED-CS, and LEED-EB projects as part of the registration process. When the project is prepared for consideration of documentation submitted for LEED certification, a certification fee must be paid. The current fee structures for registration and certification are indicated in Tables 3.4 and 3.5. Note that the project team has the option of submitting all documentation either at the conclusion of the construction process or in two phases: (1) a Design Review, in which all credits that have been completely addressed by the

TABLE 3.4

LEED Registration Fees

USGBC Membership	Registration Fee
Members	$450.00
Nonmembers	$600.00

TABLE 3.5

Certification Fees for LEED-NC, LEED-CI, and LEED-CS, and LEED-EB

	Less Than 50,000 Sq Ft Fixed rate	50,000–500,000 Sq Ft Based on sq. ft.	More Than 500,000 Sq Ft Fixed rate
Design Review			
Members	$1,250	$0.025/sq ft	$12,500
Nonmembers	$1,500	$0.03/sq ft	$15,000
Construction Review			
Members	$500	$0.01/sq ft	$5,000
Nonmembers	$750	$0.015/sq ft	$7,500
Combined Design and Construction Review			
Members	$1,750	$0.035/sq ft	$17,500
Nonmembers	$2,250	$0.045/sq ft	$22,500
LEED-EB	**Fixed rate**	**Based on sq ft**	**Fixed rate**
Initial Certification Review			
Members	$1,250	$0.025/sq ft	$12,500
Nonmembers	$1,500	$0.030/sq ft	$15,000

design team are put forward for review, and (2) a Construction Review, in which all the remaining credits are reviewed. The advantage of the two-phase review process is that it speeds the certification process and allows the project team to decide on and act on appeals far earlier in the process.

The Green Globes Building Assessment Protocol

Green Globes is a building rating protocol with roots in Canada that is making inroads in the United States as an alternative to LEED.[8] It provides a rating of one to four *Green Globes,* depending on the percentage of the maximum 1,000 points allowed that the project actually achieves (see Table 3.6). The Green Building Initiative describes the Green Globes building assessment system as a revolutionary green management tool that includes an assessment protocol, a rating system, and a guide for integrating environmentally friendly design into commercial buildings. When the assessment protocol has been completed, it also facilitates recognition of the project through third-party verification. It is designed to be an interactive, flexible, and affordable approach to environmental design.

The Green Globes environmental assessment and rating system represents more than 9 years of research and refinement by a wide range of prominent international organizations and experts. The genesis of the system was BREEAM, which was brought to Canada in 1996. The Canadian Standards Association published BREEAM Canada for Existing Buildings. In 2004, the Green Building Initiative (GBI) acquired the rights to distribute Green Globes in the United States. The GBI committed to continually refining the system to ensure that it reflects changing opinions and ongoing advances in research and technology, as well as involving multiple

TABLE 3.6

Green Globes Ratings

Rating	Globes	Description
85–100%	oooo	Reserved for select building designs which serve as national or world leaders in energy and environmental performance. The project introduces design practices that can be adopted and implemented by others.
70–84%	ooo	Demonstrates leadership in energy and environmental design practices and a commitment to continuous improvement and industry leadership.
55–69%	oo	Demonstrates excellent progress in achieving eco-efficiency results through current best practices in energy and environmental design.
35–54%	o	Demonstrates movement beyond awareness and commitment to sound energy and environmental design practices by demonstrating good progress in reducing environmental impacts.

stakeholders in an open and transparent process. In 2005, GBI became the first green building organization to be accredited as a standards developer by the American National Standards Institute (ANSI) and began the process of establishing Green Globes as an official ANSI standard. The GBI ANSI technical committee was formed in early 2006.

THE GREEN GLOBES PROCESS

The Green Globes system is questionnaire driven. At each stage of the design process, users are walked through a logical sequence of questions that guide their next steps and provide guidance for integrating important elements of sustainability (see Figure 3.2). The construction documents questionnaire is the basis for the rating system. However, to benefit fully from the value-added design assistance features of the system—and to obtain a preliminary self-assessment of a building—the project should be registered and the preliminary and subsequent questionnaires should be completed. However, a building cannot be promoted as a Green Globes certified building without third-party verification from a GBI-authorized verifier. (The Green Globes website is www.thegbi.org.)

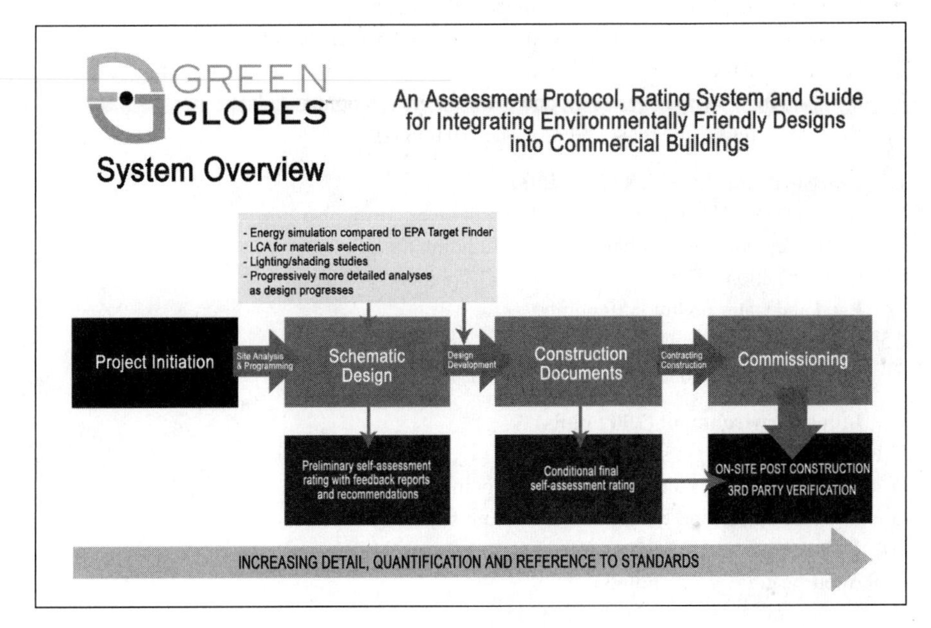

Figure 3.2 Overview of the Green Globes assessment standard showing the assessment activities at each of the design stages. (Diagram courtesy of The Green Building Initiative, Inc.)

STRUCTURE OF GREEN GLOBES

The structure of Green Globes is shown in Table 3.7. It is similar to LEED-NC in many respects but addresses some additional issues: project management, emergency response planning, durability, adaptability, deconstruction, life-cycle assessment (LCA), and noise control.

TABLE 3.7

Structure of the Green Globes v.1 Rating System

A. Project Management—Policies and Practices (50 points)
A.1 Integrated design (20 points)
A.2 Environmental purchasing (5 points)
A.3 Commissioning (20 points)
A.4 Emergency response plan (5 points)

B. Site (115 points)
B.1 Site development area (45 points)
B.2 Reducing ecological impacts (40 points)
B.3 Enhancement of watershed features (15 points)
B.4 Site ecology improvement (15 points)

C. Energy (300 points)
C.1 Energy consumption (Paths A and B) (110 points)
C.2 Energy demand minimization (Paths A and B) (135 points)
C.3 "Right-sized" energy-efficient systems (Path B only) (110 points)
C.4 Renewable sources of energy (Paths A and B) (45 points)
C.5 Energy-efficient transportation (Paths A and B) (70 points)

D. Water (100 points)
D.1 Water (40 points)
D.2 Water-conserving features (40 points)
D.3 Reducing off-site treatment of water (20 points)

E. Resources, Building Materials, and Solid Waste (100 points)
E.1 Materials with low environmental impact (40 points)
E.2 Minimized consumption and depletion of material resources (30 points)
E.3 Reuse of existing structures (10 points)
E.4 Building durability, adaptability, and disassembly (10 points)
E.5 Reduction, reuse, and recycling of waste (10 points)

F. Emissions and Effluents (75 points)
F.1 Air emissions (15 points)
F.2 Ozone depletion and global warming (30 points)
F.3 Contamination of sewers or waterways (12 points)
F.4 Land and water pollution (9 points)
F.5 Integrated pest management (4 points)
F.6 Storage for hazardous materials (5 points)

G. Indoor Environment (200 points)
G.1 Effective ventilation system (60 points)
G.2 Source control of indoor pollutants (45 points)
G.3 Lighting design and integration of lighting systems (40 points)
G.4 Thermal comfort (35 points)
G.5 Acoustic comfort (25 points)

GREEN GLOBES VERIFICATION AND CERTIFICATION

For a project to be certified, the project team must fill out the Green Globes v.1 questionnaire, which is done at the various stages in the project's design and construction. If the project can potentially achieve at least 35 percent of the available points, the project is eligible for formal certification. A third-party Verifier visits the project and the project team at the building site, interviews project team members, and reviews documentation for each of the points claimed by the project team. The Verifier is an experienced construction industry professional who has been trained on the Green Globes protocol and who has been monitored and mentored by other Verifiers prior to becoming an independent Verifier. In the USGBC LEED process, the project team completes documentation online and submits it via LEED-Online, to be reviewed by a review team that is not at any time in direct contact with the project team. Unlike LEED, the Green Globes system requires a Verifier to actually visit the project, interact directly with the team, and physically examine the project. At the end of this stage, the Verifier sends his or her recommendation to GBI concerning the appropriate certification level.

One other important feature of Green Globes is that if points are not available to a project, they do not count in the total of potentially achievable points. In effect, LEED penalizes projects that, for example, do not build on a brownfield or that are not near bus stops, even if they are in an urban setting. In LEED the available number of points is fixed, while in Green Globes the total potential number of points is adjusted, depending on the project's location. Green Globes could be said to rate the work of the project team and does not address issues that are outside of their control—for example, the location of the building, an owner issue. LEED attempts to rate both the project team and the owner, and the final certification is a reflection of their joint efforts.

International Building Assessment Systems

There are several significant building assessment systems that are used in other countries and that provide other perspectives on how to approach the problem of determining how environmentally friendly a given building design may be. In the following subsections, three building assessment systems are described: BREEAM (United Kingdom), CASBEE (Japan), and Green Star (Australia). GBTool, a building assessment method that is used by countries participating in the Green Building Challenge series of conferences to compare buildings using a uniform approach, is also described.

BREEAM (UNITED KINGDOM)

BREEAM is by far the oldest building assessment system and, until the advent of LEED, easily the most successful. Its development was initiated in 1988 by BRE, the national building research organization of the United Kingdom, to help transform the construction of office buildings to high-performance standards. BREEAM has also been adopted in Canada and in several European and Asian countries.[9]

BREEAM assesses the performance of buildings in the following areas:

- *Management:* Overall management policy, commissioning site management, and procedural issues
- *Energy use:* Operational energy and carbon dioxide (CO_2) issues
- *Health and well-being:* Indoor and external issues affecting health and well-being

- *Pollution:* Air and water pollution issues
- *Transport:* Transport-related CO_2 and location-related factors
- *Land use:* Greenfield and brownfield sites
- *Ecology:* Ecological value conservation and enhancement of the site
- *Materials:* Environmental implication of building materials, including life-cycle impacts
- *Water:* Consumption and water efficiency

Credits are awarded in each area according to performance. A set of environmental weightings then enables the credits to be added together to produce a single overall score. The building is then rated on a scale of Pass, Good, Very Good, or Excellent, and a certificate is awarded that can be used for promotional purposes.

BREEAM covers primarily offices, homes, and industrial units, with assessment methods for each general type of building: BREEAM Office version 2002, BREEAM/New Industrial Units, and BREEAM EcoHomes. In 2003, a new version, BREEAM/Retail, was issued to address the design, construction, and operation of retail stores.

CASBEE (JAPAN)

The Japan Sustainable Building Consortium, composed of academic, industrial, and government entities, is cooperating to develop a building assessment system, CASBEE, designed specifically for Japan and Japanese cultural, social, and political conditions. CASBEE is a suite of assessment tools for the various phases of the building being evaluated: planning, design, completion, operation, and renovation (see Tables 3.8 and 3.9).[10]

The key concept in CASBEE is Building Environmental Efficiency (BEE), which is an attempt to describe the eco-efficiency of the building. The World Business Council on Sustainable Development (WBCSD) defines eco-efficiency as maximizing economic value while minimizing environmental impacts:

$$\text{Eco-efficiency} = \frac{\textit{Value of Products or Services}}{\text{Environmental Loadings for Products or Services}}$$

BEE is simply a modification of the concept of eco-efficiency for application to buildings:

$$\text{BEE} = \frac{\textit{Building Environmental Quality and Performance}}{\text{Building Environmental Loadings}}$$

Building Environmental Quality and Performance is described as the amenities provided for building users and consists of several quantities:

Q1: Indoor environment

Q2: Quality of service

Q3: Outdoor environment on-site

Similarly, the Building Environmental Loadings consist of several different categories:

L1: Energy

L2: Resources and materials

L3: Off-site environment

TABLE 3.8

CASBEE Assessment Tools and Applicable Phases

Name	Title	Applicable Phases
Tool-0	Predesign Assessment Tool	Planning, design
Tool-1	Design for the Environment (DfE) Tool	Design, completion
Tool-2	Eco-labeling Tool	Completion, operation
Tool-3	Sustainable Operation and Renovation Tool	Operation, renovation

The BEE rating is a number, generally in the range of 0.5 to 3, that corresponds to a building class, from class S (highest for BEE of 3.0 or higher) to classes A (BEE of 1.5 to 3.0), B+ (BEE of 1.0 to 1.5), B– (BEE of 0.5 to 1.0), and C (BEE less than 0.5). The relationship of Quality (Q) to Loading (L) in CASBEE and the resulting BEE letter scores are diagrammed in Figure 3.3.

As noted earlier, CASBEE and its various tools are still under development, so it remains to be seen how it will be accepted in the Japanese marketplace for transforming the building stock to high-performance standards.

GREEN STAR (AUSTRALIA)

Green Star is a new building assessment system developed for use in the Australian building market, with the first products being directed at offices and office buildings. It will eventually have rating tools for different phases of the building life cycle (e.g., design, construction, interiors, and operation) and for different building classes (office, retail, industrial, residential, etc). Green Star builds on existing rating systems and tools in overseas markets, including the British BREEAM system and the U.S. LEED system. Green Star has established individual environmental measurement criteria with particular relevance to the Australian marketplace and

TABLE 3.9

Description of CASBEE Tools

Tool-0 Predesign Assessment Tool
For use by owners and planners for identifying the project context, selecting the proper site, and determining the basic impact of the project

Tool-1: Design for the Environment (DfE) Tool
A simple check system for designers and engineers to use in improving BEE during the design phase

Tool-2: Eco-Labeling Tool
Used to rate the building in terms of BEE after construction and to determine the basic property of the labeled building in the property market

Tool-3: Sustainable Operation and Renovation Tool
For use in informing building owners and managers how to improve the BEE of their building during its operation

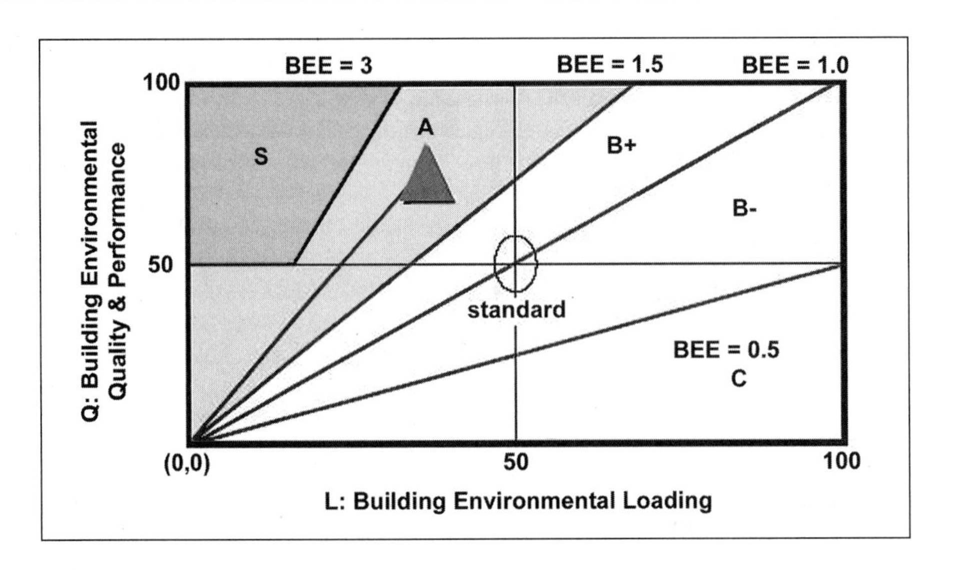

Figure 3.3 The BEE rating is determined by finding the intersection of Q (Building Environmental Quality and Performance) and L (Building Environmental Loadings). High ratings (S and A) are achieved by buildings with high environmental quality and performance and low environmental loadings. Higher resource consumption and lower environmental quality produces below-standard ratings (B– or C).

environmental context. Green Star Office Design version 1.0 covers the following categories:

- Management (12 points)
- Indoor environmental quality (27 points)
- Energy (24 points)
- Transportation (11 points)
- Water (12 points)
- Materials (20 points)
- Land use and ecology (8 points)
- Emissions (13 points)
- Innovation (5 points)

A maximum of 132 points is achievable for Green Star Office Design. It awards various numbers of *stars* to indicate the level of performance. Six stars is the highest level and is said to recognize and reward international leadership. Five stars recognizes and rewards Australian leadership, while four stars indicates best practice in building environmental initiatives. The management category illustrates some of Green Star's use of the LEED approach in its design. For example, a Green Star Accredited Professional engaged in the project earns 2 points.

GBTOOL

GBTool is a very comprehensive and sophisticated building assessment tool that was developed for the biannual international Green Building Challenge, which has been held four times to date: in 1998 (Paris), 2000 (Maastricht), 2002 (Oslo), and 2005 (Tokyo); the 2008 event is scheduled for Melbourne, Australia. In the meeting in Tokyo, national teams from 21 countries submitted entries to demonstrate the art and science of green building in their countries.

GBTool provides a standard basis of comparison for the wide range of buildings being evaluated in the Green Building Challenge. It requires a comprehensive set of information not only on the building being assessed, but also on a benchmark building for use in comparing how well the green building performs compared to the norm. GBTool requires the group using it to establish benchmark values and weights for the various impacts. The tool is implemented in the form of a sophisticated Excel spreadsheet that can be downloaded from the website of the International Initiative for a Sustainable Built Environment (iiSBE). The output from GBTool provides an assessment of the building in seven different categories: Resource Consumption, Environmental Loadings, Indoor Environmental Quality, Service Quality, Economics, Management, and Commuting Transport.

Example of a LEED-NC Platinum Building: The Audubon Center, Debs Park, Los Angeles

To illustrate the type of building that is achieving the highest USGBC rating for new construction, this section looks at the Audubon Center building in the Ernest E. Debs Regional Park, located about 10 minutes from downtown Los Angeles in an urban wilderness. It is specifically designed to serve inner-city children and educate them about the environment and its ecological system, which is populated by coyotes and 136 bird species. The Audubon Center has 5,022 square feet (467 square meters) of fully enclosed space and 2,816 square feet (262 square meters) of partially enclosed areas. The total cost of the project was $5 million, or $371 per square foot ($3,993 per square meter), including the cost of restoring the site to its original condition.

The center is an off-grid building that has all of its energy provided by three solar systems: (1) a photovoltaic array of 208 panels that generate the building's electricity and is connected to a 3- to 5-day battery backup system; (2) a 1,100-square-foot (102 square meters) array of glass vacuum tube solar collectors that provide high-temperature hot water (160°F to 180°F) (71°C to 82°C) to a solar-powered absorption chiller for air conditioning the facility; and (3) a solar hot water system for providing domestic-use hot water. An automated load-shedding system drops loads in priority fashion when the energy stored in the building's battery system is low. Advanced passive strategies are employed to minimize the need for heating and cooling. Operable windows and the building's geometry allow cross-ventilation from low to high windows. Ceiling fans create air movement to provide an enhanced cooling effect, and exposed interior concrete floors and concrete block walls provide thermal mass for storing the cooling effect.

The building's water needs are very low, and the building's hydrologic strategy employs low-flow shower heads, dual-flush toilets, and a graywater/blackwater recycling system that was installed for future use. Wastewater is treated on-site using microfilters and microorganisms. Stormwater is retained on-site and treated prior to release to assist in groundwater recharge.

Green building materials were employed extensively in the design of the Audubon Center, and 97 percent of construction waste was recycled. Some of the materials used include (1) synthetic gypsum board with 95 percent recycled content; (2) plywood, redwood, and Douglas fir pergola components certified by the Forestry Stewardship Council; (3) ceramic tiles with recycled content; (4) cast-in-place concrete with 25 percent fly ash displacing cement; (5) linoleum countertops; (6) steel reinforcing bars with 97 percent recycled content; (7) formaldehyde-free batt insulation with recycled content; (8) cabinetry and wainscoting made with organic wheatboard, urea formaldehyde–free medium-density fiberboard made from wheat and sunflower composites; and (9) carpet made of sisal fibers.

The landscape is designed to emphasize native and adapted species of plants that are drought-tolerant and fire-resistant. It also provides a setting that attracts birds and other wildlife, and features a children's garden and nature trails for experiential learning.

(A)

(B)

Figure 3.4 The Audubon Center at the Ernest E. Debs Regional Park, near Los Angeles, California. (A) The building is an off-grid facility located in the 282-acre (114-hectares) Debs Regional Park. (Photograph: Gary Leonard.) (B) Glass vacuum solar tubes (upper right) generate high-temperature hot water, which is stored in a tank (lower left) for use with an absorption chiller. (Photograph: EHDD Architecture.)

In December 2003, the Audubon Center received 53 out of a possible 69 points in the LEED-NC rating system, qualifying it for the most prestigious award level, Platinum. At present, there are just a handful of buildings in the world with this level of achievement, and the center effectively represents today's cutting edge for green building. The design team carefully considered the building's site and its natural assets, particularly solar energy, and maximized the use of these assets in the building's energy systems. The building hydrologic cycle, including potable water, wastewater, and stormwater, was thoughtfully implemented, and every opportunity for innovation was considered. The challenge for the USGBC and the overall green building movement is to use the Audubon Center and other superior building projects as the launching point from which to consider the future and how green buildings will evolve.

Beyond Today's Building Rating Systems

The basic structure of LEED-NC dates from 1998, when LEED 1.0 was issued and applied to a group of 32 new building projects to shake down the rating system. The result of the pilot was a relatively major revision, LEED 2.0, in 2000 followed by LEED-NC 2.1 and 2.2, which contained some significant changes but no new philosophical directions. There has been ample frustration with this basic framework, which is now almost 10 years old and which remains based on intuition as to what constitutes a green building, rather than shifting to approaches that are scientifically justifiable. Clearly, LEED is often confused with a plan of action when in fact it is meant to be nothing more than a measuring tool. A very promising approach suggested as a new foundation for future LEED products, is described by Alex Zimmerman below.[11]

Informing LEED with the Natural Step

In spite of its success in bringing attention and credibility to green building, LEED has been criticized for its lack of scientific robustness, particularly its lack of credits dealing directly with practices that impact climate change and its failure to address persistent organic pollutants. Others have argued that issues are weighted inequitably. In addition, it can be argued that LEED, by focusing on impacts, in effect rewards incremental solutions and does not adequately recognize major step-change or paradigm-shifting advances. Failure to address these concerns may limit the market penetration and ultimate success of LEED and of the green building movement. LEED, as an environmental rating system, also does not explicitly address economic or social issues. LEED is voluntary, has few prerequisites, and has a graduated scale of achievement. The result is that individual project owners and managers determine what makes the most economic sense for them, as determined by the particular mix of financial and market recognition drivers that apply to the project. There is an urgent need to align LEED more closely to established sustainability principles, for example The Natural Step. And this needs to occur (1) without losing the accessibility that has contributed to its market success to date (2) while keeping LEED relevant to its target market and (3) while fulfilling the stated goal of transforming the entire market.

The way forward is laid out in a seminal 2002 paper by Karl Henrik Robèrt and others that demonstrated that there is a remarkable agreement among on:

- A commonly agreed-upon physics-based definition of sustainability beyond the Brundtland Commission's definition (The Natural Step)
- A set of hierarchical principles to enable us to comprehensively plan for and work toward sustainability
- How a given framework, tool method, or system fits with other frameworks

Another key paper by Ny and others, published in 2006, proposed using hierarchical principles as the basis for incorporating sustainability into planning and management, defined as *strategic life-cycle management.* The primary objective is to identify viable investment paths toward social and ecological sustainability. Generally, the approach set forth in these papers can be used to inform the next generation of LEED in two major ways:

1. As a filter, to ensure that prerequisites and credits align with fundamental sustainability system conditions, and to fill gaps if they exist
2. As a way of restructuring or fine-tuning the language in the prerequisites and credits and, specifically, to separate Objectives and Indicators from Actions and process principles

THE ROLE OF THE NATURAL STEP

The Natural Step, developed by Karl Henrik Robèrt, is addressed in Chapter 2. As noted there, the first three System Conditions (which could be said to refer to ecological sustainability) are derived from observing the basic mechanisms by which natural life-sustaining systems can be destroyed, while the fourth System Condition (referring to social sustainability) is simply stated as the requirement to meet human needs.

In order to use The Natural Step framework, Robèrt and his colleagues propose five hierarchical system levels for use in comprehensive planning for any complex system. The five levels are:

1. Principles for the *constitution* of the system (e.g., ecological and social principles)
2. Principles for a favorable *outcome* of planning within the system (e.g., *principles for sustainability*)
3. Principles for the *process* to reach this outcome (e.g., *principles for sustainable development*)
4. *Actions,* that is, *concrete measures* that comply with the principles for the process to achieve a favorable outcome in the system (e.g., recycling and switching to renewable energy)
5. *Tools* to monitor and audit (a) the relevance of actions with reference to principles for the process (e.g., indicators of flows and key figures to comply with principles for sustainability) and/or monitoring (b) the status of the system itself, and its impacts (e.g., ecotoxicity and employment), or reduced impacts, as a consequence of strategically planned societal actions

LEVEL 1 PRINCIPLES

These principles describe human society and the physical system in which society operates. They are, in effect, a restatement of the laws of physics.

LEVEL 2 PRINCIPLES

The authors of the 2002 paper argue that The Natural Step System Conditions constitute the level 2 principles for sustainability. They further argue that Sustainability Objectives can be derived from each of the System Conditions as follows:

1. Eliminate our contribution to systematic increases in concentrations of substances from the Earth's crust. This means substituting for certain minerals that are scarce in nature others that are more abundant, using all mined materials efficiently, and systematically reducing dependence on fossil fuels.
2. Eliminate our contribution to systematic increases in concentrations of substances produced by society. This means systematically substituting for certain persistent and unnatural compounds ones that are normally abundant or break down more easily in nature and using all substances produced by society efficiently.

3. Eliminate our contribution to the systematic physical degradation of nature through overharvesting, introductions, and other forms of modification. This means drawing resources only from well-managed ecosystems, systematically pursuing the most productive and efficient use of those resources and of land, and exercising caution in all kinds of modification of nature.
4. Contribute as much as we can to meeting human needs in our society and worldwide, over and above all the substitution and dematerialization measures taken in meeting the first three objectives. This means using all of our resources efficiently, fairly, and responsibly so that the needs of all people on whom we have an impact, and the future needs of people who are not yet born, stand the best chance of being met.

Further, the authors argue that there are fundamentally only two broad mechanisms or Strategies for achieving these Sustainability Objectives, either Dematerialization or Substitution, although the particulars vary somewhat for each System Condition.

LEVEL 3 PRINCIPLES

The level 3 principles are defined as process principles that need to be followed in order to guide the actions taken and to move in the direction of sustainability. Sixteen process principles are articulated for society as a whole, ranging, for example, from the Precautionary Principle, to a good return on investment (ROI), to international agreements. The 2002 paper focuses on three key principles for informing viable investment paths or solutions in strategic life-cycle management. These principles can be posed as questions to be asked of proposed solutions or actions:

- *Direction*—is the proposed solution headed in the direction of compliance with the four System Conditions (level 2 principles)?
- *Flexible*—will the proposed solution avoid dead ends, that is, will it provide a flexible stepping stone to link to future solutions in the same direction?
- *Good ROI*—will the proposed solution (monetary, meeting market demand, foreseeing regulatory changes, etc.) yield a good return in order to seed subsequent solutions?

LEVEL 4 PRINCIPLES

Actions that need to be taken are defined as level 4 principles. We are reminded that actions should not be confused with the principles underpinning them; for example, more efficient automobiles may lead to increased consumption of fossil fuels through rebound effects.

LEVEL 5 PRINCIPLES

Tools and metrics to indicate and audit progress toward sustainability are defined as level 5. Robèrt and his colleagues agree that many well-known tools and approaches, such as Factor 10, ecological footprinting, zero emissions, natural capitalism, ISO 14000, and life-cycle assessment (LCA), can be considered as belonging to level 5. They emphasize that all these tools have different entry points in a sustainability planning framework, and therefore different primary focuses, and that these should provide opportunities for synergies rather than being viewed as being competitive.

LEED ANALYZED WITH THE PLANNING HIERARCHY

When LEED is analyzed with this set of principles and hierarchy, it can be shown that it effectively covers levels 2 to 5. This is represented schematically in Figure 3.5. Strategies and Actions are described in the Requirements of LEED, along with processes to achieve them. Impacts are embedded in the Intent statements. Indicators are largely a by-product of the Requirements and Documentation statements. Assessment is spelled out in the Documentation. Weightings are implicit in the assignment of points to prerequisites and credits. The spanning of several levels of principles helps explain the success of LEED in a marketplace that, at the time of the launch of LEED, was unsophisticated in its knowledge and understanding of the scientific issues at stake. Within one tool, sufficient definition of the impacts and rationale for appropriate action is presented, along with practical actions that can be taken, so that users can immediately understand how to move forward on their projects. Assessment is related to existing standards and guidelines, which reduces the learning curve, and the market recognition of graduated progress provides tangible incentives beyond the intrinsic satisfaction of doing the right thing.

USING THE PLANNING HIERARCHY APPROACH TO INFORM LEED

The planning hierarchy can be used to show how LEED can be better aligned with sustainability principles. One way to do this is to approach the problem in a series of steps as follows:

1. Review existing LEED prerequisites and credits from a Natural Step Systems Conditions perspective rather than solely from the perspective of Impacts.
2. Identify the System Condition toward which the prerequisite or credit is intended to move the project. This could be stated as an Objective.

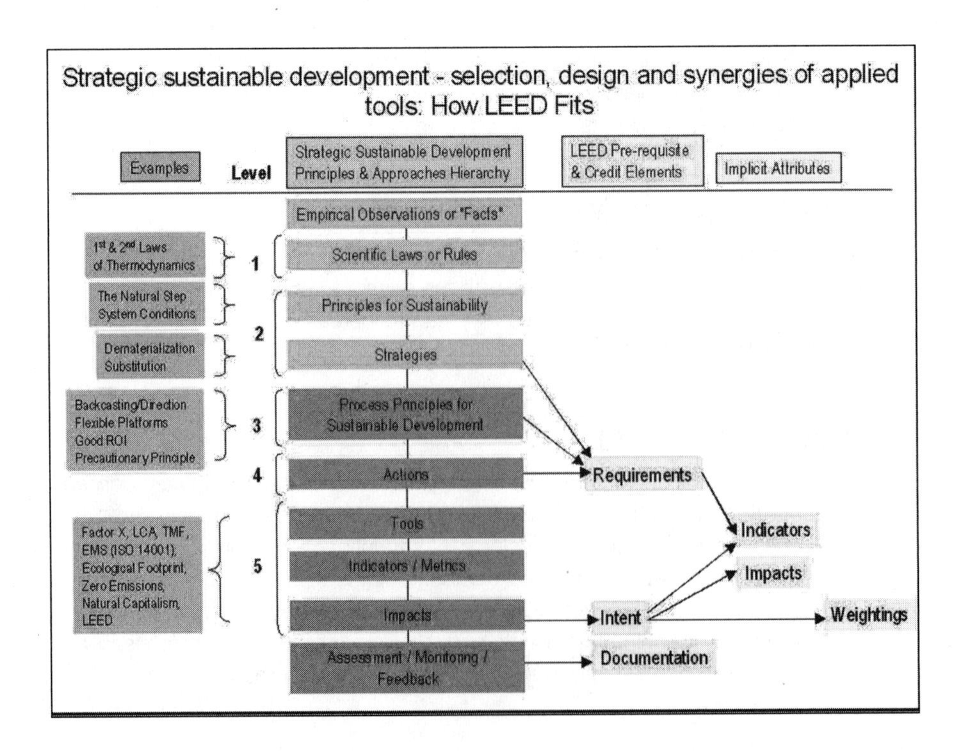

Figure 3.5 Analyzing the LEED rating system with the planning hierarchy.

3. Reword the Intent statements in relation to Dematerialization and Substitution principles.
4. Use appropriate process principles as filters for selecting the required Actions for each credit.
5. Ensure that the Indicators of success for the Action taken for each credit reflect the process principles and that they can be connected with Impacts and System Conditions to ensure measurable feedback directly to principles for sustainability.

These steps are represented schematically in Figure 3.6.

EXAMPLE CREDIT EXAMINED WITH PROPOSED PLANNING HIERARCHY

In order to illustrate the proposed process, two credits, Energy and Atmosphere Credit 2 (EAc2) and Renewable Energy and Energy and Atmosphere Credit 6 (EAc6), Green Power, are examined, and new wording and structure are proposed for a single credit to replace them.

Existing Wording EAc2

Intent: Encourage and recognize increasing levels of on-site renewable energy self-supply in order to reduce environmental impacts associated with fossil fuel energy use.

Requirements: Supply at least 5 percent of the building's total energy use (expressed as a fraction of the annual energy cost) through the use of on-site renewable energy systems.

Existing Wording EAc6

Intent: Encourage the development and use of grid-source, renewable energy technologies on a net zero pollution basis.

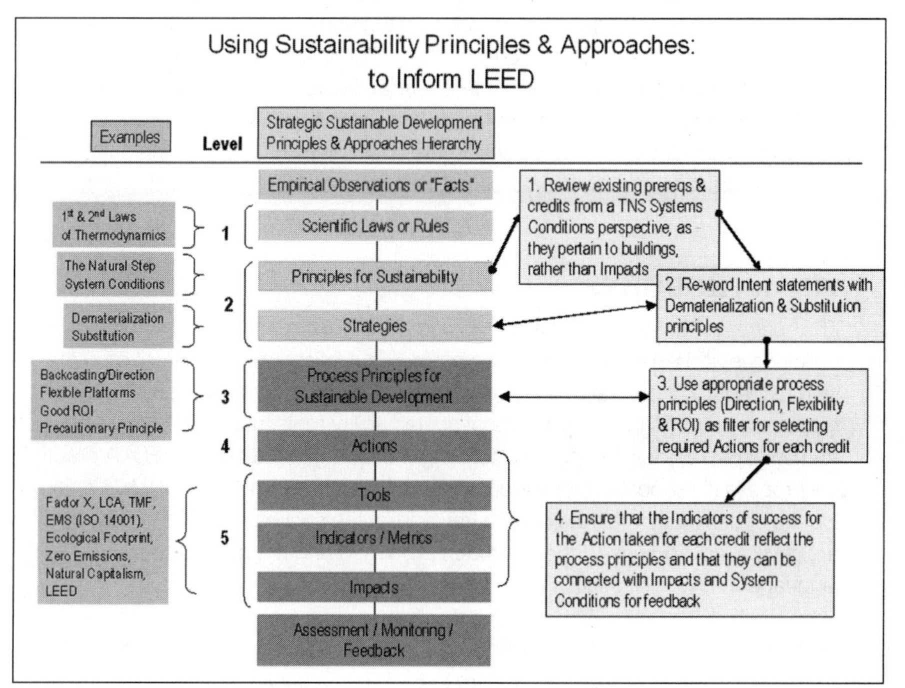

Figure 3.6 Using sustainability principles and approaches to inform the LEED rating systems.

Requirements: Provide at least 50 percent of the building's electricity from renewable sources by engaging in at least a 2-year renewable energy contract. Renewable sources are those that meet the Environment Canada Environmental Choice programs' EcoLogo requirements for green power supplies.

Proposed New Wording for a Credit to Replace EAc2 and EAc6

Objective: Eliminate humanity's contribution to systematic increases in concentration in the ecosphere of substances from the Earth's crust.

Intent: Increase the rate of substitution of fossil fuels by renewable energy from the lowest LCA source.

Indicator: Renewable energy supply, characterized by LCA impact

Requirements: Supply at least xx percent of the building's or project's total energy consumption with renewable energy, chosen from the alternative with the lowest LCA impact of those available.

There are several advantages to be obtained from using the proposed approach to inform and rework LEED:

- It provides awareness and education on a deep level to the industry about the fundamental sustainability challenges facing society.
- It allows solutions to move faster in the direction of solving or avoiding problems we are not yet aware of.
- By focusing on a first-principle, physics-based approach, the end objective is more easily understood, which should encourage projects to move beyond "point chasing."
- The credits and issues can be better integrated if they are examined with the goal of satisfying all of the system conditions and not, potentially, one at the expense of another.
- This approach is better able to withstand criticisms from vested interests, because the science is more easily defended.
- The first-principle, physics-based approach will make it relatively easier to deal with contentious issues such as the use of polyvinyl chloride (PVC) rather than deal with the end-of-pipe, allowable dosage, toxicology mindset that seems to prevail in these kinds of discussions. It would also put issues such as PVC and polybrominated diphenyl ether (PBDE) in better context with other persistent organic pollutants (POPs) and materials where low-level pollutants cause long-term toxicological effects.

FINAL THOUGHTS

The success that LEED has had to date in transforming the market for green buildings is at risk because of the increasing sophistication of its target market with respect to understanding of sustainability issues. Recent work by highly respected authors and proponents of sustainability tools, methodologies, and frameworks provides a basis for informing the next generation of LEED. A practical process is proposed that will allow greater rigor to be introduced while retaining the accessibility that has been a key part of the success of LEED so far. There are several advantages to the approach proposed and the new structure of LEED language that would result.

Alex Zimmerman is President of Applied Green Consulting, Limited, and Director and past President of the Canada Green Building Council.

Summary and Conclusions

The high-performance building movement worldwide is being propelled by the success of building assessment methods, in particular LEED in the United States and BREEAM in the United Kingdom. Both methods take complex arrays of numerical and nonnumerical data and provide a score that indicates the performance of a building according to the scoring and weighting system built into the method. LEED-NC 2009, the current version of the USGBC assessment standard for new construction, is being applied to a wide range of other types of public and private buildings, from gymnasiums to schools. New versions of LEED to address home construction, as well as to deal with renovations and existing buildings, are being produced to provide a better match of the assessment method to the building situation. Newcomers to the marketplace, such as GBI's Green Globes and NAHB's Model Green Home Guidelines, can help to bring the movement and these collective green building design concepts and strategies even further into the mainstream.

Notes

1. The Building Research Establishment (BRE) is the national building research organization for the United Kingdom and the developer of BREEAM, which is described in detail at http://products.bre.co.uk/breeam.
2. The Japan Sustainable Building Consortium developed CASBEE. A detailed description can be found at the consortium's website, www.ibec.or.jp.CASBEE/english/overviewE.htm.
3. At present, Green Star provides a series of assessment tools directed at new offices, existing offices, and office interiors. The Australian Green Building Council website is www.gbcaus.org.
4. From December 1 through December 11, 1997, more than 160 nations met in Kyoto, Japan, to negotiate binding limitations on greenhouse gases for the developed nations, pursuant to the objectives of the Framework Convention on Climate Change of 1992. The outcome of the meeting was the Kyoto Protocol, in which the developed nations agreed to limit their greenhouse gas emissions relative to the levels emitted in 1990. The United States agreed to reduce its emissions from 1990 levels by 7 percent during the period 2008 to 2012.
5. GBTool, developed by Natural Resources Canada in collaboration with a wide range of academics and practitioners worldwide, has been used by the Green Building Challenge to determine how well buildings compare to base or typical buildings in each category, for example, schools. The tool consists of an Excel spreadsheet. The most recent version is available for research and academic purposes at http://greenbuilding.ca/iisbe/gbc2k2/gbc2k2-start.htm.
6. The development of LEED during the period 1994–1998 was led by Rob Watson of the Natural Resources Defense Council (NRDC). It is Watson who is credited with coining the term Leadership in Energy and Environmental Design (LEED).
7. *Environmental Building News* provides excellent updates on the status of LEED development. "Spotlight on LEED" in the December 2003 issue reviewed the various LEED standards and progress in their production, piloting, and revision.
8. Additional information about Green Globes can be found at the Green Building Initiative website, www.thegbi.com.
9. More detailed information on BREEAM, including the system of assessment used to ensure compliance, can be found at www.breeam.com.
10. A detailed description of CASBEE can be found at the website of the Japanese Building Consortium, www.ibec.or.jp/CASBEE/english/overviewE.htm.
11. The paper by Alex Zimmerman, co-authored by Charles Kibert, is in review for publication by the *Building Research & Information Journal* (BRIJ), where it is expected to be published in its entirety in 2007. BRIJ is published by the Taylor & Francis Group (U.K.), and its website is www.tandf.co.uk/journals/titles/09613218.asp.

References

LEED Green Building Rating System 2.2. November 2005. Washington, DC: U.S. Green Building Council.

Ny, Henrik, Jamie McDonald, Göran Broman, Ryoichi Yamamoto, and Karl-Henrik Robèrt 2006. "Sustainability Constraints as System Boundaries," *Journal of Industrial Ecology,* Vol 10, Issue 1–2, pp. 61–77.

Robèrt, Karl-Henrik 2002. "Strategic Sustainable Development—Selection, Design, and Synergies of Applied Tools," *Journal of Cleaner Production,* Vol. 10, Issue 3, pp. 197–214.

"Spotlight on LEED." December 2003. *Environmental Building News,* 12(12), pp. 1, 12–17.

Chapter 4

The Green Building Process

The movement toward high-performance buildings is changing both the nature of the built environment and the delivery systems used to design and construct the facility according to a client's needs. The result has been the emergence of the high-performance green building delivery system, introduced in Chapter 1. This system is distinguishable from conventional practice by the selection of project team members based on their green building expertise; increased collaboration among the project team members; more focus on building performance than on building systems; heavy emphasis placed on environmental protection during the construction process; careful consideration of occupant and worker health throughout all phases; scrutiny of all decisions for their resource and life-cycle implications; the added requirement of building commissioning; and the emphasis placed on reducing construction and demolition waste. Some of these differences are driven by LEED requirements, while others are part of the evolving culture of green building.

This chapter more fully describes the differences between standard practice and the green building process, paying particular attention to the highly collaborative *charrette* process, probably one of the most distinguishing hallmarks of contemporary green building.

Conventional Versus Green Building Delivery Systems

Contemporary construction delivery systems in the United States fall into three major categories: *design-bid-build, construction management-at-risk,* and *design-build.* In the following subsections, these three systems are briefly described, and then compared and contrasted with the emerging high-performance green building delivery system.

DESIGN-BID-BUILD (HARD BID)

The primary objective of a design-bid-build, or hard bid, delivery system is low-cost delivery of the completed project. The design team is selected by the owner and works on the owner's behalf to produce construction documents that define the location, appearance, materials, and methods to be used in the creation of the building and its infrastructure. General contractors bid on the project, with the lowest qualified bidder receiving the job. Similarly, the general contractor selects subcontractors based on competitive bidding, and awards the specific work—for example, steel erection or masonry—to the lowest qualified bidder. Although the project is theoretically delivered at the lowest cost to the owner, conflicts among the parties to the contract (owner, design team, general contractor, subcontractors, materials suppliers) are frequent, and emotional tension and miscommunication generally permeate the process, often resulting in higher costs from change orders, repairs, and lawsuits.

CONSTRUCTION MANAGEMENT-AT-RISK (NEGOTIATED WORK)

In the construction management-at-risk system, the owner contracts separately with the design team and the contractor, or *construction manager,* who will work on the owner's behalf. This system is also referred to as *negotiated work* since the construction manager negotiates a fee for management services with the owner. Early in the design process, the construction manager is usually required to guarantee that total construction cost will not exceed a maximum price, referred to as the *guaranteed maximum price* (GMP). Ideally, both the construction manager and the design team are selected at the start of the project. The construction manager can then provide preconstruction services such as cost analysis, constructability analysis, value engineering, and project scheduling.

Working together, the parties produce construction documents that meet the owner's requirements, schedule, and budget, and prevent physical conflicts among systems, missing information, and other products of miscommunication often found in the construction documents produced for hard bid projects. Using a bidding process, the construction manager selects subcontractors based on their capabilities and the quality of their work, not merely the lowest bid. Accordingly, the level of conflict in negotiated work is much lower because of the closer relationship among the parties to the contract. Additionally, construction management firms undertaking negotiated work understand that the primary source of future income will be current and past clients. Consequently, client satisfaction becomes a primary objective.

DESIGN-BUILD

Although negotiated work reduces the frequency and intensity of conflicts present in a hard bid construction delivery system, the classic tension between the design team and the construction manager still exists, albeit to a lesser degree. *Design-build* is a method of project delivery in which one entity (the design-builder) forges a single contract with the owner to provide for architectural/engineering design services and construction services.[1] Design-build is also known as *design/construct* and provides the owner with single-source responsibility. In the typical design-bid-build project, the owner commissions an architect or engineer to prepare drawings and specifications under a design contract and subsequently selects a construction contractor by competitive bidding (or negotiation) to build the facility under a construction contract. In contrast, the design-build delivery system provides the owner with a single contractual relationship with an entity that combines both design and construction services. This entity may be a firm that possesses in-house design and construction capabilities or a partnership between a design firm and a construction firm. Thus, the design-build delivery system is more likely to reduce typical design-construction conflicts, provide a lower price for the owner, improve quality, speed the project to completion, and facilitate improved communication among the project team members. The design-build delivery system is very compatible with the green building concept, and due to its emphasis on a high degree of collaboration between the design and construction phases, it is very consistent with the design approach required to produce high-performance buildings.

HIGH-PERFORMANCE GREEN BUILDING DELIVERY SYSTEM

The evolving high-performance green building delivery system is a variant of the negotiated work system, but with additional responsibilities for the project team. Most notably, it requires much greater communication among the project team members. Consequently, initial team building, which engages the widest possible range of stakeholders, ensures that everyone understands the project's goals and the unique specifications. This delivery system also demands special qualifications from its participants, especially an understanding of, and commitment to, the concept of green building and, in the case of projects to be certified using LEED, strong familiarity with this standard and its requirements. The team members should also have experience with the charrette process and be especially willing to engage a wide range of stakeholders, including some who are traditionally not included in building projects. An example would be the inclusion of community members in the charrette for the design of a corporate facility.

Due to its adversarial nature, the hard bid delivery system is exceptionally difficult to employ for a green building project. The collaborative spirit needed for a successful high-performance green building project would be difficult to develop in this adversarial climate. The design-build delivery system has significant potential to deliver green buildings because, like negotiated work, it is designed to minimize adversarial relationships and simplify transactions among the parties. However, unlike conventional construction, the checks and balances provided by transparent interaction between the design team and the construction entity are virtually absent. And, as with other aspects of sustainable development, transparency is an important characteristic of green building projects. In spite of this potential problem, several successful green building projects have been executed using design-build, for example the Orthopaedics and Sports Medicine Building at the University of Florida, completed in September 2004 by a design-build team of URS and Turner Construction.[2]

Executing the Green Building Project

Because the high-performance green building delivery system is distinctly different in many ways from conventional delivery systems, the project team needs to be aware of these differences and where they occur in the building design and construction process. After the programming and budgeting of the proposed building project have been accomplished by or on behalf of the owner, the execution of a high-performance green building project has the following phases:

1. Setting priorities for the green building project by the owner in collaboration with the project team.
2. Selection of the project team: the design team and the construction manager or the design-build firm.
3. Implementing an Integrated Design Process (IDP): orienting the project team to the concept of IDP and how it will be implemented during the design and construction processes. IDP is described in more detail below.
4. Conduct a charrette to obtain input for the project from a wide variety of parties, including the project team, the owner and users, the community, and other stakeholders.
5. Execution of the design process, consisting of schematic design, advanced schematic design, design development, construction documents, and documentation of green building measures for a project that is to be certified, all conducted using IDP. This involves full use of IDP in the development of the design, marked by extensive interdisciplinary interaction to maximize design synergies.
6. Construction of the building, to include implementing green building measures that address soil and erosion control, minimizing site disturbance,

protecting flora and fauna, minimizing and recycling construction waste, ensuring building health, and documenting the construction phase of green building measures.

7. Final commissioning and handover to the owner.

OWNER ISSUES IN HIGH-PERFORMANCE GREEN BUILDING PROJECTS

The decision to produce a high-performance green building brings with it a number of unique issues that have to be resolved by the owner prior to initiating the design and construction of the building. Among the questions that must be answered are the following:

- Does the owner want the building to be a certified green building? Although the LEED approach is the predominant method for producing a green building, a green building based on a different philosophical and technical approach may be desirable. For example, the Green Globes building assessment protocol is an alternative approach that may be a good choice in some situations. In at least one state, Florida, there is a commercial green building assessment standard that can be used in lieu of the national standard.[3]
- If the building is to be certified, what level of certification is desired (Platinum, Gold, Silver, Certified for LEED or the number of Green Globes)? The building's owner may have a preconceived idea of the level of certification desired for the facility, in which case the task of the project team will be to design and build the facility to meet the owner's goals. Often the project team will have to address the cost/benefit issues involved in achieving different certification levels and provide LCC analyses for each level to give the owner the data needed to make a decision.
- If the building need not be certified, what design criteria should be followed by the design team? The LEED and Green Globes standards each provide a consistent framework that contains virtually all the criteria needed to produce a green building. If LEED or Green Globes is not to be the basis for creating the green building, the owner will have to provide the project team with a detailed description of the criteria the team members are to use in their work.
- What are the desired qualifications of the design team and construction manager with respect to the high-performance building? In the case of a design-build project, what background and training should the designers and construction professionals have? It is certainly advantageous for the owner to hire project team members who have green building experience. If certification is desired, significant documentation of numerous aspects of the project will be required. For example, if one of the credits being addressed is the recycled content of materials used in the project, the construction manager will have to obtain information from most of the subcontractors about the quantity of recycled materials in the products they are using in the building, and then compile the data from all the subcontractors to determine the overall percentage of recycled content in the project.
- What level of capital investment, beyond that required for conventional construction, will the owner provide to make the facility a high-performance green building? And is the owner willing to consider trading off lower operational costs for higher front-end capital costs? Green buildings are specifically designed to have lower operational costs, which are often accompanied by higher front-end capital costs. An LCC analysis will provide a breakdown of costs versus savings on an annual basis and indicate where the breakeven point, in years, for the investment occurs. It is up to the owner to decide whether the

breakeven point is satisfactory and, based on this information, whether the additional capital cost is warranted.

SETTING PRIORITIES AND MAKING OTHER KEY INITIAL DECISIONS

When the decision has been made to create a high-performance green facility, the owner must decide on the priorities for the building. For example, in water-short areas of the United States, water issues may be so important that the owner may decide to focus heavily on the building's hydrologic cycle (water conservation, water reuse, rainwater harvesting, graywater systems, and employment of reclaimed water) rather than, for example, to make an exceptional effort to reduce energy consumption. Another owner may opt for implementing an extensive and exceptional system of daylighting and lighting controls due to its energy-conserving possibilities and potential health benefits and, conversely, undertake minimal water conservation measures.

Another priority to be set and a decision to be made concern the financial investment the owner is willing to make in a high-performance building. Green buildings normally involve systems not commonly used in conventional buildings; for example, rainwater harvesting systems, with their associated piping, pumps, and cisterns, entail additional design effort. Many state governments are forced by law to operate within strict per-square-foot cost guidelines. As a result, very simple, cost-effective measures must be considered. Other types of organizations may have revolving funds that can be used to invest in high-performance options that will pay back the fund over time. Harvard University, for example, has a $3 million revolving fund that can be used for investing in higher-capital projects that are repaid out of the savings. The federal government requires LCC to be employed to justify building investment decisions, a requirement that works in favor of high-performance building decisions. Private sector owners have considerably more leeway, and their decisions can be based on LCC, as is the case for the federal government. Certified green buildings will have additional documentation requirements, requirements for commissioning, fees for registration and certification review, and other costs that must be allocated in the building budget.

SELECTING THE GREEN BUILDING TEAM

When an owner has decided to produce a high-performance green building, the next order of business is to select the design and construction teams. The actual selection process proceeds in the conventional fashion with the issuance of a Request for Proposals (RFP) or Request for Qualifications (RFQ) by the owner to announce the upcoming selection of the architect and construction manager. The RFP/RFQ should specify the additional qualifications required of the architect, interior designers, landscape architects, civil engineers, structural engineers, electrical engineers, and mechanical engineers supporting the design. One of the challenges in writing an RFP for a high-performance building is to ensure that the architects and construction managers understand the owner's green goals. To facilitate this effort, the Committee on the Environment (COTE) of the AIA has produced a guide to writing RFPs and RFQs for green buildings, *Writing the Green RFP*.[4]

After reviewing the submissions by the architect and construction management firms or design-build firms that respond to the RFP/RFQ for the project, the owner typically creates a list of three to five firms in each category, then organizes presentations by the shortlisted design firms and construction management companies. The final selection is based on experience, qualifications, previous work, and demonstrated understanding of the owner's program and requirements, the building site, and the firm members' ability to work with other project team members. The architect and construction manager or design-build firm should be selected prior to the start of the design so that both will be on board during the entire project.

Clearly, it is important that the architect and engineers have a detailed understanding of the concept of green building and a commitment to investing creativity and energy to produce an exceptional building. At this point in the evolution of high-performance green buildings, even though the movement is relatively new, there are a large number of design professionals who have already engaged in the design of one or more green buildings. Detailed knowledge of the USGBC LEED building assessment standard or the Green Globes building assessment protocol is absolutely essential if the owner decides that the goal is green building certification. It is also important to note that there are some outstanding architects who have experience creating high-performance buildings that have not been submitted for USGBC certification; thus, the owner must judge the ability of these firms to meet the owner's requirements.

If the building is to be certified, the construction manager should have great familiarity with, or staff trained in, the requirements of the LEED or Green Globes standard. The certification process imposes enormous responsibility on the construction manager; lack of experience with the standards could compromise the certification of the project.

ROLE OF THE LEED ACCREDITED PROFESSIONAL IN A LEED PROJECT

For both the design team and the construction manager, the USGBC has a training and testing program that, if successfully passed, designates the individual as a LEED Accredited Professional (LEED-AP). This designation provides the building owner with a high degree of assurance that the requirements of the USGBC certification programs will be understood and that the extensive documentation required for certification will be provided. The LEED-AP Examination, based on LEED-NC 2.2, is intended to test the individual's knowledge of green building principles, as well as familiarity with LEED requirements. In February 2007, the USGBC announced LEED-AP examinations for LEED-EB and LEED-CI. There are no requirements for work experience or educational background, although the USGBC does indicate that applicants who consider taking the exam should have industry work experience. The following are the points covered on the LEED-AP Examination:[5]

- In-depth familiarity with the LEED building assessment system
- Understanding of LEED project registration/technical support/certification processes
- Knowledge of LEED documentation requirements
- Demonstrated knowledge of design and construction industry standards and processes
- General understanding of the various standards referenced in LEED
- Understanding of green and sustainable design strategies and practices, and corresponding credits in the LEED rating system
- Familiarity with key green and sustainable design resources and tools

One of the other benefits of having a LEED-AP on the project team is that one credit is awarded for the certification of the project. One of the drawbacks of the current system of awarding this credential is that it does not require in-depth knowledge of the building design and construction process, nor does it require professional experience. One of the challenges for the USGBC is to create a rigorous accreditation process with requirements either for periodic recertification or for continuing education to maintain currency on green building issues and the LEED system.

The Integrated Design Process

Although it is true that excellent teamwork is required for any building project, the level of interaction and communication needed to ensure the success of a green building project is significantly higher. Green buildings are a new concept to the industry, and it is generally necessary to orient all members of the project team to the goals and objectives of the project that are related to issues such as resource efficiency, sustainability, certification, and building health, to name a few. This orientation can serve three purposes. First, it can fulfill its primary purpose of informing the project team about all project requirements. Second, it can familiarize the project team with the owner's priorities for the high-performance green building aspects of the project. Third, it can provide an opportunity to accomplish team building in the form of group exercises for familiarizing the group with the building, the building program, and the building's green building issues.

Integrated building design or *integrated design* is the name given to the high levels of collaboration and teamwork that help differentiate a green building design from the design process found in a conventional project. According to the U.S. Department of Energy, integrated design is

> [a] process in which multiple disciplines and seemingly unrelated aspects of design are integrated in a manner that permits synergistic benefits to be realized. The goal is to achieve high performance and multiple benefits at a lower cost than the total for all the components combined. This process often includes integrating green design strategies into conventional design criteria for building form, function, performance, and cost. A key to successful integrated building design is the participation of people from different specialties of design: general architecture, HVAC, lighting and electrical, interior design, and landscape design. By working together at key points in the design process, these participants can often identify highly attractive solutions to design needs that would otherwise not be found. In an integrated design approach, the mechanical engineer will calculate energy use and cost very early in the design, informing designers of the energy-use implications of building orientation, configuration, fenestration, mechanical systems, and lighting options.[6]

The integrated design process is characterized by early significant collaboration in the design process. In conventional design, the team begins their joint effort at the start of schematic design, whereas in a green building project employing integrated design, the collaboration starts at the very beginning of the project and all team members have input on design decisions during the entire cycle of design (see Figure 4.1). The earlier integrated design is implemented, the greater the benefits (see Figure 4.2).

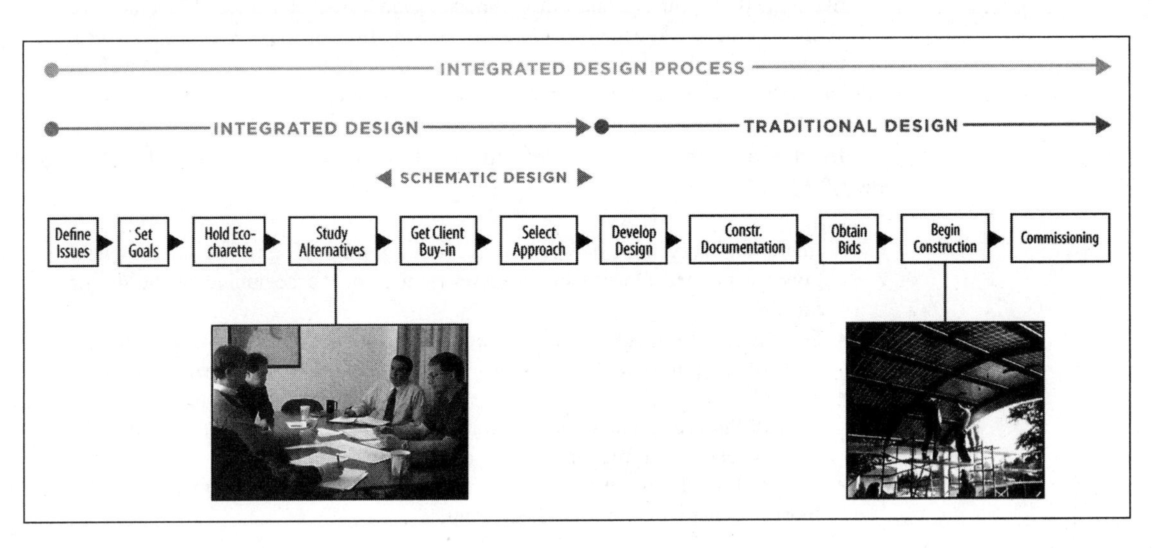

Figure 4.1 In green design, the integrated design starts much earlier in the project development process compared to conventional design, involving interaction with the owner to define issues and set goals prior to schematic design and continuing through construction and commissioning. (Diagram courtesy of Interface Engineering, Inc.)

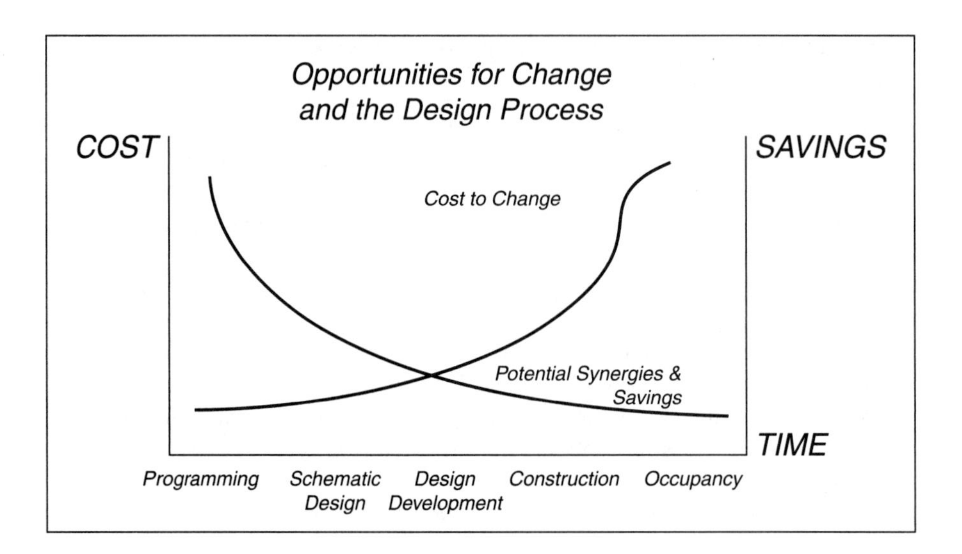

Figure 4.2 The earlier an integrated design process is implemented, the greater the potential savings and the lower the cost of changes to the building design.

There are numerous potential areas for integrated design in any building project: the building envelope, the daylighting scheme, green roofs, minimization of light pollution, indoor environmental quality, and the building hydrologic cycle, to name but a few. The Green Globes building assessment protocol spells out the requirements for integrated design in its Project Management section, where a team can achieve 20 points for demonstrating that they have indeed implemented integrated design in the process. In addition to appointing a Green Design Coordinator, the team must demonstrate how they interacted by documenting the results of their collaboration in the form of the minutes of goal-setting meetings and lists of items on which the team worked jointly for resolution.[7]

Another term that describes integrated design is *integrated design process* (IDP). Some of the foundational work on developing IDP occurred in Canada, and perhaps the most thorough definition was a result of a National Workshop on IDP held in Toronto in 2001:[8]

> IDP is a method for realizing high performance buildings that contribute to sustainable communities. It is a collaborative process that focuses on the design, construction, operation and occupancy of a building over its complete life-cycle. The IDP is designed to allow the client and other stakeholders to develop and realize clearly defined and challenging functional, environmental and economic goals and objectives. The IDP requires a multi-disciplinary design team that includes or acquires the skills required to address all design issues flowing from the objectives. The IDP proceeds from whole building system strategies, working through increasing levels of specificity, to realize more optimally integrated solutions.

In addition to this extensive definition of IDP, the main elements of the IDP were identified as:

- Interdisciplinary work between architects, engineers, costing specialists, operations people and other relevant actors right from the beginning of the design process;
- Discussion of the relative importance of various performance issues and the establishment of a consensus on this matter between client and designers;
- The addition of an energy specialist, to test out various design assumptions through the use of energy simulations throughout the process, to provide relatively objective information on a key aspect of performance;
- The addition of subject specialists (e.g., for daylighting, thermal storage, etc.) for short consultations with the design team;

- A clear articulation of performance targets and strategies, to be updated throughout the process by the design team.
- In some cases, a Design Facilitator may be added to the team, to raise performance issues throughout the process and to bring specialized knowledge to the table.

It was also noted that it may be useful to launch the IDP with a charrette, described in more detail in the following section.

Traditional design could be said to have three steps:

Step 1: The client and architect agree to a design concept that includes the general massing of the building, its orientation, its fenestration, and probably its general appearance and basic materials.

Step 2: The mechanical and electrical engineers are engaged to design systems based on the building design concept agreed to in Step 1. The civil engineer and landscape architect develop a concept for landscaping, parking, paving, and infrastructure based on the building design concept and the owner's wishes.

Step 3: Each phase of design (schematic, design development, and construction documents) is carried out employing the same pattern, with minimal interaction between disciplines, little or no interdisciplinary collaboration, and attention to the speed and efficiency of executing each discipline's design.

The result of traditional design is a linear, noncollaborative process in which no goals are set and the performance of the building is purely random and not optimized, albeit according to code. Each discipline functions in isolation, with interdisciplinary communications kept to a minimum. As is the case with every other system, optimizing each subsystem of the project results in a suboptimal building. The most likely outcome is not only an unoptimized project, but also a range of other potential problems caused by a lack of strong coordination among disciplines.

In contrast to traditional design, the point of IDP is to optimize the entire building project, and the requirements for communication are intense, nonstop, and at all stages of the project, from design through construction, commissioning, turnover to the owner, and postoccupancy analysis. Integrated design starts prior to the actual design process, with the project team articulating goals for the project and determining the opportunities for synergies in which design solutions have multiple benefits for the project. The following is a typical sequence of events that are indicative of integrated design:

- The project team establishes performance targets for a broad range of parameters, to include energy, water, wastewater, landscape performance, heat island issues, indoor environmental quality, and construction and demolition waste generation, to name a few. In conjunction with establishing these performance targets, the project team develops preliminary strategies to achieve the targets. IDP should bring engineering skills and perspectives to bear at the concept design stage, thereby helping the owner and architect to avoid becoming committed to a suboptimal design solution. It should also involve all members of the team bringing their skills to bear on designing the optimal building. Mechanical engineers are better placed in terms of their background in thermodynamics than the architect, and it makes sense to engage them in the design of the building envelope.
- The team should minimize heating and cooling loads and maximize daylighting potential through orientation, building configuration, an efficient building envelope, and careful consideration of the amount, type, and location of fenestration. A potentially wide variety of plug loads should be addressed due to the effects of large numbers of computers, printers, fax machines, sound systems,

and other equipment on the performance of the building. Minimizing these loads and selecting equipment with the lowest possible energy consumption is needed so that the intent of the high-performance building is not compromised by neglecting to account for this consumption. The broad range of indoor environmental quality issues should be addressed, to include air quality, noise, lighting quality and daylighting, temperature and humidity, and odors. The team should also collaborate on site issues to maximize the use of natural systems, minimize hardscape, use trees to assist heating and cooling of the building, and integrate rainwater harvesting, graywater systems, and reclaimed water into the design of the building's hydrologic cycle.

- The team should maximize the use of solar and other renewable forms of energy, and use efficient HVAC systems, while maintaining performance targets for indoor air quality, thermal comfort, illumination levels and quality, and noise control.
- The result of the process should be several concept design alternatives, employing energy, daylighting, and other simulations to try out the alternatives, and then the selection of the most promising of these for further development.

The earlier IDP is instituted, the greater its effect on the design process. The maximum benefit occurs when the decision to employ IDP is made prior to the start of the design process and the project team has the opportunity to set goals for the project that guide the design process.

The result of IDP should be a full understanding of the potential design synergies and the connection of the project goals to the resulting building design. A truly

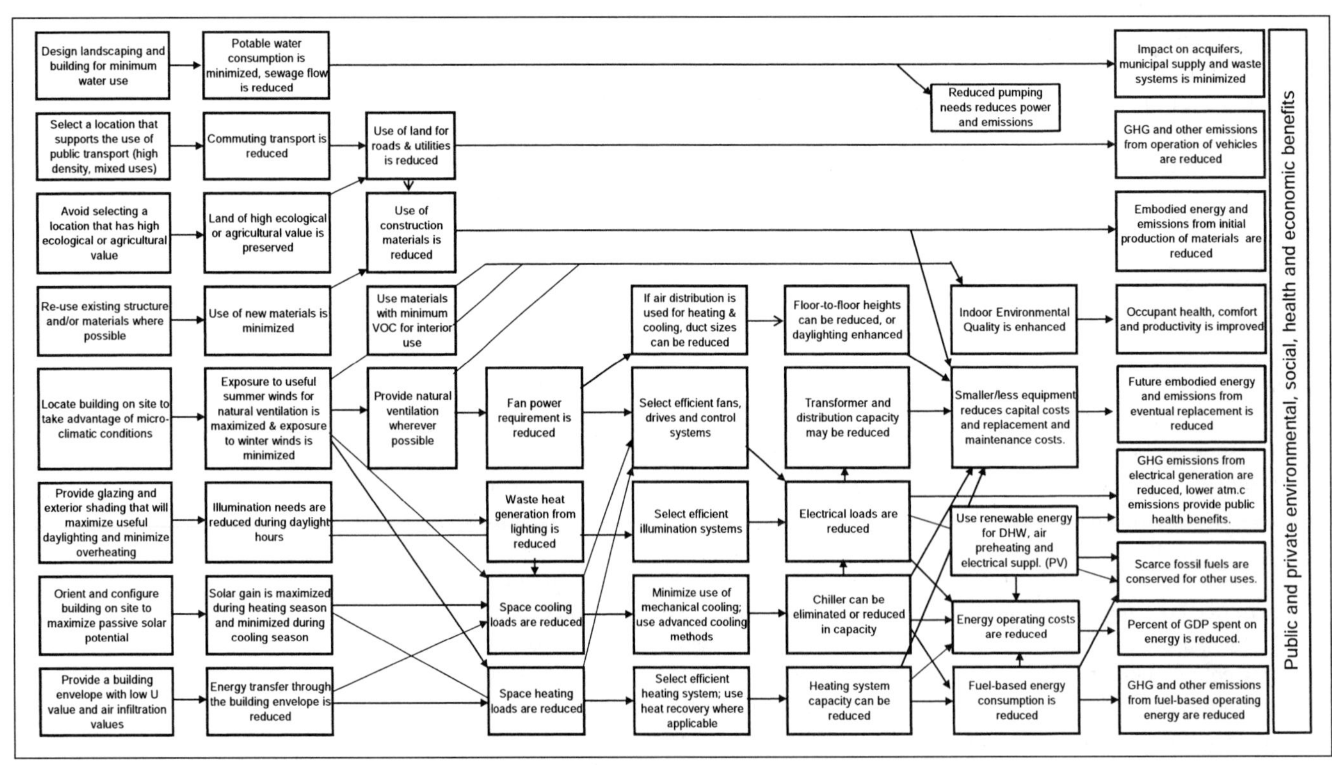

Figure 4.3 The integrated design process can assist in achieving design synergies by stimulating interdisciplinary collaboration that results in green strategies listed in the leftmost column of this example project being translated into benefits for the building owner and occupants as well as for the global environment. (Illustration courtesy of Nils Larsson, Natural Resources Canada, and the United Nations Environmental Program.)

collaborative process will use these project goals as the basis for wide-ranging, dynamic interaction among the project team members to capitalize on the potential for reducing resource consumption, reducing environmental impacts, and restoring the site to its maximum ecological potential. Figure 4.3 is a schematic that demonstrates how project goals can be used in conjunction with IDP to produce a wide range of benefits, both for the project and for the environment.

Another term related to integrated design is *whole building design,* a concept advocated by the National Institute of Building Sciences (NIBS) and described as consisting of two components: an *integrated design approach* and an *integrated team process.*[9] Whole building design has been adopted by a group of federal agencies as the core concept of high-performance green buildings, and the emphasis is on collaboration and life-cycle performance. The concept of collaboration is extended outside the project team to include all stakeholders in the building process. In the integrated team process, the design team and all affected stakeholders work together throughout the project phases to evaluate the design for cost, quality of life, future flexibility, and efficiency; overall environmental impact; productivity and creativity; and effect on the building's occupants. Whole building design, as described by NIBS, draws from the knowledge pool of all the stakeholders across the life cycle of the project, from defining the need for a building, through planning, design, construction, building occupancy, and operations. The process does not conclude at the end of construction and handover to the owner. During operation, the building should be evaluated to ensure that it has met its high-performance design objectives. Furthermore, the building should be recommissioned periodically to maintain its high-performance character throughout its life cycle.

Role of the Charrette in the Design Process

Creating a green, sustainable building implies that the widest range of possible stakeholders will be engaged in the process because buildings ultimately affect a large variety of people and, in fact, affect future buildings. As the predominant artifacts of modern society, and due to their relative longevity, buildings are important cultural symbols; hence, they impact enormous numbers of people every day. Passersby are affected either positively or negatively by the appearance of a building based on its design, materials, color, location, and function. The stakeholders in a building will vary widely, depending on its type and its ownership. For example, a public building such as a library or city administration building will affect not only the employees who will directly use the building, but virtually all persons in the local jurisdiction, who, as taxpayers, have contributed to its realization. In the case of a corporate building, although its impact may not be as widespread, a savvy owner would nevertheless engage a wide range of users, customers, local government, and citizens to obtain the maximum input. The process of gathering this input is referred to as a *charrette.* A general overview of the charrette concept is provided here. The detailed integration of the charrette into the design process is covered in the next section.

The word *charrette* is derived from the French term meaning "little cart." As noted in Chapter 1, the concept has its roots in French architectural education when proctors at the École des Beaux Arts in nineteenth-century Paris collected student projects on wheeled carts, literally pulling the drawings from the students' hands at the end of their final frenzied efforts on a design project. Today, the term is used to refer to an effort to create a plan. The National Charrette Institute (NCI) states that there are four guiding principles for a charrette (see Table 4.1). Note that these principles are meant to apply to a community planning charrette, not specifically to the design of a single building. Consequently, they are presented here in a modified form from the actual NCI guiding principles.[10]

TABLE 4.1

Four Guiding Principles for a Built Environment Charrette

1. *Involve everyone from the start.* The identification and solicitation of stakeholders to provide input to a project is of the utmost importance because the participants in the charrette process will feel a sense of ownership for the outcomes. The broader the range of input, the more likely the project is to be successful and accepted by the community. It is also important to note that people or organizations that may potentially play a role in blocking a project should be invited to participate.

2. *Work concurrently and cross-functionally.* All disciplines engaged in a project should work together at the same time during the charrette and with the other stakeholders to generate alternative designs under the guidance of a facilitator. The level of design detail that emerges from the charrette will be a function of the time available and the complexity of the project. In general, a building design charrette produces a wide range of potential solutions and approaches that not only cover green issues, but also address the function of the building and its relationship to the community. For larger, more complex projects, the participants can divide into groups to tackle specific issues, then return to a caucus or plenary meeting for each group to share its progress with other groups and to make decisions on how to proceed.

3. *Work in short feedback loops.* For a building project, proposed solutions and measures are laid out in a brainstorming session during which the participants, guided by a facilitator, cover all aspects of the building, its infrastructure, and its relationship to the community. This approach produces far more alternatives and engages far more creativity than a conventional design process. This is an advantage in that many more ideas and options are presented. That said, the information must also be processed efficiently and rapidly to provide useful input to the actual design process. The result of the brainstorming sessions must be distilled to the essential outcomes, and duplications must be eliminated and priorities established. For example, it would certainly be advantageous if all buildings had photovoltaics, but few owners have the resources at present to incorporate them into their facilities. The feedback loops between initial brainstorming sessions and design decisions should be as rapid as possible so that more than one iteration is possible during the charrette.

4. *Work in detail.* The more detail in a charrette the better. Alternatives for building appearance, orientation, massing, and electrical and mechanical systems should be sketched out in as much detail as possible. The NCI recommends working on problems at different scales during the charrette. Larger-scale issues of drainage, paving, and relationships to other buildings and the street should be addressed, as should details such as entrance location, window selection, and roof type.

The NCI has also proposed a four-step charrette process that, although designed for a community planning charrette, is also applicable to a building project charrette. These steps are outlined in Table 4.2.[11]

At the conclusion of the charrette, it is the responsibility of the project team to transform the results into a report that can be used to guide the design of the project. A final review of the outcome of the brainstorming sessions should be conducted to ensure that the measures selected for implementation meet cost and other criteria that may be important. Communications may need to be established with entities or groups external to the charrette to ensure that they act to maximize the high-performance aspects of the project. For example, Rinker Hall, a LEED Gold building at the University of Florida in Gainesville that was certified in 2004, is connected to a central plant that provides its heating and cooling. The project team decided that the LEED Energy & Atmosphere point for eliminating hydrochlorofluorocarbon (HCFC) use could be justified only by obtaining a commitment from the university to implement a program to replace its older, HCFC-based chillers with efficient hydrofluorocarbon (HFC)

TABLE 4.2

Four Steps for a Built Environment Charrette

1. *Startup.* In the context of a building project, the startup for a charrette is very simple. It involves determining who the stakeholders are, engaging the stakeholders in the process, establishing the goals for the charrette, determining the time and place for holding the charrette, and notifying the participants of the details.

2. *Research, education, and concepts.* Prior to the charrette, the building owner, the charrette facilitator, and design team members should discuss the information needs for the charrette. The owner's directions, the building program, site details, utility information, and other pertinent data should be gathered and readied for the charrette. Information on specific technologies may be useful. For example, if a fuel cell is a strong-candidate technology for the project, technical information about the device, issues of connecting the fuel cell to the grid, and information about fuel and emissions should be gathered for use during the process. In some cases, the process of gathering information for the charrette may highlight the need to engage other organizations in the process. In the example of a fuel cell, the local utility company could provide valuable input as to how best to incorporate the fuel cell into the project. The location of the charrette should be selected to best facilitate its conduct. Generally, it is best to hold the charrette at the owner's location if adequate space and facilities are available. A large room with blackboards or whiteboards, space for large-paper tripods, projector and projector screen should be available.

3. *The charrette.* Generally, the charrette should be conducted by a facilitator familiar with the green building process. A typical building charrette might occur over several days and continue in phases until complete. The first step should be an effort to educate all the participants on the owner's requirements and the concept of high-performance green building. The second step would be to review the building program, previously generated architectural schemes, building siting, proposed construction budget, and construction schedule. The third step would be to lay out the goals of the project with respect to its green high-performance aspects. The owner may desire a specific level of certification, for example a LEED Gold certification, that will affect many of the decisions made during the charrette. When these steps have been completed and the project team and stakeholders understand the context of the project, the actual charrette begins. The facilitator conducts a guided brainstorming session that draws out input from the group about every aspect of the project, with a special emphasis on the sustainability of the building. During the conduct of the charrette, the team should keep a running scorecard on how the decisions made during the process are impacting the building assessment score. The economics of each decision also need to be taken into account, and the construction manager should ensure that enough data are available to provide a conceptual cost estimate for review by the owner.

4. *Review, revise, and finalize.* After the charrette is complete, the design team reviews the results with the owner, makes any appropriate adjustments and changes, and then produces a report of the charrette to guide the balance of the design process.

refrigerant chillers. Another point, for maintaining open space, was acquired by obtaining a letter from the university administration stating that specific property contiguous to Rinker Hall would be maintained as open space for the life of the building. In the private sector, cooperation of municipal officials may be necessary to obtain points for proximity of mass transit.

The final version of the charrette report becomes one of the guiding documents for the launch of the schematic design phase of the project, and ultimately serves to help steer the project through design development, construction documents, and the actual construction process.

CONDUCT OF THE GREEN BUILDING CHARRETTE

The charrette typically marks the launch of the design phase for a green building. As noted previously, the stakeholders in a green building are drawn from the widest possible range of people who will be affected by its construction. For a typical university building, for example, the stakeholders would include the campus planning department, the physical plant building operators, the architect, the design engineers, students, faculty, and anyone else impacted by the design and construction of the building.

Although there is not yet a standard charrette process, the steps in a typical charrette for a building that will use the LEED or Green Globes building assessment system can be summarized as follows:

1. Provide the stakeholders and participants with an overview of the goals of the client, the building program, the budget, the project schedule, and other pertinent information. This should be done prior to the charrette and reviewed by the charrette facilitator at the start of the charrette.
2. Describe the LEED or Green Globes building assessment system to the group, thoroughly covering the point system and any goals of the owner/client with respect to achieving certification. Because the participants may not be familiar with building assessment, a review of the system being used for the project is essential to a successful outcome of the charrette.
3. Conduct an open, uncritical brainstorming session with the goal of generating as many ideas as possible for making the proposed facility a high-performance building. Typically, the measures proposed by the participants are recorded on charts situated in the room where the brainstorming takes place.
4. Organize the results of the brainstorming into major categories, including site, water, energy, materials, indoor environmental quality, and innovations. Post the measures organized by category on charts in the brainstorming room.
5. Have the participants select their top measures from the brainstorming session. A typical method is to provide each participant with three to five stars that they use to indicate their favorite green building measures.
6. Sort the measures by order of interest by identifying those suggestions that receive the most attention—that is, the most stars.
7. List the top measures by order of interest and begin the process of determining how to achieve them.
8. Compare the results to the LEED or Green Globes structure to determine how many credits are achievable based on the results of the charrette.
9. Determine the cost of the entire project based on the owner's program and directions and the results of the charrette.
10. If the number of points and the costs are acceptable, the result of the charrette can be used as input to the design process. If the owner/client has predetermined a level of performance—for example, a minimum of a LEED Gold building—then the participants must determine if this criterion has been met.
11. If the process has failed to achieve the desired level for certification, additional brainstorming may be needed to ensure that the owner/client's desired level of achievement will be met.
12. If the process has failed to meet budgetary constraints, the owner has the choice of directing the project team to review the results of the charrette to

keep the project within budget or to allow higher capital costs so that the results of the charrette at this point can be implemented.

The last point in the charrette process is crucial. The construction manager on the project team who is providing preconstruction services must be able to assess costs rapidly during the process. To that end, the manager may employ a software tool. For example, DPR Construction, Inc., uses such a tool developed by Paul Shahriari of Green Mind Technologies, Inc. It can be used to track the costs associated with various LEED measures during a charrette. It provides capital costs, life-cycle paybacks, and other data to assist the decision-making process during a rapidly moving charrette. A cash-flow analysis derived from the use of this tool is shown in Figure 4.4.

The charrette, as the first stage in the process that brings together the design and construction team with the client in an active participatory process, offers an opportunity for all stakeholders to come away with a detailed understanding of the project. In addition, the charrette provides education about green building issues, enables team building within the project team, and generates an unprecedented level of momentum for launching the effort.

Following the charrette process, the results should be examined by the building team to determine which measures will be implemented. Calculating the cost of implementing some of the measures may be necessary to determine the feasibility of implementation. A typical issue is whether to incorporate a photovoltaic system, which, although desirable because it uses renewable energy, may not be achievable due to its cost.

One of the dangers in the initial stages of the design process is that measures identified as lowering the environmental, resource, and health impacts of the building may be eliminated later on in a *value engineering* (VE) process. VE is a process wherein the owner/client engages a consultant in a formal review of the design with the goal of maximizing the so-called value of the project while at the same time reducing costs in order to meet the budgetary goals of the building.

There are many instances of innovative green building measures being eliminated during the VE process. This occurs because of a lack of sound cost data on the

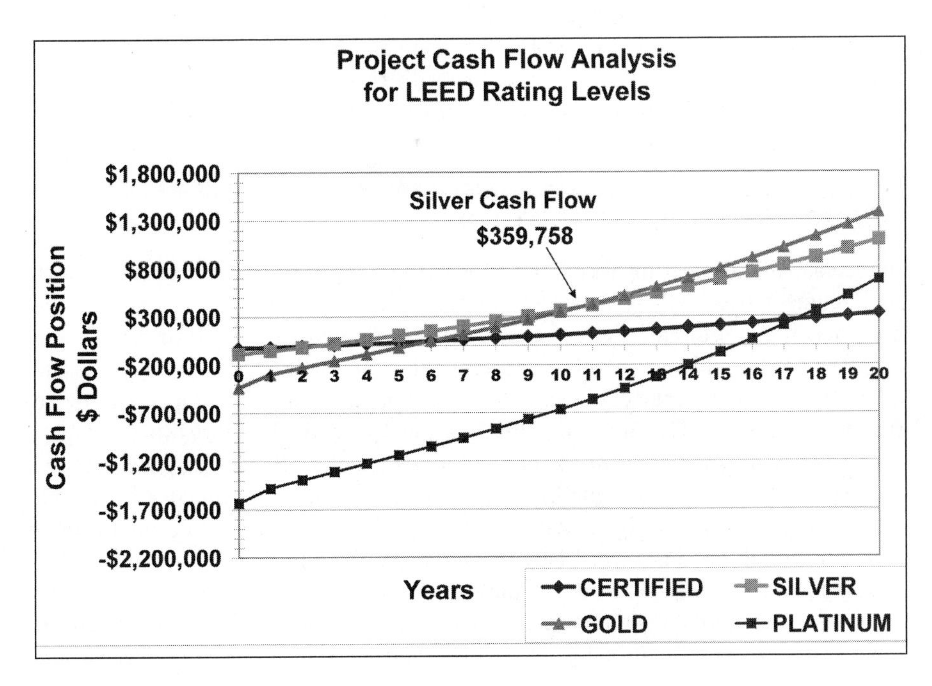

Figure 4.4 A spreadsheet tool used by DPR Construction, Inc., during green building charrettes helps provide a rapid assessment of the costs and benefits of various green building options, as well as the overall costs and benefits for various levels of LEED certification. As the chart indicates, the higher the LEED rating, the higher the first, or construction, costs. However, the payback line is steeper as the rating level increases from Certified to Silver, Gold, and Platinum. (Courtesy of Paul Shahriari of Greenlight Strategies, Inc.)

measures being implemented at the start of the project. The life-cycle impacts of the various measures must be well known in order to understand their capital costs versus their savings. For example, most energy-conserving features require additional capital up front but have correspondingly lower operational costs. The cost tolerance of the owner/client for these measures must be understood at the start of the project to ensure that they are not eliminated later in the project.

Some measures that may gain LEED points cost significantly more on a unit basis, and have no payback, but are justifiable solely on the basis of lowering environmental or health impacts. For example, a new product for replacing pressed board in millwork, comprised of compressed wheat husks, may be justified as being far more environmentally friendly than the alternative. However, it may cost $200 per sheet for the millwork compared to $40 per sheet for the conventional alternative. Another example is paints that have zero emissions but may cost about $5 more per gallon than conventional alternatives. Likewise, protecting ductwork during the construction process to ensure that it remains uncontaminated entails additional costs. Innovative construction waste recycling measures, too, may cost the construction manager far more to implement—for example, the recycling of gypsum wallboard. A rainwater harvesting system, with its cistern, pumps, and controls, will cost additional capital to implement. Thus, the owner/client must understand these costs and be willing to pay for the more environmentally friendly alternatives. The fact that the owner/client is seeking LEED certification is an indication that many of these measures will be acceptable, but it is the responsibility of the building team to make these costs clear to the owner during the design process, in particular after the charrette has taken place.

Green Building Documentation Requirements

To certify a green building using one of the major building assessment systems requires that a great deal of attention be paid to gathering information throughout the course of the design and construction of the project that ultimately will be reviewed as part of the certification process. The two main building assessment systems in the United States, LEED and Green Globes, have different approaches to how the documentation is ultimately reviewed, and the project team should carefully review the requirements for each approach.

USGBC LEED DOCUMENTATION

In the case of a project for which the owner is seeking USGBC certification, careful documentation of the efforts to achieve credits is needed. As noted earlier, the documentation requirements for the first versions of the LEED building assessment standard were relatively complex and difficult. For LEED-NC Version 2009 and other LEED products, the documentation requirements, while far simpler, are by no means easy to meet. The advent of LEED-Online has made the entire process paperless. The documentation may be submitted in two batches: the design phase and the construction phase. The design phase submission is for those credits that are essentially complete during design and do not require any documentation during the construction phase. For instance, LEED Materials & Resources (MR) Prerequisite 1 requires that a space be set aside for the storage and collection of recyclables in the building. The required documentation is a drawing that shows this area and the location of the containers required for recycling. This prerequisite is completed during design and can be submitted with other design phase credits for review by the USGBC via LEED-Online. Most credits are documented at least in part by means of the aforementioned LEED Letter Templates. A LEED Letter Template is to be filled out for each credit the project team is claiming for the building. For example, to demonstrate that the LEED Prerequisite 1 for Construction Activity Pollution Prevention in the Sustainable Sites (SS) category has been adequately addressed, the civil engineer or other responsible party must fill out the LEED Letter Template designated for this purpose, stating that the project followed U.S. Environmental Protection Agency (USEPA) Document No. EPA 832/R-92-005 (September 1992), "Storm Water Management for Construction Activities," or local erosion and sedimentation control standards and codes, whichever is more stringent. A brief list of the measures actually implemented must also be provided, along with a description of how they meet or exceed the local or USEPA standards. The effort to document that this prerequisite has been met should be factored into the overall design and construction process to ensure that all the documentation has been prepared by the completion of construction.

Another example of required documentation is Materials and Resources (MR) Credit 4.1 (MRc4.1), Recycled Content of 10 percent. This credit is achieved if the project team can demonstrate that 10 percent of the value of the nonmechanical and nonelectrical materials in the building have a combination of postconsumer and preconsumer recycled content. Only one-half of the preconsumer content can be included in the calculation. For MRc4.1, the architect, owner, or other responsible party must state that these requirements have been met and include details about products, product value, postconsumer and preconsumer recycled content, and the resulting overall recycled content for the project.

The project team also must decide at the start of the project how information will flow among the various parties and who will actually compile and produce the information for the appropriate LEED Letter Template. For MRc 4.1, the calculations provided with the Letter Template must clearly demonstrate that a requirement has been met by indicating the product or material, its value, and its postconsumer and preconsumer recycled content. The final computation should demonstrate that at least 10 percent of the total value of the materials, excluding mechanical and electrical systems, is recycled content, counting postconsumer content at its full percentage and preconsumer at half of its percentage in each product. This requirement can be challenging for products such as glass and aluminum storefronts, where part of the aluminum components may have recycled content but the glass will not. Additionally, because the product is likely to be assembled by a local glass subcontractor, that firm must research this information for its product. The contractor then compiles the information on the recycled content for all products to produce a final picture of the total recycled content of the project. Finally, either the contractor or the architect submits this data along with the MRc4.1 Letter Template for the project at LEED-Online.

It should also be noted that the USGBC audits submissions, meaning that much more extensive backup information may be required to verify the assertions made in the Letter Templates. Therefore, it is good practice to ensure that full documentation is maintained throughout the design and construction processes and that all assumptions are clearly stated in the backup materials.

GREEN GLOBES DOCUMENTATION

Green Globes relies on an online questionnaire that the project team should utilize to guide the green aspects of the design and construction process. A careful review of the questionnaire should alert the team that, for example, as an indicator of Integrated Design, meetings should be held and documented to demonstrate that Integrated

Design was indeed being fostered. (See Section A1.4 of the Green Globes questionnaire in Appendix E.) Another indicator of integrated design is the appointment of a Green Design Coordinator who must be assigned duties such as to (Section A1.1):

- Outline the overall green design framework for the project
- Communicate the client's/user's intentions to the project team
- Develop measurable green design performance requirements
- Assist in evaluating responses against the green design objectives

A careful review of the questionnaire will provide the project team with valuable information about the required documentation and what the Verifier, who audits the project documentation in an on-site visit at the conclusion of construction, will be reviewing to determine if the documentation is adequate.

Summary and Conclusions

The process of green building and its delivery system are unique in that they provide not only improved buildings to owners but also an improved process. In a short time, this movement has developed several key elements that will undoubtedly find their way into mainstream construction, among them better teamwork among project team members, the use of the charrette to maximize input and creativity at the start of the design process, and the extensive use of building commissioning as a tool for ensuring that owners receive precisely the buildings they anticipated. In effect, this delivery system is based on the conventional construction management-at-risk delivery system, with significant improvements in the areas of collaboration and communication among the project team members. The design-build delivery system can also be modified to a green building delivery system by selecting a team with green building familiarity and an orientation to environmentally friendly design and construction practices. The end result in either case should be a vastly superior end product, not only in its environmental attributes but also in the quality of design and construction, due to the improved working atmosphere fostered by the green building concept.

Notes

1. The definition of design-build is from the website of the Design Build Institute of America, www.dbia.org.
2. As of January 2007, there were 18 green building projects at the University of Florida registered or certified by the USGBC LEED building rating system. More on these projects can be found at www.facilities.ufl.edu/sustain/index.htm.
3. The Florida Green Building Coalition Green Commercial Building Designation Standard can be found at www.floridagreenbuilding.org/standard.
4. *Writing the Green RFP* can be found at the AIA COTE website, www.aia.org/cote_rfps. The guide also provides examples of green RFPs/RFQs and the experience of people who have had a role in writing this type of document. It also contains "Sustainable Design Basics" and "Frequently Asked Questions (FAQs)" sections.
5. Current information about the LEED Accredited Professional Exam and the latest requirements can be found at the USGBC website, www.usgbc.org/LEED/Accredited_Pros/professionalaccred.asp.
6. As found in the Building Toolbox section of the U.S. Department of Energy's Building Technology Program at www.eere.energy.gov/buildings/info/design/integratedbuilding.
7. The potential Project Management points of Green Globes can be found at www.thegbi.com/greenglobes/pdf/1_ProjectManagement_053106.pdf.

8. The National Workshop on the Integrated Design Process was held in Toronto in October 2001. An excellent document describing a Canadian perspective on IDP is "Integrated Design Process Guide," written by Alex Zimmerman in 2006, available at www.cmhc-schl.gc.ca/en/inpr/bude/himu/coedar/upload/article_design_guide_en_aug23.pdf.
9. The concept of whole building design and an online reference, "The Whole Building Design Guide," can be found at www.wbdg.org.
10. Adapted from the "Four Guiding Principles," proposed by the National Charrette Institute, available at www.charretteinstitute.org.
11. Adapted from the four-step charrette process proposed by the National Charrette Institute, available at www.charretteinstitute.org.

Chapter 5

Ecological Design

The key to creating a high-performance green building is the ability of the design team to understand and apply the concept of *ecological* or *green design.* Although a design rooted in ecology and nature should be integral to creating a green building, ecological design is in the early stage of evolution, and it will take considerable time and experimentation before a robust version matures. Meanwhile, designers often must use their best judgment when making decisions from among the myriad choices available. Where materials and product selection are concerned, the best choices can be far from obvious. In addition to the environmental implications, performance and cost criteria must be addressed in the selection process.

One benefit of the building assessment standards such as LEED or Green Globes is that knowledge and application of ecological design are not absolutely necessary to produce a high-performance green building. A design team can simply utilize a building assessment standard as a checklist and be able to create a certified building, regardless of the underlying philosophy and without ever having studied or pondered the diverse and complex issues of the construction industry's environmental impact. LEED's commonly accepted approach to green building design is a major advantage that has rapidly increased the penetration of green buildings in the marketplace. Yet, simple adherence to a LEED checklist without deeper thinking and innovation could ultimately result in building stereotypes that stagnate rather than advance the art of green building, certainly an outcome never intended by the USGBC when fashioning LEED. Nonetheless, commitment to a design approach that is rooted in an understanding of natural systems and in the behavior of ecosystems, and that is concerned with resource conservation, will undoubtedly produce a high-performance building of higher economic and aesthetic value.

In the brief history of the green building movement, several philosophical approaches have been articulated and various terms have evolved, including *ecological design, environmental design, green design, sustainable design,* and *ecologically sustainable design.* Fundamentally, each approach seeks to acknowledge, facilitate, and/or preserve the interrelationship of natural system components and buildings. In doing so, particular questions and problems recur, such as:

- What can be learned from nature and ecology that can be applied to buildings?
- Should ecology serve as model or metaphor for green buildings?
- How can natural systems be directly incorporated to improve the functioning of the built environment?
- How can the human-nature interface best be managed for the benefit of both systems?

These profound questions have no easy answers, yet responses to them are critical to the evolution of truly sustainable buildings. Clearly, the progress of green building requires greater understanding and consideration of the environmental and human impact of the built environment, as well as incorporation of nature's lessons into the

Figure 5.1 The San Francisco Federal Building exemplifies ecological design by employing local natural forces such as the prevailing winds and sunlight to provide cooling and daylighting. Detailed analysis of natural airflows induced by wind and thermal processes was accomplished using sophisticated computational fluid dynamics (CFD) modeling. (Illustration courtesy of Morphosis Architects.)

building process. The striking lack of understanding of ecology itself among design and construction professionals is less surprising when one considers that the green building movement was not created by ecologists, but rather by building professionals and policymakers with only glancing familiarity with the dynamic discipline of ecology. Yet, without greater understanding of ecology and ecological theory, green buildings may cease to evolve beyond merely fanciful, intuitive structures that are green in name only. With this in mind, this chapter reviews certain fundamental principles of ecological or green design, and explores the philosophy and rationale of practitioners and academics whose life's work has centered on these issues. An overview of the history and current efforts to connect ecological thinking to buildings provides a starting point; further study of ecology, industrial ecology, and related fields is recommended.

Design Versus Ecological Design

According to Sim Van Der Ryn and Stuart Cowan, the authors of *Ecological Design,* in its simplest form design can be defined as ". . . the intentional shaping of matter, energy, and process to meet a perceived end or desire."[1] This broad definition means that literally everyone is a designer because we are all using resources to achieve some end; consequently, the responsibility for design does not rest solely with those who might be called the design professionals, the most prominent of whom are architects. The world we design collectively is a rather simple one compared to the design of nature. In our world, we use a limited number of models and templates to produce an impoverished urban and industrial landscape largely devoid of true imagination and creativity. It is clear that this human-designed and -engineered landscape often replaces the natural landscape with unrecyclable and toxic products produced by wasteful industrial processes that were implemented with little regard for the consequences for humans or ecological systems. It is often said that the environmental problems we face today, such as climate change and biodiversity loss, reflect a failure of design. The disconnection of human design from nature is precisely the problem that high-performance green building, through the application of ecological design, seeks to redress.

In contrast to their definition of *design,* Van Der Ryn and Cowan define *ecological design* as that which transforms matter and energy using processes that are compatible and synergistic with nature and that are modeled on natural systems. Thus, unlike design that destroys landscapes and nature, ecological design, in the context of the built environment, seeks solutions that integrate human-created structures with nature in a symbiotic manner, that mimic the behavior of natural systems, and that are harmless to humans and nonhumans in their production, use, and disposal. Some would widen the concept of ecological design to an even broader concept, that of *sustainable design,* which would address the triple bottom-line effects of creating buildings: environmental impacts, social consequences, and economic performance. Clearly, the larger context and impacts of building design and construction need to be kept in mind by all the players in the process. Ecological design, sometimes referred to as *green design,* focuses on the human-nature interface and uses nature rather than the machine as its metaphor.

The key problem facing ecological design is a lack of knowledge, experience, and understanding of how to apply ecology to design. Complicating the issue is that there are several major approaches to understanding ecology, even among ecologists. Systems ecology, for example, focuses on energy flows, whereas proponents of adaptive management study processes.[2] Nature functions across scales and time horizons that are virtually unimaginable to human designers, who continue to struggle to apply even relatively simple ecological concepts such as resilience and adaptability to their work. An even deeper flaw is that building professionals have little or no background or education in ecology; hence, any application of so-called ecological or green design is likely to be shallow and perhaps even trivial. Equally problematic is that an enormous legacy of machine-oriented design is in place in the form of buildings and infrastructure; and the industrial products comprising buildings are still being created based on concepts, design approaches, and processes that have their roots in the Industrial Revolution. Thus, contemporary ecological designers are engaged in a struggle on several fronts in their attempt to shift to a form of thinking that would reconnect humans and nature. These "fronts" can be itemized as follows:

1. Understanding ecology and its applicability to the built environment
2. Determining how to use nature as the model and/or metaphor for design
3. Coping with an industrial production system that operates using conventional thinking
4. Reversing at least two centuries of design that used the machine as its model and metaphor

The classic approach to building design has been for the architect to define and lead the design effort, with input from the building owner but with scant input from other entities affected by the project. Contemporary ecological design changes this thinking dramatically by engaging a wide range of stakeholders in the charrette process from the onset of the effort. The key point of the charrette is to obtain the maximum amount of input from as many parties to the project as possible.

BENEFITS OF ECOLOGICAL DESIGN

For green buildings to be successful, the benefits of designing them must be known to those purchasing construction services and facilities. Because sustainability addresses a broad range of economic, environmental, and social issues, the benefits of ecological or sustainable design are potentially enormous. A list of these benefits recently published by the Federal Energy Management Program (FEMP) provides an overview of the promise of a shift to sustainable design (see Table 5.1).[3]

TABLE 5.1

Benefits of Sustainable Design

	Economic	Societal	Environmental
Siting	Reduced costs for site preparation, parking lots, roads	Improved aesthetics, more transportation options for employees	Land preservation, reduced resource use, protection of ecological resources, soil and water conservation, restoration of brownfields, reduced energy use, less air pollution
Water Efficiency	Lower first costs, reduced annual water and wastewater costs	Preservation of water resources for future generations and for agricultural and recreational uses; fewer wastewater treatment plants	Lower potable water use and reduced discharge to waterways; less strain on aquatic ecosystems in water-short areas; preservation of water resources for wildlife and agriculture
Energy Efficiency	Lower first costs, lower fuel and electricity costs, reduced peak power demand, reduced demand for new energy infrastructure	Improved comfort conditions for occupants, fewer new power plants and transmission lines	Lower electricity and fossil fuel use, less air pollution and fewer carbon dioxide emissions, lowered impacts from fossil fuel production and distribution
Materials and Resources	Decreased first costs for reused and recycled materials, lower waste disposal costs, reduced replacement costs for durable materials, reduced need for new landfills	Fewer landfills, greater markets for environmentally preferable products, decreased traffic due to the use of local/regional materials	Reduced strain on landfills, reduced use of virgin resources, better-managed forests, lower transportation, energy and pollution, increase in recycling markets
Indoor Environmental Quality	Higher productivity, lower incidence of absenteeism, reduced staff turnover, lower insurance costs, reduced litigation	Reduced adverse health impacts, improved occupant comfort and satisfaction, better individual productivity	Better indoor air quality, including reduced emissions of volatile organic compounds, carbon dioxide, and carbon monoxide

TABLE 5.1 *(Continued)*

Benefits of Sustainable Design

	Economic	Societal	Environmental
Commissioning; Operations and Maintenance	Lower energy costs, reduced occupant/owner complaints, longer building and equipment lifetimes	Improved occupant productivity, satisfaction, health, and safety	Lower energy consumption, reduced air pollution and other emissions

Historical Perspective

Although the green building movement, marked by the rise of the USGBC, is a relatively recent phenomenon, it has its roots in the work and thinking of several previous generations of architects and designers, dating back at least to the end of the nineteenth century. In the American context, several key figures laid the foundation for today's ecological or green design, among them R. Buckminster Fuller, Frank Lloyd Wright, Richard Neutra, Lewis Mumford, Ian McHarg, Malcolm Wells, and John Lyle. A brief introduction to each of these thinkers is presented here. The following section, "Contemporary Ecological Design," covers the synthesis of this foundational thinking about ecological design into an emerging, coherent process for green building design. To articulate today's thinking, the efforts of William McDonough, Ken Yeang, Sim Van Der Ryn, Stuart Cowan, and David Orr are described.

R. BUCKMINSTER FULLER

Perhaps more than any other figure, R. Buckminster Fuller (1895–1983) laid the foundation for the green building revolution in the United States. His list of accomplishments is long, among them the design of the aluminum Dymaxion car in 1933; the design of the autonomous Dymaxion House in the 1920s, one of which was built in Wichita, Kansas, in 1946; and, of course, the creation of the geodesic dome in the 1950s. Fuller has been called an inventor, architect, engineer, mathematician, poet, and cosmologist. He was, at heart, an ecologist. His designs emphasized resource conservation: the use of renewable energy in the form of sun and wind; the use of lightweight, ephemeral materials such as bamboo, paper, and wood; and the concept of design for deconstruction. His geodesic dome has been called the lightest, strongest, and most cost-effective structure ever devised.

Fuller is also credited with originating the term *Spaceship Earth* to describe how dependent humans are on the planet and its ecosystems for their survival and how the waste we create ends up in the biosphere, to the peril of everyone. His Dymaxion Map and World Game were designed to allow players to observe world resources and create strategies for solving global problems by matching human needs with the planet's resources. Fuller understood the issue of renewable and nonrenewable resources, and his research showed that all energy needs could be provided by renewables. In the United States, he showed that, at the time, wind energy alone could provide three and a half times the country's total energy needs.[4] His work influenced many of today's green building movement participants, so much so that he is sometimes referred to as the "father of environmental design."

Figure 5.2 The R. Buckminster Fuller postage stamp was issued by the U.S. Postal Service in July 2004 to commemorate the 50th anniversary of Fuller's patent for the geodesic dome, said to be the lightest, strongest, and most cost-effective structure ever devised. (Stamp Designs © 2004 United States Postal Service. Displayed with permission. All rights reserved.)

Figure 5.3 Buckminster Fuller's Dynamic Maximum Tension, or Dymaxion House, in Wichita, Kansas, was the first serious attempt to create an autonomous house. It was designed for mass production, weighed just 3,000 pounds (1,364 kilograms) (compared to the 150 tons (137 metric tons) of a typical house), featured a built-in wind turbine for generating power, and had a graywater system. (Photograph courtesy of the Buckminster Fuller Foundation.)

Fuller was also a prolific author; he is credited with writing 28 books, among them *Operating Manual for Spaceship Earth* (1969), in which he imagines humans as the crew of the planet, all bound together by a shared fate on what amounts to a tiny spaceship in an infinite universe. The question he posed to his fellow planetary inhabitants was: How do we contribute to the safe operation of Spaceship Earth? In the book, he describes many of his basic concepts, two of which are synergy and ephemeralization. Another notable book by Fuller was *Critical Path* (1981), in which he explored social issues, marking him as one of the first people to connect the issues of environment, economics, and humans, labeled many years later by Lester Brown as *sustainability.* In *Critical Path,* Fuller analyzes how humanity has found itself at the limits of the planet's resources and facing political, economic, environmental, and ethical crises. Fuller, labeled "the planet's friendly genius," was an extraordinary member of the planet's "crew."

FRANK LLOYD WRIGHT

Frank Lloyd Wright (1867–1958) is well known as an important figure in architecture. Less well known is that he developed some of the first ideas about nature and building, ideas that directly influence what is happening today. His early exposure to nature had a profound effect on both his life and his architecture. Under the tutelage of his mother, who employed Friedrich Froebel's nature-based training, he learned about nature's forms and geometries. His architecture reflects this influence, building on the underlying structure of nature. Wright's goal was to create buildings that were, as he put it, integral to the site, to the environment, to the life of the inhabitants, and to the nature of the materials.

RICHARD NEUTRA

Richard Neutra (1892–1970), a pupil of Wright's, recognized how flawed were the products of human creation compared to those of nature. He noted that human artifacts were static and unable to self-regenerate or self-adjust, unlike nature's creations, which are dynamic and self-replicating. He observed that nature's form and function emerge simultaneously, whereas humans must first create a building's form and then allow it to function. Neutra was one of the first to recognize the concept of *biophilia,* the need or craving of humans to be connected to nature, a concept that has been expounded on more recently by E.O. Wilson and Stephen Kellert.[5]

Neutra advocated the close connection of living spaces to the "green world of the organic." According to Neutra, imitating nature is not simply flattery on the part of humans; it is the copying of systems that function in an extraordinarily successful fashion. He was also one of the first architects to recognize the connection between human health and nature and the need to consider this relationship in building design. In designing what became know as the Health House, a Los Angeles residence for Dr. P.M. Lovell, a naturopath, or integrated medical practitioner, Neutra explored the health relationship between nature and structure. In today's green buildings, health issues are of paramount importance, and connections between nature and health are again being explored in a wide variety of building experiments.

LEWIS MUMFORD

Lewis Mumford (1895–1990) was renowned for his writings on cities, architecture, technology, literature, and modern life. His long-term connection with the built environment was forged over a 30-year stint as architectural critic for *The New Yorker* magazine. He was also a cofounder of the Regional Planning Association of America, which advocated limited-scale development and the region as significant for city planning. He wrote *The Brown Decades* in 1931 to detail the architectural achievements of Henry Hobson Richardson, Louis Sullivan, and Frank Lloyd Wright. Mumford was particularly critical of technology, and in *The Myth of the Machine,* written in 1967, he argued that the development of machines threatened humanity itself, citing, for example, the design of nuclear weapons. He argued in *Values for Survival,* written in 1946, for the restoration of organic human purpose and for humankind to exert ". . . primacy over its biological needs and technological pressures" and to ". . . draw freely on the compost from many previous cultures." Mumford advocated the implementation of *ecotechnics,* technologies that rely on local sources of energy and indigenous materials in which variety and craftsmanship add ecological consciousness, as well as beauty and aesthetics. He drew his conclusions from observations of how cities evolved, from preindustrial cities that respected nature to post–Industrial Revolution metropolises that sprawled and destroyed compact urban forms, caused resources to be wasted, and that virtually no connection to nature.

IAN McHARG

The disconnect between buildings and nature in the Industrial Age was also noted and articulated by Ian McHarg (1920–2001), particularly the lack of a multidisciplinary effort to produce a built environment that was responsive to nature. He decried the lack of environmental consideration in planning; the lack of interest on the part of scientists in planning; and the absence of consideration of life itself in many of the sciences such as geology, meteorology, hydrology, and soil science. According to McHarg, the compartmentalization and specialization of disciplines have created conditions that at present may make truly ecological design difficult or impossible to achieve.

McHarg's 1969 book, *Design with Nature,* is a modern classic, especially for the discipline of green building. McHarg called for environmental planning on a local level and advocated taking everything in the environment (such as humans, rocks, soils, plants, animals, and ecosystems) into account when planning the built environment. He was also one of the first people to realize that the best way to preserve open space is to sustain urban areas, which contain existing resources (such as sewer systems and streets) to handle human growth. He also noted that it was critical that everyone have an ecological education in order to be able to make the best-informed decisions about growth and development.

MALCOLM WELLS

Malcolm Wells (1926–) is generally critical of architects for failing to be aware of or moved by the biological foundations of both life and art. In his 1981 work *Gentle Architecture,* he asked a key question: "Why is it that every architect can recognize and appreciate beauty in the natural world yet fail to endow his own work with it?" His solution was a simple but very effective one: leave the surface of the planet alone and submerge the built environment underground so that the Earth's surface can continue to provide unimpeded services. Wells' approach was to tread gently on the Earth, minimize the use of asphalt and concrete, and use local natural resources and solar energy as the primary resources for the built environment. He is known as the father of *gentle architecture* or of *earth-sheltered architecture,* and although he claims that his work has not had the effect he had hoped for, his thinking has significantly influenced today's green building movement. He suggests that buildings should consume their own waste, maintain themselves, provide animal habitat, moderate their own climate, and match nature's pace—all notions that are frequently presented in the increasing number of green building forums throughout the United States.

JOHN LYLE

Landscape is perhaps the most neglected and underrated issue in green design, but one man, John Lyle (1934–1998), pursued the goal of creating *regenerative* landscapes. His book, *Design for Human Ecosystems,* originally published in 1985, is his classic text. In it he explores methods of designing landscapes that function in the sustainable ways of natural ecosystems. The book provides a framework for thinking about and understanding ecological design, highlighted by a wealth of real-world examples that bring Lyle's key ideas to life. Lyle traces the historical growth of design approaches involving natural processes, and presents an introduction to the principles, methods, and techniques that can be used to shape landscape, land use, and natural resources in an ecologically sensitive and sustainable manner. He articulates the problems inherent in imposed and artificial infrastructures, which are part of a linear industrial system in which materials extracted from nature and the Earth end up as useless waste.

Unlike its natural counterpart, the urban landscape does not produce food; store, process, or treat stormwater; or provide diverse habitat for wildlife; is not part of an ecological system; and does not contribute to biological diversity. And the artificial landscape is not sustainable because it is highly dependent for its survival on fossil fuel, chemicals, and large quantities of water. In contrast, Lyle's regenerative landscape is characterized by the qualities of locality, fecundity, diversity, and continuity. A regenerative landscape grows out of a particular place (locality) in a manner unique to that place. It is fertile and continually grows and renews itself through reproduction, the heart of regeneration (fecundity). The regenerative landscape is composed of a wide variety of plants and organisms, each occupying a niche in its environment (diversity). And the regenerative landscape is not fragmented; it changes gradually over space and time (continuity).

Contemporary Ecological Design

The influence of these architects, designers, and philosophers on today's green building movement has been profound. In addition to establishing the foundations for ecological design, they influenced a large number of today's practitioners. Even though ecological design is still in development, the green building movement is driving efforts to refine its meaning and to explore in detail the connection between ecology and the built environment. Today's green building movement builds on the thoughts and work of figures like Fuller, Wright, Neutra, Mumford, Lyle, and McHarg. To a few voices on the subject of ecological design prior to 1990 are now added the intellectual capital and professional output of thousands of individuals, organizations, and companies. The process of discovery and implementation will be a long but exciting journey as design, practices, materials, methods, and technologies adapt to a world that is truly in need of a refined approach to the built environment.

Perhaps the first step in describing where ecological design is today is to sort through the terminology being used in association with this concept. Christopher Theis, professor of architecture at Louisiana State University, in a paper published in 2002 on the website of the Society of Building Science Educators, suggests that we first have to deal with several differing sets of nomenclature floating around in the building community.[6] A variety of terms, including those already introduced in this book, are being used to describe the approach to delivering high-performance buildings: *sustainable design, green design, ecological design,* and *ecologically sustainable design.* Theis advocates the use of *ecological* to describe the design strategy needed to produce a high-performance green building. Although using the word *sustainable* to describe this design strategy may be more comprehensive, doing so leads to levels of complexity that are not resolvable in designing a building, because it is necessary to consider the three major aspects of sustainability: social, economic, and environmental. This is a nearly impossible task for the building team because their task is to take projects awarded to them by an owner or client and meet the requirements spelled out in their contract. This is not to say that the building team should be unaware of sustainability issues, and as much as possible, they should consider the ramifications of all their decisions with respect to sustainability. In fact, the building team can exert a powerful influence on owners by educating them about these broad issues, both directly, through an articulation of their philosophy, and indirectly, by their approach to building design.

As for ecological design itself, Peter Wheelwright, chair of the Department of Architecture at the Parsons School of Design, described two often contradictory and conflicting approaches to ecological design currently proffered in schools of architecture: the *organic* one, which combines an activist social agenda with a "Wrightian" design ethic, and the *technological* one, which is "futurist in orientation and scientific in method." In fact, they coexist, with designers seeking to create solutions rooted in nature, yet applying technology as appropriate.

KEY GREEN BUILDING PUBLICATIONS: EARLY 1990S

The early 1990s marked the start of the green building movement in the United States. Three publications of this era provided an early articulation of green building design: the *Hannover Principles* in 1992, *The Local Government Sustainable Buildings Guidebook* in 1993, and *The Sustainable Building Technical Manual* in 1996. In addition, in 1992, *Environmental Building News,* the first and still the most authoritative publication on green building issues, was launched and featured a checklist for green design. Each of these key publications is briefly reviewed below.

TABLE 5.2

The Hannover Principles

1. Insist on the rights of humanity and nature to coexist
2. Recognize interdependence
3. Respect relationships between spirit and matter
4. Accept responsibility for the consequences of design
5. Create safe objects of long-term value
6. Eliminate the concept of waste
7. Rely on natural energy flows
8. Understand the limitations of design
9. Seek constant improvement by the sharing of knowledge

The Hannover Principles

In 1992, the city manager of Hannover, Germany, Jobst Fiedler, commissioned William McDonough, one of the early major figures in the emergence of green buildings, to work with the city to develop a set of principles for sustainable design for the 2000 Hannover World Fair. The principles were not intended to serve as a how-to for ecological design but as a *foundation* for ecological design. One of the contributions that emerged from this relatively early attempt to articulate principles for the green building movement was a definition of *sustainable design* as the "conception and realization of ecologically, economically, and ethically responsible expression as part of the evolving matrix of nature." These principles, commonly known as the Hannover Principles, are listed in Table 5.2.[7]

The Local Government Sustainable Buildings Guidebook and *The Sustainable Building Technical Manual*

In the 1990s, several publications attempted to provide an orientation to the current era of ecological design, especially as driven by the emergence of the LEED building assessment system. Two of the first publications on the subject of designing a green building were produced by Public Technology, Inc. (PTI): *The Local Government Sustainable Buildings Guidebook,* in 1993, and *The Sustainable Building Technical Manual,* in 1996. At the time of their publication, the USGBC was a very new organization and the first drafts of the LEED standard were just beginning to emerge from its committees.

The Local Government Sustainable Buildings Guidebook reveals some of the very first thoughts on the direction of the U.S. green building movement. A number of the guiding principles noted in the guidebook are shown in Table 5.3.[8]

In contrast to the guidebook, *The Sustainable Building Technical Manual* was in essence a stopgap measure to serve the rapidly growing interest in green building.

TABLE 5.3

Design Considerations and Practices for Sustainable Building

- Resources should be used only at the speed at which they naturally regenerate, and should be discarded only at the speed at which local ecosystems can absorb them.
- Material and energy resources must be understood as a part of a balanced human/natural cycle. Waste occurs only to the extent that it is incorporated back into that cycle and used for the generation of more resources.
- Site planning should incorporate resources naturally available on the site, such as solar and wind energy, natural shading, and drainage.
- Resource-efficient materials should be used in construction of the building and in furnishings to lessen local and global impact.
- Energy and materials waste should be minimized throughout the building's life cycle from design through reuse or demolition.
- The building shell should be designed for energy efficiency.
- Material and design strategies should strive to produce excellent total indoor environmental quality, of which indoor air quality is a major component.
- The design should maximize occupant health and productivity.
- Operation and maintenance systems should support waste reduction and recycling.
- Location and systems should optimize employee commuting and customer transportation options and minimize the use of single-occupancy vehicles. These include using alternative work modes such as telecommuting and teleconferencing.
- Water should be managed as a limited resource.

The manual provides a list of areas that should be considered in designing a green building. These are summarized in Table 5.4.[9] The manual emphasizes the need for an integrated, holistic approach to design, with the building being considered a system rather than an assemblage of parts. This marked one of the first public statements of this key aspect of green building. As noted previously, the notion of a systems approach has emerged as one of the dominant themes of green building, even though in practice it is difficult to achieve due to the large quantities of information being processed, the many actors involved, and the same difficulties in communication that occur in conventional design.

TABLE 5.4

Overview of Building Design Issues as Stated in *The Sustainable Building Technical Manual*

Passive Solar Design
Daylighting
Building envelope
Renewable energy
Building Systems and Indoor Environmental Quality
HVAC, electrical, and plumbing systems
Indoor air quality
Acoustics
Building commissioning
Materials and Specifications
Materials
Specifications

Environmental Building News

The most prominent U.S. publication on green building is *Environmental Building News* (EBN), a monthly newsletter/journal dedicated to the subject of high-performance buildings. Periodically, it has featured checklists on various subjects related to green building, among them one for environmentally responsible design. Although not considered a philosophical approach, it does provide an overview of the major issues that should be considered in designing green buildings. Table 5.5 presents this checklist.[10]

TABLE 5.5

EBN Checklist for Environmentally Responsible Design

- *Smaller is better.* Optimize use of interior space through careful design so that the overall building size—and the resources used in constructing and operating it—are kept to a minimum.
- *Design an energy-efficient building.* Use high levels of insulation, high-performance windows, and tight construction. In southern climates, choose glazings with low solar heat gain.
- *Design buildings to use renewable energy.* Passive solar heating, daylighting, and natural cooling can be incorporated cost-effectively into most buildings. Also consider solar water heating and photovoltaics—or design buildings for future solar installations.
- *Optimize material use.* Minimize waste by designing for standard ceiling heights and building dimensions. Avoid waste from structural overdesign (use optimum-value engineering/advanced framing). Simplify building geometry.
- *Design water-efficient, low-maintenance landscaping.* Conventional lawns have a high impact because of water use, pesticide use, and pollution generated from mowing. Landscape with drought-resistant native plants and perennial groundcovers.
- *Make it easy for occupants to recycle waste.* Make provisions for storage and processing of recyclables—recycling bins near the kitchen, undersink compost receptacles, and the like.
- *Look into the feasibility of graywater.* Water from sinks, showers, or clothes washers (graywater) can be recycled for irrigation in some areas. If current codes prevent graywater recycling, consider designing the plumbing for easy future adaptation.
- *Design for durability.* To spread the environmental impacts of building over as long a period as possible, the structure must be durable. A building with a durable style ("timeless architecture") will be more likely to realize a long life.
- *Avoid potential health hazards—radon, mold, pesticides.* Follow recommended practices to minimize radon entry into the building and provide for future mitigation if necessary. Provide detailing to avoid moisture problems, which could cause mold and mildew growth. Design insect-resistant detailing to make minimizing pesticide use a high priority.

KEY CONTEMPORARY PUBLICATIONS ABOUT ECOLOGICAL DESIGN

In addition to the publications just described, in the mid-1990s, two landmark books on the subject of contemporary ecological design were published: *Designing with Nature,* written in 1995 by Ken Yeang, a Malaysian architect, and *Ecological Design,* authored by Sim Van Der Ryn and Stuart Cowan, in 1996. Although there are several other volumes on the subject of designing buildings in a manner that employs either the metaphor or model of nature, these two are particularly noteworthy for their deeper thinking on the subject of ecological design.

Designing with Nature: Ken Yeang (1995)

Designing with Nature was perhaps the first publication to attempt to tackle the tremendous challenge of how to apply ecology directly to architecture. Yeang uses the terms *green architecture* and *sustainable architecture* interchangeably, defining them as "designing with nature and designing with nature in an environmentally responsible way." He approaches this problem by making several important assumptions:[11]

- The environment must be kept biologically viable for people.
- Environmental degradation by people is unacceptable.
- Destruction of ecosystems by humans must be minimized.
- Natural resources are limited.
- People are part of a larger closed system.
- Natural system processes must be considered in planning and design.
- Human and natural systems are interrelated and essentially one system.
- Changing anything in the system affects everything else.

Yeang also suggests several premises or bases for ecological design (see Table 5.6).[12]

TABLE 5.6

Bases for Ecological Design as Suggested by Ken Yeang

1. Design must be integrated not only with the environment, but also with the ecosystems that are present.
2. Because Earth is essentially a closed system, matter, energy, and ecosystems must be conserved and the biosphere's waste assimilation capacity considered.
3. The context of the ecosystem, that is, its relationship with other ecosystems, must be considered.
4. Designers must analyze and use each site for its physical and natural structures to optimize the design.
5. The impact of the design must be considered over its entire life cycle.
6. Buildings displace ecosystems, and the matter-energy impacts must be considered.
7. Due to the complex impacts of built environments on nature, design must be approached holistically rather than in a fragmented manner.
8. The limited assimilative capacity of ecosystems for human-induced waste must be factored into design.
9. Design should be responsive and anticipatory, and as much as possible result in beneficial effects for natural systems.

In the actual implementation of ecological design, Yeang suggests that there are three major steps:

1. Define the building program as an ecological impact statement (analysis).
2. Produce a design solution that comes to grips with the probable environmental interactions (synthesis).
3. Establish the performance of the design solution by measuring inputs and outputs throughout the life cycle (appraisal).

Yeang continues his efforts to develop his concept of ecological design, and he is particularly well known for his work on tall greening buildings. He has written several other books on the subject of ecological design and in 2004 published an updated work on the general subject of ecological design called *EcoDesign: A Manual for Ecological Design.*[13]

Ecological Design: Sim Van Der Dyn and Stuart Cowan (1996)

Sim Van Der Ryn and Stuart Cowan also delved deeply into the subject of ecological design in their book by the same name. *Ecological Design* was written to provide a context for green design rather than specific details. The main feature of the book is the articulation of five ecological design principles:

1. *Solutions grow from place.* Each location has its own character and resources; hence, design solutions are likely to differ accordingly. Solutions should also take advantage of local style, whether it be the adobe architecture of New Mexico or the cracker architecture of Florida. Sustainability has to be embedded in the process so that choices can be made about how a project can interact with local ecosystems and, ideally, improve on the conditions that presently exist—for example, to clean up contaminated industrial sites or brownfields for productive uses.
2. *Ecological accounting informs design.* For true ecological design to take place, the impact of *all* decisions must be taken into account. These include the effects of energy and water consumption; solid, liquid, and gaseous wastes; and toxic materials use and waste. Moreover, materials selection should support the design of facilities that minimize resource consumption and environmental effects. In regard to materials selection, LCA is appropriate to determine the total resource consumption and emissions over the entire life of the building and to find the solution with the minimum total impact.
3. *Design with nature.* Ecological design should foster a collaboration with natural systems, and the result should be buildings that coevolve with nature. Buildings should mimic nature, where, for example, there is essentially no waste because in nature, waste equals food. Buildings are one stage in a complex industrial system that has to be redesigned with this strategy in mind to ensure that waste is minimized and closed-loop behavior rather than large-scale waste is the result. A synergistic relationship with nature is desirable, one in which matter-energy flows across the human-nature interface and is beneficial to both subsystems, human and natural. The heating and cooling systems in buildings can be assisted by landscaping; waste can be processed by wetlands; trees can take up vast quantities of stormwater; and waste generated by a building's occupants can provide nutrients for the landscape.
4. *Everyone is a designer.* The participatory process is emerging as a key ingredient of ecological design; that is, including a wide array of people affected by a building provides more creative and interesting results. Schools of architecture need to be reinvigorated, reoriented to teach about building holistically, and to include ecological design as a foundation for the curriculum. A

new ecological design discipline should be created to address not only issues that may be connected to the built environment but also issues such as industrial product design and the materials supply chain.

5. *Make nature visible.* Having lost their connection with nature, humans have forgotten details as simple as where their water and food originate and how they are processed and moved to humans for consumption. Ecological design should reveal nature and its workings as much as possible, celebrate place, and reverse the trend from denatured cities to urban spaces with life and vitality. Drainage systems, normally hidden, might be exposed. The disposal areas for waste, sewage systems, wastewater treatment plants, and landfills should be located closer to the human waste generators to expose them to the consequences of wasteful behavior. By the same token, the elegant and complex behavior of natural systems in the form of natural wetlands that treat effluent can serve to educate people about integration with nature. As part of the design and construction process, the regenerative approaches advocated by John Lyle can be employed to restore areas once damaged by human activities to their natural state.

Van Der Ryn and Cowan provide a framework for designers—that is, everyone—for creating a nature- and ecology-based process that is flexible, adaptable, and useful for the building project and the place. Again, their framework does not give details on how to accomplish this process, because the details would be immense in scope and volume. Rather, it provides a strong philosophical underpinning for high-performance green building design that, if faithfully followed, will produce human-made structures that cooperate rather than compete with nature.

The Nature of Design: Ecology, Culture, and Human Intention: David Orr (2002)

More recently, in 2002, David Orr addressed ecological design in his book, *The Nature of Design: Ecology, Culture, and Human Intention.* Orr takes a much broader view, addressing the full array of human interaction with nature, to include how we acquire and use food, energy, and materials and what we do for a living. Although he is not a professional in a built environment discipline, Orr has made a significant impact on today's green building movement by virtue of his ability to clearly elucidate a vision of ecological design. Orr broadens our thinking about ecological design by comparing it to the Enlightenment of the eighteenth century, with its connections to politics and ethics. He describes ecological design as an emerging field that seeks to recalibrate human behavior to, in effect, synchronize it to nature and connect people, places, ecologies, and future generations in ways that are fair, resilient, secure, and beautiful. According to Orr, changing the behavior of both the public and private sectors is badly needed to transform our production and consumption patterns.

In addition to his work as an author and as a proponent of environmental literacy, Orr successfully raised funds for what is perhaps the most important green building project of the later 1990s: the Lewis Center for Environmental Studies at Oberlin College. The Lewis Center was designed by an elite team of architects and other professionals, among them William McDonough, one of the leading green building architects, and John Todd, creator of the Living Machine, a waste treatment system that uses natural processes to break down the components of the building's wastewater stream. Orr sees buildings as contributing to a pedagogy for environmental literacy and cites numerous examples of how designers can create structures that teach as well as function. For example, buildings can teach us how to conserve energy, recycle materials, integrate with nature, and contribute rather than detract from their surroundings. The landscape around the Lewis Center, by virtue of its design, helps teach ecological competence in horticulture, gardening, natural systems agriculture, forestry, and aquaculture, as well as techniques to preserve biodiversity and ecological restoration. As Orr notes, we need a national effort to engage students of every discipline in ecological design because our current system of production and consumption is poorly designed. This is perhaps the key challenge facing us: understanding how nature can inform design of all types, including that of buildings.

Future Ecological Design

At present, sustainable construction is constrained by an inability to come to grips with a more precise notion of what ecological design is and what can and cannot be achieved through its application. A wide variety of hypotheses about ecological design have been presented in addition to those mentioned in the previous discussion of the history of ecological design. Some of the major hypotheses suggested by designers, industrial ecologists, and others are as follows:

1. General management rules for sustainability (Barbier 1989, Daly 1990)
2. Design principles for industrial ecology (Kay 2002)
3. The golden rules for ecodesign (Bringezu 2002)
4. Adaptive management (Peterson 2002)
5. Biomimicry (Benyus 1998) (briefly described in Chapter 2)
6. Factor 4 and Factor 10 (von Weizsäcker, Lovins, and Lovins 1997) (briefly described in Chapter 2)
7. Cradle to cradle (McDonough and Braungart 2002)
8. The Natural Step (Robert 1989) (described in Chapter 2)
9. Natural capitalism (Hawken, Lovins, and Lovins 1999)

In the following sections, these major contributions to ecological design are presented as the basis for a future more robust and more refined version of ecological design that can serve as both a philosophical and technical basis for sustainable construction.

GENERAL MANAGEMENT RULES FOR SUSTAINABILITY

Proponents of ecological economics have formulated several pragmatic rules for "managing" sustainability.[13] According to the first rule, the use of renewable resources should not exceed the regeneration rate. In order to operationalize this demand, one has to consider that the use of either naturally or technically renewable materials always requires some inputs of nonrenewables (e.g., mineral fertilizer for the loss of nutrients due to leaching in agriculture, and the requirements for materials and energy for recycling processes). As a consequence, the total life cycle of products has to be checked for the use of renewables and nonrenewables. The former will have to be distinguished according to criteria on sustainable modes of production in agriculture, forestry, and fishery. An example in the construction sector would be the origin of timber products from sustainable cultivation.

The second rule states that nonrenewable resources may be used only if physical or functional substitutes are provided—for example, investments in solar energy systems from gains from fossil fuels. Here the basic assumption is that man-made capital may be substituted for natural capital (*weak sustainability*). The central requirement from an economic perspective is that the sum of natural and man-made capital is not reduced (Pearce and Turner 1990). However, from a natural systems perspective, it may be argued that there are minimum requirements of nature that may not be depleted without risk for life-support functions. Therefore, man-made capital should not be substituted (permanently) for natural capital (strong sustainability). Under this assumption, the second rule would require minimization of the use of nonrenewables.

The third rule states that the release of waste matter should not exceed the absorption capacity of nature. This can be operationalized by comparing *critical loads* of water, soil, and air compartments with actual levels of emission rates. After measures have been successfully applied to reduce pollution problems, the *after-end-of-pipe* approach to limit critical loads is also important. The implementation of the third rule is usually based on substance-specific analyses. This approach has some limitations. Generally, we must acknowledge that we are aware of only the tip of the iceberg with respect to the potential future impacts of all materials and substances released to the environment. Many natural functions react in a nonlinear manner. The complex interactions of natural substances like carbon dioxide, not to mention thousands of synthetic chemicals, cannot be foreseen in total.

From experience, we know that the effects of certain emissions become obvious *after* release and after the change of the environment takes place. There is a huge time lag between the scientific finding, public perception, and political reaction. Thus, the chances for comprehensive and precautionary materials management are extremely limited. A long-term effective implementation of the third rule should begin before the end-of-pipe and should aim to minimize the environmental impact potential of anthropogenic material flows. This impact potential is generally determined by the volume of the flow times the specific impacts per unit of flow. The second term is unknown for most materials released to the environment. The first term, the volume or weight used or released in a certain time period, can be made available for nearly every material handled. It may be used to indicate a generic environmental impact potential. As long as detailed information on specific impacts is lacking, it may be assumed that the impact potential is growing with the volume of the material flow. The overall volume of outputs from the anthroposphere can only be reduced when the inputs to this system are diminished. This is especially important for construction material flows with large scale and significant retention time within the anthroposphere. Starting from a situation in which the assimilation capacity of nature is overloaded by a variety of known substances, the long-term implementation of the third rule requires a reduction of the resource inputs of the anthroposphere in order to lower the throughput and ultimate output to the environment.

Another rule that has not yet attracted sufficient attention may be derived from the relation of inputs and outputs of the anthroposphere. Currently, the input of resources exceeds the output of wastes and emissions in industrialized as well as developing countries. As a consequence, the economies of these countries are growing physically (in terms of new buildings and infrastructure). The stock of materials in the anthroposphere is therefore increasing. In Germany, for example, the rate of net addition to stock was about 10 tons per capita annually in the mid-1990s. Associated with this accumulation of stock is an increase in built-up land area and a consequent reduction in reproductive and ecologically buffering land. Keeping in mind the limited space on our planet, this development cannot continue infinitely. Thus, a flow equilibrium between input and output must be expected. However, a question naturally arises: When will the economy stop growing physically and to what physical level?

DESIGN PRINCIPLES FOR INDUSTRIAL ECOLOGY (JAMES J. KAY)

James Kay (2002), the late ecologist from the University of Waterloo, proposed a set of principles that would govern the production-consumption system.[14] They are based on the premise that all man-made systems should contribute to the survival of natural systems.

1. *Interfacing:* The interface between societal systems and natural ecosystems reflects the limited ability of natural ecosystems to provide energy and absorb waste before their survival potential is significantly altered, and the fact that the survival potential of natural ecosystems must be maintained.
2. *Bionics:* The behavior of large-scale societal systems should be as similar as possible to that exhibited by natural systems.
3. *Appropriate biotechnology:* Whenever feasible, the function of a societal system should be carried out by a subsystem of a natural biosphere.
4. *Nonrenewable resources:* Nonrenewable resources are used only as capital expenditures to bring renewable resources on line.

The interfacing and appropriate biotechnology principles are related to intermediate ecological design in that they call for natural systems to interface with human systems in a synergistic manner to the benefit of both systems. Natural systems could provide services that would otherwise be performed by expensive engineered systems, such as stormwater control and waste processing. The bionics principle is closely related to strong ecological design but notably for large-scale functions. The nonrenewable resources principle has its roots in ecological economics where investing limited nonrenewables in transitioning to renewable resources is a key tenet. In effect, Kay's design principles are a mix of various levels of several types of ecological design, and he does not state that one version is most preferable.

THE GOLDEN RULES FOR ECODESIGN

To assist engineers, architects, and planners in the production of an environmentally benign built environment, Stefan Bringezu of the Wuppertal Institute suggested five "Golden Rules of ecological design":[15]

1. Potential impacts to the environment should be considered on a life-cycle basis (from cradle to cradle).
2. The intensity of use of processes, products, and services should be maximized.
3. The intensity of resource use (material, energy, and land) should be minimized.
4. Hazardous materials should be eliminated.
5. Resource input should be shifted toward renewables.

The first Golden Rule aims to avoid shifting problems between different processes and actors. For instance, if the energy requirements for heating or cooling during the use phase of buildings were not considered in the planning phase, the options with the highest potential for energy efficiency would be neglected. And if one considers only the direct material inputs for construction, the environmental burden associated with the upstream flows will be hidden.

The second Golden Rule reflects the fact that most building products are not used much of the time. For a considerable part of each day and each week, homes, offices, and public buildings are essentially unoccupied. Nevertheless, economic, environmental, and probably also social costs have to be paid for maintenance. Multifunctionality and more flexible models of use may reduce the demand for additional construction and contribute to lower costs for the users. The model of car sharing may also be applied for construction. Part-time employees already share the same office. And there is even potential for more efficient building use beyond normal working hours.

The third Golden Rule may be specified with the Factor 4 to 10 target for material requirements, including energy carriers, and should be applied to average products and services. In order to reach these goals, it seems essential to invest more intellectual power in the search for alternative options to provide the services and functions demanded by users.

The fourth Golden Rule calls for the elimination of hazardous substances, at face value a very sensible rule, but very difficult to implement from the perspective of today's economy. The use of nuclear energy violates this rule, and self-replicating nanomachines or genetically modified organisms may also be considered hazardous according to some criteria.

The fifth and final Golden Rule is a restatement of a key concept of ecological economics, namely, that supplies of nonrenewables will clearly diminish over time as they are consumed. For example, recent studies of copper consumption in the United States indicate that only one-third of the original dowry of copper ore exists today. The logic is that as these resources disappear, a shift to renewable resources must occur, and that in fact the consumption of nonrenewables should support the development of renewable resources. In the case of copper, a substitute renewable material may not be easy to develop.

ADAPTIVE MANAGEMENT

Ecology, like other fields, has several distinct schools. One of them is *adaptive management,* as articulated by Gary Peterson, who described it as an approach to ecosystem management that argues that ecosystem functioning can never be totally understood.[16] As Peterson notes, ecosystems are continually changing due to internal and external forces. Internally, ecosystems change due to the growth and death of individual organisms, as well as fluctuations in population size, local extinction, and the evolution of species traits. Ecosystems are also changed by external events such as the immigration of species, alterations in disturbance frequency, and shifts in the diversity and amount of nutrients entering the ecosystems. To cope with these changes, management must continually adapt. Management becomes adaptive when it persistently identifies uncertainties in human-ecological understanding and then uses management intervention as a tool to strategically test the alternative hypotheses implicit in these uncertainties. Consequently, basing the design of human systems on ecosystem function means creating materials, products, and processes using models that are not very well understood. Clearly, this means that it is probably impossible to implement strong ecological design in other than one-dimensional, virtually trivial applications.

Adherents to this line of thinking are also responsible for posing the fundamental and crucial question: "Why are systems of people and nature not just ecosystems?"[17] As noted in the Chapter 2 discussion of ethics and sustainability, the qualities of humankind that make them the only forward-looking and thinking species on this planet can result in humans thinking of themselves as "apart" from nature rather than "a part" of nature. Coupled with the ability to infer the laws of nature and physics and the ability to create materials and products that have no precedent in nature, the challenge is how to address the results of human inventiveness.

BIOMIMICRY

Janine Benyus (1997) described biomimicry as the conscious emulation of life's genius.[18] In her popular book on the subject, she states that "'Doing it nature's way' has the potential to change the way we grow food, make materials, harness energy, heal ourselves, store information, and conduct business." She goes on to say, "In a biomimetic world, we would manufacture the way animals and plants do, using sun and simple compounds to produce totally biodegradable fibers, ceramics, plastics, and chemicals." Farms would be modeled on prairies, new drugs would be based on plant and animal chemistry, and even computers would use carbon- rather than silicon-based structures. Proponents of biomimicry point to the 3.8 billion years of research and development that nature has invested in evolving a wide range of materials and processes that could benefit humans. Benyus also laid out 10 lessons for corporations that are based on the emulation of nature as the model for human-designed systems:

1. Use waste as a resource.
2. Diversify and cooperate to fully use the habitat.
3. Gather and use energy efficiently.
4. Optimize rather than maximize.
5. Use materials sparingly.
6. Don't foul the nest.
7. Don't draw down resources.
8. Remain in balance with the biosphere.
9. Run on information.
10. Shop locally.

Benyus also suggests "four steps to a biomimetic future":

1. *Quieting:* immerse ourselves in nature.
2. *Listening:* interview the flora and fauna of our own planet.
3. *Echoing:* encourage biologists and engineers to collaborate, using nature as a model and measure.
4. *Stewarding:* preserve life's diversity and genius.

With respect to step 3, Echoing, she provides 10 questions for testing innovation or technology for its acceptability, and all 10, according to Benyus, should be answered affirmatively.

1. Does it run on sunlight?
2. Does it use only the energy it needs?
3. Does it fit form to function?
4. Does it recycle everything?
5. Does it reward cooperation?
6. Does it bank on diversity?
7. Does it utilize local expertise?
8. Does it curb excess from within?
9. Does it tap the power of limits?
10. Is it beautiful?

In the area of materials, Benyus states that nature has four approaches:

1. Life-friendly manufacturing processes
2. An ordered hierarchy of structures
3. Self-assembly
4. Templating of crystals with proteins

As she points out, nature does produce a wide range of complex and functional materials. Abalone (twice as tough as high-tech ceramics), silk (five times stronger than steel), mussel adhesive (works underwater), and many other natural materials are remarkable in their performance. Each is created out of the local environment and biodegrades back to the environment in a harmless manner at the end of its useful life.

Biomimicry has many drawbacks when it is applied to the design of products and materials in the human sphere. Nature manufactures its products at a built-in evolved rate that is a function of information and local resources. In contrast, humans have learned to make products at an astoundingly rapid pace and, over time, to dematerialize and deenergize their production systems. Humans can and do observe nature and natural phenomena and apply their observations to create all manner of products, not all of them beneficial. The strength of biomimicry is that it provides us with a deeper appreciation for the elegant designs of nature and instructs us about how to design systems that are materials and energy conserving, that largely close materials loops, that use renewable energy, and that are niche players in complex ecosystems. The value of biomimicry as a teacher is probably far greater than as a provider of specific information about the chemical composition and structure of materials, and in this regard it should be part of the toolbox of ecological design.

CRADLE-TO-CRADLE DESIGN (McDONOUGH AND BRAUNGART)

The concept of *cradle-to-cradle* design describes approaches that contrast to designs that employ a *cradle-to-grave* approach or mentality. More recently, this concept has been popularized in *Crade to Cradle: Remaking the Way We Make Things,* by McDonough and Braungart (2002).[19] In laying the foundation for the cradle-to-cradle concept, they suggest that people and industry should set out to create the following:

- Buildings that, like trees, produce more energy than they consume and purify their own wastewater
- Factories that produce effluents that are drinking water
- Products that, when their useful life is over, do not become useless waste but can be tossed on the ground to decompose and become food for plants and animals and nutrients for soil; or, alternatively, that can return to industrial cycles to supply high-quality raw materials for new products
- Billions, even trillions, of dollars' worth of materials accrued for human and natural purposes each year
- A world of abundance, not one of limits, pollution, and waste

McDonough and Braungart suggest that the solution is to follow nature's model of *eco-effectiveness.* This entails separating the materials we use in human activity into *biological* substances (which can be returned to the natural ecosystem, where they can benefit other creatures as nutrients) and *technical* substances (which can, with proper design, be 100 percent recollected and recycled or even upcycled, producing, in second use, products of greater value than their original use, with zero waste). Carpets and shoes, for example, could be made of two layers—a biological outer layer that abrades over time, whose fibers could serve as nutrients in the soil or compost, and a much more durable technical inner layer that would be 100 percent recyclable, after its long life, into another identical product. A *biological nutrient* is a material or product that is designed to return to the biological cycle. McDonough and Braungart state that packaging, for example, can be designed as biological nutrients, so that, at the end of its use it can be, as they put it, thrown on the ground or compost heap. A *technical nutrient* is a material or product that is designed to be returned to the technical cycle, the industrial metabolism from which it comes. The authors also define a class of materials they refer to as the *unmarketables,* which are neither technical nor biological nutrients.

The cradle-to-cradle approach has a number of shortcomings that make it difficult to implement. Biological nutrients, for example, are not easily defined. Is a biopolymer, produced from corn or cellulose and biodegradable, a biological nutrient? Is a biodegradable synthetic material a biological nutrient or a technical nutrient? The fact is that biomaterials such as biopolymers use natural materials as feedstock but result in alterations to the basic feedstock and produce materials that have no precedent in nature. Furthermore, the consequences of their biodegradation are not well known. Whether or not biodegradation results in nutrients or waste has not been firmly established.

McDonough and Braungart suggest implementing changes to products and systems based on Five Steps to Eco-Effectiveness:

Step 1. Get rid of known culprits. These include X substances: materials that are bioaccumulative: mercury, cadmium, lead, PVC.

Step 2. Follow informed personal preferences. Prefer ecological intelligence, being sure that a product or substance does not contain or support substances or practices that are blatantly harmful to human and environmental health. This also includes admonitions to prefer respect and prefer delight, celebration, and fun.

Step 3. Create a "passive positive" list concerning harm in manufacture or in use. This is the X list, involving the X substances in Step 1. It includes substances that are carcinogens or problematic as defined by the International Agency for Research on Cancer (IARC) and Germany's Maximum Workplace Concentration (MAK) list. MAK defines two lists of substances, the gray list and the P list. The gray list includes problematic substances not urgently in need of phaseout. The P list consists of benign substances.

Step 4. Activate the positive list. Redesign products focusing on the P list substances.

Step 5. Reinvent. Totally reinvent products such as the automobile to be "nutri-vehicles."

Dave Pollard describes this process more elegantly in his blog:[20]

1. Free ourselves from the need to use harmful substances (e.g., PVC, lead, cadmium, and mercury).
2. Begin making informed design choices (materials and processes that are ecologically intelligent, respectful of all stakeholders, and which provide pleasure or delight).
3. Introduce substance triage: (a) phase out known and suspected toxins, (b) search for alternatives to problematic substances, and (c) substitute for them "known positive" substances.
4. Begin comprehensive redesigns: to use only "known positives," separate materials into biological and technical, and ensure zero waste in all processes and products.
5. Reinvent entire processes and industries to produce "net positives"—activities and products that actually *improve* the environment.

Cradle-to-cradle design provides an interesting framework for designing materials and products, and focuses attention on waste and on the proliferation of toxic substances used in the production system. Clearly, these are important issues that deserve significant attention when selecting building systems and products for a high-performance built environment.

THERMODYNAMICS: LIMITS ON RECYCLING AND THE DISSIPATION OF MATERIALS

One of the notions repeatedly suggested by McDonough is that human designs should behave like natural systems. One of his oft-stated principles is, "There is no waste in nature," with the implication that human systems should be designed to eliminate the concept of waste. In fact, zero-waste systems are not possible due to the laws of physics, more specifically the laws of thermodynamics. Georgescu-Roegen (1971) dealt with the implications of the entropy law and the second law of thermodynamics for economic analysis.[21] He described the important difference between primary factors of production (energy and materials) and the agents (capital and labor) that transform those materials into goods and services. The agents are produced and sustained by a flow of energy and materials that enter the production process as high-quality, low-entropy inputs and ultimately exit as low-quality, high-entropy wastes. This restricts the degree to which the agents of production (capital and labor) can substitute for depleted or lower-quality stocks and flows of energy and material inputs from the environment. Thermodynamics can inform us about ultimate limits. There are irreducible thermodynamic minimum amounts of energy and materials required to produce a unit of output that technical change cannot alter. In sectors that are largely concerned with processing and/or fabricating materials, technical change is subject to diminishing returns as it approaches these thermodynamic minimums. Ruth (1995) uses equilibrium and nonequilibrium thermodynamics to describe the materials-energy-information relationship in the biosphere and in economic systems.[22] In addition to illuminating the boundaries for material and energy conversions in economic systems, thermodynamic assessments of material and energy flows, particularly in the case of effluents, can provide information about depletion and degradation that are not reflected in market prices.

What are the implications of thermodynamics and the entropy law for materials recycling? Georgescu-Roegen argued that materials are dissipated in use, just as energy is, so complete recycling is impossible. He elevated this observation to a fourth law of thermodynamics—or Law of Matter Entropy—describing the degradation of the organizational state of matter. The bottom line for Georgescu-Roegen is that due to material dissipation and the generally declining quality of resource utilization, materials in the end may become more crucial than energy. However, Georgescu's fourth law has been criticized by a number of analysts in both economics and the physical sciences.

A recent paper by Reuter et al. (2005) addresses the dissipation of materials in recycling by examining the technical feasibility of an EU mandate for 95 percent end of vehicle life (ELV) recycling by 2015, with an intermediate goal of 85 percent by 2006.[23] One of the conclusions is that while the 85 percent target is achievable, the basic constraints of thermodynamics make it virtually impossible to reach the 95 percent goal. Consequently, at least 5 percent of the automobile mass dissipates into the biosphere. This is true of all recycling activities; the materials being recycled are dissipating to background concentrations, as dictated by the second and perhaps the fourth (according to Georgescu) laws of thermodynamics. Indeed, the dissipation of materials in the recycling process begs a number of questions; among them is: What are the health and ecological impacts of recycling as practiced and as envisioned for a sustainable future?

A 1998 U.S. Geological Survey report by Michael Fenton indicated some of the practical problems with so-called cradle-to-cradle strategies.[24] Steel and iron scrap, for which there is high demand, is not recycled at a very impressive rate. Fenton's report stated that in 1998, an estimated 75 million metric tons of steel and iron scrap was generated. The recycling efficiency was 52 percent, and the recycling rate was 41 percent.

In short, materials will be lost in recycling processes and, due to entropy, will naturally seek to return to background concentrations for naturally occurring substances and to very low concentrations for synthetic materials. Cradle-to-cradle and other approaches do not address this potentially difficult issue when suggesting that recycling of technical nutrients is desirable. Again, recycling, like most other issues involved in improving materials cycles, is a matter of ethics, risk, and economics.

NATURAL CAPITALISM

The concept of *natural capitalism* was articulated by Hawken et al. (1999) in its most recent form in a book with the same name.[25] Implementing natural capitalism entails four basic shifts in business practice:

Shift 1: Radical resource productivity. Dramatically increase the productivity of natural resources.

Shift 2: Ecological redesign. Shift to biologically inspired models.

Shift 3: Service and flow economy. Move to solutions-based business models.

Shift 4: Investment in natural capital. Reinvest in natural capitalism.

Each of these shifts is echoed in the other previously mentioned sets of principles and approaches. Relative to Shift 1, the productivity of natural resources can certainly be increased. However, natural renewable resources have little role in the creation of buildings, the vast bulk of which are made of human-designed materials. The authors claim that the industrial manufacturing system converts 94 percent of extracted materials into waste, with just 6 percent becoming product. It is unclear how accurate these numbers are or if they reflect the actual situation. The ultimate goal is to reduce resource extraction, which can be accomplished in several ways:

1. Dematerialization of products
2. Increasing the recycling rate of products at the end of their life cycle
3. Increasing the durability of products

If the industrial system were to double each of these factors, a Factor 8 increase in resource productivity would occur. And each of these is achievable over the short term.

Shift 2, to biologically inspired models, is also echoed time and again, and focuses on developing systems with closed-loop behavior. However, as pointed out by Reuter et al. (2005), the laws of thermodynamics and separation efficiency dictate that closed loops are not closed loops at all; that some fraction of the materials being recycled will dissipate into the environment; and that ultimately, after many recycling loops, materials will, for all practical purposes, be totally dissipated.

Shift 3 to a service and flow economy is a proposal that has been made numerous times over the past decade and has received little serious attention. Having manufacturers retain ownership of building components and maintain responsibility for reusing or recycling them makes good sense on paper. However, maintaining the link between manufacturer and product, even after decades of use, would be extremely difficult, and the logistics system that would be required to dismantle buildings and return materials to their originators would be enormously complicated.

Shift 4, reinvesting in natural capital, is an important point, and its implementation in the built environment context can be strongly reinforced. It is indeed possible to restore damaged sites and to ensure that the net ecological value of many sites is greater than it was prior to the alterations caused by building.

BIOLOGICAL MATERIALS, BIOMATERIALS, AND OTHER NATURE-BASED MATERIALS

One of the shifts advocated by many of the approaches described above is a shift from nonrenewable to renewable resources. Natural capitalism, The Natural Step, and cradle-to-cradle design, for example, suggest that this shift is fundamental for sustainability in general. A shift to renewable resources implies a shift in the materials sector to biological materials, biomaterials, and other natural or nature-based materials. Biological materials and biomaterials are two distinct classes of materials. *Biological materials* are natural systems products such as wood, hemp, and bamboo, while *biomaterials* are materials with novel chemical, physical, mechanical, or "intelligent" properties, produced through processes that employ or mimic biological phenomena.[26] Biomaterials include several emerging classes of biopolymers such as polylactic acid (PLA) and olyhydroxyalkanoate (PHA). Long chain molecules synthesized by living organisms, such as proteins, cellulose, and starch, are natural biopolymers. Synthetic biopolymers are generated from renewable natural sources, are often biodegradable, and are not toxic to produce. Synthetic biopolymers can be produced by biological systems (i.e., microorganisms, plants, and animals) or chemically synthesized from biological starting materials (e.g., sugars, starch, natural fats or oils). Biopolymers are an alternative to petroleum-based polymers (traditional plastics). (Bio)polyesters have properties similar to those of traditional polyesters. Starch-based polymers are often a blend of starch and other plastics (e.g., polyethylene [PE]), which allows for enhanced environmental properties.

Biological materials, such as wood pulp and cotton, can pose environmental problems. Unsound agricultural or silvicultural practices can quickly turn a fertile tract into a disaster area. Because biological resources are renewable, there is a tendency to think of them as unlimited. Nothing could be further from the truth. If cultivated carefully, crops can be planted in perpetuity. But if the land is pushed past its carrying capacity or otherwise abused, permanent damage can be done.[27]

A widespread shift to biological materials for both energy and materials has other implications because large quantities of land may be required to provide ethanol, biological materials, and the feedstock for biomaterials such as biopolymers. An ethical debate is shaping up over taking excess land from food production and shifting it to these other applications, causing increases in food prices and impacting the poor and hungry of the world.

The fact that these materials are biodegradable and compostable means that they are recyclable via a biological route. However, there is a great deal of uncertainty about the quality and utility of the degraded materials and the logistics for effectively using these nutrients of unknown quality in agriculture or the support of natural systems.

Finally, there is little evidence that biologically based materials can replace the synthetic materials that have become common in construction, especially structural materials such as steel and concrete, not to mention copper and aluminum wiring, glass, and the wide variety of polymers used in myriad applications.

SYNTHESIS

After the range of principles and approaches that describe how to create an environmentally sound and sustainable built environment have been examined, and taking into account the orientation of the human species toward the future, the development and deployment of new materials and products will likely be based on ethics, risk, and economics. Clearly, many lessons have been learned about the introduction of toxins and estrogen mimickers into the environment, the impacts of emissions on human and natural systems, the effects of extraction on the environment and human communities, the impacts of waste, and all the other well-known negatives of the production system. Changing the decision system, screening all substances for a broad range of impacts, is badly needed to ensure that the risks to nature and humans are minimized. Certainly, nature's materials and processes provide inspiration for human-designed materials and products, and the behavior of natural systems can inform human systems. But many novel materials and products will continue to be produced, and a systematic approach to examining the extraction, production, use, recycling, and disposal of these resources is needed. This would include LCA, but with application of toxicology and other screens to produce a fuller understanding of the risks associated with the entire life cycle of materials. Beyond the question of materials is responsibility for products and ensuring their potential for disassembly. In the context of the built environment, one other level of disassembly, that of the whole building, must be considered for closing materials loops. Economics, underpinned by policy in the form of taxes that penalize negative behavior in the production and consumption system, will also help dictate the future. In the final analysis, ethics will have to govern the decision system. It must also address how humans use knowledge of potential negative impacts and, ideally, require detailed screening of all new chemicals and processes to ensure that their effects are well understood. Knowing this would allow risk assessment and the ultimate decision as to whether the benefits outweigh the costs.

An Emerging View: A Shift from Green Design to Regenerative Design

Bill Reed, an internationally known architect and thinker, suggests that today we are at the beginning of a shift in thinking about the design of human systems that ultimately needs to be restorative and regenerative, that we are faced with the necessity of actually having to help revive nature after the enormous damage done by human activities over centuries.[28] Clearly, the very survival of humanity is dependent on just such a radical rapid shift. The process of shifting is sketched out in Figure 5.4, where at the lowest rung lies conventional design, a relatively high-energy, fragmented approach. At present, we are shifting to lower-impact, lower-energy approaches. In this framework the next stage is green design, representing building rating systems such as LEED, followed by sustainable design, which is nothing more than a "less bad" approach as it is currently interpreted by most practitioners. Beyond sustainable design lies restorative design, in which the first steps in reviving nature will occur, leading finally to regenerative design, where humans coevolve with nature once again. Some of the terms Reed (2006) uses to describe these stages are defined below.

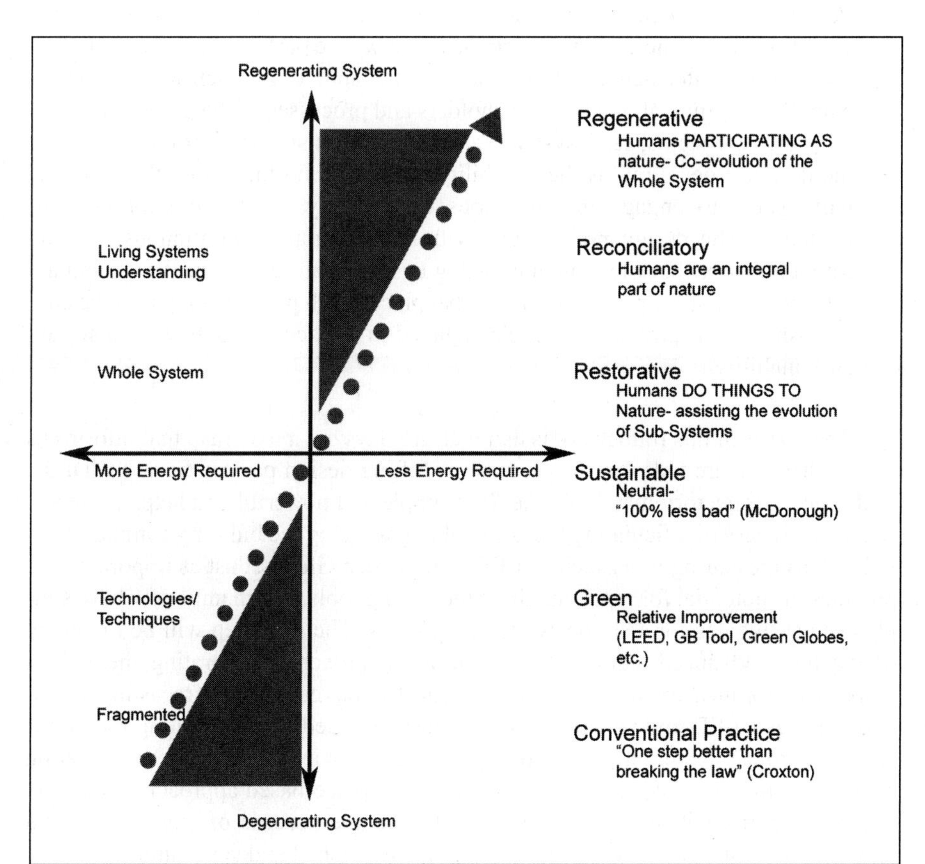

Figure 5.4 A diagram developed by Bill Reed of Integrative Design Collaborative and Regenesis provides a thought-provoking view of the ongoing shift in design from conventional approaches that barely meet building codes to regenerative design that actively engages humans in a synergistic relationship with natural systems. (Diagram courtesy of the Integrative Design Collaborative.)

ISSUE-BASED APPROACHES

A "Limiting the Damage" Approach

High-performance design: Design that realizes high efficiency and reduced impact in the building structure, operations, and site activities. This term can imply a more technical efficiency approach to design and may limit an embrace of the larger natural system benefits.

"Neutral" Approaches

Green design: A general term implying a direction of improvement in design—i.e., continual improvement towards a generalized ideal of doing no harm. Some people believe this is more applicable to buildings and technology than to other activities.

Sustainable design: Green design with an emphasis on reaching a point of being able to sustain the health of the planet's organisms and systems over time.

LIVING SYSTEM APPROACHES

"Restoration" Approaches

Restorative design: In this approach, designers think in terms of using the activities of design and building to restore the capability of relatively independent local natural systems to a healthy state of self-organization.

Reconciliation design: This is a design process that acknowledges that humans are an integral part of nature and that human and natural systems are one.

The "Regeneration" Approach

Regenerative design: This is a design process that engages and focuses on the evolution of the whole of the system of which we are part. Logically, our place—community, watershed, and bio-region—is the sphere in which we can participate. By engaging all the key stakeholders and processes of the place—humans, other biotic systems, earth systems, and the consciousness that connects them—the design process builds the capability of people and the "more than human" participants to engage in continuous and healthy relationship through co-evolution. The design process draws from and supports continuous learning through feedback, reflection, and dialogue so that all aspects of the system are integral parts of the process of life in that place. Such processes tap into the consciousness and spirit of the people engaged in a place, the only way to sustain sustainability.

The power of this framework is that it clearly lays out a road map that informs us about where we are today in the process of shifting design paradigms and what the evolutionary trajectory may look like. It is simple and powerful and helps us understand the process of articulating the key challenges facing the building community as it develops measuring tools such as LEED and Green Globes. Just as importantly, it provides the potential for replacing these measuring tools, which until now have simply been surrogates for a seriously lacking process. The next step will be to clearly define these advanced, nature-oriented design approaches, separating them from tools meant for measurement. This would allow for the redesign and recasting of systems such as LEED and Green Globes for their intended purpose: rating the owner and the project team on how closely the built environment they are creating approaches the ideal of restoring nature. This is a place-based approach. Indicators and benchmarks will ultimately be derived from the process of engagement and understanding of how life works in each unique watershed and community.

Summary and Conclusions

Clearly, a shift is afoot in the design of buildings in the United States, and perhaps as many as 2–3 percent of all commercial and institutional buildings were using LEED as the basis for their greening in 2009. This is a remarkable turn of events, considering that this approach was virtually unknown in 1998. At a minimum, it can be said that LEED has been a tremendous success. Whether LEED can be said to be driving building design to be environmentally friendly and resource-efficient is another question. Certainly, considering the rapid deterioration of our planet's fragile health, any measures that help reduce the destruction of its ecological systems, minimize waste, and use resources more effectively are helpful.

At the very least, LEED can be said to be making deep inroads in addressing the disproportionate impact of the built environment on the Earth. In particular, LEED makes design and construction professionals more aware of their ethical responsibilities for providing high-quality, healthy buildings that have the potential for complementing and working synergistically with nature.

Notes

1. As defined by Van Der Ryn and Cowan (1996).
2. Coincidentally, the founder of systems ecology, Howard T. Odum (1924–2002), and the founder of adaptive management, Crawford (Buzz) Holling (1930–), worked together at the University of Florida for an extended period of time in the last two decades of the twentieth century. Odum's book, *Systems Ecology* (1983), and Holling's book, *Adaptive Environmental Assessment and Management* (1978), are landmark works that redefined how scientists think about ecological systems.
3. Excerpted from *The Business Case for Sustainable Design,* published by the Federal Energy Management Program (FEMP) in 2003.
4. Excerpted from the Buckminster Fuller Institute website, www.bfi.org.
5. Wilson and Kellert (1993) addressed the strong and fundamental connection between humans and nature.
6. Theis's 2002 paper can be found at the website of the Society of Building Science Educators, www.sbse.org.
7. *The Hannover Principles: Design for Sustainability* is available at the McDonough and Partners website, www.mcdonough.com/principles.pdf.
8. Excerpted from *The Local Government Sustainable Buildings Guidebook* (1993).
9. Excerpted from the *Sustainable Building Technical Manual* (1996).
10. The checklist is also available at the BuildingGreen, Inc., website, www.buildinggreen.com, as part of a paid subscription service. The website also provides access to all of the back issues of *Environmental Building News* (EBN), the oldest and most prominent source of information in the United States on green building issues.
11. These assumptions are paraphrased from Yeang (1995), chapter 1.
12. The bases for ecological design are paraphrased from Chapter 1 of Yeang (1995).
13. From Daly (1990) and Barbier (1989).
14. Kay (2002).
15. Bringezu (2002).
16. Peterson (2002).
17. Westley, Carpenter, Brock, Holling, and Gunderson (2002).
18. Benyus (1997).
19. McDonough and Braungart (2002).
20. http://blogs.salon.com/0002007/2006/02/12.html (February 12, 2006).
21. Georgescu-Roegen (1971, 1979).
22. Ruth (1995).
23. Reuter, van Schaik, Ignatenko, and de Haan (2005).
24. Fenton (1998).

25. Hawken, Lovins, and Lovins (1999).
26. As described at the U.S. Department of Agriculture website, www.nal.usda.gov/bic/bio21/gloss.html.
27. Hayes (1978).
28. Bill Reed is President of the Integrated Design Collaborative, which has a website at www.integrativedesign.net with links to many of his papers and articles. His paper "Shifting Our Mental Model—'Sustainability' to Regeneration" is the source of his thinking on shifting from conventional design to regenerative design. It was presented at Rethinking Sustainable Construction 2006 (RSC06), a conference held in Sarasota, Florida, in September 2006.

References

Barbier, E.B. 1989. *Economics, Natural Resource Scarcity and Development. Conventional and Alternative Views.* London: Earthscan.

Benyus, J. 1997. *Biomimicry: Innovation Inspired by Nature.* New York: William Morrow.

Bottero, Maria. December 2002. "Richard Buckminster Fuller," *Your Private Sky.* Available at www.designzine.com.

Bringezu, Stefan. 2002. "Construction Ecology and Metabolism," in *Construction Ecology: Nature as the Basis for Green Building,* C.J. Kibert, J. Sendzimir, and G.B. Guy, eds. London: Spon Press.

Daly, H.E. 1990. "Towards Some Operational Principles of Sustainable Development," *Ecological Economics* 2, pp. 1–6.

Fenton, M.D. 1998. *Iron and Steel Recycling in the United States in 1998.* U.S. Geological Survey Open File Report 01-224.

Fuller, R. Buckminster. 1969. *Operating Manual for Spaceship Earth,* Carbondale: Southern Illinois University Press.

——— 1981. *Critical Path.* New York: St. Martin's Press.

Georgescu-Roegen, N. 1971. *The Entropy Law and the Economic Process,* Cambridge, MA: Harvard University Press.

——— 1979. "Energy Analysis and Economic Valuation," *Southern Economic Journal* 45, pp. 1023–1058.

Hawken, P., A. Lovins, and H. Lovins. 1999. *Natural Capitalism.* New York: Little, Brown.

Hayes, D. 1978. *Repair, Reuse, Recycling: First Steps Toward a Sustainable Society.* Worldwatch Paper 23. Washington, DC: The Worldwatch Institute.

Holling, Crawford S. 1978. *Adaptive Environmental Assessment and Management.* London: John Wiley & Sons.

Kay, James. 2002. "Complexity Theory, Exergy, and Industrial Ecology," in *Construction Ecology: Nature as the Basis for Green Building,* C.J. Kibert, J. Sendzimir, and G.B. Guy, eds. London: Spon Press.

Lyle, John Tillman. 1985. *Design for Human Ecosystems: Landscape, Land Use, and Natural Resources.* New York: Van Nostrand Reinhold. (Republished by Island Press, 1999.)

——— 1994. *Regenerative Design for Sustainable Development.* New York: John Wiley & Sons.

McDonough, William. 1992. *The Hannover Principles: Design for Sustainability.* Charlottesville, VA: William McDonough+Partners. Available at www.mcdonough.com/principles.pdf.

McDonough, William, and Michael Braungart. 2002. *Cradle to Cradle: Remaking the Way We Make Things.* New York: North Point Press.

McHarg, Ian. 1969. *Design with Nature.* Garden City, NY: Natural History Press.

Mumford, Lewis. 1946. *Values for Survival: Essays, Addresses, and Letters on Politics and Education.* New York: Harcourt, Brace.

——— 1955. *The Brown Decades: A Study of the Arts in America.* New York: Dover Publications.

——— 1967. *The Myth of the Machine.* New York: Harcourt, Brace & World.

Odum, Howard T. 1983. *Systems Ecology.* New York: John Wiley & Sons.

Office of Energy Efficiency and Renewable Energy (EERE). 2003. *The Business Case for Sus-*

tainable Design in Federal Facilities. Washington, DC: Federal Energy Management Program (FEMP), U.S. Department of Energy. Available at www.eere.energy.gov/femp/technologies/sustainable_federalfacilities.cfm.

Orr, David W. 2002. *The Nature of Design: Ecology, Culture, and Human Intention.* New York: Oxford University Press.

Pearce, David W. and R. Kerry Turner. 1990. Economics of Natural Resources and the Environment, Hemel Hempstead, U.K.: Harvester Wheatsheaf.

Peterson, Gary. 2002. "Using Ecological Dynamics to Move Toward an Adaptive Architecture," in *Construction Ecology: Nature as the Basis for Green Building,* C.J. Kibert, J. Sendzimir, and G.B. Guy, eds. London: Spon Press.

Reed, Bill. 2006. "Shifting Our Mental Model—'Sustainability' to Regeneration," *Proceedings of Rethinking Sustainable Construction 2006* (*RSC06*), September 19–22, 2006, Sarasota, FL.

Reuter, M.A., A. van Schaik, O. Ignatenko, and G.J. de Haan. 2005. "Fundamental Limits for the Recycling of End of Life Vehicles." *Minerals Engineering.* Draft available at www.doc.tms.org/ezMerchant/prodtms.nsf/ProductLookupItemID/JOM-0408-39/$FILE/JOM-0408-39F.pdf?OpenElement#search=%22Reuter%20van%20Schaik%20recycling%22.

Ruth, M. 1995. "Thermodynamic Implications for Natural Resource Extraction and Technical Change in U.S. Copper Mining," *Environmental and Resource Economics* 6, pp. 187–206.

Sustainable Building Technical Manual: Green Design, Construction and Operations. 1996. Washington, DC: Public Technology, Inc. Available from the USGBC website, www.usgbc.org.

The Local Government Sustainable Building Guidebook. 1993. Washington, DC: Public Technology.

Theis, Christopher C. 2002. "Prospects for Ecological Design Education," Society of Building Science Educators. Available at www.sbse.org.

Van Der Ryn, Sim, and Stuart Cowan. 1996. *Ecological Design.* Washington, DC: Island Press.

von Weizsäcker, Ernst, Amory Lovins, and L. Hunter Lovins. 1997. *Factor Four: Doubling Wealth, Halving Resource Use.* London: Earthscan.

Wells, Malcolm. 1981. *Gentle Architecture.* New York: McGraw-Hill.

Westley, F., S. Carpenter, W. Brock, C.S. Hollins, and L.H. Gunderson. (2002). "Why Systems of People and Nature Are Not Just Social and Ecological Systems," in *Panarchy,* L.H. Gunderson and C.S. Holling, eds. Covelo, CA: Island Press.

Wheelwright, Peter M. April 2000. "Environment, Technology and Form: Reaction." Paper delivered to the Architecture League of New York in response to the symposium Environment, Technology, and Form.

——— December 2000. "Texts and Lumps: Thoughts on Science and Sustainability," *ACSA NEWS,* pp. 5–6.

Wilson, E.O., and Stephen Kellert, eds. 1993. *The Biophilia Hypothesis.* Washington, DC: Island Press.

Yeang, Ken. 1995. *Designing with Nature: The Ecological Basis for Architectural Design.* 1995. New York: McGraw-Hill.

——— 1996. *The Skyscraper Bioclimatically Considered: A Design Primer.* London: Academy Editors.

——— 1999. *The Green Skyscraper: The Basis for Designing Sustainable Intensive Buildings.* Munich: Prestel.

——— 2004. *EcoDesign: A Manual for Ecological Design.* London: John Wiley & Sons.

Part II

Green Building Systems

Since its emergence in 1998, the USGBC's LEED building assessment system for new construction (LEED-NC) has been the most important method for rating and certifying buildings. Since then, LEED has morphed into a suite of standards including LEED-NC 2009, the latest version of the rating system for new construction and major renovations. A second building assessment standard, Green Globes, appeared in 2006 as a potential competitor to LEED-NC 2009 and in some respects has advanced the state of the art of assessing buildings for their environmental performance. For example, Green Globes has provisions for the LCA of materials, addresses noise issues, directly addresses integrated design, and awards points for emergency planning for natural disasters. Part II of this book covers the major categories of issues covered by most building assessment systems, including LEED and Green Globes. These categories include site and landscaping, energy systems, materials and products, the building hydrologic cycle, and indoor environmental quality.

Part II contains the following chapters:

Chapter 6: Sustainable Sites and Landscaping

Chapter 7: Energy and Atmosphere

Chapter 8: The Building Hydrologic System

Chapter 9: Closing Materials Loops

Chapter 10: Indoor Environmental Quality

These chapters correspond to the heart of the LEED-NC 2009 and Green Globes schemes for rating buildings. It should be noted that Green Globes does contain several additional categories, namely, Project Management Policies and Practices (Section A) and Emissions and Other Impacts (Section G). However, most of the issues covered in these sections are discussed in the five chapters of Part II. At the end of each chapter, the LEED-NC and Green Globes approaches to addressing these issues are considered.

Chapter 6, "Sustainable Sites and Landscaping," parallels the corresponding LEED-NC and Green Globes categories that include issues such as locating the building near mass transit, siting the building on a brownfield instead of a greenfield, minimizing the ecological footprint of the construction process, and other measures designed to ensure that the building is sited to have the lowest possible environmental impact. This category also covers the potential for enhancing ecosystems as a component of developing green buildings. Stormwater management and alternatives to conventional practices are addressed. The problem of urban heat islands and measures to reduce temperature buildup in urban areas are considered. Light pollution—a health, safety, and environmental problem—is covered, and techniques for preventing excessive light from affecting the surrounding areas are presented. At the end of Chapter 6, the connection to LEED-NC and Green Globes is briefly covered to show how these building assessment standards address site and landscaping issues.

Chapter 7, "Energy and Atmosphere," covers a range of energy issues including passive design, design of the thermal envelope, equipment selection, renewable energy systems, green power, and emerging technologies, all of which can help achieve a very low building energy consumption profile. Energy-efficient lighting systems and lighting controls that sense occupancy and can be tied into the lighting system are covered. Innovative practices such as radiant cooling and ground coupling are used as examples of cutting-edge methods for addressing energy issues in green buildings. Smart buildings and building energy management systems are described to show how technology can be used in an effective manner to reduce a building's energy profile. Chapter 7 also concludes by making the connection to LEED-NC and Green Globes.

Chapter 8, "The Building Hydrologic System," focuses on minimizing potable water use, water recycling and reuse, and provisions for minimizing off-site stormwater flows. A strategy for the design of an effective building hydrologic cycle is provided. Technologies are described that can help provide alternative sources of water when potable water is not absolutely necessary. A wastewater strategy for high-performance buildings is also provided to minimize the need to move wastewater off-site for processing. Water-efficient landscaping is described, and its role in a green building hydrologic strategy is covered. Chapter 8 ends by briefly delineating the LEED-NC and Green Globes points distributions for this category.

The selection of environmentally friendly construction materials is addressed in Chapter 9, "Closing Materials Loops." It covers the use of recycled content materials, used components, embodied energy due to transportation of material, and the minimization of construction waste. Defining green building materials remains the most difficult problem for designers of contemporary green buildings. For example, recycled content materials are in principle green building materials, but many contain industrial and agricultural waste, so it is not clear that recycling these byproducts into the built environment is the best solution. Consequently, one objective of this chapter is to promote an understanding of the broad range of issues and problems connected to building materials and products. The chapter also covers the topic of LCA, a method for analyzing the resources, waste, and health effects associated with the entire life of a product or material, from its extraction as raw materials to its ultimate disposal. David Hobbs, President of Interface FLOR Commercial, Inc., provides a case study of his company's journey to develop sustainable products. Chapter 9, like the other chapters in this part, ends with a brief delineation of the points available from LEED-NC and Green Globes in this category.

Indoor environmental quality (IEQ), the last of the five major categories for achieving points in Green Globes and LEED-NC, is covered in Chapter 10. The various types of health-related building problems are described. Selecting low-emissions materials, protecting HVAC systems during construction, monitoring indoor air quality (IAQ), and issues surrounding the health of the construction workforce and future building occupants are explored. Lighting quality, access to daylight and views, and noise as IEQ issues are covered. Best practices for providing building IAQ are also addressed. At the end of Chapter 10, the connections of this category to LEED-NC and Green Globes are covered.

Additional best practices and checklists that provide assistance in achieving building assessment system points or that address issues not covered in these systems are also provided. Experiences with the first LEED-NC buildings emerging from the process demonstrate how the standard is being applied in practice. In short, this part addresses the core issues of the technical side of sustainable construction and discusses approaches that can be employed to limit resource depletion, negative environmental consequences, and impacts on human health that are too often the result of the creation, operation, and disposal of the built environment. Future buildings should contribute to the restoration and regeneration of ecological capacity, recycle water and discharge potable water, generate the energy needed for their operation, contribute to the health of their human occupants, and serve as materials resources for future generations rather than as a disposal headache.

Chapter 6

Sustainable Sites and Landscaping

Land use and landscape design are closely coupled; arguably, they offer the greatest opportunity for innovation in the application of the resources needed to create the built environment. Carefully designed and executed work by architects, landscape architects, civil engineers, and construction managers is required to produce a building that optimizes the use of the site; that is highly integrated with the local ecosystem; that carefully considers the site's geology, topography, solar insolation, hydrology, and wind patterns; that minimizes impacts during construction and operation; and that employs landscaping as a powerful adjunct to its technical systems. Other members of the project team must also have a voice in the decisions made about land and landscape. The location of the facility on the site, the type and color of exterior finishes, and the materials used in parking and paving all affect the thermal load on the building, and hence the design of the heating and cooling systems by the mechanical engineer. Minimizing the impact of light pollution requires the electrical engineer to carefully design exterior lighting systems to eliminate unnecessary illumination of the building's surroundings. Providing access to mass transportation, encouraging bicycling and alternative-fuel vehicles, or the capability to refuel alternative-fuel vehicles ensures that the context of the building is not neglected. Collaboration among all these players marks high-performance green building as a distinct delivery system and is essential to make optimal use of the site and landscape.

Site and landscape also provide the opportunity to move beyond mere greening to the potential restoration of the land as an integral part of the building project. Until the advent of the green building movement, scant attention was paid to the impacts of construction on the environment, particularly on the land. Buildings alter the ecology, biodiversity, fecundity, and hydrology of the site, leaving it in a degraded state. Contemporary green building approaches call for the reuse of land, its cleanup in the case of contaminated land, and increasing density to minimize the need for greenfield development.

In the context of green buildings, the role of the landscape architect should perhaps be redefined from that of simply providing exterior amenities for the project to serving as the integrator of ecology and nature within the built environment. Because they are probably the best-equipped members of the project team to deal with natural systems, landscape architects should also provide expertise to the rest of the project team on the relationship between buildings and natural systems.

Historically, there has not necessarily been a connection between landscape architecture and the environment. As noted by Robert France in a 2003 critique of landscape architecture, "[T]he desire of planners to make their personal mark on the landscape, and of ecologists to understand the workings of nature, can be at odds with a desire to preserve, protect, and restore environmental integrity."[1] It might even be useful at this point in the evolution of the green building delivery system for members of this profession to review the term *landscape architect* and consider a more appropriate one, perhaps *ecological architect.* At present, there is no professional on the conventional project team with the knowledge of buildings, ecology, and the flow of matter and energy across the human-nature interface. New, emerging topics for landscape design include stormwater uptake, wastewater treatment, food production, and assisting in heating and cooling buildings. New approaches that include a robust role for natural systems in buildings are at the cutting edge of high-performance building and point to areas where their design must eventually evolve.

The appropriate use of land is a major issue in green building, if for no other reason than that a building designed and constructed to the most exacting green building standards will be badly compromised if the users or occupants must drive long distances to reach it. Other land issues include building on environmentally sensitive property, in flood-prone areas, or on greenfields or agricultural land instead of on land already affected by human activities. Putting formerly contaminated land or brownfields back into productive use in a building project has the dual advantage of improving the local environment and recycling land, as opposed to employing greenfields. Contemporary green building approaches also require far more care in the use of the building site. In a green building project, the construction footprint is typically minimized, and the construction manager plans the construction process to minimize soil compaction and the destruction of plants and animal habitat. Erosion and sedimentation control are emphasized, and detailed planning of systems to minimize soil flows during construction is part of the green building delivery system. The potential for so-called heat islands, caused by the use of energy-storing materials in the building and on the site, is addressed. Likewise, the issue of light pollution from buildings is addressed in the design of a high-performance green building.

Land and Landscape Approaches for Green Buildings

Buildings require several categories of resources for their creation and operation: materials, energy, water, and land. Land, obviously, is an essential and valuable resource, so its appropriate use is a prime consideration in the development of a high-performance building. There are several general approaches to land use that fit in with the concept of high-performance green buildings:

- Building on land that has been previously utilized instead of on land that is valuable from an ecological point of view
- Protecting and preserving wetlands and other features that are key elements of existing ecosystems
- Using native and adapted, drought-tolerant plants, trees, and turf for landscaping
- Developing brownfields, properties that are contaminated or perceived to be contaminated
- Developing grayfields, areas that were once building sites in urban areas
- Reusing existing buildings instead of constructing new ones
- Protecting key natural features and integrating them into the building project for both amenity and function
- Minimizing the impacts of construction on the site by minimizing the building footprint and carefully planning construction operations
- Minimizing earth moving and compaction of soil during construction
- Fully using the sun, prevailing winds, and foliage on the site in the passive solar design scheme
- Maintaining as much as possible the natural hydroperiod of the site

- Minimizing the impervious areas on site through appropriate location of the building, parking, and other paving
- Using alternative stormwater management technologies, such as pervious pavement, bioretention, rainwater gardens, and others, which assist on-site or regional groundwater and aquifer recharge
- Minimizing heat island effects on the site by using light-colored paving and roofing, shading, and green roofs
- Eliminating light pollution through careful design of exterior lighting systems
- Using natural wetlands to the maximum extent possible in the stormwater management scheme and minimizing the use of dry-type retention ponds
- Using alternative stormwater management technologies such as pervious concrete and asphalt for paved surfaces

These approaches cover a wide range of possibilities. Their general purpose is to integrate nature and buildings, reuse sites that have already been impacted by human activities, and minimize disturbances caused by the building project.

Land Use Issues

The selection of a building site is generally the purview of the building owner, but often it may be affected by input from members of the project team. Rinker Hall, a recently completed USGBC Gold LEED-certified green building at the University of Florida, was originally slated for construction in an open greenspace on the campus that had previously provided environmental amenity and recreation. However, following interaction between the project user group and the university administrators, the building was relocated to a plot of used land—in this case, a parking lot. The general population of the university benefited by this move in that it did not lose the environmental amenity of the greenspace; and, as it turned out, Rinker Hall's new location was a far more prominent site than its original location. One of the most important green measures in siting a new building is to locate it where the need for automobiles is minimized. Consequently, urban locations reasonably close to mass transit are highly desirable. In some cases, additional discussion with local government and the local transit authority may be required to articulate the need for bus service to what would otherwise be a good location for the facility.

In this section, several issues related to land use and siting are covered: the loss of prime farmland; building in 100-year flood zones; using land that is habitat for endangered species; and reusing brownfields, grayfields, and blackfields. These topics are addressed in the USGBC LEED-NC and the Green Globes building assessment standard.

LOSS OF PRIME FARMLAND

In addition to addressing concerns over the loss of ecosystems, the green building effort considers the loss of agricultural land that, although impacted by human activities, is an important renewable resource. Of the various categories of agricultural land, prime farmland is especially important to preserve. The U.S. Department of Agriculture (USDA) defines prime farmland as follows:[2]

> Prime farmland is land on which crops can be produced for the least cost and with the least damage to the resource base. Prime farmland has an adequate and dependable supply of moisture from precipitation or irrigation and favorable temperature

> and growing season. The soils have acceptable acidity or alkalinity, acceptable salt and sodium content, and a few rocks. They are not excessively eroded. They are flooded less often than once in two years during the growing season and are not saturated with water for a long period. The water table is maintained at a sufficient depth during the growing season to allow cultivated crops common to the area to be grown. The slope ranges mainly from 0 percent to 5 percent. To be classified as prime, land must meet these criteria and must be available for use in agriculture. Land committed to nonagricultural uses is not classified as prime farmland.

In its publication "Farming on the Edge," published in 1997, the American Farmland Trust made the following observations about the impacts of development on the nation's farmland.[3]

- Every single minute of every day, America loses 2 acres of farmland. From 1992 to 1997, more than 6 million acres of agricultural land were developed, an area the size of Maryland.
- Farm and ranch land were lost at a rate 51 percent faster in the 1990s than in the 1980s. The rate of loss for 1992 to 1997, 1.2 million acres per year, was 51 percent higher than from 1982 to 1992.
- The best land, the most fertile and productive, is being lost the fastest. The rate of conversion of prime land was 30 percent faster, proportionally, than the rate for nonprime rural land from 1992 to 1997. This results in *marginal* land, which requires more resources, like water, being put into production.
- Food is increasingly in the path of development: 86 percent of U.S. fruits and vegetables, and 63 percent of our dairy products, are produced in urban-influenced areas.
- Wasteful land use is the problem, not growth itself. From 1982 to 1997, the U.S. population grew by 17 percent, while urbanized land grew by 47 percent. Over the past 20 years, the acreage per person for new housing almost doubled; and since 1994, 10-plus acre housing lots have accounted for 55 percent of the land developed.
- Every state is losing some of its best farmland. Texas leads the nation in high-quality acres lost, followed by Ohio, Georgia, North Carolina, and Illinois. And for each of the top 20 states, the problem is getting worse.

Redirecting development away from prime farmland is addressed in the USGBC LEED-NC and Green Globes building assessment standards, indicating that preserving this valuable resources is high on the priority list for green building projects.

Figure 6.1 In the United States, farmland is being lost at the rate of 2 acres (0.8 hectares) per minute, with the most fertile, productive land being lost most rapidly. Farms abutting urban areas, as shown here, are especially threatened by land development and urban sprawl.

GREENFIELDS, BROWNFIELDS, GRAYFIELDS, AND BLACKFIELDS

Greenfields are properties that have experienced little or no impact from human development activities. Greenfields can also be defined to include agricultural land that has had no activity other than farming. Like recycling in general, recycling of land is an important objective in creating high-performance green buildings. *Land recycling* refers to reusing land impacted by human activities instead of using greenfields. There are at least three identifiable categories of potentially recyclable land: brownfields, grayfields, and blackfields.

The U.S. Environmental Protection Agency (USEPA) defines *brownfields* as abandoned, idled, or underused industrial and commercial facilities where expansion or redevelopment is complicated by real or perceived environmental contamination.[4] The official definition of a brownfield site, according to Public Law 107-118 (H.R. 2869), the Small Business Liability Relief and Brownfields Revitalization Act, signed into law January 11, 2002, is as follows: "With certain legal exclusions and additions, the term 'brownfield site' means real property, the expansion, redevelopment, or reuse of which may be complicated by the presence or potential presence of a hazardous substance, pollutant, or contaminant." The key word in the first definition is *perceived;* the key phrase in the second is *potential presence.* Former industrial properties are often thought to be contaminated because of the activities that occurred on the site—for example, metal plating or leather tanning. In fact, not infrequently these properties are fairly clean, requiring minimal cleanup. In many U.S. cities, brownfields are now valuable real estate because of their proximity to extensive infrastructure and a potential workforce. Industries formerly fleeing to greenfields outside urban areas, thereby causing impoverishment of minority communities due to job loss, are returning to former industrial sites because the economics dictate the return to the city. A prime example of the potential success of a well-developed brownfields strategy is Chicago's Brownfields Initiative, which since 1993 has been assisting in the cleanup and transfer of 12 major former industrial sites in the city. An interesting aspect of the Chicago strategy has been to emphasize the return of these zones to industrial use, thus bringing jobs back into the city.[5]

Both the LEED-NC 2009 and Green Globes building assessment standards provide a credit for the use of a former brownfield as a building site. According to a USGBC credit ruling on brownfields, the project team can consider a site not officially designated a brownfield by the USEPA if the team can convince the USEPA that the site fulfills the requirements of a brownfield and the USEPA agrees in writing.

Grayfields, another form of urban property, can be defined as blighted or obsolete buildings sitting on land that is not necessarily contaminated. The term *grayfield* is actually an expanded definition of *brownfield.* The State of Michigan, for example, embeds the term *grayfield* in the concept of *core community,* areas that are economically blighted and need investment to restore them to economic health. A grayfield could be a former machine shop where the building is obsolete, perhaps because it lacks a fire suppression system; had a septic system and old fuel tanks; or contains asbestos. In general, boarded-up housing is often an indication of a grayfield. The Congress of New Urbanism (CNU) points out that former or declining malls can be classified as grayfields because they occupy impacted land that can be returned to productive use.[6] Declining malls are caused by a number of factors: population shifts, increasing numbers of big-box stores, changing demographics, and a failure of developers to reinvest in upgrades and modernization of older malls. Changes in the retail environment are also affecting the big-box stores as they continue to increase the size of their facilities. In June 2004, Wal-Mart Corporation listed 394 properties for sale, ranging in size from 2,700 to 162,000 square feet (251 to 15,050 square meters).[7] Larger abandoned big-box stores are now referred to as *ghostboxes.* Some of the strategies communities are using to deal with these types of properties are:[8]

- *Adaptive reuse:* Turning ghostboxes into office space, entertainment space, or space for light manufacturing.
- *De-malling:* Reversing storefronts to face the street; converting the property to give it a "Main Street" look; and making connections to nearby housing, using pedestrian-friendly planning.
- *Razing and reuse:* Older malls are being demolished to make room for new retail developments.
- *Passing community ordinances to prevent future grayfields and ghostboxes:* Some communities are setting a maximum size for big-box stores or requiring that an escrow account covering future demolition costs be established for the construction of a big-box store.

Both grayfields and brownfields are becoming valuable properties because of the presence of good infrastructure in urban areas; a trend toward urban living prompted by a perceived higher quality of life; and incentives offered by local and state governments in the form of tax rebates, tax credits, tax increment district financing, and other innovative strategies. In addition to access to infrastructure and a willing workforce for business, cities ultimately receive far greater tax revenues, creating a true win-win scenario. Though grayfields are not explicitly addressed in either LEED-NC or Green Globes, credits and points are awarded for building in a dense urban environment.

Yet another category of blighted land is *blackfields.* These properties are abandoned coal mines and are found in former coal mining areas such as eastern Pennsylvania, where abandoned strip mines and subsurface mines comprise an area three times the size of Philadelphia and which will require an estimated $16 billion to clean up. Surface waters in these zones have a very low pH and are contaminated with iron, aluminum, manganese, and sulfates. The term *blackfields* also can be considered as an expanded definition of *brownfields.* There is a potential for obtaining LEED-NC points for using one of these properties for a building project.[9]

Figure 6.2 Grayfields are urban properties that are underperforming or declining in value for technological, economic, or social reasons. Strip malls throughout the United States, such as the one shown here, often become outmoded and their tenants move on to larger facilities or to more profitable locations. A potential outcome is blighted areas and impacts on the local economy and property values, raising major challenges for local government.

BUILDING IN 100-YEAR FLOOD ZONES

Clearly, buildings should not be constructed in flood-prone areas due to the high potential for disasters that result not only in human suffering, but also in enormous environmental and resource impacts caused by the cycle of destruction and rebuilding. This is such a vital matter that the Federal Emergency Management Agency (FEMA) has become deeply involved in issues of flood mapping and insurance. Specifically, in support of the National Flood Insurance Program (NFIP), FEMA has undertaken a massive program of flood hazard identification and mapping to produce Flood Hazard Boundary Maps, Flood Insurance Rate Maps, and Flood Boundary and Floodway Maps. Several areas of flood hazards are commonly identified on these maps. One of these areas is the *special flood hazard area* (SFHA), which is defined as an area of land that would be inundated by a flood having a 1 percent chance of occurring in any given year (previously referred to as the *base flood* or *100-year flood*). The 1 percent annual chance standard was decided after considering various alternatives. The standard constitutes a reasonable compromise between the need for building restrictions to minimize potential loss of life and property and the economic benefits to be derived from floodplain development. Development may take place within the SFHA, provided that it complies with local floodplain management ordinances, which must meet the minimum federal requirements. Flood insurance is required for insurable structures within the SFHA to protect federally funded or federally backed investments and assistance used for acquisition and/or construction purposes within communities participating in the NFIP.[10]

Before continuing with this discussion, it is important to point out that the term *100-year flood* is misleading. It is not the flood that will occur once every 100 years; rather, it is the flood *elevation* that has a 1 percent chance of being equaled or exceeded each year. Thus, the 100-year flood could occur more than once in a relatively short period of time. The 100-year flood, which is the standard used by most federal and state agencies, is also used by the NFIP as its standard for floodplain management and to determine the need for flood insurance. A structure located within an SFHA shown on an NFIP map has a 26 percent chance of suffering flood damage during the term of a 30-year mortgage.

ENDANGERED SPECIES

Passed in 1973 and reauthorized in 1988, the Endangered Species Act (ESA) regulates a wide range of activities affecting plants and animals designated as endangered or threatened. By definition, an *endangered species* is an animal or plant listed by regulation as being in danger of extinction. A *threatened species* is any animal or plant that is likely to become endangered within the foreseeable future. A species must be listed in the Federal Register as endangered or threatened for the provisions of the act to apply.

The ESA prohibits the following activities involving endangered species:

- Importing into or exporting from the United States
- Taking (which includes harassing, harming, pursuing, hunting, shooting, wounding, trapping, killing, capturing, or collecting) within the United States and its territorial seas
- Taking on the high seas
- Possessing, selling, delivering, carrying, transporting, or shipping any such species unlawfully taken within the United States or on the high seas
- Delivering, receiving, carrying, transporting, or shipping in interstate or foreign commerce in the course of a commercial activity
- Selling or offering for sale in interstate or foreign commerce

The ESA also provides for:

- Protection of critical habitat (habitat required for the survival and recovery of the species)
- Creation of a recovery plan for each listed species

As of February 2003, the following were the statistics for endangered and threatened species in the United States, according to the U.S. Fish and Wildlife Service:

- 517 U.S. species of animals are listed.
- 745 U.S. species of plants are listed.
- 30 U.S. species of animals are currently proposed for listing.
- 5 U.S. species of plants are currently proposed for listing.

As is the case with construction within a 100-year flood zone, LEED-NC 2009 Sustainable Site Credit 1 (SSc1) is not achievable if the project site is on land identified as habitat for species that are on state or federal lists of threatened or endangered species.

EROSION AND SEDIMENT CONTROL

Sediment is eroded soil that is suspended, transported, and/or deposited by moving water or wind. *Erosion* is the process of displacing and transporting soil particles by the action of gravity. Some general principles and best management practices that should be used in sediment and erosion control are indicated in Table 6.1.

For high-performance green buildings, care must be taken to ensure that soil loss is minimized. The construction manager and subcontractors must pay attention to soil loss in the form of airborne dust and due to stormwater runoff. Additionally, the contemporary green building delivery system requires that measures be put in

TABLE 6.1

Principles and Best Practices for Sedimentation and Erosion Control

- Design the project to fit the site's context: its topography, soils, drainage patterns, and natural vegetation.
- Minimize the area of construction disturbance and limit the removal of vegetative cover.
- Reduce the duration of bare-area exposure by scheduling construction such that bare areas of the site are exposed only during the dry season or for as short a time as possible.
- Decrease the amount of bare area exposed at any one time.
- Shield soil from the impact of rain or runoff by using temporary vegetation, mulch, or groundcover on exposed areas.
- Divert run-on and runoff water away from exposed areas.
- Prevent offsite runoff from entering the site.
- Inspect and maintain the erosion and sediment control practices that have been put in place.
- Use vegetative buffer strips, mulching and temporary seeding, surface roughening, erosion control blankets, permanent vegetation, and gravel-surfaced construction areas for erosion control.
- Use silt fences, fiber wattles, and logs; check dams in swales, sediment traps and basins, detention/retention ponds, and silt filters/inlet traps for sedimentation control.
- Where high winds are likely to transport soil, use sand or wind fences as a barrier to soil movement.

place to prevent the sedimentation of both stormwater systems and receiving water bodies. An erosion and sedimentation control plan is a prerequisite for certification under LEED-NC, meaning that this plan is required for the building to be considered for even the lowest level of certification. Green Globes awards a range of points for erosion and sedimentation control measures in Part B of this building assessment standard.

Sustainable Landscapes

The advent of high-performance green buildings is causing noteworthy changes to the traditional notion of the constructed landscape. Landscape design has typically been an afterthought in the conventional building delivery system, and in many cases it is given very low priority. As funding for a project becomes tighter near the end of construction it will likely be the budget for the constructed landscape that will be reduced to the bare minimum. The outcome of such conventional thinking is that landscape design is given short shrift, treated *apart* from the building rather than *integral* to it. Today, the role of landscape design in high-performance building is in a state of transition; some projects treat it conventionally, while others are assigning it new roles. Among these new roles are to assist building heating and cooling, help control stormwater and eliminate stormwater infrastructure, treat waste, and provide food.

The concept of *sustainable landscape* predates the contemporary high-performance green building movement. The term emerged in the vocabulary of landscape architecture in 1988, when the Council of Educators in Landscape Architecture defined it as landscapes that contribute ". . . to human well-being and at the same time are in harmony with the natural environment. They do not deplete or damage other ecosystems. While human activity will have altered native patterns, a sustainable landscape will work with native conditions in its structure and functions. Valuable resources—water, nutrients, soil, etc.—and energy will be conserved, diversity of species will be maintained and increased."[11] The movement to reconsider the role of landscape architecture was initiated by the late John Tillman Lyle with the publication of his 1985 book, *Design for Human Ecosystems: Landscape, Land Use, and Natural Resources.* It was almost a decade, however, before more was heard on the subject of sustainable landscapes. In 1994, two volumes appeared, coincidentally at the onset of the American green building movement: Robert Thayer's *Gray World, Green Heart: Technology, Nature and the Sustainable Landscape,* and another book by Lyle, *Regenerative Design for Sustainable Development.*

In *Design for Human Ecosystems,* Lyle considered how landscape, land use, and natural resources could be shaped to make the human ecosystem function in the sustainable ways of natural ecosystems. He suggested that designers must understand ecological order and how it operates at a wide variety of scales, from minute to global. The understanding of ecological order has to be linked with human values in order to develop solutions that are long-lasting, beneficial, and responsible.

In *Gray World, Green Heart,* Thayer notes that landscape is the place where ". . . the conflict between technology and nature is most easily sensed." A sustainable landscape, according to Thayer, would have the following properties:

- An alternative landscape where natural systems are dominant.
- A landscape where resources are regenerated and energy is conserved.
- A landscape that allows us to see, understand, and resolve the battle between the forces of technology and nature.
- A landscape where essential life functions are undertaken, revealed, and celebrated.
- A landscape where the incorporated technology is sustainable, the best of all possible choices, and can be considered part of nature.

- A landscape that counters the frontier ethic of discovery, exploitation, exhaustion, and abandonment with one where we plant ourselves firmly, nurture the land, and prevent ecological impoverishment.
- A landscape that responds to the loss of place with reliance on local resources, celebration of local cultures, and preservation of local ecosystems.
- A landscape that responds to the view that landscape is irrelevant by making the physical landscape pivotal to our existence.

Thayer admits that this vision is utopian but suggests that such a vision is needed to give us direction. He goes on to provide five characteristics of a sustainable landscape that are based on the function and organization of natural landscapes:

1. Sustainable landscapes use primarily renewable, horizontal energy[12] at rates that can be regenerated without ecological destabilization.
2. Sustainable landscapes maximize the recycling of resources, nutrients, and byproducts, and produce minimum waste or conversion of materials to unusable locations or forms.
3. Sustainable landscapes maintain local structure and function and do not reduce the diversity or stability of the surrounding ecosystems.
4. Sustainable landscapes preserve and serve local human communities rather than change or destroy them.
5. Sustainable landscapes incorporate technologies that support these goals and treat technology as secondary and subservient, not primary and dominant.

As a cautionary note, Thayer also tells us that "Without sustainable values, landscapes designed to be sustainable will be misused, become unsustainable, and fail." Contemporary American culture does not have a sense of, nor does it value, place, and it is oriented toward consumption, profit, and waste. Creating a sustainable landscape in the face of these values is challenging but necessary to at least launch a countermovement that puts value on nature and ecosystems and that helps increase human awareness of their role in daily life.

In *Regenerative Design for Sustainable Development,* Lyle introduced designers of the built environment to the concept of regenerative landscape, reminding them, as John Dewey did in 1916, that ". . . the most notable distinction between living and inanimate things is that the former maintain themselves by renewal."[13] He maintains that the developed landscape, the one created and built by humans, should be able to survive within the bounds of local energy and materials flows and that, in order to be sustainable, it must be *regenerative,* which, in the case of landscape, means being capable of *organic self-renewal.* Landscapes must be created using regenerative design, that is, design that creates cyclical flows of matter and energy within the landscape. According to Lyle, a regenerative system is one that provides for continuous replacement, through its own functional processes, of the energy and materials used in its operation. A regenerative system has the following characteristics:

- Operational integration with natural processes and, by extension, with social processes
- Minimum use of fossil fuels and man-made chemicals, except for backup applications
- Minimum use of nonrenewable resources, except where future reuse or recycling is possible and likely
- Use of renewable resources within their capacities for renewal
- Composition and volume of wastes within the capacity of the environment to reassimilate them without damage

Lyle gained considerable experience with regenerative landscapes as a professor at the 1-acre Center for Regenerative Studies that he founded at California Polytech-

(A)

(B)

Figure 6.3 (A) A sustainable landscape renovation project at the Garfield School in Seattle, Washington. Working with University of Washington students, the high school students restored the site by designing and planting a self-maintaining sustainable landscape. In addition to embodying Robert Thayer's vision of a sustainable landscape, this project provided an important learning experience for all the participants. (B) A sustainable landscape designed by Alan Kettler re-creates a meadow and uses eastern red cedar trees as its centerpiece. (Photograph courtesy of Alan Kettler, Kettler Ecological Design Studio.)

nic University in Pomona, where faculty and students worked with regenerative landscapes and technology to try to solve the daily problems of providing shelter, food, energy, and water and dealing with waste. He and his students took what was then a compacted cow pasture within sight of a large landfill and created what a former center director, Joan Stafford, described as a landscape that ". . . now yields armfuls of scented, exuberant lavender, sage, [and] rosemary, growing from rejuvenated soils."

In *Sustainable Landscape Construction: A Guide to Green Building Outdoors,* J. William Thompson and Kim Sorvig provide a set of principles to guide landscape design and construction for green buildings. These principles are outlined in Table

TABLE 6.2

Principles of Sustainable Landscape Construction

Principle 1: Keep sites healthy. Ensure that biologically productive sites with healthy ecosystems are not harmed by the building project. Special attention must be paid to utility installation and road construction, which can be especially destructive to natural systems.

Principle 2: Heal injured sites. Using grayfields, brownfields, or blackfields reduces pressures on biologically productive sites and can result in restoration of blighted properties to productive ecosystems.

Principle 3: Favor living, flexible materials. Slope erosion can be controlled with living structures rather than artificial physical structures. Greenwalls, artificial structures that provide a support system for living matter, may be needed in especially steep terrain. Living materials on roofs create eco-roofs that provide additional green area and assist heating and cooling.

Principle 4: Respect the waters of life. Water bodies, including wetlands, should be protected and even restored. Rainwater can be harvested from roofs, stored in cisterns, and used for nonpotable applications. Landscape irrigation should be minimized and landscape designed to be durable and drought-tolerant.

Principle 5: Pave less. Paving destroys natural systems and should be minimized. Stormwater should be quickly infiltrated through the use of porous concrete and asphalt paving and through the use of pavers. Heat islands should be minimized by appropriate landscaping.

Principle 6: Consider the origin and fate of materials. Minimize the impact of landscape materials by carefully analyzing their embodied energy and other effects. Emphasize reused and recycled materials and avoid toxic materials.

Principle 7: Know the costs of energy over time. Landscape construction requires considerable energy in the form of work by machinery, the embodied energy of materials. The total energy consumption for all purposes, including maintenance, should be minimized.

Principle 8: Celebrate light, respect darkness. Landscape lighting should be accomplished such that plants are unaffected by lighting schemes, and lighting should be energy-efficient. Lighting should not spill over to areas where it is not wanted. Low-voltage lighting, fiber-optic lighting, and solar lighting should be considered.

6.2).[14] In general, they are fairly straightforward and parallel the logic of LEED-NC, which addresses many of these issues.

Some of the innovations emerging in today's high-performance green buildings include the application of landscaping directly to buildings in the form of green roofs, or living or eco-roofs, and the use of vertical landscaping, especially for skyscrapers. These two emerging landscaping concepts are described in the following subsections.

GREEN, OR LIVING, ROOFS

A *green,* or *living,* roof is nothing more than an updated version of the ancient sod roof used in Europe that is making a comeback in today's green building movement. An alternative term used by some practitioners is *eco-roof.* The city of Portland, Oregon, provides tax breaks to motivate the creation of eco-roofs.[15] It approved a regulation in January 2001 that allows developers to expand their building plans if those plans include an eco-roof; it also waives certain code requirements for buildings with green roofs. Though Portland is the only city in North America to offer such an incentive, in other cities, including Chicago, Toronto, and Seattle, gardens are grown on the roofs of city halls and courthouses. And in Dearborn, Michigan, the Ford Motor Company has a 10-acre living roof on its Rouge Center Assembly Plant; county buildings in Anne Arundel County, Maryland, also have grassy roofs.

But Portland goes much further than most jurisdictions thanks to the financial commitment the city has made to eco-roofs, in the form of tax breaks, grants, building code waivers, and the variety of private and public buildings with rooftop gardens.[16] Portland provides such incentives for living roofs because they have been found to reduce building energy costs by 10 percent and to decrease summer roof temperatures by 70°F (21°C); furthermore, these roofs can reduce storm runoff by 90 percent and delay the flow of stormwater for several hours, thereby reducing the probability of stormwater and sewer system overflow. In an area like Portland, which suffers chronic stormwater and sewage system overflows that affect both the Willamette and Columbia rivers, an extensive array of eco-roofs on buildings may help mitigate this problem. Living roofs can also filter pollution and heavy metals from rainwater and help protect the regional water supply.

An eco-roof can fulfill several distinct roles: as an aesthetic feature, to help the building blend into its environment, and to help support climatic stabilization. An eco-roof is particularly useful in wet, snowy areas but has more limited potential in dry climates. Green roofs must be built on a sufficiently strong frame with carefully applied waterproofing, because it is very difficult to locate leaks once the growing medium is in place. The living aspect of the roof is a compost-based system, usually composed of a base of straw that is left to decompose, within which native or introduced plants can then take root. As might be expected, a living roof requires ongoing tending; another disadvantage is that it could be a fire hazard in hot, dry climates. In contrast, it is advantageous in that it protects the waterproofing from damage by ultraviolet radiation, and it precludes the need for tiles or other shingles.

According to the Living Systems Design Guild, green roofs are generally classified as either *extensive* or *intensive:*

> *Extensive eco-roof systems.* Extensive landscaped roofs are defined as low-maintenance, drought-tolerant, self-seeding vegetated roof covers that incorporate colorful sedums, grasses, mosses, and meadow flowers that require little or no irrigation, fertilization, or maintenance. The types of plants suitable for extensive landscaping are those native mainly from locations with dry and semi-dry grassy conditions or with rocky surfaces, such as an alpine environment. Extensive systems can be placed on low-slope and pitched roofs with up to a 40 percent slope.
>
> *Intensive eco-roof systems.* If there is adequate load-bearing capacity, it is possible to create actual roof gardens on many buildings. This type of eco-roof system may include lawns, meadows, bushes, trees, ponds, and terraced surfaces. Intensive systems are far more complex and heavy than extensive eco-roof systems and hence require far more maintenance.

Eco-roof systems are made up of 6 to 10 individual components, as shown in Table 6.3.[17] The soil substrate differentiates the extensive from the intensive system. The extensive system has a soil substrate of 4 to 6 inches (10 to 15 centimeters) of formulated, lightweight growing medium, whereas an intensive system may have as much as 18 to 24 inches (46 to 61 centimeters) of a heavier soil mix.

As should be obvious from this discussion, an eco-roof is far more complex than a conventional roof and requires significantly more research and planning. Additionally, eco-roofs generally cost twice as much as conventional roofs, or \$10 to \$15 per square foot (\$107 to \$162 per square meter). However, the payback due to energy savings alone can be fairly rapid, and the benefits of reduced stormwater infrastructure and natural water cleaning make eco-roofs an attractive option.

TABLE 6.3

Components of Eco-Roof Systems

- *Plants:* Extensive eco-roof systems include shallow root systems; regenerative qualities; and resistance to direct radiation, drought, frost, and wind. A much larger variety of plant selections are available for intensive roofscapes due to their greater soil depths.
- *Soil mix:* The planting mix is a specially formulated, lightweight, moisture-retaining mix that is enriched with organic material.
- *Filter fabric:* The filter fabric prevents fine particles from being washed out of the substrate soil, ensuring efficiency of the drainage layer.
- *Water retention layer:* This layer provides mechanical protection and retains moisture and nutrients. Profiled drainage elements retain rainwater for dry periods in troughs or cups on the upper side of this layer.
- *Drainage layer:* Eco-roofs must have a drainage layer to carry away excess water; on very shallow, extensive eco-roofs, the drainage layer may be combined with the filter layer.
- *Root barrier:* The root barrier prevents roots from affecting the efficiency of the waterproofing membrane in case it is not root-resistant.
- *Waterproof membrane:* An eco-roof system may consist of a liquid-applied membrane or a specially designed sheet membrane.
- *Insulation layer:* An insulation layer is optional and prevents water stored in the eco-roof system from extracting heat in the winter or cool air in the summer.

VERTICAL LANDSCAPING

Skyscrapers are not normally thought of as candidates for landscaping. However, Ken Yeang, a Malaysian architect, has been advocating what he calls *vertical landscaping* to at least in part render these very large structures green. He also advocates vertical landscaping for reducing energy consumption, stating that a 10 percent increase in vegetated area can produce annual cooling load savings of 8 percent. He describes vertical landscaping as "greening the skyscraper," which he says involves introducing plants and ecosystem components at a high level, in addition to the ground-level landscaping.[18]

The vertical landscape creates a microclimate at the façade on each floor, can be used as windbreaks, absorbs carbon dioxide and generates oxygen, and improves the well-being of the occupants by providing greenery throughout the building. This strategy also helps counterbalance the enormous mass of concrete, glass, and steel with plants and soil. In addition to these benefits, a vertical landscape that is well integrated with the building can provide architectural visual relief from otherwise uninteresting, nondescript surfaces. In order for the vertical landscaping to make visual sense, Yeang suggests that a series of stepped and linked planter boxes be designed into the building. The use of trellises also allows for vertical growth and interaction of the landscape from ground level to the roof. But because wind speeds at roof level will often be twice their ground level speed, plants at upper levels may need protection, which can be provided by side louvers that allow the landscape to be seen, yet deflect the wind from around the plants.

(A)

(B)

Figure 6.4 (A) Cross section of an extensive eco-roof structure that provides structure and drainage for eco-roofs. (B) Extensive and intensive 8,000-square-foot (743 square meters) roof garden on the Vancouver City Library, Vancouver, British Columbia. (Photographs courtesy of American Hydrotech, Inc.)

Enhancing Ecosystems

A desired outcome of any building project would be a landscape and an ecosystem that are regenerated and improved as a consequence of the project. *Environmental Building News* (EBN) provides a checklist for owners and designers to use in helping restore the vitality of natural ecosystems, (see Table 6.4).[19] Although directed primarily at enhancing the presence of wildlife on a site, it is very useful for general ecosystem restoration.

Stormwater Management

Transforming the natural environment by development dramatically affects the quantity and flows of stormwater across the surface of the Earth. Covering natural landscapes with buildings and infrastructure replaces largely pervious surfaces with impervious materials, thereby increasing the volume and velocity of horizontal water flows. Moreover, ecosystems, most prominently wetlands, that have a function of absorbing pulses of stormwater and returning it in a controlled manner to bodies of water and aquifers, are subject to modification or destruction by these same construction activities. One of the functions of green building is to address the issue of stormwater management by protecting ecosystems and the pervious character of the

TABLE 6.4

EBN Checklist for Wildlife Habitat Enhancement of Developed Land

1. **Research and Planning**
   Hire a qualified consultant specializing in natural landscaping and ecosystem restoration.
   Test soils for contaminants.
   Inventory existing ecosystems.
   Research ecosystems that may have been on the site prior to European settlement.
   Inventory current landscape management practices.
   Develop an ecosystem restoration plan.

2. **Ecosystem Restoration**
   Reduce turf area.
   Eliminate invasive plants.
   Establish native ecosystems.
   Ensure diversity in plantings.
   Provide wildlife corridors.
   Use bioengineering for erosion control.

3. **Enhancements for Wildlife**
   Select native plant species that attract wildlife.
   Encourage birds to "plant" seeds of species they like.
   Provide edible landscaping.
   Provide "edge" areas.
   Establish a bird feeding program, if desired.
   Provide bird nesting boxes and platforms.
   Provide bat houses.
   Provide water features.
   Avoid chemical usage in the landscape.

4. **Helping People Appreciate Natural Areas and Wildlife**
   Provide wildlife viewing areas.
   Provide easy and inviting access to the outdoors.
   Provide for easy management of bird feeders and nesting boxes.
   Provide clear signage in public spaces.
   Provide features that will get people outside.

TABLE 6.5

Checklist for Stormwater Management

Reduce the Amount of Stormwater Created
1. Minimize the impact area in a development.
2. Minimize directly connected impervious areas.
3. Do not install gutters unless rainwater is collected for use.
4. Reduce paved areas through cluster development and narrower streets.
5. Install porous paving where appropriate.
6. Where possible, eliminate curbs along driveways and streets.
7. Plant trees, shrubs, and groundcovers to encourage infiltration.
Keep Pollutants Out of Stormwater
8. Design and lay out communities to reduce reliance on cars.
9. Provide greens where people can exercise pets.
10. Incorporate low-maintenance landscaping.
11. Design and lay out streets to facilitate easy cleaning.
12. Control high-pollution commercial and industrial sites.
13. Label storm drains to discourage dumping of hazardous wastes into them.
Managing Stormwater Runoff at Construction Sites
14. Work only with reputable excavation contractors.
15. Minimize the impact area during construction.
16. Avoid soil compaction.
17. Stabilize disturbed areas as soon as possible.
18. Minimize slope modifications.
19. Construct temporary erosions barriers.
Permanent On-Site Facilities for Stormwater Control and Treatment
20. Rooftop water catchment systems
21. Vegetated filter strips
22. Vegetated swales for stormwater conveyance
23. Check dams for vegetated swales
24. Infiltration basins
25. Infiltration trenches
26. Dry detention ponds
27. Retention ponds
28. Constructed wetlands
29. Filtration systems

landscape, as well as to carefully consider how to affect as little as possible the natural hydroperiod of the site. EBN provides a useful checklist for dealing with stormwater issues; it is presented in Table 6.5.[20]

Heat Island Mitigation

An issue that is not normally considered in site and landscape design but that is a matter for consideration in high-performance green buildings is the *urban heat island effect.* Temperatures in cities are substantially higher than those in surrounding rural areas, usually in the range of 2°F to 10°F (1°C to 6°C) hotter (see Figure 6.6). The result is that cooling requirements for buildings in urban areas will be

Figure 6.5 Natural and constructed wetlands, such as this constructed wetland in the Blackwater River watershed in West Virginia, serve as buffers for stormwater flows, reducing or eliminating the need for stormwater infrastructure. Wetlands can also interface with the built environment to break down waste and absorb a wide variety of undesirable and even toxic materials, such as heavy metals.

higher than those in a rural setting. The additional energy required to support the higher cooling loads results in more air pollution, greater resource extraction impacts, and higher costs. Reducing or mitigating urban heat islands can counter these negative effects and result in a more pleasant urban lifestyle.

Heat islands are caused by the removal of vegetation and its replacement with asphalt and concrete roads, buildings, and other structures. The shading effect of trees and the evapotranspiration, or natural cooling effect, of vegetation are replaced by human-made structures that store and release solar energy.

In addition to their negative energy impacts, heat islands are problematic for the following reasons:[21]

- Heat islands contribute to global warming by increasing fossil fuel consumption by power plants.
- Heat islands increase ground-level ozone pollution by increasing the reaction rate between nitrogen oxides and volatile organic compounds (VOCs).
- Heat islands adversely affect human health, especially that of children and older people, by increasing temperatures and ground-level ozone levels.

Heat island effects can be reduced by several measures:

- Installing highly reflective (or high-albedo) and emissive roofs that reflect solar energy back into the atmosphere
- Planting shade trees near homes and buildings to reduce surface and ambient air temperatures
- Using light-colored construction materials where possible to reflect rather than absorb solar radiation

The USEPA launched the Urban Heat Island Pilot Project in 1998 to quantify the potential benefits of reducing heat islands. For the city of Sacramento, California, a Lawrence Berkeley National Laboratory study showed the following:

- Citywide energy bill reduction of $26.1 million per year, assuming high penetration of reduction measures

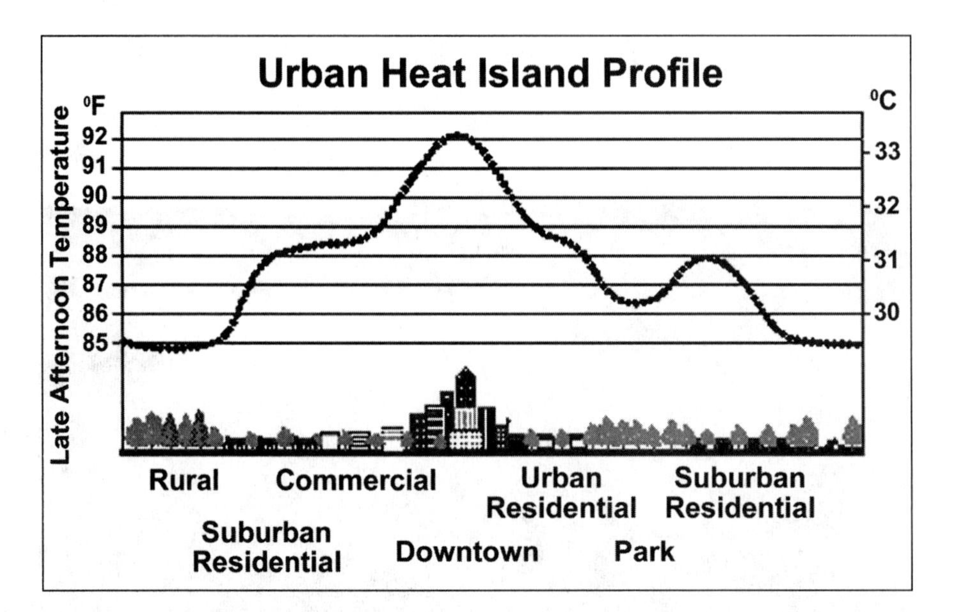

Figure 6.6 The removal of vegetation in urban areas and its replacement with buildings and infrastructure produces a *heat island effect* and results in urban temperatures being 2° F to 10°F (1°C to 5° C) higher than those in nearby rural areas. (Drawing by Bilge Çelik.)

- Savings of 468 million watts (MW) of peak power and 92,000 tons (83,600 metric tons) of carbon annually
- An improvement in air quality caused by a decrease in ozone of 10 parts per billion
- Cooling-energy savings of 46 percent and peak power savings of 20 percent by increasing roof albedo, or reflectivity, on two school buildings

The LEED-NC 2009 standard provides up to 2 points for mitigating heat islands in the Sustainable Site category. For nonroof areas, Credit 7.1 allows 1 point for creating shade or reducing heat islands for the site's impervious surfaces such as parking lots, walkways, and plazas. Sustainable Sites Credit 7.2 provides 1 point for providing a high-albedo (high-reflectivity) or vegetated roof.

Similarly, Green Globes Part B, which addresses site issues, awards points on a sliding scale, depending on how much of the project hardscape and roof area include heat island mitigation measures.

Light Trespass and Pollution Reduction

Exterior lighting systems on buildings frequently emit light that, in addition to performing their primary role of illuminating the buildings and their walkways and parking areas, illuminate areas off-site. This condition is sometimes referred to as *light trespass,* defined as unwanted light from a neighboring property. This unwanted light poses a number of problems, ranging from being a nuisance to causing safety problems when it "blinds" pedestrians and automobile drivers. Nuisance light can also negatively affect wildlife, as well as human health, because it can interrupt normal daily light cycles that are needed for the average person's well-being. For example, chicken farmers have discovered that 24-hour lighting disturbs the growth of chicks. Bright lights can affect the migration patterns of birds and baby sea turtles.

Another negative lighting condition is *light pollution,* which prevents views of the night sky by the general population and astronomers. The solution to both light trespass and pollution is proper lighting system design. The location, mounting height, and aim of exterior luminaires must all be taken into account to ensure that lighting energy is used efficiently and for its intended purposes. To prevent light pollution:

- Parking area and street lighting should be designed to minimize upward transmission of light.
- Exterior building and sign lighting should be reduced or turned off when not needed.
- Computer modeling of exterior lighting systems should be used to design exactly the level and quality of lighting needed to meet the project's needs without straying off-site and causing undesirable conditions.

The LEED-NC 2009 Sustainable Site Credit 8 addresses light pollution, awarding up to 1 point for compliance with this requirement. Similarly, Green Globes offers 7 points on its scale for preventing light pollution under Section B.2.4

Connection to LEED-NC

The structure of the Sustainable Sites category of the USGBC LEED building assessment system is addressed in this section, including an overview of the credits and requirements. For detailed information on how to document points properly for LEED certification, consult the LEED-NC 2009 Reference Manual.[23] (Note that each version of LEED has its own specific reference manual.) What follows is a detailed explanation of each credit, the required LEED Letter Template, and other miscellaneous documentation.

The Sustainable Sites (SS) category of LEED has a single prerequisite and a maximum of 26 points that can be achieved by employing measures that make the siting of the building as environmentally responsible as possible. LEED requires that all prerequisites be met before a building becomes eligible for LEED certification. Table

Figure 6.7 The exterior lighting system for Rinker Hall, a LEED-NC 2.1 Gold-certified building at the University of Florida, was designed to minimize light pollution. The result is a pleasant evening view of the building that enhances the experience of passersby. (Photograph courtesy of Gould Evans Associates and Timothy Hursley.)

6.6 lists the SS credits and points available under LEED-NC 2009. Note that the descriptions of the LEED-NC credits and points in this section are abbreviated, and the appropriate LEED Reference Manual should be consulted for detailed information about how to achieve the points.

SS PREREQUISITE 1: CONSTRUCTION ACTIVITY POLLUTION PREVENTION

The purpose of this prerequisite is to minimize the environmental impacts of erosion on the environment. The primary requirement is the design and implementation of an Erosion and Sedimentation Control (ESC) Plan that prevents soil loss via water or wind and sedimentation of stormwater infrastructure and receiving bodies of water.

SS CREDIT 1 (SSc1): SITE SELECTION (1 POINT MAXIMUM)

The selection of a site with minimal environmental or ecological system impact is a very important feature of a high-performance green building. This credit requires that buildings, roads, or parking areas on portions of sites must not be built on:

1. Prime farmland
2. Previously undeveloped land whose elevation is lower than 5 feet above the elevation of the 100-year flood
3. Land that is specifically identified as habitat for any species on federal or state threatened or endangered lists
4. Within 100 feet (30 meters) of any wetlands, or according to state or local regulations if they require greater setback distances from wetlands
5. Previously undeveloped land within 50 feet (15 meters) of a body of water (seas, lakes, rivers, streams, and tributaries that do or could support fish, recreation, or industrial use)

TABLE 6.6

Sustainable Sites (SS) Credits and Points Under LEED-NC 2009

Prerequisite/Credit	Name of Prerequisite/Credit	Maximum Points
SS Prerequisite 1	Construction Activity Pollution Prevention	NA
SS Credit 1	Site Selection	1
SS Credit 2	Development Density & Community Connectivity	1
SS Credit 3	Brownfield Redevelopment	1
SS Credit 4.1	Public Transportation Access	1
SS Credit 4.2	Bicycle Storage and Changing Rooms	1
SS Credit 4.3	Low Emitting & Fuel Efficient Vehicles	1
SS Credit 4.4	Parking Capacity	1
SS Credit 5.1	Protect or Restore Habitat	1
SS Credit 5.2	Maximize Open Space	1
SS Credit 6.1	Stormwater Design: Quantity Control	1
SS Credit 6.2	Stormwater Design: Quality Control	1
SS Credit 7.1	Heat Island Effect: Non-Roof	1
SS Credit 7.2	Heat Island Effect: Roof	1
SS Credit 8	Light Pollution Reduction	1
	Total SS Points Available	**14**

6. Land that, prior to acquisition for the project, was public parkland, unless land of equal or greater value as parkland is accepted in trade by the public landowner

The key to earning the point associated with this credit is to not use sites that include sensitive site elements and restrictive land types. A suitable building location should be selected and the building designed with a minimal footprint to minimize site disruption.

SS CREDIT 2 (SSc2): DEVELOPMENT DENSITY & COMMUNITY CONNECTIVITY (5 POINTS MAXIMUM)

Along with reusing disturbed land in preference to greenfields, it makes sense to increase the density of existing development consistent with maintaining or increasing the quality of life of the area. There are two options for earning this credit. Option 1 is to build on a previously developed site that is located within an existing minimum development density of 60,000 square feet per acre (13,800 square meters per hectare) (two-story downtown development). Option 2 is for the project to be on a previously developed site that is within half a mile of a residential zone or neighborhood with an average density of at least 10 units per acre (25 units per hectare). It must also have pedestrian access to (be within half a mile of) 10 so-called Basic Services (bank, cleaners, day-care facility, pharmacy, post office, and fitness center, to name a few).

SS CREDIT 3 (SSc3): BROWNFIELD REDEVELOPMENT (1 POINT MAXIMUM)

Land that has already been impacted by human activities is preferable for a building project to land that is a greenfield. Although brownfields are generally urban sites with access to excellent infrastructure, there are numerous issues with respect to remediating or cleaning up these properties. This is a complex and potentially costly process; hence, it applies to only a very small number of building projects. A site can be designated as a brownfield via an Environmental Site Assessment or a local Voluntary Cleanup Program, or by a federal, state, or local government agency.

SS CREDIT 4 (SSc4): ALTERNATIVE TRANSPORTATION (12 POINTS MAXIMUM)

The overall purpose of this credit is to reduce dependence on conventional fossil fuel–powered automobiles. SS Credit 4 actually consists of four different subcredits, each with a maximum of 1 point, as follows:

SS Credit 4.1: Public Transportation Access

SS Credit 4.2: Bicycle Changing and Storage Rooms

SS Credit 4.3: Low Emitting and Fuel Efficient Vehicles

SS Credit 4.4: Parking Capacity

The requirements for each of these subcredits are described below.

SS Credit 4.1 (SSc4.1): Public Transportation Access (6 Points Maximum)

For a building to be truly green, it should be in a location where there is ready access to mass transportation. For the purpose of LEED-NC, this credit requires the building project to be within one-half of a mile walking distance (0.8 kilometer) of a commuter rail, light rail, or subway station, or within one-fourth of a mile walking distance (0.4 kilometer) of two or more public or campus bus lines usable by building occupants.

SS Credit 4.2 (SSc4.2): Bicycle Storage and Changing Rooms (1 Point Maximum)

Another aspect of green buildings that helps reduce the impacts of their operations is to facilitate the use of bicycles by the occupants. To earn this credit, the building should provide secure bicycle storage with convenient changing/shower facilities [within 200 yards (183 meters) of the building] for 5 percent or more of regular building occupants. For residential buildings, in lieu of changing/shower facilities, covered storage facilities for securing bicycles for 15 percent or more of building occupants must be provided.

SS Credit 4.3 (SSc4.3): Low Emitting and Fuel Efficient Vehicles (3 Points Maximum)

Another approach to reducing the impacts associated with traveling to and from the building is to facilitate the use of alternative-fuel vehicles. This credit can be achieved by providing alternative-fuel vehicles for 3 percent of building occupants *and* preferred parking for these vehicles, by providing preferred or discounted parking for low-emitting and fuel-efficient vehicles for 5 percent of the parking capacity of the site by installing an alternative fuel refueling station, or by providing access to a fuel efficient vehicle sharing program.

SS Credit 4.4 (SSc4.4): Parking Capacity (2 Points Maximum)

This credit emphasizes the reduction of parking capacity for automobiles to the bare minimum needed to meet local zoning requirements. To earn the point associated with this credit, for nonresidential projects the parking capacity must be sized to meet, but not exceed, minimum local zoning requirements *and* provide preferred or discounted parking for carpools or vanpools capable of serving 5 percent of the building occupants; or add no new parking for rehabilitation projects *and* provide preferred parking for carpools or vanpools capable of serving 5 percent of the building occupants. For residential projects, the parking capacity must be sized to meet, but not exceed, minimum local zoning requirements and facilitate shared vehicle usage through carpool dropoff areas, designated parking for vanpools, car-share services, ride boards, and shuttle service to mass transit.

SS CREDIT 5 (SSc5): SITE DEVELOPMENT (2 POINTS MAXIMUM)

Site clearing, earthwork, compaction, temporary roads and structures, and other construction operations involving earth movement and construction can have significant environmental impact. This credit has two subcredits, each offering 1 point, for measures that reduce disturbance to the site during construction:

SSc5.1: Protect or Restore Habitat

SSc5.2: Maximize Open Space

The requirements for each of these subcredits are described below.

SS Credit 5.1 (SSc5.1): Protect or Restore Habitat (1 point)

The idea behind this credit is to minimize the impacts of the construction process on natural systems by requiring minimal site disturbance during construction. Option 1 for this credit is that for greenfield sites, site disturbance, including earthwork and clearing of vegetation, must be limited to 40 feet beyond the building's perimeter; 10 feet beyond surface walkways, patios, surface parking, and utilities less than 12 inches in diameter; 15 feet beyond primary roadway curbs, walkways, and main utility branch trenches; and 25 feet beyond constructed areas with permeable surfaces (such as pervious paving areas, stormwater detention facilities, and playing fields) that require additional staging areas in order to limit compaction in the constructed area; or, on previously developed sites, a minimum of 50 percent of the site area (excluding the building footprint) must be restored by replacing impervious surfaces with native or adaptive vegetation.

SS Credit 5.2 (SSc5.2): Maximize Open Space (1 point)

This credit emphasizes the inclusion of requirements for conserving open space and restoring damaged areas to productive ecosystems. Option 1: The point associated with this credit can be achieved by reducing the development footprint (defined as entire building footprint, access roads, and parking) to exceed the local zoning's open space requirement for the site by 25 percent. Option 2: For areas with no local zoning requirements (e.g., some university campuses and military bases), open space area adjacent to the building that is equal to the development footprint must be designated. Option 3: If a zoning ordinance exists but there is no requirement for open space, vegetated open space must equal at least 20 percent of the project's site area. Vegetated roofs can contribute to all three options for projects that earn SSc2 (Development Density & Community Connectivity), as can wetlands if the side slope gradients are an average of 1:4 (vertical to horizontal) or less and are vegetated. Pedestrian-oriented hardscapes for projects that can earn SSc2 credit can contribute to the open space requirement as long as a minimum of 25 percent of the open space counted is vegetated.

SS CREDIT 6 (SSc6): STORMWATER DESIGN (2 POINTS MAXIMUM)

Stormwater management is required largely because of the significant reduction in pervious surfaces caused by buildings and their associated parking and paving. This credit comprises two subcredits:

SS Credit 6.1 (SSc6.1): Quantity Control

SS Credit 6.2 (SSc6.2): Quality Control

The requirements for each of these subcredits are described below.

SS Credit 6.1 (SSc6.1): Quantity Control (1 Point Maximum)

The goal of this credit is to ensure that the imperviousness of the building site does not increase. In cases where there is significant imperviousness, it should be decreased. This can be accomplished by increasing the area of pervious pavement, by using vegetative roofs or eco-roofs, and by other measures that increase the infiltration of water back into the soil. Another approach is to capture stormwater and use it for nonpotable water purposes such as flushing of sanitary fixtures and landscape irrigation. This credit requires that if existing imperviousness is less than or equal to 50 percent, a stormwater management plan must be implemented that prevents the postdevelopment 1.5-year, 24-hour peak discharge rate from exceeding the predevelopment 1.5-year, 24-hour peak discharge rate. Or, if existing imperviousness is greater than 50 percent, a stormwater management plan must be implemented that results in a 25 percent decrease in the rate and quantity of stormwater runoff.

SS Credit 6.2 (SSc6.2): Quality Control (1 Point Maximum)

Treating stormwater by using simple approaches and natural systems reduces infrastructure and energy for moving and treating large volumes of water. Mechanical or natural treatment systems such as constructed wetlands, vegetated filter strips, grass swales, bioswales, detention ponds, and filtration basins can be designed to collect and treat the site's stormwater. This credit requires these systems be designed to remove 80 percent of the average annual postdevelopment total suspended solids (TSS) and 40 percent of the average annual postdevelopment total phosphorus (TP) based on the average annual loadings from all storms less than or equal to the 2-year/24-hour storm.

SS CREDIT 7 (SSc7): REDUCING HEAT ISLAND EFFECTS (2 POINTS MAXIMUM)

The air temperature in urban areas can be 2° to 10°F (1°C to 6°C) higher than in the surrounding countryside, a consequence of solar energy absorption and reradiation by components of the built environment, particularly dark, nonreflective surfaces used for paving and roofing. This increase in air temperature means that significantly more energy is needed for cooling and even that distinct microclimates are created in the affected areas. Reducing the heat island effect can markedly reduce summertime energy use. This credit comprises two subcredits, each carrying a maximum of 1 point:

SS Credit 7.1 (SSc7.1): Non-Roof Heat Islands

SS Credit 7.2 (SSc7.2): Roof Heat Islands

The requirements for each of these subcredits are described next, together with the documentation requirements for each.

SS Credit 7.1 (SSc7.1) Heat Island Effect: Non-Roof (1 Point Maximum)

The heat island effects of nonroof surfaces can be reduced by providing shade or using light-colored (high-albedo) materials for parking and paving. Using open grid pavement is another appropriate option for reducing thermal energy buildup, as is locating parking structures underground. Two other heat island reduction strategies are using trees and other vegetation to shade structures and using architectural shading devices where planting vegetation is not feasible. To earn the point associated with this credit, at least 50 percent of the site hardscape must be shaded within 5 years of occupancy, paving materials must have a Solar Reflectance Index (SRI) of at least 29, or an open grid pavement system can be used. Optionally, 50 percent of parking spaces can be put undercover (for example in underground parking), but the roof must have an SRI of at least 29.

SS Credit 7.2 (SSc7.2) Heat Island Effect: Roof (1 point maximum)

Using eco-roofs, vegetative roofs, or light-colored (high-albedo), highly reflective, Energy Star–compliant roofs can greatly reduce the heat island effect associated with this building component. Roofing materials must have an SRI equal to or greater than 78 (for a roof slope ≤ 2:12) or 29 (for a roof slope ≥ 2:12). Optionally, a vegetated roof covering at least 50 percent of the roof area can satisfy the requirements for this credit. A combination of high-albedo roof and vegetated roof can also meet the requirements for this credit.

SS CREDIT 8 (SSc8): LIGHT POLLUTION REDUCTION (1 POINT MAXIMUM)

Light pollution is a complex problem that can be caused by both exterior and interior lighting. It can be addressed by adopting site lighting criteria to maintain safe light levels while avoiding off-site lighting and night-sky pollution. Site lighting should be minimized where possible and should be designed using a computer model. Technologies to reduce light pollution include full cutoff luminaires, low-reflectance surfaces, and low-angle spotlights. Exterior illumination should not exceed 80 percent

TABLE 6.7

Outline of Green Globes v.1, Part B (Site)

Section	Description	Points
B.1	**Site Development Area**	**45**
B.1.1	Site location	30
B.1.2	Avoiding sensitive locations	15
B.2	**Reduce Ecological Impacts**	**40**
B.2.1	Leaving slopes greater than 15% undisturbed	2
B.2.2	Limiting construction site activities	2
B.2.3	Protection of trees	1
B.2.4	Erosion control Best Management Practices	8
B.2.5	Heat island mitigation, site hardscape	10
B.2.6	Heat island mitigation, roof	10
B.2.7	Light trespass and skyglow minimized	7
B.3	**Enhancement of Watershed Features**	**15**
B.3.1	Controlling stormwater from damaging the project or waterways	10
B.3.2	Roof runoff control	5
B.4	**Site Ecology Improvement**	**15**
B.4.1	Native landscape vegetation	6
B.4.2	Minimizing turf	5
B.4.3	Design to avoid bird collisions	4
	Total Points Available	**115**

of the lighting power densities for exterior areas and 50 percent for building façades and landscape features as defined in ASHRAE/IESNA Standard 90.1-2004. Interior lighting must be designed such that the angle of maximum candela from interior luminaires does not exit the windows.

Connection to Green Globes v.1

Part B of the Green Globes v.1 building assessment standard addresses site issues and carries a maximum of 115 points. The structure of Part B and the issues covered in it are outlined in Table 6.7. Unlike the LEED-NC approach, Green Globes uses a questionnaire to ask if certain actions have been taken. Consequently, obtaining the number of points allocated to an issue requires the team to develop documentation for review by the Verifier during the Postconstruction Assessment. Note that there are no prerequisites in Green Globes, and situations that do not apply are not included in the maximum points totals.

Summary and Conclusions

The most exciting and underutilized resources for creating high-performance green buildings are natural systems, and they should be employed as more than superficial components of the project. The ultimate green building will undoubtedly feature a much deeper integration of ecosystems with buildings, and exchanges of matter-energy between human systems and natural systems, in ways that are beneficial to both. The need to dramatically reduce building and infrastructure energy consumption will motivate designers to better understand the processing of waste by natural or constructed wetlands, which contribute to their sustainability and to that of the human systems with which they cooperate. Natural systems can shade and cool buildings, yet allow sunlight through for heating during appropriate seasons. They can also provide calories and nutrition and may be able to take up large quantities of stormwater, thus allowing the downsizing of conventional stormwater handling systems.

The high-level integration of ecosystems and the built environment is, at present, only a concept. But a future of high energy costs will inevitably force changes that decentralize many of the waste-processing functions, which currently are performed at distant wastewater treatment plants to which building waste must be pumped, often through miles of piping, with motive energy provided by a series of lift stations. By integrating buildings with ecosystems, an alternative framework can be designed to ensure a future with a low energy profile. Though today's green building designers make only a minimal effort to use natural systems for anything other than amenities, in the future they will have a much more detailed knowledge of ecology and ecological systems, enabling them to successfully weave nature into the built environment.

Notes

1. From France (2003). The author provides an insightful analysis of how landscape architecture must change to participate in ecological design. He points out the possibility of landscape as "functional art," most prominently in the form of wetlands that, in addition to being pleasing to the human eye, provide numerous services such as stormwater uptake and wastewater processing. He adds that the shift to multifunctional wetlands is a success story for sustainable landscape architecture.
2. As defined by the U.S. Department of Agriculture and listed on the website of the American Farmland Trust, www.farmland.org.
3. Excerpted from "Farming on the Edge" (1997).
4. The USEPA brownfields website is www.epa.gov/brownfields.
5. The Chicago Brownfields Initiative is a partnership of private and public sector institutions that advocates and assists in the conversion of formerly contaminated industrial zones to productive use. The Chicago Department of the Environment hosts the website for this initiative at www.ci.chi.il.us/Environment/Brownfields/index.html.
6. The Congress of New Urbanism website is www.cnu.org.
7. Data on Wal-Mart stores is from the Wal-Mart Realty Company website, www.wal-martrealty.com.
8. Excerpted from an excellent article on the issue of grayfields, "Grayfields and Ghostboxes" (2003).
9. The Eastern Pennsylvania Coalition for Abandoned Mine Reclamation (EPCAMR) has an excellent website (www.orangewaternetwork.org) that describes the extent of the problem with blackfields or abandoned mine properties.
10. Detailed information about the NFIP, SFHA, and flood mapping can be found at the FEMA website, www.fema.gov/plan/prevent/fhm/fq_term.shtm.
11. As defined in Thayer (1989). The *Landscape Journal*'s website is www.wisc.edu/wisconsinpress/journals/journals/lj.html.
12. According to Robert Thayer, horizontal energy is low-intensity, widely dispersed, renewable energy in the form of sunlight, wind, water moving by tides or gravity, and energy fixed by plants. Horizontal energy is limited by its location and the rate of its natural generation, and landscape must exist within the limits of its availability.
13. From Dewey (1916).
14. Excerpted from Thompson and Sorvig (2000).
15. Portland's eco-roof program is described at the city's Bureau of Environmental Services website, www.portlandonline.com/shared/cfm/image.cfm?id=53987.
16. Excerpted from Flaccus (2002).
17. As described on the website of Living Systems Design Group, LLC, www.lsdg.net.

18. From Yeang (1996).
19. The "Checklist for Wildlife Habitat Enhancement of Developed Land" is excerpted from the February 2001 issue of EBN, pp. 8–12. The original checklist provides a detailed description of each of the points in the table. EBN is a publication of Building Green, Inc., www.buildinggreen.com.
20. The "Checklist for Stormwater Management Practices" is excerpted from the September/October 1994 issue of EBN, p. 1 and pp. 8–13. The original checklist provides a detailed description of each of the points in the table.
21. From the USEPA Heat Island Effect website, http://yosemite.epa.gov/oar/globalwarming.nsf/content/ActionsLocalHeatIslandEffect.html.
22. Lawrence Berkeley National Laboratory has a website devoted to heat island issues, eetd.lbl.gov/HeatIsland/.
23. Each LEED product has or will have a Reference Manual that describes in detail the technical and documentation requirements for each credit. The Reference Manual for LEED-NC Version 2.2 is in its second edition (November 2006) and is published by the USGBC in Washington, D.C.

References

Campbell, Craig, and Michael Ogden. 1999. *Constructed Wetlands in the Sustainable Landscape.* New York: John Wiley & Sons.

Dewey, John. 1916. *Democracy and Education.* New York: Free Press.

"Farming on the Edge: Sprawling Development Threatens America's Best Farmland." 1997. American Farmland Trust. Available as an Adobe Acrobat file from the American Farmland Trust website, www.farmland.org.

Flaccus, Gillian. 2002. "Portland at Forefront of Eco-Friendly Roof Trend." Available online at www.evesgarden.org/archives/2002/11/17/portland_at_forefront_of_ecofriendly_roof_trend.

France, Robert. 2003. "Grey World, Green Heart?" *Harvard Design Magazine,* no. 18, pp. 30–36.

"Grayfields and Ghostboxes: Evolving Real Estate Challenges. May 2003. *Let's Talk Business,* Issue 81. Available online at www.uwex.edu/ces/cced/lets/0503ltb.html.

LEED-NC 2.2 Reference Manual, 2d Edition 2006. Published by the USGBC in Washington, D.C. The Reference Manual can be ordered online at www.usgbc.org. LEED 2009 reference manual not yet available at time of publication.

Lyle, John T. 1985. *Design for Human Ecosystems: Landscape, Land Use, and Natural Resources.* Washington, DC: Island Press.

———. 1994. *Regenerative Design for Sustainable Development.* New York: Wiley & Sons.

Thayer, Robert. 1989. "The Experience of Sustainable Design," *Landscape Journal,* 8, pp. 101–110.

———. 1994. *Gray World, Green Heart: Technology, Nature, and the Sustainable Landscape.* New York: John Wiley & Sons.

Thompson, Robert, and Kim Sorvig. 2000. *Sustainable Landscape Construction: A Guide to Green Building Outdoors.* Washington, DC: Island Press.

Yeang, Ken. 1996. *The Skyscraper Bioclimatically Considered.* London: Academy Editions.

Chapter 7

Energy and Atmosphere

Creating a low energy profile is a major challenge for designers of high-performance green buildings. The environmental impacts of extracting and consuming nonrenewable energy resources such as fossil fuels and nuclear energy are profound. Pronounced land impacts from coal and uranium mining, acid rain, nitrous oxides, particulates, radiation, ash disposal problems, and long-term storage of nuclear waste are just some of the consequences of energy consumption by the built environment. Building energy consumption in the United States is at about the same scale as energy consumption by automobiles, with about 40 percent of primary energy being consumed by buildings and about the same quantity by transportation.[1] In fact, much automotive energy consumption is caused by the placement of buildings on the landscape in the planning process.

Toward the end of the current decade, humankind will face the aforementioned rollover point in petroleum production,[2] the point at which the extraction rate is predicted to peak, and considerable additional energy and financial resources will be needed to extract the remaining oil resources. At the same time, economies around the world continue to grow, all of them dependent on abundant, cheap energy, none of them more so than the United States. H.T. Odum, the eminent ecologist who founded the branch of ecology known as *systems ecology,* forecasted that, at the rollover point, the energy required to extract the oil would be greater than its energy value.[3] Sounding another warning note, Odum and his colleagues calculated that some key technologies suggested as substitutes for a predominantly fossil fuel–powered energy system, among them photovoltaics and fuel cells, require more energy to produce than they themselves will ever generate. The point is, the technological optimists who believe that a technical solution will always be found to solve our energy, water, or materials problems have no truly viable substitute for fossil fuel–derived energy. For the built environment, truly dramatic reductions in building energy consumption, accompanied by tremendous progress in passive design, will be needed to meet a potentially costly energy future.

As we approach this day of reckoning, when energy costs are likely to rise dramatically as a result of fierce international demand and competition, we still have time to make some very important decisions with respect to how we live and the types of buildings we create. The green building movement and allied efforts to improve building energy performance are attempting to influence a major shift in the way buildings are designed. It is a fundamental transformation that must take place, one that does not just reduce energy consumption by a small percentage but that involves a total rethinking of building design. Advocates of just such a radical change believe that buildings should be *energy-neutral* or even net *exporters* of energy. Advancing the use of solar energy, ground coupling, radiant cooling, and other radical approaches may indeed enable buildings to generate at least as much energy as they consume. In the interim, however, we must learn how to cut building energy use by a marked quantity, perhaps as by much as 90 percent—a daunting challenge, to be sure.

Figure 7.1 Rinker Hall is a high-performance LEED-certified Gold building at the University of Florida that incorporates advanced energy strategies. (Photograph: M.R. Moretti.)

Building Energy Issues

Energy consumption remains the single most important green building issue, not only because of its environmental impacts but also because of the probability of significantly higher future energy costs. In 2002, 80 million buildings in the United States consumed 33 quads (1 quad equals 1 quadrillion BTUs, or 1 million billion BTUs), about 36 percent of the country's primary energy. Lighting consumed 31 percent of commercial building energy, heating accounted for 22 percent, and space cooling for 18 percent. Impacts from electrical power generation include global climate change, acid rain, ground-level ozone creation, and a wide range of health effects caused by the emission of particulates. Energy consumption by buildings in 2002 contributed to:[4]

- 47 percent of U.S. sulfur dioxide emissions
- 22 percent of nitrogen oxide emissions
- 35 percent of carbon dioxide emissions

In the future, this situation will only worsen, as 18.4 million new homes and 21.5 billion square feet of new commercial buildings are forecasted to be built between 2002 and 2010. An ambitious federal government program, Buildings for the 21st Century, proposes to reduce energy consumption by 50 percent in new homes and commercial buildings and by 20 percent in existing homes and buildings by 2010. It calls for reductions of another 50 percent and 20 percent for new and existing homes and buildings, respectively, by 2020.[5] To accomplish this feat will require technology breakthroughs, better building design tools and simulations, improved construction techniques, and a wholesale shift in the attitude of building owners with respect to investing in high-performance green buildings at the front end to produce greatly reduced operating costs over the life of a building.

One general school of thought with respect to sustainability is that energy consumption worldwide should be reduced by a factor of 10, a concept that has taken root in the development of energy policy in the European Community. (See Chapter 2 for a discussion of Factor 4 and Factor 10.) A nominal U.S. commercial office building consumes on the order of 100,000 BTUs per square foot per year (292 kWh per square meter/year). Today's U.S. green buildings typically reduce this energy requirement to less than 50,000 BTUs per square foot per year (146 kWh per square meter per year). But a Factor 10 building would use just 10,000 BTUs per square foot annually (292 kWh per square meter/year). Based on today's assumptions about comfort and aesthetics, it is challenging to create a truly energy-responsible building in the spirit of Factor 10.[6]

Recent programs in Germany indicate that buildings can be designed to use far lower levels of energy than even the most ambitious U.S. high-performance buildings. As part of a 10-year demonstration program that ended in 2005, a group of 23 office buildings throughout Germany were designed, built, and monitored with a goal of using 34,000 BTUs per square foot (100 kWh per square meter) of primary energy annually. Primary energy is the source energy for the energy delivered to the building; consequently, the electrical energy used in the building was multiplied by a factor of 3 to account for the primary energy. In comparison, as noted above, U.S. office buildings consume on the order of 100,000 BTUs per square foot (262 kWh per square meter) per year of purchased energy. For a U.S. all-electric building, this would equate to 300,000 BTUs per square foot (780 kWh per square meter) of primary energy each year. For a building that derives 50 percent of its energy from electricity and the remainder from natural gas, the primary energy would be about 200,000 BTUs per square foot (520 kWh per square meter) annually. A U.S. energy-efficient, high-performance building would have to use one-fifth to one-seventh of the energy of a conventional U.S. building to match today's best practices in Germany.[7] Best U.S. practices result in the typical office building using at least twice the primary energy of a German building, pointing to a need for dramatic changes in the way buildings are designed in the United States.

A green building would ideally use very little energy, and renewable energy would be the source of most of the energy needed to heat, cool, and ventilate it. Today's green buildings include a wide range of innovations that are starting to change the energy profile of typical buildings. Many organizations are committed to investing in innovative strategies to help create buildings with Factor 10 performance, notably the U.S. federal government, which has been the leader in requiring LCC analysis as the basis for decision making with respect to building procurement. Some state governments have followed suit, notably those of Pennsylvania, New York, and California; in contrast, others, such as Florida, have passed legislation requiring decisions based solely on the capital or first cost of a particular strategy. This latter, shortsighted approach will result in enormous expenditures of energy as we approach the rollover point.

Green building advocates often note that the strategies used to heat, cool, ventilate, and light some high-performance buildings allow a significant downsizing of the mechanical plant and a parallel reduction in the overall capital costs of the building. This is clearly the ideal outcome, wherein both capital and operating costs are lower than those of a comparable base-case building. However, there are very few of these buildings in typical U.S. climactic zones for a variety of reasons, including building code constraints. LCC analysis of a building's performance is key for giving designers the creative freedom to optimize a given building's energy consumption.

High-Performance Building Energy Design Strategy

The basic steps in designing an energy-efficient building are as follows:

1. Use building simulation tools to assist designers in minimizing energy consumption.

2. Optimize the passive solar design of the building.
3. Maximize the thermal performance of the building envelope.
4. Minimize internal building loads.
5. Design an efficient HVAC system that minimizes energy use.
6. Incorporate renewable energy use to the maximum extent possible.
7. Harvest waste energy through combined heat and power (CHP) systems, cogeneration, ventilation/exhaust air energy recovery, and other means.
8. Incorporate innovative emerging strategies where appropriate—for example, ground coupling and radiant cooling.

The design of an energy-efficient building is a complex undertaking, and these steps cannot just be performed in a single sequence; they are, in effect, part of an iterative process that starts with passive design. Trade-offs inevitably must be made between these steps, including the client's requirements and budget. Designed properly, an energy-efficient building should provide greatly reduced operational costs for minimal or no increase in capital costs. In some cases, an excellent passive design and building envelope strategy can markedly reduce the costs of HVAC equipment due to the reduction in load that can occur.

GOAL SETTING FOR BUILDING ENERGY DESIGN

The design of the energy strategy for a high-performance building should involve an examination of energy targets for the building based on a combination of reviewing the performance of similar conventional buildings, an understanding of contemporary high-performance building best practices, and building energy simulations. The two major building assessment systems, LEED-NC and Green Globes, take distinctly different approaches to energy goal setting. LEED-NC relies on Illumination Engineering Society of North America/American Society of Heating Refrigeration and Airconditioning Engineers (IESNA/ASHRAE) Standard 90.1-2007 (ASHRAE 90.1-2007) for directions on how to establish the baseline building and how to compare it to the designed building. The baseline building is generally thought of as a building designed to minimal building code requirements, with no special effort made to achieve energy efficiency. Green Globes relies on the USEPA Target Finder to determine the baseline for comparison. Both of these approaches are described in more detail below.

Energy Goal Setting in LEED-NC

Prerequisite 1 (EAp1) and the Energy & Atmosphere Credit 1 (EAc1) of LEED-NC both rely on ASHRAE 90.1-2007 to provide a standard set of instructions that dictate how the Baseline Design is to be defined and how the design of the high-performance building, referred to as the Proposed Design, is to be compared to the Baseline Design. It should be noted that there are two other Options for obtaining points under EAc1; these are briefly described at the end of this chapter. Appendix G of the Standard describes the Performance Rating Method, which is a modification of the Energy Cost Budget Method. The Baseline Design is simulated for each of four orientations, with specified opaque assemblies, limits on vertical fenestration, and HVAC systems as defined in Appendix G. This approach uses energy cost as the basis for determining savings, with the cost of energy based on actual local utility rates or on state average prices published by the U.S. Department of Energy's Energy Information Administration (EIA).[8]

Energy Goal Setting in Green Globes

Green Globes uses a significantly different approach to setting targets for energy performance. Instead of developing an arbitrary baseline model based on ASHRAE

90.1-2007, Green Globes states that the building must surpass the 75 percent target as evaluated by the USEPA Target Finder, meaning that the building has to be in the top 25 percent of buildings of that type, in the specific area, as contained in the Target Finder database. The 75 percent target also designates the building as meeting Energy Star standards. This target is the threshold performance for earning points in Green Globes, earning the minimum 10 points out of the maximum 100 points that can be awarded for energy consumption. The maximum number of points is achieved for buildings in the 96th percentile or higher. The advantage of this approach is that the target is based on actual buildings and the designed building is compared to like structures in the immediate area. The drawback of Target Finder is that there is a limited range of building types listed in the database. Target Finder does have the capability of taking mixed-use buildings into account; for example, a building combining office and residential space can be analyzed to determine the appropriate target.[9]

BUILDING ENERGY SIMULATION AND DAYLIGHTING SIMULATION

Building energy simulation is an important tool in the design of a high-performance building. Contemporary building energy simulation tools allow the building to be modeled in great physical detail and to be operated on an hourly basis in a given configuration for an entire year. It is important to employ building energy simulation at a very early stage in the design process, when decisions about building shape, number of stories, and orientation are being made. Today's simulation tools allow the integration of active and passive building systems and can easily examine the interplay and trade-offs among heating and cooling systems, walls and roof choices, insulation, lighting, windows and doors, exterior and interior shading, and skylights. Perhaps the best-known whole building energy simulation tool is DOE 2.2, which now has a user-friendly interface and wizards to speed the energy simulation process.[10] Daylighting is a key component of an energy-efficient building, and performing simulations that optimize daylighting is important to understand the trade-offs between fenestration, envelope thermal resistance, and energy use for artificial lighting. Some building energy simulation tools, such as Energy-10™, allow the integrated evaluation of daylighting, passive solar heating, low-energy heating and cooling strategies, and envelope design.[11] Daylighting can also be evaluated with sophisticated software such as Radiance, developed by the Lawrence Berkeley National Laboratory. Radiance contains libraries of materials, glazings, electric lighting luminaires, and furniture to facilitate the daylighting analysis.[12] The simulation provides a quantitative check on the intuitive guesswork of the design team about the interrelationship of the building systems. Typical tools for whole-building energy simulation include E-Quest, DOE 2.2, and Energy-10™.

In a recent study of how well energy modeling represents the actual performance of buildings, Lawrence Berkeley National Laboratory conducted a study of 21 buildings certified under LEED 2.0 or LEED 2.1, with about half located in the Pacific Northwest and the others from areas throughout the United States. Part of the study separated federal from nonfederal buildings and considered only nonlaboratory buildings. The results of this study are shown in Table 7.1.[13]

The results of this study indicate a wide range of results when comparing modeling to actual performance. In some cases the modeling is quite accurate, while in others it tends to be far off. The real value of the modeling is to find the relative importance of changes to the building's envelope and energy systems; providing an accurate prediction of building energy performance is less important. The modeling of plug loads (computers, printers, fax machines, copiers, appliances) is notoriously inaccurate because the behavior of the building users is unpredictable. Actual plug loads are often substantially higher than those simulated in the energy model. Addi-

TABLE 7.1

Comparison of Energy Models of LEED Buildings and Their Associated Base Cases Compared to Actual Energy Consumption (For this study, nine federal and eight nonfederal, nonlaboratory buildings were compared.)

Building Type	Modeled Base Case*	Modeled LEED Case*	Savings of LEED Case Based on Modeling	Actual Energy Use*	Actual Energy Compared to Modeled†
Federal	131	117	21%	81	−30%
Nonfederal	105	61	42%	57	−7%

*Thousands of BTUs per square foot per year.
†A negative number indicates that the actual energy consumption was less than the modeled energy consumption.

tionally, with the continual addition of new electrically powered devices in office buildings, plug loads tend to increase over time. The issues of plug loads and techniques for reducing them are addressed later in this chapter in the subsection "Plug Load Reduction."

VERIFYING BUILDING ENERGY PERFORMANCE

The International Performance Measurement and Verification Protocol (IPMVP) provides an overview of current best practices for verifying energy efficiency, water efficiency, and renewable energy performance for commercial and industrial facilities. It may also be used by facility operators to assess and improve facility performance. Energy conservation measures (ECMs) covered in the protocol include fuel-saving measures, water efficiency measures, load shifting and energy reductions through installation or retrofit of equipment, and/or modification of operating procedures. The IPMVP is maintained with the sponsorship of the U.S. Department of Energy by a broad international coalition of facility owners/operators, financiers, energy services companies (ESCOs), and other stakeholders.

The IPMVP was first published in 1996 and contained methodologies that were compiled by a technical committee including hundreds of industry experts, initially from the United States, Canada, and Mexico. In 1996 and 1997, 20 national organizations from a dozen countries worked together to revise, extend, and publish a new version of the IPMVP in December 1997. This second version has been widely adopted internationally, and has become the standard Measurement and Verification (M&V) document in countries ranging from Brazil to Romania. Volume III of the IPMVP applies to New Construction, and its purpose is to provide a description of best practices for verifying the energy performance of new construction.[14]

The IPMVP requires the user to develop an M&V plan that includes defining the ECMs employed in the building; identifying the boundary conditions for measurement; establishing base year data; defining conditions to which all data will be adjusted for comparison; and meeting a range of other requirements that establish a standardized method for comparing information. The LEED-NC 2009 points that can be earned for M&V requires that the IPMVP be used for measuring both energy and water consumption data.

Passive Design Strategy

Due to the complexity of designing the energy systems for a high-performance green building, the starting point must be full consideration of *passive solar design*, or *passive design.* Passive design is the design of the building's heating, cooling, lighting, and ventilation systems, relying on sunlight, wind, vegetation, and other naturally occurring resources on the building site. Passive design includes the use of all possible measures to reduce energy consumption prior to the consideration of any external energy source other than the sun and wind. Thus, it defines the energy character of the building prior to the consideration of active or powered systems (chillers, boilers, air handlers, pumps, and other powered equipment). Randy Croxton, one of the pioneers of contemporary ecological design, describes a good passive design as one that allows a building to "default to nature." A building that has been well designed in a passive sense could be disconnected from its active energy sources and still be reasonably functional due to daylighting, adequate passive heating and cooling, and ventilation being provided by the chimney effect, cross-ventilation, operable windows, and the prevailing winds. A successful passive design scheme creates a truly climate-responsive, energy-conserving building offering a wide range of benefits.

Passive design has two major aspects: (1) the use of the building's location and site to reduce the building's energy profile and (2) the design of the building itself—its orientation, aspect ratio, massing, fenestration, ventilation paths, and other measures. Passive design is complex, as it depends on many factors, including latitude, altitude, solar insolation,[15] heating and cooling degree days,[16] humidity patterns, annual wind strength and direction, the presence of trees and vegetation, and the presence of other buildings. An optimized passive design can greatly reduce the energy costs of heating, cooling, ventilation, and lighting.

Some of the factors that should be included in the development of a passive design strategy are:

- *Local climate:* Sun angles and solar insolation, wind velocity and direction, air temperature, and humidity throughout the year
- *Site conditions:* Terrain, vegetation, soil conditions, water table, microclimate, relationship of other buildings
- *Building aspect ratio:* Ratio of the building's length to its width
- *Building orientation:* Long axis oriented east-west, room layout, glazing
- *Building massing:* Energy storage potential of materials, fenestration, color
- *Building use:* Occupancy schedule and use profile
- *Daylighting strategy:* Fenestration, daylighting devices (light shelves, skylights, internal and external louvers)
- *Building envelope:* Geometry, insulation, fenestration, doors, air leakage, ventilation, shading, thermal mass, color
- *Internal loads:* lighting, equipment, appliances, people
- *Ventilation strategy:* Cross-ventilation potential, paths for routine ventilation, chimney effect potential

Like any concept, passive design can be improperly applied to building design. Its success is highly dependent on the wide range of factors just listed, and its application differs widely from New York to California, Colorado, or Florida. For example, using thermal mass as a passive design strategy, an excellent choice in the high desert altitudes found in New Mexico, with its abundant sunlight and wide daily temperature swings, would not be an appropriate choice in a hot, humid climate with generally narrow daily temperature differences, as would be found in Tampa, Florida.

The optimum building orientation, the location and types of windows, the use of daylighting, and many other decisions must be based on a careful examination of the situation found in each locale.

SHAPE, ORIENTATION, AND MASSING

The classic passive design approach to orienting a building on its site is to locate the long side on a true east-west axis to minimize solar loads on the east and west surfaces, particularly during the summer. The *aspect ratio* is the ratio of a building's length to its width, which is an indicator of the general shape of a building. Passive design dictates that a building in the northern United States should have an aspect ratio of close to 1.0; that is, it should be virtually square in shape. For buildings in the warmer southerly latitudes, the aspect ratio increases, with the building becoming longer and narrower. The reasoning behind this shift in aspect ratio is that a square building will have the minimum skin surface area compared to its volume. It is important in colder climates to minimize the surface area through which heat can be transmitted. Temperature differentials for heating are generally much greater than for cooling; thus, the total skin area of the building is more important in heating situations. The long, narrow building favored by passive design experts for warmer climates minimizes the relative exposure of east and west surfaces that experience the greatest sun load. Windows on east and west surfaces are typically minimized to eliminate as much as possible the potential high morning and afternoon solar loads. South-facing walls will experience a variable sun load during the day, and windows are easily protected from solar loads through the use of roof overhangs, shading devices, or recessed windows.

Thermal mass is an important aspect of passive design. In cases where passive solar heating is desired, the geometry of the building should be arranged to allow materials with high heat capacity and significant mass to store solar energy during the day. Materials such as brick, concrete masonry, concrete, and adobe, used for floors and walls, can absorb solar energy during the day and release it in the evening, when internal temperatures begin to drop. For passive solar cooling, buildings in climates such as that of Florida should have minimal mass for storing energy, and should generally be lightweight and well insulated. Preventing solar energy transmission into the structure is the desired strategy for passive cooling. The ideal design, which would consider both passive heating and passive cooling, could provide heating in winter and promote cooling in summer. This requires careful consideration of orientation, fenestration, shading, and massing.

Because large commercial and institutional buildings are complex and are often restricted with respect to siting, trying out various passive design approaches using computer simulation is necessary to sort through the wide array of possibilities. The integration of landscaping with the building also has enormous potential for contributing to natural heating and cooling by shielding windows during the summer and allowing solar energy through in winter.

DAYLIGHTING

Using natural light or daylight for illumination is one of the hallmarks of a high-performance building. In addition to the benefits of supplying substantial light for free, natural lighting has been shown to provide great physical and psychological benefits to the building occupants. The first comprehensive scientific studies of the benefits of daylighting were conducted by the Pacific Gas & Electric Company in California in the late 1990s for two general types of buildings: retail stores and schools.[17] Daylighting in stores was shown to increase sales per square foot of retail space from 30 to 50 percent, while the learning rate of students was 20 to 26 percent higher in classrooms with daylighting compared to those with only artificial lighting.[18] Clearly, daylighting produces a win-win situation, marked by lower energy costs and better performance in classrooms. Most likely, the same is true in offices. Although not yet proven by scientific methods, it is thought that a 10 to 15 percent increase in office worker productivity can be expected as a consequence of daylighting. A 10 percent increase in employee productivity due to decreased illness and absenteeism or an improved sense of well-being translates into savings that far exceed the energy costs of a typical office building. If the connections between daylighting and human health could be proven with a high degree of certainty, this alone would cause an enormous transformation in the way buildings are designed and built. At present, productivity and health effects are not fully taken into account in the LCC analysis of high-performance buildings. However, if and when science catches up with speculation and the benefits are verified, daylighting will leap past its use as a green building strategy to near-universal incorporation. (Chapter 13 addresses LCC of green buildings in more detail.)

Developing an effective daylighting strategy can, however, be a complex undertaking due to the trade-offs that must occur between admitting light and cooling the building. The cost of windows, skylights, light shelves, and other features that function to transmit light, versus conventional construction where daylighting is not much of an issue, must also be factored in. Fortunately, experience with daylighting is growing at an exponential rate, along with the green building movement itself; consequently, the information from these efforts is becoming available to a wider audience of designers and owners. A list of key ideas for assessing daylight feasibility from Lawrence Berkeley National Laboratory is shown in Table 7.2.[19] An excellent checklist for daylighting from *Environmental Building News* (EBN) is shown in Table 7.3.[20]

The daylighting strategy for Rinker Hall at the University of Florida was to maximize the energy and health benefits of daylighting. The building's site and purpose dictated that the building would be on a true north-south axis, an orientation that is not generally favorable for passive design. The daylighting simulations indicated that through the use of louvers, low-emissivity glass, and slanted ceilings in the classroom, energy-efficient and effective daylighting could be created. A complete daylighting control system on the building's east and west façades is provided by large exterior windows, spectrally selective glazing, a shaped ceiling geometry, photosensor-controlled electric lighting, upper daylighting louvers, and lower vision-panel blinds with 97 percent reduced transmission in the closed position. A central skylight-covered atrium provides the open public stairways with dynamic beam daylight, marking solar noon each day. The daylighting simulations indicated that

TABLE 7.2

Key Ideas for Daylight Feasibility

- *Windows must see the light of day.* A high-density urban site may make daylighting difficult if the windows will not see much sky.
- *Glazing must transmit light.* A strong desire for very dark glazing generally diminishes the capacity to daylight in all but very sunny climates.
- *Install daylight-activated controls.* To save energy, lights are dimmed or turned off with controls. Automated lighting controls in a daylighted building can have other cost-saving applications (occupancy, scheduling, etc.) and benefits.
- *Design daylight for the task.* If the occupants require very bright light, darkness, or a highly controllable lighting environment, tailor the design to meet their needs.
- *Assess daylight feasibility for each portion of the building.* Spaces with similar orientation, sky views, ground reflectance, and design can be treated together. Within a single building, the feasibility and cost effectiveness of daylighting may vary greatly.

TABLE 7.3

Checklist for Daylighting

General Daylighting
1. Provide a daylighting scheme that will work under the range of sky conditions expected at that location.
2. Orient building on an east-west axis.
3. Brighten interior surfaces.
4. Organize electric lighting to complement daylighting.
5. Provide daylight controls on electric lighting.
6. Commission the daylight controls.
Perimeter Wall Daylighting
1. Provide perimeter daylight zones.
2. Extend windows high on perimeter walls.
3. Provide light shelves on south-facing windows.
4. Minimize direct-beam sunlight penetration into work spaces.
5. Choose the right glazing.
6. Arrange interior spaces to optimize the use of daylighting.
Roof Daylighting
1. Provide roof apertures for daylighting.
2. Optimize skylight spacing.
3. Consider extending skylight performance with trackers.
4. Use reflective roofing on sawtooth clerestories.
5. Diffuse daylight entering the building through roof apertures.
Core Daylighting
Provide a central well or atrium for daylighting.

through the use of louvers, low-emissivity glass, and slanted ceilings in the classroom, energy-efficient and effective daylighting could be created. Some of the techniques used to optimize the daylighting for Rinker Hall are shown in Figure 7.2A–D. Utilizing the upper room volume and diffusion off a sloping ceiling in conjunction with optimized daylight louvers allows the building occupants to comfortably "experience a day in nature." Half of the solar day is a no-glare exterior condition, and half is a full solar traverse from solar noon to sunset (west) or sunrise to solar noon (east). The design team's hour-by-hour analysis showed that the design case achieves a 30 percent overall increase in usable light (with a 15° or higher incidence to glass) duration (4,153 hours, compared to 3,171 hours in the base case). Further, the design case achieves a 48 percent increase in low-angle light (1,325 hours, compared to 686 hours in the base case). The emerging consensus is that learning environments that are enlivened by the subtle and natural variation of light intensity, color, and direction as the day passes are healthier environments that support higher productivity.[21]

PASSIVE VENTILATION

Providing ventilation to building occupants is normally accomplished by using fans, dampers, and controls to move outside air into the building while at the same time removing an equal amount of interior air to the outside. In more advanced designs, an economizer cycle uses outside air for cooling, providing significant savings. Ventilation air using natural forces to move the air, rather than mechanical systems, can also be provided, greatly reducing the energy needed to move air. Passive ventilation can

Figure 7.2 The daylighting strategy for Rinker Hall, University of Florida, employs (A) skylights, (B) window louvers, (C) tilted ceilings to reflect daylighting deep into rooms, and (D) low-emissivity glass to facilitate deep daylighting and control solar thermal radiation. (Photographs: M.R. Moretti.)

be accomplished by using a thermal chimney effect, whereby air normally rises due to heating, inducing airflow in a generally vertical direction; or a Venturi effect, whereby air movement is induced by the development of a low-pressure area created by wind flow.

The Jubilee Campus of the University of Nottingham in England, designed by Sir Michael Hopkins and Partners and built in 1999, has one of the most advanced passive ventilation strategies among modern buildings. Windcatchers are used to position the air exhaust stacks for optimal ventilation. The windcatchers automatically turn in the direction of the wind, creating suction behind them and driving the ventilation system for the buildings. Cool, clean air is brought in at a high level and fanned down to the floor levels, where it starts to rise with the sunlight, body heat, and equipment. This intricate pattern of environmental cause and effect is echoed throughout the building's staircases and corridors. Thermal wheels are used in conjunction with the windcatchers to exchange energy between exiting exhaust air and

Figure 7.3 Windcatcher (upper right) on the Jubilee Campus Building, University of Nottingham, United Kingdom. The windcatcher pivots in the wind, with the vane indicating the direction of airflow. Wind flowing past the vane induces the convection of air through the structure. (Photograph courtesy of Hopkins Architects and Ian Lawson.)

incoming fresh air. The innovations in this design resulted in the Jubilee Campus winning the Royal Institute of British Architects (RIBA) sustainability award in 2001. Figures 7.3 and 7.4 show this passively ventilated building on the Jubilee Campus and depict the ventilation pattern through the building.

In a typical European passive ventilation design, the quantity of air required for ventilation is first determined. In England, the Chartered Institute of Building Services Engineers (CIBSE) publishes standards and guidelines requiring the following ventilation rates:

- Classrooms: 2 to 4 air changes per hour
- Offices: 4 to 6 air changes per hour
- Theaters: 6 to 10 air changes per hour
- Storage areas: 1 to 2 air changes per hour

The outside wind speed, which is generally in the range of 3 to 19 feet per second (1 to 6 meters/second) in England, is factored into the design, and the number of passive ventilation stacks required to move the calculated amount of ventilation air are designed into the structure. In the base of the stack, dampers connected to the building's energy management system, and possibly to carbon dioxide, humidity, and/or temperature sensors, control the rate of ventilation. Diffusers at ceiling level introduce the ventilation air into the occupied spaces. In some passive ventilation stacks, solar tubes that bring in light, as well as air, are incorporated.

In contrast to Europe, which has a wide range of examples of passive ventilation systems, the concept has not had much success yet in this country. One of the best U.S. examples is the Federal Building in San Francisco, for which a sample computation fluid dynamics (CFD) simulation is shown in Figure 7.5.

PASSIVE COOLING

Earlier in this chapter, it was noted that today's German office buildings achieve substantially better energy performance than their U.S. high-performance counterparts. An obvious question is, how do the Germans achieve such exceptional energy performance in their buildings? The answer is that they are changing some of the basic

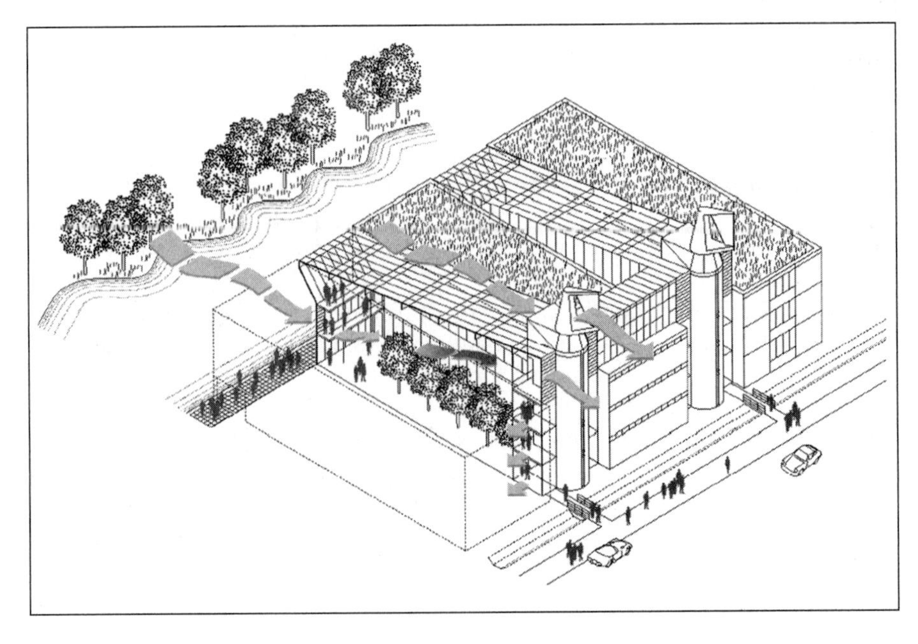

Figure 7.4 Schematic of the natural ventilation strategy for the Jubilee Campus Building, University of Nottingham. Air flows from a low level at the rear of the building, moves gradually upward, and then exits through the pivoting windcatchers on the front of the building. A *Venturi effect,* or induced airflow, caused by low pressure, is created in the windcatcher by the wind flowing past the vanes. (Illustration courtesy of Hopkins Architects.)

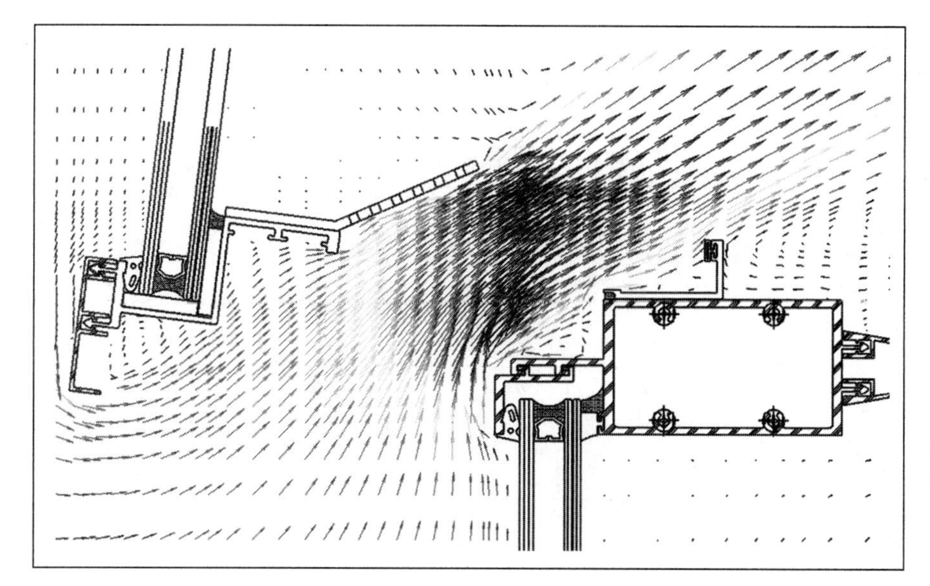

Figure 7.5 Design of passive ventilation systems requires the use of tools not traditionally used in building design, such as the computational fluid dynamics (CFD) modeling of wind and airflows around the San Francisco Federal Building. The illustration shows a simulation of the design of an air deflector for the windows of the building that helps accelerate airflows, propelling them deep into the building's spaces. (Illustration courtesy of Natural Works.)

assumptions of the past several decades about how buildings should operate. Rather than completely isolating the building occupants from outdoor conditions, they now assume moderate interaction by means of natural ventilation, daylighting, and passive cooling. This concept, called *lean building*, results in smaller building service equipment for heating and cooling. In the German context, passive cooling is the interaction of all measures that reduce heat gains and render natural heat sinks—night air and the ground—accessible (see Figure 7.6). Heat loads are transferred to the surrounding environment with some time delays, and heat storage in the building mass itself is substantial. The main design priority is to restrict the amplitude and dynamics of external heat gains. Limiting glazing while maintaining daylighting is the key to this strategy, and the ratio of glazing to façade area is less than 43 percent for the set of 23 German demonstration buildings mentioned above. Almost all buildings use external adjustable sun-shading devices, and total solar energy transmittance is kept below 15 percent. Cooling is accomplished by using night ventilation in which the building mass is cooled using earth-to-air heat exchangers, which are simply underground metal ducts through which the air is brought into the building, or by slab cooling in which ground-

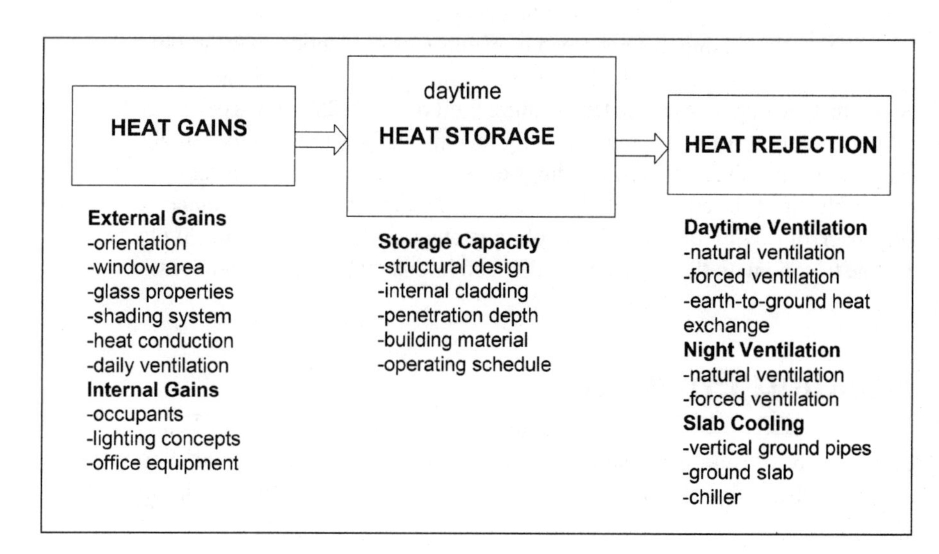

Figure 7.6 Passive cooling strategies use heat gain avoidance to minimize external thermal loads, minimize internal gains from occupants and electrical equipment, and use the building structure for storing residual heat gains, which are then removed by a combination of natural and forced ventilation with ground coupling.

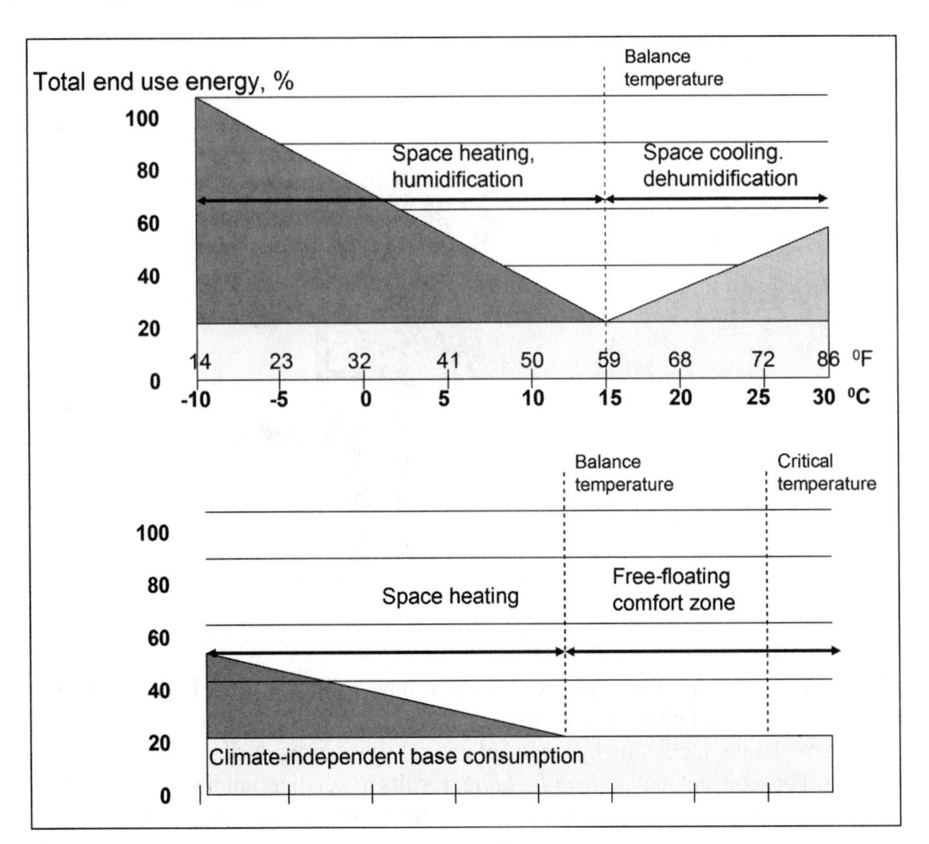

Figure 7.7 Low-energy German buildings use passive ventilation and cooling to eliminate much or all of the need for conventional mechanical cooling. Studies indicate that this strategy results in interior temperatures that rarely exceed acceptable space conditions for offices. As a consequence, buildings that use 100 kWh of primary energy per square meter (31,700 BTUs per square foot) annually are achievable, a fraction of the energy of today's U.S. high-performance green buildings.

water is pumped through cavities in the slab. The coefficient of performance (COP) for mechanical and hybrid night ventilation ranges between 4.5 and 14, far higher than that of conventional cooling.[22] The earth-to-air heat exchangers have extremely high COPs, ranging from 20 to 280. Note that today's best-performing chillers, the heart of many air conditioning systems, have a maximum COP of about 8. Figure 7.7 illustrates this strategy graphically. Eliminating conventional cooling systems gives the project the resources to accomplish the technical analysis to design a lean building appropriate to the bioregion, one that transfers daytime internal energy to the structure and minimizes the intrusion of external heat energy into the building. Even if the outdoor conditions vary, the result is that indoor conditions remain within a well-defined comfort zone, meeting the needs of the occupants.[23]

The result of using this approach is an enormous reduction in the cooling capacity typically needed for office buildings. In monitoring these buildings, the researchers found that the upper desirable temperature limit of 77°F (25°C) was exceeded less than 10 percent of the working hours. During the unusually warm summer of 2002 in Germany, the naturally ventilated buildings exceeded the temperature criterion only 5 percent of the time, the equivalent of 1 hour every 2½ days, a remarkable outcome. The one drawback of relying on a passive cooling strategy is that the mechanical plant will be unable to cope with the extreme weather conditions that can occur on occasion.

Building Envelope

After passive design is considered to minimize the need for external energy inputs, energy transmission through the building skin should be minimized through a tight, thermally resistant envelope. The building envelope must control solar heat gain, conduction or direct heat transmission, and infiltration or leakage heat transmission. The three major building envelope issues that need to be addressed are thermal resistance of the walls, window selection, and roof strategy. These are covered in the following subsections. (The environmental impacts of materials selection are covered in Chapter 9.)

WALL SYSTEMS

The thermal conductance, or *U-value*, of building walls is an important factor in building energy efficiency because walls are generally the dominant component of the envelope. Maximum U-values are set by state building energy codes and by ASHRAE 90.1-2004. The maximum U-value is a function of the number of heating degree days (HDDs) and cooling degree days (CDDs). Both HDD and CDD are measures of how much heating or cooling will probably be required in a given climatic zone. In general, wall thermal resistance becomes more important the farther north the building is located in the United States. Two other considerations in selecting wall systems are the thermal mass of the exterior surface that receives direct sunlight during the day and the placement of insulation with respect to the building façade. Placing insulation to the exterior nearest the outdoor conditions and having the thermal mass closest to the interior provides ideal conditions for using the mass beneficially and for minimizing the thermal loads transmitted into the building's interior that must be removed by air conditioning. In southern climates, it is generally important to design energy-shading façades that will reflect energy or that are ventilated to carry away energy that is absorbed on the building's skin.

WINDOW SELECTION

Windows play a variety of roles in the building envelope. They allow light into the room spaces, permit the occupants to admit air into the space in the case of operable windows, and provide a thermally resistant layer to energy movement. Windows must be installed so as to balance the amount of light admitted into the structure with the control of solar heat gain and conduction of energy through the window assembly. Window performance is a combination of several factors: the *solar heat gain coefficient* (SHGC), the *visible transmittance* (VT), the thermal conductance or U-value of the window assembly, and the infiltration or leakiness character of the window assembly.

Solar heat gain is largely a function of where windows are placed in the building and the types of glass used. SHGC and VT are used to express the performance of windows in the building envelope. SHGC, with a value between 0 and 1, is the fraction of solar heat that enters the window and becomes heat; it includes both directly transmitted and absorbed solar radiation. The lower the SHGC, the less solar heat the window transmits through the glazing from the exterior to the interior and the greater its shading capability. In general, south-facing windows in buildings designed for passive solar heating should have windows with a high SHGC to allow beneficial solar heat gain in the winter. East- or west-facing windows encounter high levels of solar energy in the morning and afternoon and should generally have lower SHGC assemblies.

The VT, ranging in value from 0 to 1, refers to the percentage of the visible spectrum (380 to 720 nanometers) that is transmitted through the glazing. When daylight in a space is desirable, high-VT glazing would be the logical choice. However, low-VT glazing may be more applicable for office buildings or where reduced interior glare is desirable. A typical clear, single-pane window has a VT of 0.90, meaning that it admits 90 percent of the visible light.

The ratio of SHGC to VT, known as the *light-to-solar gain ratio* (LSG), provides a gauge of the relative efficiency of different glass types in transmitting daylight while blocking heat gains. The higher the LSG, the brighter the room is without adding excessive amounts of heat. Table 7.4 shows average values of SHGC, VT, and LSG for typical windows.[24]

Low-emissivity (low-e) and reflective coatings usually consist of a layer of metal a few molecules thick. The thickness and reflectivity of the metal layer (low-e coating) and the location of the glass to which it is attached directly affect the amount of solar heat gain in the room. Most window manufacturers now use one or more layers of low-e coatings in their product lines.

Any low-e coating is roughly equivalent to adding an additional pane of glass to a window. Low-e coatings reduce long-wave radiation heat transfer by 5 to 10 times. The lower the emissivity value (a measure of the amount of heat transmission through the glazing), the better the material reduces the heat transfer from the inside to the outside. Most low-e coatings also slightly reduce the amount of visible light transmitted through the glazing relative to clear glass. Representative emissivity values for different types of glass are as follows:

- Clear glass, uncoated: 0.84
- Glass with single hard-coat low-e: 0.15
- Glass with single soft-coat low-e: 0.10

Increasing the window area to maximize daylighting has the effect of replacing a highly thermally resistant wall with far less thermally resistant glass, creating an opportunity for the infrared or heating component of light to enter the envelope and also creating the potential for infiltration around the window frame. In trading off daylighting versus optimizing the thermal envelope, controlling solar heat gain through windows is critically important. Prior to the development of today's window glazing and film technologies, 75 to 85 percent of infrared energy could pass through typical single- or double-paned glass.

A standardized national system for rating windows is important to enable performance comparisons of these important components of the building envelope. The National Fenestration Ratings Council (NFRC) operates a uniform national rating system for measuring the energy performance of fenestration products, including windows, doors, skylights, and similar products.[25] The key to the rating system is a procedure for determining the thermal transmittance (U-factor) of a product. The U-factor rating procedure is supplemented by procedures for rating products for solar

TABLE 7.4

Typical Values of SHGC, VT, and LSG for Total Window and Center of Glass (in parentheses) for Different Types of Windows

Window Type	Glazing	SHGC	VT	LSG
Single-glazed	Clear	0.79 (0.86)	0.69 (0.90)	0.97 (1.04)
Double-glazed	Clear	0.58 (0.86)	0.57 (0.81)	0.98 (1.07)
Double-glazed	Bronze	0.48 (0.62)	0.43 (0.61)	0.89 (0.98)
Double-glazed	Spectrally selective	0.31 (0.41)	0.51 (0.72)	1.65 (1.75)
Triple-glazed	Low-e	0.37 (0.49)	0.48 (0.68)	1.29 (1.39)

Figure 7.8 Low-e glazing on Rinker Hall, University of Florida. The advent of high-technology glazing allows the design of buildings that admit visible light for daylighting but reflect infrared radiation. DOE 2.1 and daylighting simulations confirmed that daylighting and low-e glass produced greater savings than focusing solely on the thermal resistance of the building envelope. (Photograph: M.R. Moretti.)

heat gain coefficient, visible transmittance, air leakage, and annual energy performance. Together, these rating procedures, as set forth in documents published by NFRC, are known as the NFRC Rating System. This system is expected to be supplemented by additional procedures for rating energy performance characteristics, including long-term energy performance and condensation resistance. The NFRC Rating System employs computer simulation and physical testing by NFRC-accredited laboratories to establish energy performance ratings for fenestration products and product lines. The system is reinforced by a certification program under which a window and door manufacturer may label and certify its products to indicate those energy performance ratings.

ROOF SELECTION: THERMAL RESISTANCE AND COLOR

The roof of a high-performance building is especially important because it is a major area for heat transmission due to its generally large area and exposure to the sun. According to Cool Communities, a nonprofit organization based in Rome, Georgia, the roofs of structures such as shopping malls, warehouses, and office buildings can reach 150°F (83°C) in the summer, enough to affect whole neighborhoods.[26] Using surfaces with high albedo (a measure of the reflectivity of solar radiation) for roofing can reduce the ambient air temperature so that the entire area is cooler. Light-colored roofs have high albedo, or high reflectivity, which helps reduce the thermal load on the building as well as the surrounding neighborhood. Lawrence Berkeley National Laboratory and the Florida Solar Energy Center both estimate that buildings with light-colored, reflective roofs use 40 percent less energy than similar buildings with dark roofs.

A new rating system, the Solar Reflectance Index (SRI), which measures how hot materials are in the sun, has been developed and is being applied to roofing materials. A building with light-colored shingles with an SRI of 54 would reflect 54 percent of incident solar energy, and would be very cool relative to a building with conventional dark shingles. Manufacturers have recently developed clean, "self-washing" white shingles with an even higher SRI, up to 62 percent. This is useful because the labor costs of maintaining the white color and high reflectivity of a conventional roof may exceed the worth of the energy being saved; consequently, a self-washing roof system will significantly reduce maintenance costs and improve energy performance. The

Figure 7.9 Rinker Hall has a white-colored thermoplastic roof with very high reflectivity. Research by national laboratories has demonstrated that light-colored roofs with a high SRI use 40 percent less energy than dark roofs. (Photograph: M.R. Moretti.)

reflectance of commonly used roofing materials is shown in Table 7.5.[27] As can be seen, dark colored roofs have a tendency to absorb solar radiation, and they can be as much as 90°F (50°C) hotter than the air just above the roof. Because heat transmission is a function of temperature difference, a dark-colored, hot roof will have proportionately more heat conduction than a light-colored, relatively cool roof.

TABLE 7.5

Reflectance of Roof Materials and Air Temperatures Above Roof

Material	Solar Reflectance	Temperature of Roof Over Air Temperature (°F/°C)
Bright white coating (ceramic, elastomeric) on smooth surface	80%	15°/8°
White membrane	70–80%	15°–25°/8°–14°
White metal	60–70%	25°–36°/14°–20°
Bright white coating (ceramic, elastomeric) on rough surface	60%	36°/20°
Bright aluminum coating	55%	51°/28°
Premium white shingle	35%	60°/33°
Generic white shingle	25%	70°/39°
Light brown/gray shingle	20%	75°/42°
Dark red tile	18–33%	62°–77°/34°–43°
Dark shingle	8–19%	76°–87°/42°–48°
Black shingle or materials	5%	90°/50°

Internal Load Reduction

Excellence in passive design and in the design of a high-performance building envelope needs to be combined with a significant effort to address the internal heat loads of the building. This is achieved in part by a good daylighting strategy that has the dual benefit of reducing energy consumption for lighting and removing the lighting power saved from the total building cooling load. People constitute a major fraction of the building's internal heat load, and we can generally assume that reducing the number of people in a building is not a viable strategy. Reducing loads due to computers, peripherals, copiers, and other miscellaneous equipment is a promising strategy because it has been found that these loads constitute a substantial fraction of a building's energy consumption. Increasing wiring sizes beyond those required by code has the benefit of reducing energy losses in the wiring system and proportionately reducing the impact of these heat losses on the building's cooling system.

PLUG LOAD REDUCTION

Designers of high-performance buildings typically do not closely examine one of the major internal building loads, *plug loads,* a name for devices plugged into electrical outlets around the building that not only consume substantial energy, but also increase cooling loads due to their heat emissions. The U.S. Department of Energy estimates that office equipment loads comprise about 18 percent of a U.S. commercial building's electrical load, exceeded only by HVAC and lighting loads. In a study of plug loads for their newly renovated 6,700-square-foot (622-square-meter) office building, IDeAs Z^2, a consulting engineering firm, estimated that plug loads would consume in excess of 40,000 kWh per year, or almost 7 kWh per square foot (75 kWh per square meter) each year. In many office buildings the largest plug loads are due to desktop computers, typically averaging about 160 watts per unit.[28] One alternative that should be considered for reducing desktop computer plug loads is to replace them with laptops designed for energy efficiency to maximize battery life. Laptops generally consume 40 watts, or about 25 percent of the power of desktops. The economics of replacing desktops with laptops will depend largely on the processing power required. In the case of IDeAs Z^2, which requires high-speed processors and large amounts of software to run their computational and graphics software, the cost difference between a desktop and a laptop with equivalent processing speed and storage was about $1,835. Their analysis indicated that the additional cost for high-end laptops would be greater than the cost of photovoltaics to offset the additional load; that is, it would be cheaper to buy photovoltaics than to replace desktops with laptops. In cases where exceptional computing capability is not needed, the cost differential and payback will be more favorable. Purchasing liquid crystal display (LCD) screens instead of cathode ray tube (CRT) screens results in a 50 percent energy savings, and LCD screens take up less space.

MISCELLANEOUS PLUG LOADS

Electrical loads for printers, scanners, copiers, and fax machines also contribute to higher building energy consumption. Energy Star–rated equipment costs 20 to 100 percent more than existing equipment, with energy savings of less than 10 percent. A typical high-efficiency, USEPA Energy Star–rated refrigerator saves 53 percent over a standard refrigerator but costs about $4.83 per kWh saved annually. There appears to be a very high premium for the highest-efficiency Energy Star–rated equipment. IDeAs Z^2 concluded that immediate replacement of existing devices was not cost effective. Instead, their strategy is to replace old equipment with Energy Star–rated devices as much as possible and to evaluate the situation on an individual basis. They also researched energy-efficient dishwashers and concluded that unlike refrigerators, which run 365 days a year, dishwashers run infrequently; therefore, the energy savings resulting from purchasing a high-efficiency dishwasher would be minimal. A final item that was evaluated was the existing coffee maker, which stays on "warm" all day (and occasionally all night), holding half a pot of coffee. The firm will purchase a single-cup coffee maker for daily staff use, which will heat one cup at a time and has no warming element. The old coffee maker will only be used in conjunction with a thermos for large meetings.

PLUG LOAD CONTROL

Several types of control strategies are employed to reduce plug loads. Some equipment needs to be left on 24 hours a day, 7 days a week. This includes fax machines, main servers, and security systems. In these cases, high energy efficiency will be a key criterion when selecting equipment. Control of "phantom loads" in office equipment is another key strategy for conserving power. For equipment that has infrequent duty cycles, such as microwave ovens, the energy consumed during long hours of standby can be more than the energy consumed while in use. A second group of items, including printers, plotters, and copiers need to be turned on only during working hours. These items typically have a long startup time and would be inconvenient to turn on prior to each use, so they cannot be turned off between uses. However, there is no reason for these items to continue to run when the office is unoccupied, regardless of whether they are active or in sleep mode. IDeAs Z^2 found that the worst case was a laser plotter that consumed 1,440 watts when plotting, 30 watts in the sleep mode, and 25 watts when manually switched to standby. Oddly enough, there is no true Off switch. To ensure that this equipment is not left on, the security system automatically turns off the electrical circuits to it when armed and turns on the circuits the next day when disarmed, reducing phantom loads. Occupancy sensor-controlled surge protectors are used at each workstation to turn off the power to task lights, computer monitors, speakers, and other nonessential peripherals when a user leaves his or her desk. Desktops are routinely left on all day but are set to go into sleep mode when not in use. Sleep mode saves energy and allows for fast restart times compared to hibernate mode. However, if power is lost, data will also be lost. Hibernate mode saves data to the hard drive, so if power is lost, data will not be lost. However, computers in hibernate take much longer to restart when they come back to active mode. IDeAs Z^2 is currently working with USEPA-sponsored researchers and experimenting with personal computer settings and individual occupancy sensors to determine how best to minimize energy consumption without significantly reducing productivity or creating inconvenience. There is a debate within IDeAs Z^2 as to whether it is wise to automatically shut off power to personal computers when the building is unoccupied. The argument against it is that risking the loss of work left unsaved on the computer is not worth the attempt to save a few watts. However, most software has a built-in auto-save feature, reducing the potential to lose a significant amount of work. IDeAs Z^2 is currently experimenting to determine if using the security system to turn off computer circuits is worth saving additional phantom losses.

UPSIZED ELECTRICAL WIRING

All circuits lose small amounts of energy through resistance as power flows through the wiring. Wire sizes recommended by code are based on keeping the heat generated from wiring losses below temperatures that would damage wire insulation. If wires are upsized, resistance in the wires is lower and losses are reduced. IDeAs Z^2 estimated losses for sample circuits and compared the value of saved electricity with the additional cost to increase wire sizes one size above code recommendations. The paybacks were as low as 4 years for circuits that were highly loaded. All branch circuits carrying large, continuous loads were upsized to reduce wiring losses. In addition to reducing the electrical energy consumption of a circuit, this reduced the cooling loads associated with those losses.

Active Mechanical Systems

After the passive solar design of the building is optimized, the internal loads in the building should be minimized. The load of some buildings will be people-dominated; that is, the bulk of the load will be due to the number of people in the facility, so little can be done to reduce the load. A classroom building at a university is a good example of this situation. In other buildings, the load may be dominated by equipment, lighting, and other powered devices. In this situation, energy-efficient appliances, lighting, computers, and other energy-efficient systems can contribute to a significant reduction in cooling load. Office buildings may have equipment-dominated loads if they have relatively large quantities of powered devices such as computers and copy machines and a moderate to low population.

A very wide variety of HVAC systems can be used to meet the needs of a facility's occupants. The type of system selected is a function of the size of the building, the climatic conditions, and the load profile of the building. A typical building HVAC system will have an air side that delivers conditioned air into the spaces and a fluid side that creates chilled and hot water for use in the HVAC system, so equipment with the highest possible efficiency should be selected for all roles. The following subsections contain information about selecting some of the major types of equipment in a HVAC system: chillers, air distribution system components, and energy recovery systems.

CHILLERS

According to Lawrence Berkeley National Laboratory's Energy Environmental Technologies Division, chillers are the single largest energy users in commercial buildings, consuming 23 percent of total building energy. Chillers also have the unfortunate characteristic of increasing their power consumption during the day, contributing to peak demand and forcing utilities to build new power plants to meet high daytime power demand. Consequently, chillers are responsible for large portions of peak power charges for commercial customers. In addition to these problems, most chiller plants tend to be oversized during the design process. Chillers operate at peak efficiency when they operate at peak load. However, chillers tend to operate at part load during much of the day. Even those that are correctly sized operate most of the time at low, part-load efficiencies.

Four types of chillers are commonly available today (note: 1 ton equals 12,000 BTU/hour of cooling capacity, or 3.4 kW):

- Centrifugal, primarily large tonnage above 1,000 kW or 300 tons
- Screw (50 to 400 tons) (170 to 1360 kW)
- Scroll (up to 50 tons) (170 kW)
- Reciprocating (up to 150 tons) (510 kW)

Figure 7.10 The ERM manufactured by Greenheck, Inc., houses a desiccant wheel that rotates between fresh and exhaust airstreams, exchanging energy and humidity and providing enormous energy conservation benefits. (Photograph courtesy of Greenheck, Inc.)

Manufacturers of chillers have been working to produce high-efficiency chillers that meet the needs of high-performance green buildings. For example, the Trane CVHE/F EarthWise centrifugal chiller was awarded the USEPA's Environmental Protection Award (see Figure 7.10). Rated at 0.45 kWh per ton of cooling, the EarthWise centrifugal chiller has the highest efficiency in this major category of HVAC equipment.

Chiller plant efficiency can be improved by more than 50 percent while improving reliability by combining new technologies such as direct digital control (DDC) and variable-frequency drives with improved design, commissioning, and operation. California tends to lead the nation in developing energy performance standards, and the latest version of the California Title 24 Energy Efficiency Standards has substantially increased requirements for chiller efficiency.[29] Several different chiller technologies can be considered for a building. In general, water-cooled rotary screw or scroll chillers have the highest COP of all types of chillers. COP is the ratio of cooling power delivered by a chiller to the input power. A COP of 3.0, for example, indicates that the chiller provides 3 kWh of cooling for 1 kWh of input energy. Note that a high-capacity screw or scroll water-cooled chiller has a COP of over 6, a very high level of performance and more than double the COP of 2.50 for an electrically operated, air-cooled chiller. One significant disadvantage of a water-cooled chiller is the need to provide a cooling tower to reject the energy absorbed from the building.

Absorption chillers tend to have a comparatively low COP, normally less than 1.0, which appears to indicate a very low level of performance. However, absorption chillers can use heat energy that would normally be wasted to provide cooling. A steam-driven screw chiller could reject its waste energy to an absorption chiller to provide additional cooling, thus increasing the COP of the overall system. Absorption chillers can also use relatively low-temperature heat to produce chilled water and thus work well with solar thermal energy and waste heat from some varieties of fuel cells such as the phosphoric acid fuel cell (PAFC).

Table 7.6 describes the characteristics of a high-performance chiller plant,[30] and Table 7.7 provides several design strategies for achieving a relatively low-cost, high-efficiency chiller plant.[31]

AIR DISTRIBUTION SYSTEMS

Another major consumer of energy in modern buildings is the air distribution system, comprised of air handlers, electric motors, ductwork, air diffusers, registers and grilles, energy and humidity exchangers, control boxes, and its control system. The air distribution system should be designed using much the same approach as for the chiller plant to deliver the precise capability needed and to do so efficiently across a wide range of operating conditions. According to *Greening Federal Facilities*, design options for improving air distribution efficiency include (1) variable air volume (VAV) systems, (2) VAV diffusers, (3) low-pressure ductwork design, (4) low-face

TABLE 7.6

Characteristics of a High-Efficiency Chiller Plant

- *An efficient design concept.* Selecting an appropriate design concept that is responsive to the anticipated operating conditions is essential to achieving efficiency. Examples include using a variable-flow pumping system for large campus applications and selecting the quantity, type, and configuration of chillers based upon the expected load profile.
- *Efficient components.* Chillers, pumps, fans, and motors should all be selected for stand-alone as well as systemic efficiency. Examples include premium-efficiency motors, pumps that have high efficiency under the anticipated operating conditions, chillers that are efficient with both full and partial loads, and induced-draft cooling towers.
- *Proper installation, commissioning, and operation.* A chiller plant that meets the first two criteria can still waste a lot of energy—and provide poor comfort to building occupants—if it is not installed or operated properly. For this reason, following a formal commissioning process that functionally tests the plant under all modes of operation can provide some assurance that the potential efficiency of the system will be realized.

TABLE 7.7

Design Strategies for a High-Efficiency Chiller Plant
1. Focus on Chiller Part Load Efficiency. Three methods for improving chiller plant load efficiency are: specify a chiller that can operate with reduced condenser water temperatures, specify a variable-speed drive (VSD) for the compressor motor, and select the number and size of chillers based on anticipated operating conditions. For example, it may be better to size one chiller at one-third peak load and the other at two-thirds peak load for a two-chiller system.
2. Design Efficient Pumping Systems. Energy use in pumping systems may be reduced by sizing pumps based upon the actual pressure drop through each component in the system, as well as the actual peak water flow requirements, accurately itemizing the pressure losses through the system and then applying a realistic safety factor to the total. Additionally, fluid velocities should be kept low; temperature differentials across the chiller should be as high as possible; the piping system should be kept simple; unnecessary devices, such as valves, should be minimized; the possibility of using a variable-flow system should be evaluated; and high-efficiency pumps and motors should be specified.
3. Properly Select the Cooling Tower. Proper sizing and control of cooling towers is essential to efficient chiller operation. Cooling towers are often insufficiently sized for the task. An efficient cooling tower should be specified based on using realistic wet-bulb sizing criteria; an induced draft tower, if space permits; intelligent controls; and sequences of operation that minimize overall energy use.
4. Integrate Chiller Controls with Building EMS. Although modern chillers are computer-controlled and have considerable intelligence to assist their operations, they should be integrated with the building's energy management system (EMS) to provide the capability to optimally operate the entire building energy plant. To accomplish this integration, the designers should specify an "open" communications protocol; use a hardware gateway; measure the power of ancillary equipment; and analyze the resultant data.
5. Commission the System. Even an efficient chiller plant may fail to reach its potential due to installation problems, poor control system programming, or lack of coordination between the design team and the contractor. Commissioning a chiller system—that is, functionally testing it under all anticipated operating modes to ensure that it performs as intended—can improve efficiency and reliability and ensure that the owners are getting the level of efficiency they paid for. Ideally, commissioning starts early in the design process and is performed by an independent third party (that is, an entity that is not part of the design or construction team).

velocity air handlers, (5) proper fan sizing with variable frequency drive (VFD) motors, and (6) positive displacement ventilation systems. Each of these design options is described in more detail in the following list.[32]

1. *VAV systems.* A VAV system delivers the precise volume of air needed to meet the actual load of a given building space. VAV systems offer better energy performance than constant volume systems. VAV systems are now virtually standard design practice, yet even greater efficiency gains can be made through careful selection of equipment and system design.

2. *VAV diffusers.* Local VAV diffusers for individual temperature control should be used. Temperatures across a multiroom zone in a VAV system can vary widely, causing space conditions farther from the thermostat and VAV box location to be uncomfortable. Local ceiling diffusers ducted from the VAV box to individual rooms can modulate the amount of conditioned air delivered to a space, eliminating the inefficient practice of overheating or overcooling spaces to ensure the comfort of all occupants. VAV diffusers require low-duct static pressure, a water column of 0.25 inch (62 Pa) or less, and thus save on fan energy.
3. *Low-pressure ductwork design.* Duct size should be increased to reduce duct pressure drop and fan speed. Resistance in the duct system can be reduced by improving the aerodynamics of the flow paths and avoiding sharp turns in duct routing. Increasing the size of ducting where possible allows reductions in air velocity, which in turn permits reductions in fan speed and yields substantial energy savings. Small increases in duct diameter can greatly lower pressure, resulting in fan energy savings, because the pressure drop in ducts is proportional to the inverse of duct diameter to the fifth power.
4. *Low-face velocity air handlers.* A low-face velocity air handler has reduced air velocity across its coils. Increasing the air handler size produces a larger cross-sectional area for the airflow, allowing the delivery of the same required airflow at a slower air speed for only a relatively small loss of floor space. The pressure drop across the coils decreases with the square of the airspeed, allowing the use of a smaller fan and smaller VFD, thus reducing the first costs of those components. Air traveling at a lower velocity remains in contact with cooling coils longer, allowing higher chilled-water temperatures. This can yield compound savings through downsizing of the chilled water plant (as long as all air-handling units in a facility are sized with these design strategies in mind).
5. *Proper fan sizing and VFD motors.* Oversized fans should be properly sized to match the calculated load. The fan motor's speed and torque should be electronically controlled to continually match fan speed with changing building-load conditions. Electronic control of the fan speed and airflow can replace inefficient mechanical controls, such as inlet vanes or outlet dampers.
6. *Displacement ventilation systems.* A displacement ventilation system is designed to introduce conditioned air into an underfloor plenum formed by a raised access floor. The air then moves upward into the occupied space via grilles or registers in the access floor. This strategy slowly pushes the conditioned air into the occupied space, displacing the air in the room vertically with minimal mixing, thus providing significant potential energy savings. Displacement ventilation systems can largely eliminate the need for ducting by supplying the conditioned air through the floor plenum and using a ceiling plenum for the return air. The access floor itself provides some degree of radiation cooling because the floor materials are cooled to the conditioned air temperature and serve as a heat sink for thermal energy. As a result of the absence of mixing, the air temperature in the space becomes stratified, with the warmest air at the ceiling, where it is returned for conditioning.

ENERGY RECOVERY SYSTEMS

Fresh air requirements for buildings mean that substantial quantities of fresh air are being brought into the facility while approximately the same amount of inside air is being exhausted to the outside. ASHRAE Standard 62.1-2004 governs the quantity of fresh air that is required for building operation. The energy costs for this exchange of air can be considerable. For example, on a 90°F (32°C) summer day in New York City, with 80 to 90 percent relative humidity, the hot, humid outside air is being brought into buildings for ventilation purposes. At the same time, inside air at 72°F (22°C), with 50 percent relative humidity, is being exhausted to the outside. Clearly, it would be useful to have devices that could cool down the outside air stream in summer with the air being exhausted and, conversely, heat outside air being brought into the building during the winter by using the energy in the relatively warm air being exhausted. Another approach is to simply use outside air directly for conditioning the building when outside air conditions are just right for that purpose. Two technologies, economizers and energy recovery ventilators (ERVs), have been developed to use outside air for conditioning and to exchange energy between fresh intake air and exhaust air streams. Both approaches are described in the following subsections.

Economizers

One rather obvious way to save energy in a typical building is to use outside air to cool the building when weather conditions are appropriate. The concept is quite simple: determine when the outside air temperature and humidity are in the same range as conditioned air delivered to the space would be, and then duct the outside air to replace the conditioned airstream. The ductwork and dampers in the system are also designed so that all the return air can be exhausted from the building. Chillers and chilled water pumps can be turned off, thus saving significant energy, as much as 20 to 30 percent of the energy that would ordinarily be invested in cooling.

Unfortunately, economizers have a rather high rate of operational failure. Dampers become corroded and stick in place, temperature sensors fail, actuators fail, and linkages malfunction. Estimates of the failure rate of economizers vary widely, but the consensus of experts, according to Energy Design Resources, is that only 25 percent may be functioning properly within a few years. Malfunctioning economizers can actually cause significant energy waste. For example, a system mistakenly being operated in economizer mode in the middle of summer in a hot climate such as that of Florida or inland California can increase the cooling load by over 80 percent due to the large quantities of outside air that must be cooled.

In spite of this, economizers have huge potential if properly installed, commissioned, and maintained. Energy Design Resources gives the following recommendations in its guide to economizers:[33]

1. *Stainless-steel dampers.* Stainless-steel dampers resist corrosion much better than the galvanized-steel and aluminum dampers typically used in economizers. Though stainless-steel dampers cost about twice as much as galvanized-steel dampers, they are cheaper than the total cost (including labor) of removing and replacing a failed damper.
2. *Direct-drive actuators.* Using a direct-drive actuator greatly increases the reliability of the economizer system compared to an actuator with complex linkages. In a linkage system, an actuator moves the linkage that positions the damper open or closed. An economizer with a freely moving outside-air damper and a powerful actuator will still not work if the linkage does not function. Over time, linkages have a tendency to corrode, weaken, or loosen, seriously impairing their ability to transfer torque from the actuator to the damper. A direct-drive actuator has fewer moving parts between actuator and damper, and therefore fewer parts that can fail. In addition, a direct-drive actuator is usually easier to install than a linked actuator, leading to a reduction in installation time.
3. *Return or exhaust fans.* HVAC systems for most buildings are comprised of (1) packaged units that have all components to heat, cool, and move the air through the building; or (2) built-up HVAC systems in which the heating, cooling, and air-moving components are specified separately by the mechanical engineer and assembled by the mechanical contractor. For an economizer to work at all, some means for exhausting air from the building in concert with economizer operation must be provided. Most packaged cooling units with a built-in factory economizer come with an appropriately sized exhaust

fan that runs whenever the economizer is active; or a barometric or "gravity" damper can be provided to allow excess air to escape from the building during economizer operation. Built-up HVAC systems use either a general exhaust fan or a return fan to draw air through a ducted return system.

4. *Type of control strategy.* The type and location of the building largely determine what type of economizer setup will work best. In dry climates, a simple drybulb control strategy (which responds to temperature only) can provide good performance and comfort. In more humid climates, an enthalpy control strategy (which responds to both temperature and humidity) is generally used to prevent bringing cool but clammy air into a building. In buildings where return-air conditions vary widely over the year (perhaps because occupancy varies or because air is distributed by a variable air-volume system), differential control, either drybulb- or enthalpy-based, may be advantageous. Controllers that follow this strategy measure both the outside-air and return-air conditions and select the cooler or drier airstream to minimize the use of mechanical cooling.
5. *Type and location of the outside-air sensor.* For an economizer to work well, it must be able to accurately sense the temperature (or enthalpy) of the outside air. To do this, the sensor needs to be accurate, and it must be installed so that it is shielded from direct sun and wind.
6. *Sensor array.* For larger HVAC systems, installing just one temperature sensor may not be sufficient. This is because hot and cold air may stratify in larger ducts, and a single sensor will not reflect the average air temperature. Installing a temperature-sensing array, which includes multiple temperature sensors and provides an average reading, will provide better performance.
7. *Location of outdoor-air intake.* To maximize economizer effectiveness, the outside-air intake should be located away from building exhaust sources (for example, general exhaust, kitchen exhaust, and toilet exhaust). And it should not be too close to cooling towers, or hot and humid air will be drawn into the building.
8. *Integrated or nonintegrated.* An economizer that cannot work in concert with the mechanical cooling system is referred to as *nonintegrated.* With such systems, economizer operation is an all-or-nothing situation. When conditions are cool and dry, the economizer provides all of the necessary cooling, but when conditions are warmer, the damper moves to its minimum position and mechanical cooling is engaged. An *integrated* economizer, on the other hand, can use 100 percent outside air to provide as much cooling as possible and then use mechanical cooling to make up the difference. In more humid climates, an enthalpy control strategy (which responds to both temperature and humidity) is generally used to prevent bringing cool but clammy air into a building. To maximize economizer effectiveness, the outside-air intake should be located away from building exhaust sources (for example, general exhaust, kitchen, and toilet exhaust).

Energy Recovery Ventilators (ERVs)

Properly integrated desiccant dehumidification systems have become cost-effective additions to many innovative high-performance building designs. An ERV is an energy and humidity exchanger that employs desiccant technology for its functioning. ERV devices are placed between fresh air and exhaust air streams, moving energy and humidity between the two streams to save significant quantities of energy. Additionally, indoor air quality can be improved by higher ventilation rates, and desiccant systems can help to increase fresh air make-up rates economically. In low load conditions, outdoor air used for ventilation and recirculated air from the building must be dehumidified more than cooled.

Desiccants are materials that attract and hold moisture, and desiccant air-conditioning systems provide a method of drying air before it enters a conditioned space. With the high levels of fresh air now required for building ventilation, removing moisture has become increasingly important. Desiccant dehumidification systems are growing in popularity because of their ability to remove moisture from outdoor ventilation air while allowing conventional air-conditioning systems to deal primarily with control temperature (sensible cooling loads).

The Air-Conditioning and Refrigeration Institute (ARI) has developed a standard for ERVs, ARI Standard 1060 (2001), Rating Air-to-Air Heat Exchangers for Energy Recovery Ventilation Equipment, and ERV manufacturers should provide performance data in accordance with this standard. A typical ERV is shown in Figure 7.10. The device consists of a metal wheel coated with desiccant that rotates between the intake fresh air and exhaust air streams. In summer, it dries and cools the hot, humid intake air with the cool, dry exhaust air from the building, saving significant quantities of energy, especially because the removal of moisture is accomplished via the desiccant, a very energy-efficient strategy.

VENTILATION AIR AND CARBON DIOXIDE SENSORS

A healthy indoor environment is an important goal of green buildings. Creating a healthy interior requires that fresh outside air be brought into the building to dilute the buildup of potentially toxic components of indoor air. These toxic components include carbon dioxide from respiration, carbon monoxide from incomplete combustion of fuel, volatile organic compounds (VOCs) from building materials, and potentially others. The quantity of outside air required by ASHRAE 62.1 for ventilation air is significant, and it must be either heated or cooled to allow it to remix with the supply air stream.

Contemporary U.S. buildings have two basic methods for providing fresh or ventilation air for their occupants. First, the system can be designed to provide a constant quantity of fresh air based on a conservative evaluation of the number of occupants and the building's operating conditions. This approach has the advantage of being fairly simple, but the problem is that in a building with a variable population, substantial quantities of energy are wasted to condition the fresh air. A better approach would be to determine how many people are in the building and introduce the appropriate quantity of ventilation air based on the number of occupants. The concentration of carbon dioxide provides an indicator of how many people are in the building. Carbon dioxide is used as a surrogate ventilation index for diagnosing ventilation inefficiency or distribution problems. As the number of people in the space or the level of activity increases, so will the carbon dioxide concentration. Increased concentration of carbon dioxide in a space is also linked to discomfort and an increased perception of odors. Sensors are now available to detect the concentration of carbon dioxide in building spaces, and the data can be used as a surrogate for indoor air quality. The precise quantity of ventilation air needed to dilute the carbon dioxide to an appropriate level can be admitted to the space based on the measured carbon dioxide concentration. Buildings with populations that vary greatly can benefit from the use of this sensor technology because they can admit the exact amount of ventilation air needed, not the large quantities that would otherwise be required without this detection system.

WATER-HEATING SYSTEMS

In some types of buildings, water heating can consume large amounts of energy. In facilities with kitchens, cafeterias, health club facilities, or residences, there will be heavy demand for hot water. Solar water heating and tankless water heaters are technologies that can be used to reduce the hot water demand; these are described in the following subsections.

Solar Water-Heating Systems

An estimated 1 million residential and 200,000 commercial solar water-heating systems have been installed in the United States. Although there are many different types of solar water-heating systems, the basic technology is very simple. Sunlight strikes and heats an *absorber* surface within a *solar collector* or an actual storage tank. Either a heat-transfer fluid or the actual potable water to be used flows through tubes attached to the absorber and picks up the heat from it. Systems with a separate heat-transfer-fluid loop must utilize a heat exchanger to heat the potable water. The heated water is stored in a separate preheat tank or a conventional water heater tank until needed. If additional heat is needed, it is provided by electricity or fossil-fuel energy by the conventional water-heating system. By reducing the amount of heat that must be provided by conventional water heating, solar water-heating systems directly substitute renewable energy for conventional energy, reducing the use of electricity or fossil fuels by as much as 80 percent.

Today's solar water-heating systems are well proven and reliable when correctly matched to climate and load. The current market consists of a relatively small number of manufacturers and installers that provide reliable equipment and quality system design. A quality assurance and performance rating program for solar water-heating systems, instituted by a voluntary association of the solar industry and various consumer groups, makes it easier to select reliable equipment with confidence.

Solar water-heating systems are most likely to be cost-effective for facilities with water-heating systems that are expensive to operate or with operations such as laundries or kitchens that require large quantities of hot water. A need for hot water that is relatively constant throughout the week and throughout the year, or that is higher in the summer, is also helpful for solar water-heating economics. Conversely, hard water is a negative factor, particularly for certain types of solar water-heating systems, because it can increase maintenance costs and cause those systems to wear out prematurely.

Although solar water-heating systems all use the same basic method for capturing and transferring solar energy, they use a wide variety of technologies. Systems can be either active or passive, direct or indirect, pressurized or nonpressurized. As a rough guide, the solar system should have 10 square feet (1 square meter) of collector area for every 14 gallons (50 liters) of daily hot water usage, and the storage tank should have 1.4 gallons per square foot (50 liters per square meter) of collector area. This corresponds to 40 square feet (4 square meters) of collector for every apartment suite in multiunit residential buildings and 10 square feet (1 square meter) of collector for every five office workers in an office building.

Tankless (Instantaneous) Water-Heating Systems

Tankless or instantaneous water heaters eliminate the need for hot water storage by supplying energy at the point of demand to heat water as it is being used. Clearly, this takes high energy input, either electric or gas, at the point of use, but energy losses from storage tanks are eliminated. Unlike storage water heaters, tankless water-heating systems can theoretically provide an endless supply of hot water. The actual maximum hot water flow is limited by the size of the heating element or thermal input of the gas heater.

Demand water heaters, common in Japan and Europe, began appearing in the United States about 25 years ago. Unlike conventional tank water heaters, tankless water heaters heat water only as it is used, or on demand. A tankless unit has a heating device that is activated by the flow of water when a hot water valve is opened. Once activated, the heater delivers a constant supply of hot water. The output of the heater limits the rate of flow of the heated water.

Gas tankless hot water units typically heat more gallons per minute than electric units, but in either case, the rate of flow is limited. Electric tankless heaters should use less energy than electric storage systems. But gas-fired tankless heaters are only available with standing pilot lights, which lower their efficiency. In fact, the pilot light can waste as much energy as is saved by eliminating the storage tank.

Tankless heaters have either modulating or fixed output control. The modulating type delivers water at a constant temperature, regardless of flow rate. The fixed type adds the same amount of heat, regardless of flow rate and inlet temperature.

Electrical Power Systems

In addition to the building's air-conditioning and heating systems, the lighting system and electric motors are major consumers of electrical energy. Major advances have been made in lighting fixture and lighting control technologies that can dramatically reduce energy consumption. Because electric motors in buildings drive fans, pumps, and other devices, using the most energy-efficient motor can result in substantial energy savings. The following subsections describe advances in lighting and motor technology that can produce substantial energy savings in buildings.

LIGHTING SYSTEMS

Lighting is a voracious consumer of electrical energy; thus, a primary goal of all designs should be to reduce dependence on artificial light and to maximize the use of daylighting. These efforts should become an integrated strategy, that is, combining natural lighting and powered lighting to provide high-quality, low-energy illumination for the building's spaces. These are the subjects covered next.

Fluorescent Lighting

Fluorescent lighting is the best source for most building lighting applications because it is very efficient and can be switched and controlled easily. Modern linear fluorescent lamps have good color rendering and are available in many styles. Lamps are classified by length, form (straight or U-bend), tube diameter (for example, T-8 or T-5), wattage, pin configuration, electrical type (rapid or instant-start), Color Rendering Index (CRI), and color temperature. When specifying a lighting system, it is important that the lamp and ballast be electrically matched and the lamp and fixture optically matched.

Fluorescent lamp diameters are measured in ⅛-inch (0.3 centimeters) increments—for example, T-12s are 12/8 inch (3.2 centimeters) or 1½ inch (3.8 centimeters) in diameter, and T-8s are 1 inch (2.5 centimeters) in diameter. Typical linear fluorescent lamps are compared in Table 7.8; note that efficacy (lumens per watt) is higher with smaller-diameter lamps.[34]

TABLE 7.8

Fluorescent Light Fixture Characteristics

Lamp type	T-12	T-12 ES	T-8	T-5*
Watts	40	34	32	54
Initial lumens	3,200	2,850	2,850	5000
Efficacy (lumens/watt)	80	84	89	93
Lumen depreciation†	10%	10%	5%	5%

*High-output T-5 in metric length.
†Change from initial lumens to design lumens.

T-5 lamps are designed to replace T-8 fluorescent lamps. The T-5 lamp operates exclusively with electronic ballasts and offers continuous dimming. It has an efficacy of about 93 lumens per watt, compared to the 89 lumens per watt achievable with T-8 lamps. Most manufacturers use internal protective shield technology to minimize light depreciation to a predicted 5 percent over the life of the lamp. This technology has also made it possible to reduce the mercury content of lamps to about 3 milligrams, compared to the previous 15 milligrams.

Color rendering of fluorescent lamps is very important. Modern, efficient fluorescent lamps use rare-earth phosphors to provide good color rendition. The CRI describes how a light source affects the appearance of a standardized set of colored patches under standard conditions. A lamp with a CRI of 100 will not distort the appearance of the patches in comparison to a reference lamp, while a lamp with a CRI of 50 will significantly distort colors. T-8 and T-5 lamps are available only with high-quality phosphors that provide CRIs greater than 80. The minimum acceptable CRI for most indoor applications is 70; levels above 80 are recommended.

Color temperature influences the appearance of luminaires and the general "feel" in the space. Low color temperature (e.g., 2,700 K) provides a warm feel similar to that of light from incandescent lamps; 3,500 K provides a balanced color; and 4,100 K emits "cooler," bluish light. Standardizing the color temperature of all lamps in a room or facility is recommended. Electronic ballasts with linear fluorescent lighting should be specified. These are significantly more energy-efficient than magnetic ballasts and eliminate the hum and flicker associated with older fluorescent lighting. Dimming electronic ballasts are also widely available.

Luminaires should be selected based on the tasks being performed. Reflectorized and white industrial fixtures are very efficient and good for production and assembly areas but are usually inappropriate for office applications. Lensed fluorescent fixtures (*prismatic lens* style) typically result in too much reflected glare off computer screens to be a good choice for offices. In areas with extensive computer use, the common practice is to install *parabolic* luminaires, which minimize high-angle light that can cause reflected glare off computer screens; however, these may result in unpleasant illumination in the presence of dark ceilings and walls. Instead, for tall ceilings, over 9 feet (2.7 meters) in height, direct/indirect pendant luminaries should be used. For lower ceilings, 8 feet 6 inches (2.6 meters) in height, parabolic luminaires with semispecular louvers should be considered.

Luminaires should not be selected solely on the basis of efficiency. A very-high-efficiency luminaire can have inferior photometric performance. The most effective luminaires are usually not the most efficient, but they deliver light where it is most needed and minimize glare. The Luminaire Efficiency Rating (LER) used by some fluorescent fixture manufacturers makes it easier to compare products. Since the LER includes the effect of the lamp and ballast type, as well as the optical properties of the fixture, it is a better indicator of the overall energy efficiency than simple fixture efficiency. An LER of 60 is good for a modern electronically ballasted T-8 fluorescent fixture; 75 is very good and is close to state of the art.

Fiber-Optic Lighting

Fiber-optic lighting utilizes light-transmitting cable fed from a light source in a remote location. A fiber-optic lighting system consists of an illuminator (light source), fiber-optic tubing, and possibly fixtures for end-emitting uses. When light strikes the interface between the core and cladding of the cable, total internal reflection occurs and light bounces or reflects down the fiber within the core. Two types of fiber are used: small-diameter strands bundled together or a solid core (the latter being more limited in application). The lighting source is generally a halogen or metal halide lamp. Fiber-optic lighting is generally energy-efficient and provides illumination over a given area. The only electrical connection needed for the system

Figure 7.11 Fiberstars' EFO fiber-optic lighting, shown in the Trammel Crow's Houston offices, is low-energy, lightweight, and ultrasafe because it does not conduct electricity. The manufacturer claims that a single EFO lamp using 68 watts and replaces about 400 watts of halogen lamps. (Photograph courtesy of Fiberstars, Inc.)

is at the illuminator. No wiring or electrical connection is required along any part, either at the fiber-optic cable or at the actual point source fixture.

Fiber-optic lighting systems provide many benefits and eliminate many problems encountered with conventional lighting systems. Infrared and ultraviolet wavelengths produced by a given light source are undesirable by-products, and fiber-optic systems can filter these out, eliminating the damaging effect of ultraviolet and infrared radiation. Fiber-optic lighting requires no voltage at the fixture, is completely safe, emits no heat, and is virtually maintenance-free. This lighting technology is especially useful for retail settings, supermarkets, and museums because it emits no heat or ultraviolet radiation.

LED Lights

Light-emitting diodes (LEDs) for lighting systems are evolving very rapidly, and white-light LEDs are now being produced that can be used in many building applications. LEDs are based on semiconductors that emit light when current is passed through them, converting electricity to light with virtually no heat generation. Until the early 1990s, red, yellow, and green LEDs were being produced. In the early 1990s, blue LEDs and then white LEDs were developed.

LEDs produce 29 to 45 lumens per watt, compared to about 12 lumens per watt for incandescent bulbs. A few premium models deliver 52 to 69 lumens per watt. At present, fluorescent lights produce 75 lumens per watt, and to be competitive, LEDs will have to reach this level. The latest laboratory versions are producing 150 lumens per watt. LEDs are also very tough and durable, able to absorb large shocks without malfunctioning. At present, however, LEDs are point sources of light, which limits their applications. With a projected lifetime of at least 50,000 hours, LEDs last 20 times longer than incandescent light bulbs and 2 to 3 times longer than fluorescent lights. Some forecasts are that within 20 years, up to 25 to 30 percent of incandescent lighting applications will be replaced by this technology. For general building applications, LEDs are an emerging technology, and it is expected that room lighting products at competitive prices will be widely available by 2010.

Figure 7.12 An intelligent LED lighting system by Color Kinetics and 4Wall Entertainment illuminates the exterior of the Hard Rock Hotel & Casino in Las Vegas, cutting its annual energy costs from $18,000 to $1,900. (Photo courtesy of 4Wall Entertainment.)

LIGHTING CONTROLS

Ideally, lighting controls should comprise an integrated system that performs two basic functions:

1. Detects occupancy and turn lights on or off in response to the presence or absence of occupants
2. Throttles lights up and down or turns lights on and off to compensate for levels of natural light provided by the daylighting system

Research has shown that daylight-linked electrical lighting systems—such as automatic on/off and continuous dimming systems—have the potential to reduce the electrical energy consumption in office buildings by as much as 50 percent.

There are two basic types of daylighting control systems: dimming and switching. Dimming controls vary the light output over a wide range to provide the desired light level. Switching controls turn individual lamps off or on as required. In a conventional two-lamp fixture, there are three settings: both lamps off, one lamp on, both lamps on. The same strategy can be used with three- and four-lamp fixtures. Dimming systems, which require electronic dimmable ballasts, are more expensive than switching systems; however, they achieve greater savings and do not have the abrupt changes in light level characteristic of switching systems. Dimming systems are best suited to offices, schools, and those areas where deskwork is being performed. Switching systems can be used in areas with high natural light levels (e.g., atria and entranceways) and where noncritical visual tasks are being performed (e.g., cafeterias and hallways).

Of course, neither system is appropriate in nondaylit areas. The lighting control zones and number of sensors need to be carefully designed. At least one sensor is required for each building orientation. The lighting control zone should only be as deep into the building as is effectively daylit—about 16 feet (5 meters) from windows in conventional office plans. Light shelves can extend the daylit zone deeper into the building's interior.

In addition to energy savings, electric-light dimming systems offer two other advantages over conventional lighting systems. First, conventional lighting systems are typically designed to overilluminate rooms to account for the 30 percent drop in lighting output over time. Electric-light dimming systems automatically com-pensate for this reduced output to give a constant light level over time. Second, daylighting controls can be adjusted to give the desired light level for any space. Thus, when floor plans are changed, it is easy to adjust the light levels to meet the lighting needs of each area, provided that the system is properly zoned and has adequate lighting capacity.

The cost of switching controls is quite modest, and these systems should be considered in all applications where changes in light level can be tolerated. Dimming lighting controls are approximately twice the price of switching controls and require electronic dimmable ballasts.

ELECTRIC MOTORS

Electric motors are important components of modern buildings, as they drive fans, pumps, elevators, and a host of other devices. Over half of all electrical energy in the United States is consumed by electric motors. Motors typically consume 4 to 10 times their purchase cost in energy each year, so energy-efficient models often make economic sense. For example, a typical 20-horsepower, continuously running motor uses almost $8,000 worth of electricity annually at 6 cents per kWh, about nine times its initial purchase price. Improving the efficiency of electric motors and the equipment they drive can save energy and reduce operating costs.

The construction materials and the mechanical and electrical design of a motor dictate its final efficiency. Energy-efficient motors utilize high-quality materials and employ optimized design to achieve higher efficiencies. Large-diameter copper wire in the stator and more aluminum in the rotor reduce resistance losses of the energy-efficient motor. An improved rotor configuration and an optimized rotor-to-stator air gap reduce stray load losses. An optimized cooling fan design provides ample motor cooling with a minimum of windage loss. Thinner and higher-quality steel laminations in the rotor and stator core allow the energy-efficient motor to operate with substantially lower magnetization losses. High-quality bearings result in reduced friction losses (see Table 7.9).

When buying energy-efficient electric motors, there are three factors to keep in mind:

- Not all high-efficiency motors are the same. The National Electrical Manufacturers Association (NEMA) requires that the average nominal efficiency test method be listed on the nameplate. International Electric Wire and Cable, Inc. (IEWC) test method 34-2 and Japanese Electrotechnical Committee (JEC) test method 37 result in slightly higher efficiencies than the Institute of Electrical and Electronics Engineers (IEEE) 112 method. Current NEMA codes require that the motor nameplate carry both the efficiency and test standards.
- Motors perform best at full load. An underloaded motor, energy-efficient or not, is less efficient than a fully loaded motor.
- Energy-efficient motors are most attractive economically when energy costs and/or operating hours per year are high.

TABLE 7.9

Efficiency Comparison of Energy-Efficient and Standard Electric Motors

		Load			
Size	**Motor Type**	**Full**	**75%**	**50%**	**25%**
	Efficient	95.3	95.6	95.2	92.7
100 hp	Standard	92.9	92.5	91.2	86.5
	Spread	2.4	3.1	4.0	6.2
	Efficient	90.8	91.6	91.1	86.8
10 hp	Standard	87.0	87.9	86.6	79.9
	Spread	3.8	3.7	4.5	6.9
	Efficient	84.7	84.8	82.6	74.6
1 hp	Standard	77.2	74.0	69.2	54.7
	Spread	7.5	9.9	13.4	19.9

Innovative Energy Optimization Strategies

At least partially because of the green building revolution, a wide variety of innovations in building systems are emerging. Four of the more innovative approaches are described here: radiant cooling, ground coupling, renewable energy systems, and fuel cells. Each of these is a cutting-edge strategy that can have a marked effect on energy consumption if used properly in a building.

RADIANT COOLING

In the United States, cooling is generally delivered to conditioned spaces using air that is pressurized by fans and delivered via ductwork to the various spaces. Air has a very low heat capacity, and the result is that rather large quantities of air must be delivered to a space to provide the needed cooling effect. Additionally, air, a compressible medium, is relatively energy-intensive to move, compared to water, which is incompressible, has very high heat capacity, and can be moved comparatively cheaply via pumping. That is why, in Europe, radiant cooling is frequently used for cooling spaces. These systems use water, which has 3,000 times the energy transport capacity of air, as the medium for delivering cooling to the space. In Germany, radiant cooling systems have become the new standard.

Radiant cooling systems circulate cool water through tubes in ceiling, wall, or floor elements or panels. The water temperature does not differ noticeably from the room temperature, so care must be exercised to ensure that the temperature of the circulated water does not reach the dewpoint of the air in the space. Otherwise, condensation will occur, resulting in moisture problems. The cost of a radiant cooling system is approximately the same as that of a VAV system but the life-cycle savings are 25 percent higher compared to those of a VAV system. Moreover, the energy required for circulating water is only about 5 percent of the energy needed to circulate a comparable capacity of air.

There are three main types of radiant cooling systems:

- *Concrete core:* Plastic tubes are buried in concrete floor and ceiling slabs.
- *Metal panels:* Metal tubes are connected to aluminum panels.
- *Cooling grids:* Plastic tubes are embedded in plaster or gypsum.

The metal panel system is the most commonly used radiant cooling system and, due to its metal construction, has a relatively fast response time to changing conditions. Cooling grids are generally the choice for retrofit projects because the grid of plastic cooling tubes is readily placed in plaster or gypsum in existing walls. As a guide to system sizing, the total heat transfer rate (combined radiation and convection) is about 11 watts per square meter for each degree Celsius ($w/m^2/°C$) (0.7 watts per square foot per degree Fahrenheit) temperature difference for cooled ceilings.

Design guidelines for radiant cooling systems are as follows:

1. The building should be well sealed.
2. In humid areas, the intake fresh air should be dehumidified prior to its entry into conditioned spaces.
3. Radiant cooling requires a large surface area due to the relatively small temperature difference between the cooling surface and the room air.
4. The setpoints for cooling and heating must be carefully considered to deliver maximum conditioning without causing moisture problems. For instance, for a typical system in Germany, during the cooling season, the room temperature setpoint is about 80°F (27°C), with cold or chilled water entering the radiant cooling panels at 61°F (16°C) and leaving at 66°F (19°C). For heating, the room setpoint is 68°F (20°C), with heated water delivered at 95°F (35°C) and leaving the radiant panels at 88°F (31°C).
5. Humidity sensors should be used to detect when the temperature of the supply water is approaching the dewpoint to activate valves that will prevent condensation from occurring.

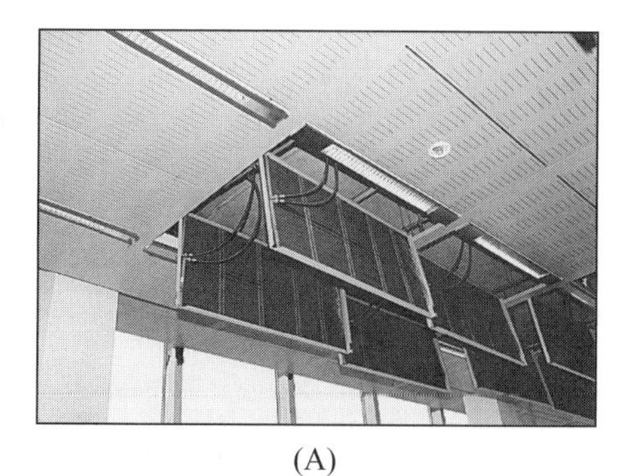

(A)

(B)

Figure 7.13 Radiant cooling panels provide a low-energy solution for cooling, requiring only a fraction of the energy of a conventional system based on air handling and ductwork. (A) A dropped-panel installation showing a radiant cooling panel installation and ease of maintenance. (B) Installing a cooling mat: grids of plastic tubing carrying chilled water are placed under the ceiling drywall. (Photographs courtesy of Juan Rudek, Karo Systems.)

GROUND COUPLING

One innovative method for reducing energy consumption in a building is ground coupling, in which the thermal characteristics of the earth and groundwater in the vicinity of the building are used for cooling and heating purposes. There are two major methods for applying ground coupling for building conditioning: direct and indirect. In the direct approach, groundwater is employed in radiant cooling systems and fresh air is cooled through ground contact. The indirect approach employs heat pumps in conjunction with the ground or groundwater to move heating and cooling energy between the building and the earth. It is feasible, for example, to use groundwater in the 60°F (16°C) range in a radiant cooling system for a building and virtually eliminate the need for a chilled water plant. The following subsections describe these two approaches.

Ground Source Heat Pumps

Ground source heat pump (GSHP) systems use the ground as a heat source in the heating mode and as a heat sink in the cooling mode. The ground is an attractive heat source or sink compared to outdoor air because of its relatively stable temperature. In many locations, the soil temperature does not vary significantly over the annual cycle below a depth of about 6.5 feet (2 meters). For example, in Louisiana, outdoor air temperatures may range from wintertime lows of 32°F (0°C) or lower to summertime highs of about 95°F (33°C), while the soil temperature at depths greater than 6.5 feet (2 meters) never falls below about 64°F (18°C) or rises to approximately 77°F (25°C), averaging around 68°F (20°C). A number of different methods have evolved for thermally connecting, or *coupling*, the heat pump systems with the ground, but the two major methods are vertical systems and horizontal systems. These systems depend on how the piping that makes ground contact is laid out.

The *horizontal* ground-coupling system uses plastic piping placed in horizontal trenches to exchange heat with the ground. Piping may be placed in the trenches either singly or in multiple-pipe arrangements. The primary advantage of horizontal systems is lower cost. This is a result of fewer requirements for special skills and equipment, combined with less uncertainty about subsurface site conditions. The disadvantages of the horizontal ground-coupling system are its high land area requirements, its limited potential for heat exchange with groundwater, and the wider temperature swings of the soil at typical burial depths.

Vertical ground coupling is the most common system used in commercial-scale systems. Vertical U-tube plastic piping is placed in bore holes and manifolded in shallow trenches at the surface. Vertical ground coupling has several advantages: low land area requirements, stable deep-soil temperatures with greater potential for heat exchange with groundwater, and adaptability to most sites. Among the disadvantages of vertical ground coupling are potentially higher costs, problems in some geological formations, and the need for an experienced driller/installer. The regulatory requirements for vertical boreholes used for ground-coupling heat exchangers vary widely by state. One note of caution to the designer is that some regulations, installation manuals, and/or local practices call for partial or full grouting of the borehole. The thermal conductivity of materials normally used for grouting is very low compared to the thermal conductivity of most native soil formations. Thus, grouting will tend to act as insulation and hinder heat transfer to the ground.

In addition to ground-coupled heat pumps, systems that use both surface water and groundwater have been successful. In fact, for commercial-scale applications, if groundwater is available in sufficient quantities, it should be considered as the first alternative, as it will often turn out to be the least costly.

Direct Ground Coupling for Fresh Air and Chilled Water

It is also possible to heat and cool fresh air being introduced into a building by bringing it in underground through large-diameter, 1- to 2-meter (3- to 7-foot) galvanized steel tubes, known as *earth-to-air heat exchangers*. Additionally, groundwater can sometimes be used as the source of chilled water, reducing or eliminating the need for mechanical chilled water systems. Both of these practices are becoming common in Germany, where buildings are now routinely conditioned using a comprehensive ground-coupling scheme. For example, a 50,000-square-meter (538,000-square-foot) Mercedes showroom in Stuttgart has all of its fresh air brought in through a 1.8-meter- (6-foot) diameter corrugated steel tube, the top of which is located 2 meters (6.6 feet) underground at a velocity of about 155 meters (500 feet) per minute. The ground temperature at this depth is a stable 60°F (16°C). In winter, cold outside air is warmed up to approximately the ground temperature prior to introduction into the building. In summer, hot outside air is significantly cooled prior to its introduction into the facility.

Groundwater, where permitted by local jurisdictions, can be used directly in a radiant cooling system, where the temperature is adjusted by mixing valves or by employing a relatively small heat pumping system to move energy to and from the groundwater stream. The groundwater is pumped into the radiant cooling system and discharged back to the ground with only a few degrees of temperature change.

It is also feasible to design and install a ground-coupling system that both conditions the fresh air being brought into the building and uses groundwater for a radiant cooling system. A well-designed ground-coupled HVAC system can provide significant savings by greatly reducing the requirements for equipment, ductwork, and air handlers.

RENEWABLE ENERGY SYSTEMS

Renewable energy can be generated on-site by three different techniques: photovoltaics, wind energy, and biomass. Each of these has advantages and disadvantages and varying levels of complexity. A brief summary of each is provided in Table 7.10.

Figure 7.14 Ground-coupled system showing 1.8-meter- (6-foot-) diameter air intake under a Mercedes showroom in Stuttgart, Germany. The 100-meter- (325-foot-) long galvanized steel tube heats cold air via ground contact in winter and cools hot air in summer. The tube is located in a zone under the building where the temperature is a constant 60°F (16°C).

Building-Integrated Photovoltaics

Photovoltaic (PV) cells are semiconductor devices that convert sunlight into electricity. They have no moving parts. Energy storage, if needed, is provided by batteries. PV modules are successfully providing electricity at hundreds of thousands of installations throughout the world. Especially exciting are building-integrated photovoltaic (BIPV) technologies that integrate PV cells directly into building materials, such as semitransparent insulated glass windows, skylights, spandrel panels, flexible shingles, and raised-seam metal roofing. PV elements can be fabricated in different

TABLE 7.10

Advantages and Disadvantages of Renewable Energy Systems

Renewable Energy Type	Advantages	Disadvantages
Photovoltaics (PVs)	▪ New technologies allow integration into building façade ▪ Price of PV modules is dropping as demand increases	▪ Remains relatively expensive ▪ Potential metering problems with local utility
Wind	▪ Lowest kWh cost of any renewable energy source	▪ Generally large, unsightly generators ▪ Significant annual wind speed needed
Biomass	▪ Can use local vegetation for fuel ▪ Potentially low-cost energy source	▪ Systems for buildings are not readily available

forms. They can be used on or be integrated into roofs and façades as part of the outer building cladding, or they can be used as part of a window, skylight, or shading device. PV laminates provide long-lasting weather protection. Their expected life span is in excess of 30 years. Warranties are currently available for a 20-year period.

PV systems are modular in nature; hence, they can be adapted to changing situations. They can usually be added, removed, and reused in other applications. Typical modules consist of glass laminates, plastic tedlar bounding material, and silicon cells with trace amounts of boron and phosphorus. Their disposal or recycling after the end of their life span should not create any environmental problems.

A variety of attractive BIPV products are available that allow building surfaces, such as the roof, walls, skylights, and sunshades, to double as solar collectors. Integrating these products into the building envelope creates a large solar collection area, enabling solar power to displace more of the electricity used in the building. The cost of the PV system is offset by the fact that the BIPV products displace standard building envelope components. PVs can be integrated into the roofing system through PV roof shingles, roof tiles, and metal roof products, all of which can replace the standard roof. Alternatively, framed PV modules can be incorporated into the roofing system. BIPV glazing systems are available that allow sloped and overhead glazing to capture solar energy. These glazing systems are insulated and can be specified to provide the desired level of light transmission for daylighting, typically as needed per kilowatt of capacity.

The curtainwall offers considerable potential for BIPVs. A wide variety of PV products can be used in place of architectural spandrel glass and vision glass. Sunshades and skylights, common BIPV applications, have become popular in Europe. BIPV systems are available for sunshades and skylights that are visually transparent or provide partial shading. It is an easy upgrade to substitute preengineered BIPV sunshades for conventional sunshades. The look and color of these building-integrated PV products vary with the application and the type of solar collector technology. The most efficient solar collectors are deep blue to black in color, although BIPV products are also available in dark gray and medium blue; some manufacturers also may produce custom-colored BIPV products for large orders.

Figure 7.15 BIPVs in the Solaire building, Battery Park, New York City. The PVs are the speckled surfaces between the windows. (Photograph courtesy of the Albanese Development Corporation.)

Depending on the type of collection medium, BIPV can generate approximately 5 to 10 watts of power per square foot (50 to 100 watts per square meter) of collector area in full sunlight. That means that a collector area of 100 to 200 square feet (10 to 20 square meters) typically is needed per kilowatt of capacity. The annual power output varies with the latitude and climate, as well as with the orientation of the building surface that comprises the PV material. The annual energy output ranges from 1,400 to 2,000 kWh per kW of installed system capacity.

Wind Energy

Wind energy is the fastest-growing form of energy production, with an estimated year-on-year growth of 25 percent. According to the U.S. Department of Energy's National Renewable Energy Laboratory (NREL), the cost of wind energy has declined from $0.40 per kWh in the 1980s to less than $0.05 per kWh today. In 2002, the United States doubled the 1,700 megawatts (MW) of additional wind-generating capacity brought online in 2001. By the beginning of 2002, installed capacity in this country was over 4,200 MW, reaching over 10,000 MW in 2006, with another 3,000 MW expected to be brought online in 2007. The American Wind Energy Association (AWEA) has estimated that, with the support of the government and utilities, wind energy could provide at least 6 percent of the nation's electricity supply by 2020. The AWEA estimates that 20 percent of U.S. electricity demand could ultimately be met by wind energy. Texas has the highest wind energy capacity in the United States, about 2,730 MW in late 2006, adding new capacity at the rate of about 400 MW per year. One of the reasons for the high rate of installation in

Texas is that state's subsidy program, which provides $0.019 per kWh for projects brought online by the end of 2007.

Small wind turbines (those with less than 100 kW output) suitable for building-scale applications are available, and there are innovative programs that can make their incorporation into a building project financially feasible.

Biomass Energy

The term *biomass* refers to any plant-derived organic matter available on a renewable basis, including dedicated energy crops and trees, agricultural food and feed crops, agricultural crop wastes and residues, wood wastes and residues, aquatic plants, animal wastes, municipal wastes, and other waste materials. Handling technologies, collection logistics, and infrastructure are important aspects of the biomass resource supply chain.

According to the American Bioenergy Association, the United States has the available land and agricultural infrastructure to produce adequate biomass in a sustainable way to replace half of the country's gasoline usage or all of its nuclear power without a major impact on food prices.[35] Shifting part of the $50 billion now spent for oil imports and other petroleum products to rural areas would have a profoundly positive effect on the economy in terms of jobs created (for production, harvesting, and use) and industrial growth (facilities for conversion into fuels and power). David Morris of the Institute for Local Self-Reliance refers to this as moving partway from a "hydrocarbon economy to a carbohydrate economy."[36]

FUEL CELLS

Fuel cells are devices that generate electricity in a process that can be described as the reverse of electrolysis. In electrolysis, electricity is input to electrodes to decompose water into hydrogen and oxygen. In a fuel cell, hydrogen and oxygen molecules are brought back together to create water and generate electricity. The principle behind fuel cells was discovered in 1839, but it took almost 130 years before the technology began to emerge, first in the U.S. space program and more recently in a host of new technologies and applications. Fuel cells provided power for on-board electronics for the Gemini and Apollo spacecraft and continue to provide electricity and water for the space shuttle today.

Fuel cells construction generally consists of a fuel electrode (anode) and an oxidant electrode (cathode) separated by an ion-conducting membrane. Fuel cells must take in hydrogen as a fuel, but any hydrogen-rich fuel can be processed to extract its hydrogen for fuel cell use. A device called a *reformer* is used to process nonhydrogen fuels to extract the hydrogen. This device reformulates nonhydrogen fuels such as gasoline, methane, diesel fuel, and ethanol to turn them into hydrogen. Due to their complexity, reformers are still very expensive. Some of the higher-temperature fuel cells can directly process some nonhydrogen fuels—methane, gasoline, and ethanol—without using the reformer.

There are several different types of fuel cells, including phosphoric acid, alkaline, molten carbonate, solid oxide, and proton exchange membrane (PEM) fuel cells. PEM fuel cells are of great interest because they operate at relatively low temperatures (below 200°F/93°C), have high power density, and can vary their output quickly to meet shifts in power demand. Current-generation fuel cells last anywhere from 1 to 6 years before they wear out or need an overhaul. Fuel cells are currently expensive to manufacture and depend on ongoing technological innovations to ensure their eventual economic viability. And as noted, unless hydrogen is available as a fuel cell, a reformer must be utilized to process hydrogen-rich fuels to extract the hydrogen gas, an expensive additional component and one that adds considerable complexity to the fuel cell system.

For buildings and utilities, fuel cell power plants are beginning to make economic sense. The potential for home and commercial building power systems to use fuel cells, particularly in the United States in an era of utility deregulation, is quite high. An additional positive feature is that heat produced by some types of fuel cells can be used for thermal cogeneration in building power systems.

> Fuel cells specifically designed for building use are beginning to emerge. Plug Power is developing the GenSys® fuel cell, which will produce electricity by using the hydrogen contained in natural gas or liquid petroleum gas (LPG).[37] For most building applications, this system has three major components:
>
> A reformer that extracts hydrogen from the natural gas or LPG
>
> A fuel cell that changes the hydrogen to electricity
>
> A power conditioner that converts the fuel cell's electricity to the type and quality of power required for use in the building

Smart Buildings and Energy Management Systems

In its simplest form, a building's energy management system (EMS) is a computer with software that controls energy-consuming equipment to ensure that the building operates efficiently and effectively.[38] Many EMS systems are also integrated with fire protection and security systems. A newer concept, *smart buildings*, uses the concept of information exchange to provide a work environment that is productive and flexible. In each building zone, a building automation system (BAS) and high-bandwidth cabling connect all building telecommunications; heating, ventilation, air-conditioning, and refrigeration (HVAC&R) components; fire, life, and safety systems (FL&S); lighting; emergency or redundant power; and security systems. The

Figure 7.16 An array of five PC25 phosphoric acid fuel cells (PAFCs), manufactured by United Technology Fuel Cells, Inc., powering a postal facility in Alaska. The PC25 provides about 200 kW of power, using natural gas as its fuel source. Waste heat generated by the PC25 can be used for heating applications or to create cooling using an absorption cycle chiller. (Photograph courtesy of United Technology Corporation Fuel Cells, Inc.)

smart building concept is important for consideration in green buildings because of the enormous demand for flexible layout and responsiveness, both afforded by smart buildings. A 1999 survey of building owners conducted by the Building Owners and Managers Association (BOMA) found that there were 13 systems desired by tenants of smart buildings (see Table 7.11).[39] In addition to the items on this list, also in demand today is the capability for wireless technologies to enable telecommunications and Internet connections. EMS systems can produce substantial energy savings, on the order of 10 percent of building energy consumption.

Modern smart buildings also use digital controls, referred to as *direct digital controls* (DDC), to control the growing variety of devices and control systems in the building's HVAC&R systems. In addition to controlling systems based on temperature and humidity, DDC permits the integration of information about air quality and carbon dioxide levels. Digital systems can process and store information and manage complex interrelationships between components and systems. Control of lighting systems can be accomplished with DDC systems and allow occupant control of lighting, a prime feature of smart buildings.

TABLE 7.11

Building Systems Typically Found in a Smart Building

Building Systems Typically Found in a Smart Building
Fiber-optics capability
Built-in wiring for Internet access
Wiring for high-speed networks
Local area network (LAN) and wide area network (WAN) capability
Satellite accessibility
Integrated digital services network (IDSN)
A redundant power source
Conduits for power/data/voice
High-tech, energy-efficient HVAC system
Automatic on/off sensor in the lighting system
Smart elevators that group passengers by floor designation
Automatic sensors installed in faucets/toilets
Computerized/interactive building directory

Ozone-Depleting Chemicals in HVAC&R and Fire Suppression

Of the many building systems, mechanical systems used for generating cooling and for fire protection employ the largest quantity of ozone-depleting chemicals. Removing these chemicals from building inventories and replacing systems with new products that use non-ozone-depleting chemicals are priorities. This section describes the replacement of refrigerants in air conditioning, and halon in fire protection systems, with newer technologies that do not impact the ozone layer. HVAC&R systems or equipment constitute the majority of mechanical engineering systems. In addressing ozone depletion, fire protection or fire suppression systems that employ halon must be discussed because halon is the most powerful ozone-depleting substance.

HCFC REFRIGERANTS

The chemicals known as *chlorofluorocarbons* (CFCs) were commonly used as refrigerants in chillers, mechanical devices that are used to generate cooling; CFC-11 and CFC-12 were the two most common. Then, in 1986, their release into the atmosphere was found to be a major cause of destruction of the ozone layer, and international treaties soon called for their phaseout. The impact of CFCs on the ozone layer is indicated in terms of a quantity called *ozone depletion potential* (ODP). The ODP is defined as 1.0 for CFC-11, meaning that a substance with an ODP of 10.0 depletes ozone at 10 times the rate of CFC-11. Other typical CFCs have a value of 1.0. For hydrochloroflurocarbons (HCFCs), the ODP ranges from about 0.02 to 0.11, or about 10 to 50 times less impact than caused by CFCs.

Several families of chemicals have been used to replace CFCs, among them HCFCs and hydrofluorocarbons (HFCs). Although HCFCs are a great improvement over CFCs, they still have relatively high ODPs. HFCs, on the other hand, have a zero ODP, and as a result have no impact on the ozone layer. HFC-containing equipment is available from all major manufacturers. HFC-134a has become the dominant refrigerant, replacing HCFC-123 and HCFC-22 in most chillers designed for building use. HCFC-22 is currently used in a large proportion of positive displacement compressor-based chillers and in some larger-tonnage centrifugal chillers. These uses predate the Montreal Protocol but will be phased out as part of the overall HCFC phaseout. In the United States, HCFC-22 cannot be used in new equipment after January 1, 2010.

According to the Carrier Corporation, HFC-134a has proven to be an optimal refrigerant in chiller applications because it has no chlorine molecules and does not contribute to ozone depletion. HFC-134a is a highly efficient thermodynamic refrigerant in application. Current centrifugal chillers using HFC-134a are 21 percent more efficient than chillers sold just 6 years ago and 35 percent more efficient than the chillers installed during the 1970s and 1980s. Because HFC-134a is a positive-pressure refrigerant, pressure vessels using it must conform to the American Society of Mechanical Engineers (ASME) pressure code, and every step in their construction must be inspected by third-party insurance companies. As a result of the stringent testing and applied technology, chiller leak rates can be lowered to less than 0.1 percent annually. Existing chillers have a leak rate of 2 to 15 percent. HFC-134a also has a smaller molecular mass than the past CFCs and HCFCs. This is an important feature, as it results in an overall product size that is 35 to 40 percent smaller, a size reduction that helps offset the cost of construction and facilitates the use of smaller interconnecting pipes. This advantage has led to the addition of isolation valves to the chiller piping connection so that the HFC-134a can be stored in the chiller during service. This feature gives the end user the option of never having to remove the refrigerant from the vessels once charged, a real "no emissions" feature. An additional advantage of HFC-134a chillers is their smaller size, requiring much less plant space than the CFC-11 chillers they replace.

HALON REPLACEMENT IN FIRE PROTECTION SYSTEMS

High-value electrical systems, telecommunications systems, rare books libraries, and other areas that could be badly damaged in a fire protection situation by a conventional water sprinkler system have been generally protected by a gaseous Halon 1301 fire protection system. Safe for human exposure, nonconductive, and inert, the bromine-bearing halon molecule was considered an ideal fire protection fluid, except that it is probably the most damaging of all the halogenated compounds used in building applications. When exposed to ozone, Halon 1301 has a destructive catalytic-type effect, and its phaseout has been a priority of the international community.

However, a wide variety of replacements for Halon 1301 have been proposed, and this has complicated the design and specification of systems in sensitive areas. Manufacturers of fire protection systems are providing their own, often proprietary, systems for these types of applications. Water mist, inert gases, carbon dioxide, halocarbons, and mixtures of halocarbons are offered to meet the need for a safe, inert fire suppression system. The problems is that, unlike Halon 1301, which was a generic fire suppression alternative that was well known and could be engineered by most mechanical or fire protection engineers, each of these replacements is manufacturer-specific, so their real-world performance is not well known or documented; therefore, dependence on the manufacturer for both design and equipment is necessary.

One of the outcomes of the ban on Halon 1301 is that water has been reconsidered for use in areas where it was once thought to be unfeasible. Water mist systems, in which a very fine water mist is sprayed into the fire zone, are the simplest and least problematic with respect to human exposure and cost. They are surprisingly compatible with electrical equipment because of the small quantities of water that are used. Dry-pipe systems, in which a conventional sprinkler system is used, but with no water in the fire protection system piping over the protected space, provide protection without the risk of an accidental leak. The risk of fire in modern telecommunications and computer facilities is quite low; consequently, the employment of costly alternatives to Halon 1301 may not be necessary or justified.

High-Performance Buildings and Energy: A Case Study

RIVER CAMPUS BUILDING ONE, OREGON HEALTH AND SCIENCE UNIVERSITY, PORTLAND

Medical facilities pose significant challenges and provide enormous opportunities for high-performance building project teams. While they are generally far more complex than other building types and have special requirements for controlling the movement of viruses and other pathogens that can result in the transmission of disease, they also have the potential for contributing to health and well-being by virtue of their design. In expanding their Portland, Oregon, main campus, the Oregon Health and Science University required a new 400,000-square-foot, 16-story medical office and wellness building, which was named River Campus Building One. In addition to a two-story wellness center, the building includes several different types of university operations, including biomedical research, clinical space, an outpatient surgery, and educational space. The medical offices are built atop a three-level below-grade parking structure (see Figure 7.17).

The project team included the Portland office of Interface Engineering, Inc., a multidisciplinary engineering firm that provided HVAC, plumbing, electrical, power and backup power distribution, lighting, security, energy, telecommunications, data, and fire alarm systems design, as well as all tenant improvements and basic commissioning. Interface Engineering's project team was instrumental in River Campus Building One's receiving a LEED Platinum certification from the USGBC. How this project team helped achieve this rating through a holistic approach to the design of these technical systems on an excellent case study on how to create a low-energy building.

THE PROJECT BUDGET

The total project was initially budgeted at $145.4 million, with $30 million allocated for the building's mechanical, electrical, and plumbing (MEP) systems. Interface's

Figure 7.17 The Oregon Health and Science University's River Campus Building One, shown in the late stages of construction, is a LEED-NC 2.1 Platinum-certified building. (Photograph courtesy of Interface Engineering, Inc.)

MEP design approach resulted in savings of nearly $4 million of the initial $30 million budget. What was truly remarkable in this project is that energy consumption was reduced about 60 percent compared to the baseline model, while at the same time the capital cost of the MEP systems was reduced 10 percent. Conventional wisdom is that high-performance MEP systems will cost more than their code-compliant alternatives. For River One, Interface's engineering team did indeed "tunnel through the cost barrier," as suggested was possible by Paul Hawken, Amory Lovins, and Hunter Lovins in their 1999 book *Natural Capitalism.*

THE STRATEGY: INTEGRATED DESIGN

Achieving a win-win combination of lower energy consumption and lower capital costs for energy systems is a difficult but clearly achievable strategy. Interface was able to reach this holy grail of sustainable design using an *integrated design* approach. As articulated for the River Campus Building One project, integrated design is different from conventional design in two key respects: (1) goal setting starts early in the sustainable design process, during the programming and conceptual design phases, and (2) the entire design team is involved in the process much earlier than is usually the case so that engineers can provide inputs to architectural decisions that affect energy and water consumption, as well as indoor air quality. For River Campus Building One, this meant that several disciplines were able to collaborate early in the design regarding the green roof, PV, and rainwater harvesting system. This early collaboration started with an eco-charrette in which participants and stakeholders from diverse backgrounds helped craft ambitious goals for the project. One of the goals that emerged was a 60 percent reduction in energy consumption relative to that of a comparable building (see Table 7.12).

Making key decisions early in the design process allowed the design team to focus on collaboration to ensure their implementation. The abundant rainfall in Portland and the facility's large roof area meant that rainwater could be used for nonpotable water uses, including cooling tower makeup water. Moderate temperatures allowed the use of outside air to flush and precondition the building at night. Due to Oregon's generous tax credits for renewable and alternative energy systems, the team also opted for PV panels on the south side of the building and a microturbine system in the central utility plant. Integrated design also allowed the design team to eliminate solutions that were not feasible early on—for example, roof-mounted, vertical axis wind turbines.

TABLE 7.12

The Interface Engineering Team's "Back of the Envelope" Goals for River Campus Building One

Load	Oregon Energy Code KBTU/SF/yr*	Percent	Target Savings, KBTU/SF/yr*
Heating	35	27	22
Cooling	10	7.7	5
Fans	6	4.6	2
Hot water	30	23	28
Lighting	30	23	15
Equipment	15	11.5	5
Exterior lighting	4	3	1
Totals	**130**	**100%**	**78**

*Thousands of BTUs per square foot per year.

THE DETAILS—HOW INTERFACE APPROACHED THE MECHANICAL SYSTEMS

The Interface team had two core principles guiding the mechanical design: (1) optimum health and (2) reduced energy use. To achieve this, the engineers followed the basic sustainable engineering dictum laid out by Amory Lovins: optimize the system, not the subsystems. Doing otherwise, that is, optimizing the subsystems without considering the system as a whole, will inevitably produce suboptimal results. Applied to buildings, the system includes all components of the building that affect energy consumption: the mass and orientation of the building, its envelope (thermal resistance, fenestration, roof, infiltration, shading), its plug loads (computers, printers, copiers, and other plug-in devices), its air delivery systems, the lighting system (lights and lighting controls), the cooling and heating plant, fans, motors, pumps, piping and duct sizing and layout. In many cases, it calls for challenging conventional wisdom. For example, mechanical engineers use tables that assume an embedded level of friction loss for fluids such as air and water circulated in pipes or ductwork. Lowering the acceptable friction loss may result in the use of larger-diameter pipes of larger cross-section ducts or the selection of smoother pipes with less friction per unit length.

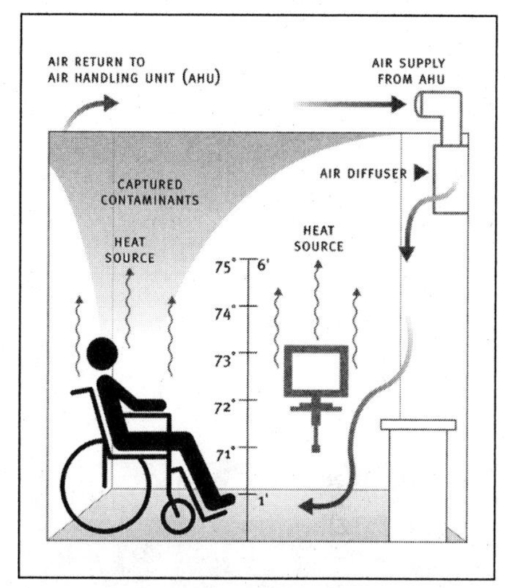

Figure 7.18 CFD modeling of patient examination rooms indicates the waterfall effect of cool incoming air falling down the walls, pooling on the floor, and then rising as it is heated by people, computers, and lights. (Illustration courtesy of Interface Engineering, Inc.)

Early on, the team examined the building's energy profile and worked with the architects to optimize the building's envelope. The team used the BetterBricks Integrated Design Lab in Portland to study year-round shading, including the shading effects of adjacent buildings. As a result, River Campus Building One was designed so that windows were shaded in the summer, allowing sunlight to warm the interior during the winter. Sunshades and building PV panels were used to assist in the shading above the fourth floor.

Plug loads from computers, printers, and other devices were examined to ensure that the selection of these components contributed to the 60 percent energy reduction goal. Similarly, all fans, water heaters, pumps, and motors were selected to support the energy-conserving goals of the team.

Computational fluid dynamic (CFD) models were used extensively to explore approaches to natural ventilation and the building's air distribution approach. A whole-building CFD model allowed the team to determine wind pressures on each face and optimize a natural ventilation strategy. CFD models were also used in making the decision to select a positive displacement ventilation system for patient examination rooms (see Figure 7.18). Similarly, the supply air temperature for the examination rooms was selected based on CFD modeling, allowing the temperature to be raised from a typical 55°F to 60°F (13°C to 16°C). In addition to lower energy costs, this permits the more extensive use of the typically temperate outside air in the Pacific Northwest to cool the building (see Figure 7.19).

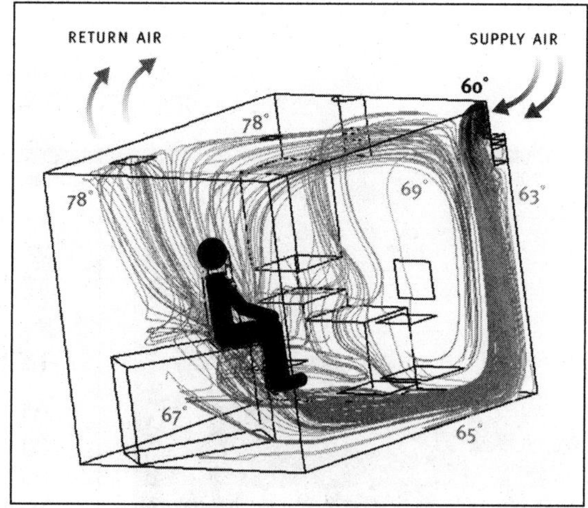

Figure 7.19 Modeling of patient examination room temperatures aided the Interface team in deciding to raise the supply air temperature to 63°F (17°C), creating a more comfortable examination room with less air movement. (Illustration courtesy of Interface Engineering, Inc.)

In short, the Interface team, by collaborating with the project architects, were able to design a downsized mechanical system. As part of this process, the engineers also bought into the notion of right-sizing the mechanical system rather than oversizing the system to accommodate hypothetical unknowns. The team accomplished this by (1) eliminating excessive safety factors; (2) calculating heating and cooling demands using basic physics rather than simply applying conventional HVAC rules of thumb; (3) assuming nothing and proving everything; (4) building in expansion capabilities rather than trying to accomplish everything at the beginning; and (5) challenging restrictive codes that add cost without benefit by making successful appeals.

Right-sizing is just one of eight design points articulated by Andy Frichtl, PE, the lead engineer for River Campus Building One. The other four design points he advocates are: (1) transfer savings in HVAC systems to other important aspects of the project; (2) use free resources such as the sun, wind, ground temperature, and groundwater to reduce building energy consumption; (3) reduce the demand for heating and cooling by superior envelope design, reduction in plug loads, and providing high-efficiency appliances and other devices; (4) shift loads from peak to off-peak periods by using energy storage strategies; (5) challenge standard practice by emphasizing comfort and health, which may also involve challenging building codes; (6) utilize radiant space conditioning, which uses radiative rather than convective heat transfer, with significantly lower energy consumption; and (7) relax comfort standards by allowing temperature and humidity setpoints to float within a specified comfort zone.

The result of applying these strategies was a variety of energy-efficient design measures to achieve the high-performance goals of the project:

- Radiant cooling of the atrium and lobby ground floor using reclaimed rainwater and groundwater in the concrete slab (see Figure 7.20)
- Radiant cooling with an overhead chilled beam (see Figure 7.21)
- High-efficiency boilers and chillers
- Double-fan VAV air handlers and VFDs on most pumps and motors
- Demand-controlled ventilation (DCV) using carbon dioxide sensors and occupancy sensors to prevent overventilating and overlighting unoccupied spaces
- Heat recovery systems including laboratory and general exhaust
- Displacement ventilation for core exam and office areas
- Load shifting using a system of hot and cold water storage to reduce peak demand
- Energy-efficient lighting fixtures and controls, incorporating daylighting where feasible
- Night-flush precooling with outside air
- Economizers for free cooling using outside air when outside temperatures permit

Figure 7.20 Construction of a radiant cooling/heating system in a ground floor slab. (Photograph courtesy of Interface Engineering, Inc.)

Chilled beams represent a potential breakthrough strategy for conditioning buildings. The HVAC systems employing this technology can be one-third the size of systems using forced air as the heat transfer medium. While relatively new to the United States, radiant cooling systems are fairly standard practice in Germany. They can function passively using only radiant effects for cooling or, with the assistance of a fan passing air through the beam, provide convective cooling. The compact size of the chilled beams allows reduced floor-to-floor heights because larger ductwork is eliminated and the space required for mechanical rooms and shafts can be reduced. Although the beams cost \$100 to \$250 a lineal foot (\$328 to \$820 per meter), the net result is reduced HVAC system costs and lower costs for architectural and structural elements.

Figure 7.21 Chilled beam systems are aluminum-finned copper assemblies through which water circulates, providing radiative cooling or heating and inducing airflow by convective effects. (Photograph courtesy of Interface Engineering, Inc.)

INTEGRATING LIGHTING AND DAYLIGHTING SYSTEMS

A properly designed lighting system for a high-performance building should integrate daylighting, lighting fixtures, and lighting controls to provide a low-energy lighting solution. For River Campus Building One, the Interface team's goal was to reduce the typical lighting system's 23 percent share of the total energy use by 50 percent. They managed to achieve a 45 percent reduction in the actual building, a savings of 16 percent in total energy use. In the exam rooms, the standard two 1 × 4 foot (0.3 × 1.2 meters) lensed fluorescent luminaires were replaced by a single 48-inch- (122 centimeters) diameter lensed skydome that mimics natural light. Combined wall switch/occupancy sensors turn on only half of the exam room lights, permitting the other half to switch on automatically when needed. Reduced lighting levels were specified for lobbies and other pass-through spaces. When there is adequate natural light, hallway daylight sensors switch off normal and emergency lighting. Outdoor lighting was significantly reduced using cutoff fixtures that also eliminate unnecessary light pollution. In the high-bay athletic club, lighting levels automatically switch down as more daylight becomes available. Occupancy sensors in stairwells switch lighting on and off to follow an occupant up or down, allowing the lighting to stay on for the minimum time needed for passage. Perimeter offices also have occupancy sensors and daylighting sensors.

INNOVATIVE SOLAR ENERGY APPLICATIONS: BIPV AND SOLAR AIR HEATER

The project team for River Campus Building One specified sunshades in the design of the south façade and used the sunshade surface for PV panels (see Figure 7.22). In addition to using renewable energy for the building, PVs are subsidized by generous federal and many state incentives such as tax credits, accelerated depreciation, and, in the case of Oregon, bonuses from the Oregon Energy Trust. These BIPV panels have a peak of 60 kW and produce about 66,000 kWh annually.

On the 15th and 16th floors of the building, the façade serves as a giant solar heater, 190 feet (58 meters) long by 32 (9.8 meters) feet high. Sheets of low-iron glass are located 4 feet (1.2 meters) from the building skin. The air between the skin and glass is warmed by solar energy and then moved by air-handling units across a heat exchanger for use in preheating water for use in bathroom sinks and exam rooms. The integrated design approach used by the project team allowed the fusion of architecture and engineering to create this innovative water heating system. This system has the added benefit of serving as a Trombe wall, warming clinic and lab spaces in winter and reducing the amount of total heating energy. It requires almost no maintenance and has no replacement costs over time.

Acknowledgment: The River Campus One Case Study is used with the permission of Interface Engineering, Inc. It is also available from Interface Engineering in a comprehensive booklet, Engineering a Sustainable World, *published in October 2005.*

Figure 7.22 The PV panels used in River Campus Building One were assembled at Benson Industries, a major supplier of curtain walls and exterior cladding systems for larger buildings. (Photograph courtesy of Interface Engineering, Inc.)

Connection to LEED-NC

The LEED-NC 2009 category Energy & Atmosphere (EA) addresses the issues of energy for high-performance buildings; it also covers several issues that connect building systems to environmental impacts on air and the atmosphere—for example, the elimination of HCFCs, which due to their presence in chillers and other mechanical equipment, are implicated in ozone depletion. The following subsections discuss LEED-NC 2009 credits and reporting requirements in the EA category. The numbers associated with these credits correspond to the numbering system used in the LEED-NC standard. Table 7.13 is a summary list of these credits and the points associated with them. Note that the three prerequisites do not carry points and must all be met for a building to even be considered for certification. Note also that the descriptions of the LEED-NC credits and points in this section are abbreviated and that the appropriate LEED Reference Manual should be consulted for detailed information about how to achieve the points.

EA PREREQUISITE 1 (EAp1): FUNDAMENTAL COMMISSIONING OF THE BUILDING ENERGY SYSTEMS

Building commissioning has become one of the hallmarks of the green building movement in the United States. According to the Building Commissioning Association, the basic purpose of building commissioning is ". . . to provide documented confirmation that building systems function in compliance with criteria set forth in the Project Documents to satisfy the owner's operational needs. Commissioning of existing systems may require the development of new functional criteria in order to address the owner's current systems performance requirements." This definition is based on the critical understanding that owners must have some means of verifying that their functional needs are rigorously addressed during design, construction, and acceptance. This prerequisite has the intent of ensuring that the installation and calibration of building elements and systems occur as designed and intended.

EA Prerequisite 1, Fundamental Commissioning of the Building Energy Systems, requires the following basic commissioning process:

TABLE 7.13

EA Prerequisites and Credits Under LEED-NC 2009

Prerequisite/Credit	Name of Prerequisite/Credit	Maximum Points
EA Prerequisite 1	Fundamental Commissioning of the Building Energy Systems	NA
EA Prerequisite 2	Minimum Energy Performance	NA
EA Prerequisite 3	Fundamental Refrigerant Management	NA
EA Credit 1	Optimize Energy Performance	19
EA Credit 2	On-Site Renewable Energy	7
EA Credit 3	Enhanced Commissioning	2
EA Credit 4	Enhanced Refrigeration Management	2
EA Credit 5	Measurement and Verification	3
EA Credit 6	Green Power	2
	Total EA Credits Available	**35**

- Designate an individual as the Commissioning Authority (CxA) to lead, review, and oversee the completion of the commissioning process. The CxA must have documented experience in commissioning at least two building projects.
- The Owner shall document the Owner's Project Requirements (OPR), the design team shall develop the Basis of Design (BOD), and the CxA shall review these documents for clarity and completeness.
- Include commissioning requirements in the construction documents.
- Develop and utilize a commissioning plan.
- Verify installation, functional performance, training, and documentation.
- Complete a summary commissioning report.

Systems that must be commissioned include HVAC&R systems, both active and passive, and their controls; lighting and daylighting controls; domestic hot water systems; and renewable energy systems.

Commissioning, a very important practice in creating a high-performance building, is covered in detail in Chapter 12.

EA PREREQUISITE 2 (EAp2): MINIMUM ENERGY PERFORMANCE

The building must be designed, at a minimum, to meet the mandatory provisions (Sections 5.4, 6.4, 7.4, 8.4, 9.4, and 10.4) of ASHRAE 90.1-2007 and the prescriptive requirements (Sections 5.5, 6.5, 7.5, and 9.5) or performance requirements (Section 11) of ASHRAE 90.1-2007. The building must comply with the mandatory provisions and either the prescriptive or Energy Cost Budget Method performance requirements of the Standard.

EA PREREQUISITE 3 (EAp3): FUNDAMENTAL REFRIGERANT MANAGEMENT

Chlorofluorocarbons (CFCs), as previously noted, are ozone-depleting substances with a long history of use in building air-conditioning equipment. This prerequisite has the intent of eliminating CFCs from buildings, thereby protecting the ozone layer. It requires zero use of CFCs in new building HVAC&R systems. For a reuse project with existing equipment, a plan to phase out the CFCs must be submitted prior to project completion.

EA CREDIT 1 (EAc1): OPTIMIZE ENERGY PERFORMANCE

Designing and building an energy-efficient building is important for sustainability reasons as well as for earning a LEED rating. By virtue of having more than half of the EA credits assigned to it, LEED-NC 2009 EA Credit 1, Optimize Energy Performance, is by far the most important credit in the EA category—in fact, in the entire LEED standard. ASHRAE 90.1-2007 is the basis for determining how well the high-performance building performs compared to the base case, that is, the building that just meets the standard's minimum requirements. To be successful in obtaining a LEED rating for the building, the design team should ensure that the requirements of ASHRAE 90.1-2007 are exceeded by a substantial margin. Of the 35 possible points in the EA category in LEED, 19 points depend on how well the building performs compared to the base case. Building energy performance is a complex issue, and can be very involved and difficult to sort through. The energy-related components of a

TABLE 7.14

Points for EAc1 Option 1 for Improvements in Building Performance Over the ASHRAE 90.1-2007 Base Case

New Building	Existing Building Renovations	Points
12	8	1
14	10	2
16	12	3
18	14	4
20	16	5
22	18	6
24	20	7
26	22	8
28	24	9
30	26	10
32	28	11
34	30	12
36	32	13
38	34	14
40	36	15
42	38	16
44	40	17
46	42	18
48	44	19

typical building are its envelope (walls, roof, floor, windows, doors), HVAC equipment, power distribution system, lighting system, and equipment such as pumps, appliances, refrigeration equipment, and elevators.

EAc1 has three options for earning points toward LEED certification, as described below. Note that as of June 2007, the USGBC was balloting a recommendation that all buildings must achieve a minimum of two points under EAc1.

Option 1: Whole Building Energy Simulation (19 Points)

The project can earn up to 19 points by running a whole building energy simulation per ASHRAE 90.1-2004 using the Performance Rating Method described in Appendix G of the Standard. A new building with a 10.5 percent improvement receives 1 point, with 1 additional point for each 3.5 percent increase, up to a maximum of 10 points for a 42 percent improvement over the base case. For an existing building, the sliding scale starts at 1 point for a 3.5 percent improvement over the base case, with 1 additional point for each 3.5 percent improvement, up to a maximum of 10 points for a 35 percent improvement over the base case (see Table 7.14).

The building must be designed, at a minimum, to meet the mandatory provisions (Sections 5.4, 6.4, 7.4, 8.4, 9.4, and 10.4) of ASHRAE 90.1-2004. The proposed design must include all energy costs of the proposed design, and the design must be compared to a baseline building as described in Appendix G of the Standard. On-site renewable energy generation is included in the modeling to show a reduction in energy demand from external sources.

Option 2: Prescriptive Compliance Path (1 Point)

A building can earn 1 point for complying with the prescriptive measures of the ASHRAE Advanced Design Guide for Small Office, Retail, or Warehouses and Self Storage Buildings.

Option 3: Prescriptive Compliance Path (1–3 points)

A building that complies with the Advanced Building Core Performance Guide and is under 100,000 square feet can earn 1 to 3 points.

EA CREDIT 2 (EAc2): ON-SITE RENEWABLE ENERGY

LEED encourages the consumption of renewable rather than nonrenewable energy for buildings and provides points for on-site or site-recovered renewable energy. Table 7.15 indicates the number of points available for various levels of renewable energy provided for the building.

TABLE 7.15

LEED-NC 2009 Points Available for Various Levels of Renewable Energy

Percent of Total Building Energy Provided by Renewable Energy Source	Points
1%	1
3%	2
5%	3
7%	4
9%	5
11%	6
13%	7

EA CREDIT 3 (EAc3): ENHANCED COMMISSIONING

LEED places considerable emphasis on building commissioning; as such, this credit provides the opportunity to gain 2 points for additional verification that building systems operations are as intended by the design. The additional point can be earned by going well beyond the minimum commissioning called for in EAp1, Fundamental Commissioning of the Building Energy Systems. Like EAp1, this credit calls for the Commissioning Authority (CxA) to have documented experience on at least two building projects. The CxA must be independent of the design team, must not be a member of the design firm or the construction manager's firm, and must report all findings directly to the Owner. In addition, the CxA shall:

- Review the Owner's Project Requirements (OPR) and Basis of Design (BOD) and the design documents prior to the middle of the Construction Documents phase.

- Review applicable contractor submittals for systems being commissioned for compliance with the OPR and BOD.
- Develop a systems manual for operators to be able to operate the commissioned systems at peak performance.
- Verify that building operators and occupants are trained.
- Review the building operation within 10 months after substantial completion with O&M staff and occupants, to include developing a plan for resolution of any outstanding commissioning-related issues.

Both the Fundamental Commissioning prerequisite and the Additional Commissioning credit are covered in detail, along with the subject of commissioning in general, in Chapter 12.

EA CREDIT 4 (EAC4): ENHANCED REFRIGERATION MANAGEMENT

EA Prerequisite 3 calls for zero use of CFC refrigerants in new building HVAC&R equipment. EAc4 provides for measures that further reduce the use of ozone-depleting refrigerants in buildings and addresses the climate change impacts of these substances. EAc4 has two options:

Option 1: Do not use refrigerants.

Option 2: Select refrigerants that minimize contributions to ozone depletion and climate change and provide a formula that takes both effects into account and determines the combined impact of the refrigerants used in the building project. The formula provides an upper limit for these combined effects:

$$\text{LCGWP} + \text{LCODP} \times 10^5 \leq 100$$

where LCGWP is the Lifecycle Direct Global Warming Potential in pounds of carbon dioxide per ton-year and LCODP is the Lifecycle Ozone Depletion Potential in pounds of CFC11 per ton-year. In both cases, *ton* refers to the unit of cooling capacity of the refrigeration systems.

EA CREDIT 5 (EAC5): MEASUREMENT AND VERIFICATION

In addition to motivating significant energy savings, the LEED-NC 2009 standard provides an incentive to measure the savings by providing 3 points for a credit that calls for a definitive system of sensors that can provide feedback on building operation. The methodology used to measure these savings is the International Performance Measurement and Verification Protocol (IPMVP), Volume III.

EA 6 (EAC6): GREEN POWER

Another approach to using renewable energy in a building is to contract for power from a utility that generates energy from renewable sources. This credit requires that the building owner engage in a 2-year contract with a source that meets the Center for Resource Solutions (CRS) Green-e products certification process and provides at least 35 percent of the building's electricity.

Green power may be procured from a Green-e certified power marketer, from a Green-e accredited utility program, through Green-e accredited Tradable Renewable Certificates, or from a supply that meets the Green-e renewable power definition. The definition of a renewable electricity product varies from state to state, but in general it can comprise one or more of the following, according to Green-e:

- Geothermal.
- Wind.
- Small hydro: Definition and types may vary in New England and California.
- Solar electric.
- Biomass: Allowed biomass and biomass emissions vary across the United States.
- Cofired fuels: Methane from landfills with natural gas varies from state to state; it does not count in Texas.
- Negawatt: A negawatt is a megawatt saved from the grid, that is, avoided energy use. It can be counted as green power, but in Pennsylvania there are specific guidelines.
- Ocean-based resources: Not yet available but could be in the near future.

Connection to Green Globes v.1

The Energy Part of Green Globes has two different Paths, depending on the size of the building. Path A applies to buildings larger than 20,000 square feet (1,860 square meters) and Part B is for buildings less than or equal to 20,000 square feet (1,860 square meters). Table 7.16 shows the sections of the Energy Part that apply to each Path. An outline of the Energy part is provided in Table 7.17.

Summary and Conclusions

As might be expected, energy receives the most emphasis in both the LEED and Green Globes building assessment systems. Clearly, the emphasis on improving building performance is in the application of passive solar design techniques that use the materials, fenestration, and orientation of the building to maximize the amount of free energy that can be used, thereby reducing the quantity of energy that must be

TABLE 7.16

Paths for Rating Energy Performance Using Green Globes v.1

PATH A Buildings over 20,000 square feet must complete the following sections:	PATH B Buildings less than or equal to 20,000 square feet must complete the following sections:
C.1 Building Energy Performance	C.1 Building Energy Performance
C.2 Energy Demand Minimization	C.2 Energy Demand Minimization
C.4 Renewable Sources of Energy	C.4 Renewable Sources of Energy
C.5 Energy-efficient Transportation	C.5 Energy-efficient Transportation
	OR
	C.2 Energy Demand Minimization
	C.3 Right-Sized Energy-efficient systems
	C.4 Renewable Sources of Energy
	C.5 Energy-efficient Transportation

TABLE 7.17

Outline of Green Globes v.1, Part C (Energy)

Section		Description	Points
C.1		**Energy Consumption**	**110**
	C.1.1	Building energy consumption vs. baseline	100
	C.1.2	Energy modeling	10
C.2		**Energy Demand Minimization**	**135**
	C.2.1	Optimization for microclimatic conditions	13
	C.2.2	Wind-mitigating measures	5
	C.2.3	Natural ventilation system	12
	C.2.4	Daylighting strategy	20
	C.2.5	Glazing strategy for daylighting	10
	C.2.6	Thermal resistance of envelope—walls	10
	C.2.7	Thermal resistance of envelope—roof	10
	C.2.8	Thermal resistance of envelope—windows	7
	C.2.9	Window solar heat gain coefficient (SHGC)	6
	C.2.10	Building air barrier	3
	C.2.11	Vapor retarder best practices	6
	C.2.12	Energy sub-metering	7
	C.2.13	Daylighting controls	4
	C.2.14	Lighting control zones	4
	C.2.15	Room occupancy lighting controls	4
	C.2.16	Building automation system	5
	C.2.17	Natural ventilation controls	3
	C.2.18	Outside air damper controls	6
	C.2.19	Vertical transport energy conservation	1
C.3		**Right-Sized Energy Systems**	**110**
	C.3.1	Lighting power density	15
	C.3.2	Energy-efficient lighting system	10
	C.3.3	High-efficiency cooling equipment	25
	C.3.4	Part-load conditioning strategy	10
	C.3.5	High-efficiency heating equipment	25
	C.3.6	High-efficiency heat pump	5
	C.3.7	Thermal zoning	5
	C.3.8	Efficient air handling system	15
C.4		**Renewable Energy**	**45**
	C.4.1	Percent renewable energy	45
C.5		**Energy-Efficient Transportation**	**70**
	C.5.1	Accessible public transit	50
	C.5.2	Car/van pool parking	6
	C.5.3	Covered secure bicycle storage	10
	C.5.4	Showers and changing facilities	4
		Total Points Available	**360**

purchased. Passive solar design addresses heating, cooling, daylighting, and ventilation of the building to minimize the employment of active mechanical and electrical systems, especially those powered by nonrenewable energy systems. The other measures called for in the EA category help round out the concept of a building that is both energy-efficient and environmentally responsible. The elimination of atmospheric ozone-depleting chemicals is a very worthwhile objective of any building rating scheme, and reducing energy consumption helps to lower the incidence of a wide range of power plant emissions.

One innovation in building assessment is the incorporation of strict requirements for building commissioning, ensuring that the building not only functions as designed but is also built to the highest quality standards. Both LEED and Green Globes also provide impetus for the development of renewable energy sources on a large scale by providing a possible credit for using energy from renewable energy power plants.

Notes

1. *Primary energy* refers to raw energy in the form of oil, coal, and natural gas that is input to a process. It does not refer to electricity leaving a generating plant, which accounts for only a fraction of the input, primary energy.
2. The oil rollover point is described in more detail in Chapter 1.
3. Systems ecology was developed into a full-fledged ecological theory by H.T. Odum during his five decades at the University of Florida. The current program in systems ecology in the Department of Environmental Engineering at Florida is described at www.enveng.ufl.edu/homepp/brown/syseco.
4. Data derived from statistics posted by the U.S. Energy Information Agency on its website, www.eia.doe.gov/emeu/consumption.
5. Buildings for the 21st Century was replaced by the EERE Building Technologies program, described at www.eere.energy.gov/buildings.
6. The Factor 10 concept is described further in Chapter 2.
7. From Löhnert, Herkel, Voss, and Wagner (2006).
8. The EIA website is www.eia.doe.gov.
9. The EPA Target Finder website is www.energystar.gov/index.cfm?c=new_bldg_design.bus_target_finder.
10. The best website for information about DOE 2.2 and the E-Quest interface is www.doe2.com.
11. Energy-10™ was developed by the National Renewable Energy Laboratory and is available from the Sustainable Buildings Industry Council under license to the Midwest Research Institute. Detailed information about Energy-10™ is available at www.sbicouncil.org/store/e10.php.
12. Detailed information about the capabilities of Radiance can be found at radsite.lbl.gov/radiance.
13. Information adapted from Diamond, Opitz, Hicks, von Neida, and Herrera (2006).
14. The January 2006 version of the IPMVP Protocol, "Concepts and Practices for Determining Energy Savings in New Construction," Volume III, Part I, plus other IPMVP references are available from the Efficiency Valuation Organization website: www.evo-world.org.
15. Insolation is a contraction of **in**coming **sol**ar radi**ation.**
16. An HDD or a CDD is a measure of the deviation of the site's temperature profile from the average temperature in a building. For heating, the average temperature is 65°F (18°C); for cooling, the average temperature used for calculations is 75°F (24°C). For example, a day with an average temperature of 60°F (16°C) would result in five Fahrenheit-based (two Celsius-based) HDDs [(65°F– 60°F) (18°C – 16°C) × 1 day]. The number of HDDs or CDDs is an indicator of how extreme the temperature profile of a site is and how much energy may be required to provide heating or cooling.
17. The study of the effects of skylights on retail sales is in the report "Skylighting and Retail Sales" (1999).
18. Data on student performance is from "Daylighting in Schools" (1999).

19. Excerpted from "Tips for Daylighting with Windows," (1997).
20. From "Daylighting: Energy and Productivity Benefits" (1999).
21. A description of the daylighting and other strategies employed to make Rinker Hall a high-performance building can be found at the American Institute of Architects (AIA), Committee on the Environment (COTE) website, www.aiatopten.org/hpb/energy.cfm?ProjectID=286.
22. COP is a measure of the performance of heat pumps and air-conditioning systems and is defined as the ratio of energy removed or added to the energy input to the system. Both energy removed and energy input must have the same units—for example, BTUs per hour or kilowatts. Unlike efficiency, which has a maximum value of 1, COP can be greater than 1 and indeed should be much greater than 1. For example, the most efficient screw chillers have a COP of 7 or higher. Another related term is *Seasonal Energy Efficiency Ratio* (SEER), which describes the ratio of energy removed in BTUs to watts of input power and is used to describe the performance of smaller residential-scale air-conditioning systems. An air-conditioning unit with a SEER 14 rating would have an equivalent COP of 4.
23. From Löhnert, Herkel, Voss, and Wagner (2006).
24. Excerpted from "Solar Heat Gain Control for Windows" (November 2006).
25. The National Fenestrations Council's website is www.nfrc.org.
26. The Cool Communities network advocates for measures that prevent urban heat islands. Its website is www.coolcommunities.org.
27. From Florida Solar Energy Center (2000).
28. Excerpted from Kaneda, Shell, Rumsey, and Fisher (2006).
29. Excerpted from "Chiller Plant Efficiency" (2000).
30. Ibid.
31. Ibid.
32. *Greening Federal Facilities* is a guide developed for use by federal government facility managers to use in greening their buildings during the course of routine operations and maintenance. It is downloadable from www.eere.energy.gov/femp/technologies/sustainable_greening.cfm.
33. Excerpted from "Economizers" (2000).
34. Source: Philips Lighting. Excerpted from *Greening Federal Facilities* (2001).
35. The American Bioenergy Association's website is www.biomass.org.
36. The Carbohydrate Economy Clearing House website (www.carbohydrateeconomy.org) is sponsored by the Institute for Local Self-Reliance (ILSR) and covers the broad range of issues associated with shifting to bio-based renewables.
37. Information about fuel cell applications can be found at www.fuelcells.org.
38. An excellent overview of building EMS is available from Energy Design Resources in the form of a Design Brief, "Energy Management Systems" (1998).
39. Excerpted from "What Office Tenants Want" (2000).

References

"Building Simulation," Energy Design Brief. 2000. Energy Design Resources. Available at www.energydesignresources.com/resource/21.

"Buildings for the 21st Century-Strategic Plan." December 1998. Office of Energy Efficiency and Renewable Energy, Department of Energy, Golden, Colorado, DOE/GO-10099-688.

"Chiller Plant Efficiency," Energy Design Brief. 2000. Energy Design Resources. Available at www.energydesignresources.com/resource/24.

"Concepts and Practices for Determining Energy Savings in New Construction." January 2006. Volume III, Part I, EVO 30000-1: 2006. Prepared by the New Construction Subcommittee of the IPMVP. Available at the Energy Valuation Organization website, www.evo-world.org.

"Daylighting: Energy and Productivity Benefits." September 1999. *Environmental Building News,* 8(9), pp. 1, 10–14.

"Daylighting in Schools: An Investigation into the Relationship between Daylighting and Human Performance, A Condensed Report." August 20, 1999. Conducted by the Heschong Mahone Group for the Pacific Gas & Electric Company. Available at www.pge.com/003_save_energy/003c_edu_train/pec/daylight/di_pubs/SchoolsCondensed820.PDF.

Diamond, Rick, M. Opitz, T. Hicks, B. Vonneida, and S. Herrera. 2006. "Evaluating the Site Energy Performance of the First Generation of LEED-Certified Commercial Buildings," *Proceedings of the 2006 Summer Study on Energy Efficiency in Buildings,* LBNL-59853. Washington, DC: American Council for an Energy Efficient Economy. Available at http://epb.lbl.gov/homepages/Rick_Diamond/LBNL59853-LEED.pdf.

"Economizers," Energy Design Brief. 2000. Energy Design Resources. Available at www.energydesignresources.com/resource/28.

"Energy Management Systems," Energy Design Brief. 1998. Energy Design Resources. Available at www.energydesignresources.com/resource/18.

Florida Solar Energy Center. 2000. "Laboratory Testing of the Reflectance Properties of Roofing Materials," by D.S. Parker et al. Available at www2.fsec.ucf.edu/en/publications/html/FSEC-CR-670-00/index.htm.

"Green Cooling: Improving Chiller Energy Efficiency." Spring 1996. *Center for Building Science News,* no. 18 (now *Environmental Energy Technologies News*), Lawrence Berkeley National Laboratory, Berkeley, CA.

Greening Federal Facilities, 2nd ed. 2001. Washington, DC: Department of Energy, DOE/GO-102001-1165. Available at www.eere.energy.gov/femp/technologies/sustainable_greening.cfm.

Hawken, P., A. Lovins, and H. Lovins. 1999. *Natural Capitalism.* New York: Little, Brown.

Kaneda, David, Scott Shell, Peter Rumsey, and Mark Fisher. 2006. "IDeAs Z^2 Design Facility: A Case Study of a Net Zero Energy, Zero Carbon Emission Office Building," *Proceedings of Rethinking Sustainable Construction 2006,* Sarasota, FL, September 18–22.

Löhnert, Günther, Sebastian Herkel, Karsten Voss, and Andreas Wagner. 2006. "Energy Efficiency in Commercial Buildings: Experiences and Monitoring Results from the German Funding Program Energy Optimized Building, ENOB," *Proceedings of Rethinking Sustainable Construction 2006,* Sarasota, FL, September 18–22.

Parker, D.S., J.E.R. McIlvaine, S.F. Barkaszi, D.J. Beal, and M.T. Anello. 2000. "Laboratory Testing of the Reflectance Properties of Roofing Material," FSEC-CR-670-00, Florida Solar Energy Center, Cocoa, FL. Available at www2.fsec.ucf.edu/en/publications/html/FSEC-CR-670-00/index.htm.

"Radiant Cooling," Energy Design Brief. 2000. Energy Design Resources. Available at www.energydesignresources.com/resource/34.

Reference Package for New Construction & Major Renovations (LEED-NC). 2005. Washington, DC: U.S. Green Building Council.

"Skylighting and Retail Sales: An Investigation into the Relationship between Daylighting and Human Performance: A Condensed Report." August 20, 1999. Conducted by the Heschong Mahone Group for the Pacific Gas & Electric Company. Available at www.pge.com/000_save_energy/003c_edu_train/pec/daylight/di_pubs/RetailCondensed820.PDF.

"Solar Heat Gain Control for Windows." November 2002. EREC Reference Briefs. Washington, DC: Office of Energy Efficiency and Renewable Energy, Department of Energy.

"Tips for Daylighting with Windows: The Integrated Approach," LBNL-39945. 1997. Berkeley, CA: Lawrence Berkeley National Laboratory.

"What Office Tenants Want: 1999 BOMA/ULI Office Tenant Survey Report." 2000. New York: BOMA International Foundation.

Chapter 8

The Building Hydrologic System

Of the various resources needed for the built environment, water is arguably the most critical. In his book *The Bioneers,* Kenny Ausubel notes that biologists occasionally refer to this resource as "Cleopatra's Water" because, like all other materials on the planet, water stays in a closed loop. The water you sip from a drinking fountain may have once been used by the Egyptian queen in her bath. The human body is 97 percent water, and water is more crucial to survival than food. It serves as a buffer in human metabolism for the transfer of oxygen at small scale, as a damper on rapid changes in the planet's environment at large scale, and as a shock absorber in cellular function at microscopic scale. Water plays a role in most of the world's spiritual traditions and religions, from baptism in the Christian faiths to sweat lodges in Native American rituals to the cleanliness traditions of the Baha'i faith. Water is the source of life for both humans and other species, yet it also has the power to destroy. It is used as a metaphor for truth and as a symbol for redemption and the washing away of sin. Water serves as habitat for a substantial fraction of the Earth's living organisms, and the remainder are totally dependent on it for their survival.

In spite of water's symbolic and practical values, water resources throughout the planet are badly stressed. During World Environment Day in 2003, Kofi Annan, the Secretary General of the United Nations, noted that one person in six is without safe drinking water, and double that number, about 2.4 billion, lack adequate sanitation. Several problems are contributing to these shortages. Of all the Earth's water, only 2.5 percent is freshwater, and of that, three-quarters is sequestered, or locked up, in glaciers and permanent snow cover. Only 0.3 percent of water is surface water found in rivers and lakes and thus readily accessible. The remainder is buried deep in the ground, and in some cases, once removed, it can be replenished only over hundreds of years. In much of the world, freshwater removed from both ground and surface sources is being used up far faster than it is being replenished. Western Asia has the most severe water supply problem in the world, with over 90 percent of its population experiencing severe water stress. In Spain, over half of the approximately 100 aquifers are overexploited. In the United States, the situation is better, but not significantly and perhaps not for long. In Arizona alone, more than 520 million cubic yards (400 million cubic meters) of water are removed from aquifers each year, double the replenishment rate from rainwater (see Figure 8.1).

Perhaps the best-known case of water supply depletion is the Aral Sea, which in the 1960s began supplying water to Soviet collective farms for the production of cotton. Formerly, it was a source of large fish; by the early 1980s, they had been virtually eliminated. By the 1990s the Aral Sea occupied half of its original area, and it had shrunk in volume by 75 percent. A once beautiful, large, rich, and deep lake with complex ecosystems had been largely destroyed in about 40 years due to human activities.

In addition to problems of water supply, public health and hygiene are important issues. Waterborne diseases, including diarrhea, typhoid, and cholera, are responsible for 80 percent of the illnesses and death in developing countries. Some 15 million children per year die from these diseases. Raw sewage and toxic materials, including industrial and chemical wastes, human waste, and agricultural waste, are dumped

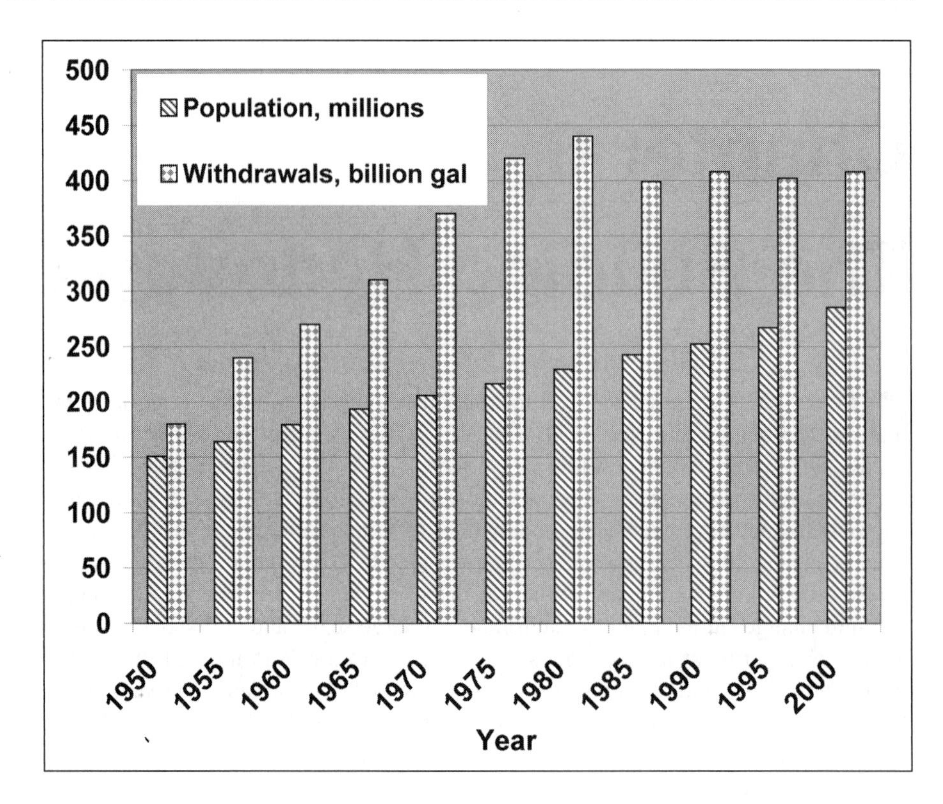

Figure 8.1 Water consumption in the United States has leveled off in the past few decades, with potable water withdrawals presently at around 400 billion gallons (1,514 liters) per year. However, shortages of potable water throughout the country indicate that, in spite of the flattening out of consumption, the current level of withdrawals is not sustainable.

into water systems at the rate of 2 million tons per day. About 300,000 gallons (1.1 million liters) of raw sewage are dumped every minute into the Ganges River in India, which is also a primary source of water for many Indians. Wastewater treatment lags in most of the world: only 35 percent is treated in Asia and approximately 14 percent in Latin America.

Agriculture is the cause of serious water supply problems because it is responsible for over 80 percent of water consumption; and 60 percent of irrigation water is

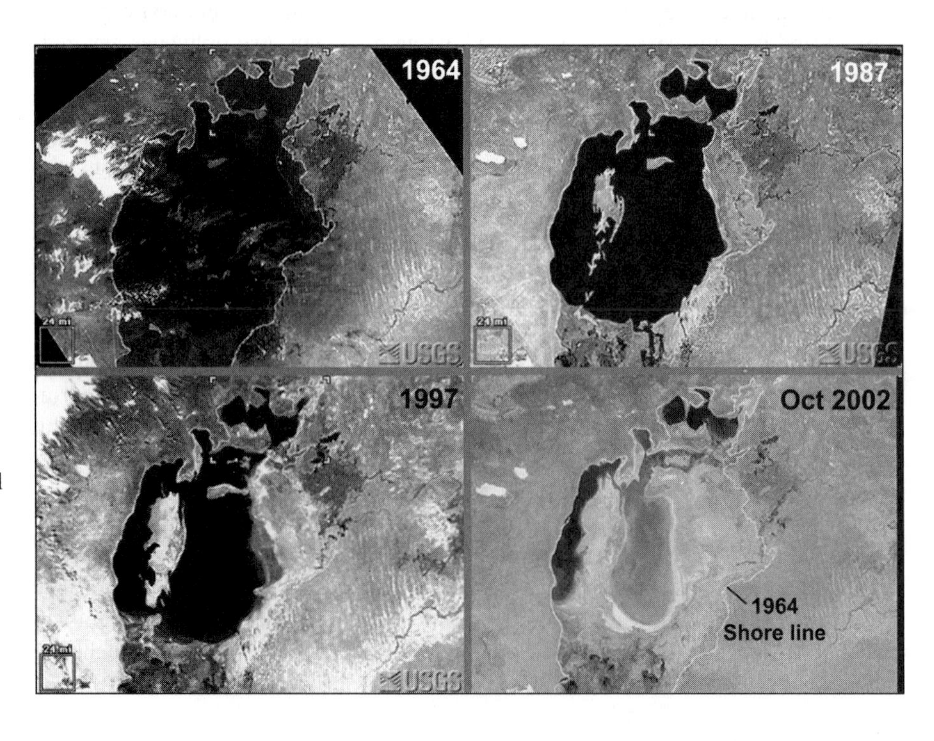

Figure 8.2 In 1964, the Aral Sea in central Asia was a very large, deep saltwater lake with diverse ecosystems (upper left). Heavy withdrawals for agriculture resulted in dramatic shrinkage of its area and volume, as shown in pictures from 1987 (upper right), 1997 (lower left), and 2002 (lower right). (Photographs: U.S. Geological Survey.)

wasted due to leaky canals, evaporation, and mismanagement. Similar problems occur in the cities of many developing countries, with about 40 percent of the water in large cities being lost to leaky systems.

It is important to note the actual amount of water needed by a population because this defines the limits of supply and consumption for a region. For bare survival, the World Health Organization (WHO) suggests that 0.5 to 1 gallon (2 to 4.5 liters) of water is needed per person for drinking and another 1 gallon (4 liters) for cooking and food preparation. The U.S. Agency for International Development (USAID) states that 26.4 gallons (100 liters) a day per person are required to maintain a reasonably good quality of life. In the United States, direct per capita daily water use is approximately four times higher, about 100 gallons (400 liters); and if agricultural and industrial water use is included, the amount per person per day is approximately 1,800 gallons (7,000 liters)—an enormous quantity of a limited and precious resource.

Although direct consumption by people in buildings is not a large fraction of total water use in the United States, water shortages in many areas of the country are having an impact on development and construction. Buildings account for about 12 percent of freshwater withdrawals. The building hydrologic cycle, characterized by the input of high-quality potable water and the release of used, contaminated water, is inefficient, wasteful, and illogical. In its more extended context, the building hydrologic cycle also includes the irrigation of landscaping and the handling of stormwater (see the discussion in Chapter 6 on stormwater, which is generally included with the general topic of the building site). As pointed out by Hawken et al. (1999), the invention of the water closet by Thomas Crapper was perhaps the start of an unfortunate trend in decision making with respect to building water use.[1] In order to dispense with the human waste generated in buildings, water closets mix high-quality potable water with disease-ridden feces and relatively clean urine for the purpose of diluting this mix. Consequently, enormous quantities of water are wasted and a potentially useful source of fertilizer is released into sanitary sewer systems to combine with industrial waste. The end result is a complex, chemically intense, energy-consuming, pollution-producing system of wastewater treatment plants. Major rethinking of the building hydrologic system is clearly needed to make better use of increasingly scarce and expensive potable water and to reduce the impact and cost of treating effluent from buildings.

In the United States, water crises are occurring almost everywhere, as is apparent in the moratoriums imposed on development and growth because of either a shortage of water supplies or insufficient wastewater treatment capacity. At present, there is active discussion of a growth moratorium in Las Vegas, currently the fastest-growing municipality in the United States. In the Diamond Valley, near Las Vegas, water levels dropped over 100 feet (30 meters) during the 1970s and 1980s and have never recovered.[2] In January 2004, in Emmitsburg, Maryland, the town commissioners passed an ordinance that invokes a growth moratorium for lots not already approved for development until the 800,000-gallon- (3 million liters) per-day maximum design capacity of the city's wastewater treatment plant is not exceeded for 180 days.[3]

Current Building Fixtures and the Energy Policy Act of 1992

One of the landmark pieces of legislation concerning potable water consumption is the Energy Policy Act of 1992 (EPAct). EPAct requires all plumbing fixtures used in the United States to meet ambitious targets for reducing water consumption; as a result, building codes now mandate these dramatically lower levels of water consumption. Table 8.1 summarizes the provisions of EPAct 1992 for plumbing fixtures.

TABLE 8.1

EPAct 1992 Maximum Flush and Flow Requirements for Plumbing Fixtures*

Water closets: 1.6 gallons (6 liters) per flush
Urinals: 1.0 gallon (3.8 liters) per flush
Showerheads: 2.5 gallons (9.5 liters) per minute at 80 psi (550 kPa) or 2.2 gallons (8.5 liters) per minute at 60 psi (410 kPa)
Faucets: 2.5 gallons (9.5 liters) per minute at 80 psi (550 kPa) or 2 gallons (7.8 liters) per minute at 60 psi (410 kPa)
Replacement aerators: 2.5 gallons (9.8 liters) per minute
Metering faucets: 0.25 gallon (0.98 liter) per cycle

*Unless otherwise indicated, flow rates correspond to a pressure of 80 psi (550 kPa).

Although these reductions are a great improvement over previous code requirements, green buildings generally attempt to improve on them by employing a combination of strategies, including specifying ultra-low-flow fixtures (ULF) that require less water than EPAct-compliant fixtures and xeriscaping for landscaping. LEED-NC 2009 requires the comparison of the building's hydrologic strategy to that of an EPAct-compliant building to demonstrate the overall reduction caused by the introduction of ULFs, the substitution of alternative water sources (rainwater, reclaimed water, and graywater) for potable water, and the use of other technologies and approaches that result in a reduction of potable water consumption.

High-Performance Building Hydrologic Cycle Strategy

High-performance green buildings incorporate novel approaches as at least partial solutions to these global water problems by addressing two major components of the building hydrologic cycle: the supply of potable water and the disposal of wastewater.

Reducing building water consumption and rethinking the wastewater strategy employed for the built environment can dramatically extend the available supply of water, improve human health, and reduce threats to ecological systems. In addition to these benefits, the Rocky Mountain Institute (RMI) suggests that water efficiency can have these other tangible and calculable benefits:[4]

- *Energy savings:* More money can be saved by reducing the energy needed to move, process, and treat water than the actual value of the saved water.
- *Reduced wastewater production:* Reducing water consumption also reduces wastewater generation, lowering costs for building owners.
- *Lower facilities services investments:* Designing water-efficient buildings reduces the costs of water and wastewater infrastructure.
- *Improved industrial processes:* Innovations in water use in production systems can result in new processes and approaches.
- *Higher worker productivity:* Facilities that incorporate resource efficiency measures are known to have a more productive workforce.
- *Reduced financial risk:* Implementing water efficiency can be accomplished as needed, thus reducing costs and risks for large facilities.
- *Environmental benefits:* Lowering water consumption results in reduced impact on natural systems.
- *Public relations value:* Protecting the environment is looked upon favorably by the general public and clients.

The building hydrologic cycle and energy use are tightly coupled, with very little of the impact being apparent to the building owner. Complex and expensive systems extract potable water from surface water and groundwater sources, then pump it for treatment and distribution, requiring large quantities of energy that are generally subsidized by the low cost of water. Similarly, wastewater must be pumped through an extensive system of sanitary sewers and lift stations to central wastewater treatment plants, consuming relatively large amounts of energy. (The term *watergy* is sometimes used to describe the tightly intertwined relationship of water and energy.) The good news is that reducing water consumption reaps numerous positive benefits, not only by reducing flows through the system but also by lowering overall energy consumption and associated pollution from energy sources.

SETTING WATER CONSUMPTION GOALS FOR BUILDINGS

Setting goals for a building's water consumption is an important first step in designing a strategy that makes sense. If the Factor 10 concept is applied to the issue of water consumption, potable water—and, by inference, wastewater—should be reduced by 90 percent for the purpose of producing a sustainable future. This means that typical per capita household consumption of potable water in this country must be reduced from 100 gallons (380 liters) per day to about 10 gallons (40 liters) per day. To accomplish this remarkable reduction requires that water must be reused and recycled at high rates. For example, per capita consumption of water is almost evenly divided between outdoor and indoor uses. If only recycled water were used outdoors for irrigating landscaping, per capita consumption of potable water would drop to 50 gallons (190 liters) per day. Indoors, almost half of the water consumed is for toilet and urinal flushing, and using only recycled water for this purpose would further reduce water consumption to 25 gallons (85 liters) per day. These relatively straightforward measures produce an immediate Factor 4 reduction. Additional measures that incorporate low-flow fixtures and electronic controls can nearly produce the desired Factor 10 reduction.

HYDROLOGIC CYCLE TERMINOLOGY

Before discussing a high-performance building hydrologic strategy, it is important to define common terms used in this context.

Blackwater. Water containing human waste. Wastewater from kitchen sinks and dishwashers also is sometimes considered blackwater because it contains oil, grease, and food scraps, which can burden the treatment and disposal processes.

Graywater. Water from showers, bathtubs, bathroom sinks, washing machines, and drinking fountains. Graywater may also include condensation pan water from refrigeration equipment and air conditioners, hot tub drainwater, pond and fountain drainwater, and cistern drainwater. Graywater contains a minimum amount of contamination and can be reused for certain landscape applications. Although the issue is still being debated by public health officials, no case of illness has ever been traced to graywater reuse. Both graywater and blackwater contain pathogens—humans should avoid contact with either—but blackwater is considered a much higher-risk medium for the transmission of waterborne diseases.

Though they are not blackwater, the following water sources should not be included in graywater that is to be used for irrigation: garden and greenhouse sinks, water softener backflush, floor drains, and swimming pool water. In buildings served exclusively by composting toilets, and thus producing no true blackwater, it may be useful to include kitchen wastewater in graywater by taking special precautions to eliminate organic matter.

Rainwater harvesting. The collection, storage, and use of rainwater. Most systems use the roof surface as the collection area and a large galvanized steel, fiberglass, polyethylene, or ferro-cement tank as the storage cistern. When the water is to be used just for landscape irrigation, only sediment filtration is typically required. When water is being collected and stored for potable uses, additional measures are required to purify the water and ensure its safety. Rainwater harvesting offers several important environmental benefits, including reducing pressure on limited water supplies and reducing stormwater runoff and flooding. It can also be a better-quality source of water than conventional sources. After purification, rainwater is usually very safe and of high quality.

Xeriscaping. A landscaping strategy that focuses on using drought-tolerant native and adapted species that require minimal to no water for their maintenance. The term is derived from the Greek word *xeri,* meaning "dry"; the strategy is also referred to as *enviroscaping.*

Reclaimed water. Water from a wastewater treatment plant that has been treated and can be used for nonpotable purposes such as landscape irrigation, cooling towers, industrial process uses, toilet flushing, and fire protection. In some areas of the United States, reclaimed water may be referred to as *irrigation quality* (IQ) water, but potential uses can extend well beyond irrigation.

STEPS FOR DEVELOPING A BUILDING HYDROLOGIC STRATEGY

The following logical steps can be used to develop a hydrologic strategy for high-performance buildings:

1. *Select water sources for each consumption purpose.* Potable water must be used only for those applications that involve human consumption or ingestion. In addition to potable water, other water sources include rainwater, graywater, and reclaimed water. These alternative sources of water can be used for landscape irrigation, fire protection, cooling towers, chilled and hot water, toilet and urinal flushing, and other applications for which valuable potable water can be minimized. In each case, the availability of each alternative water source should be analyzed to determine which mix is optimum for the particular project and its forecasted water use profile.
2. *For each purpose, employ technologies that minimize water consumption.* This strategy can include a combination of low-flow fixtures (toilets, urinals, faucets, and shower heads), no-flow fixtures (composting toilets, waterless urinals), and controls (infrared sensors). For landscaping, highly efficient drip irrigation systems use far less water and deliver the water to the plant roots with more than 90 percent efficiency. Additionally, drought-tolerant, native, and adapted species can be employed in the landscape scheme, an approach that can often eliminate the need for an irrigation system.
3. *Evaluate the potential for a dual wastewater system.* Such a system separates lightly contaminated water from sinks, drinking fountains, showers, dishwashers, and washing machines from human waste–contaminated sources such as toilets and urinals. This dual piping system separates graywater from blackwater, thus providing the capability for water recycling within the building.
4. *Analyze the potential for innovative wastewater treatment strategies.* For example, constructed wetlands or *living machines* can be employed to process effluent. These approaches are rapidly evolving and beginning to appear in more high-performance building projects each year as the practice of using nature in symbiosis with the building process becomes more refined.

5. *Apply LCC to analyze the costs and benefits of adapting practices that reduce water flow through the building and its landscape beyond the levels mandated by EPAct.* A simple LCC that examines nothing more than the cost of potable water will generally provide long payback times, perhaps in the 10- to 20-year range. Including reductions in wastewater generation and the costs associated with its treatment will provide an accelerated payback. A more liberal interpretation of costs, such as the actual energy cost of moving water and wastewater, emissions associated with energy generation, worker productivity improvements, and general environmental benefits, would also shorten the payback time of the initial investment. Finally, it can be reasonably expected that the price of potable water in most regions will increase at a greater rate than the general inflation rate and perhaps dramatically faster. Including this in the LCC evaluation, along with other indirect cost factors, should bring the paybacks into the same range as those for good energy conservation measures, namely, 7 years or less.

High-Performance Building Water Supply Strategy

The basic strategy for the water supply of a high-performance building is to reduce potable water consumption to the maximum extent possible. Thus, the first two steps in the high-performance building hydrologic cycle strategy just given also apply to the water supply strategy. The first step is to assess the potential for using nonpotable water sources to replace potable water in a wide range of applications. In this context, nonpotable water includes rainwater, graywater, and reclaimed water. When the feasibility of using each of these nonpotable sources has been assessed, the next step is to ensure that consumption of both potable and nonpotable water is minimized. A wide range of ULF fixtures are now available that provide flow rates well below the EPAct requirements. Waterless plumbing fixtures are becoming more widely available and price-competitive as manufacturers begin offering more alternatives. The strategies that can be used to minimize potable water consumption are described in detail in the following subsections.[5]

LOW-FLOW AND ULF FIXTURES

EPAct set relatively ambitious limits on water use for water fixtures. However, water use by high-performance green buildings normally exceeds the EPAct requirements. For example, the USGBC LEED-NC standard allows 1 point for a minimum 30 percent reduction in potable water consumption over the EPAct requirements, 2 points if the reduction is 35 percent, and 3 points for a 40 percent or higher reduction. The following subsections describe the main types of plumbing fixtures and their alternatives. (Note that in this context, *low flow* refers to fixtures that meet the EPAct requirements and *ultra-low flow* (ULF) refers to fixtures that exceed EPAct requirements.)

Toilets and Urinals

Toilets account for almost half of a typical building's water consumption. Americans flush about 4.8 billion gallons (18.2 billion liters) of water down toilets each day, according to the USEPA. According to the Plumbing Foundation, replacing all existing toilets with 1.6-gallon- (6-liter-) per flush models would save almost 5,500 gallons (25,000 liters) of water per person each year. A widespread toilet replacement program in New York City apartment buildings found an average 29 percent reduction in total water use for the buildings studied. The entire program, in which 1.3 million toilets were replaced, is estimated to be saving 60 to 80 million gallons (230 to

Figure 8.3 Dual-flush Cera toilet from Ifö uses 0.3 gallon (1.1 liter) for a half flush and 1.6 gallons (6 liters) for a full flush, with flush levels adjustable up and down. (Photograph courtesy of Ifö and Christer Andersson.)

300 million liters) per day. However, there is a common perception that low-flow toilets do not perform adequately. The reason is that a number of early 1.6-gallon- (6-liter) per flush gravity-flush toilets that were adapted from the 3.5-gallon- (16-liter-) per flush model (rather than being designed from the ground up to operate effectively with the lower volume) performed very poorly, and some low-flow toilets may still suffer from this problem. But studies show that most 1.6-gallon- (6-liter) per flush toilets work very well.

Several technologies of 6-gallon (6-liter) toilets are available:

- *Gravity-tank toilets:* Use basically the same design as for older toilets, but with steeper sides to allow more rapid cleaning during the flush cycle.
- *Dual-flush toilets:* Have two handles for flushing, one for minimal needs such as urine, which uses 1.0 gallon (3.8 liters) per flush, the second for maximum flow of 1.6 gallons (6 liters).
- *Flushometer toilets:* Pressure developed in the flush cycle is captured and used to assist the subsequent flush.
- *Vacuum-assisted toilets:* Use the reverse principle of a flushometer toilet by employing a vacuum, which is regenerated by flushing action, to pull the wastewater from the toilet.

For toilets, a ULF fixture would consume less than 1.6 gallons (6 liters) per flush. Where flush performance is a particular concern or where water conservation beyond that of a 1.6-gallon- (6-liter) per flush model is required, electromechanical flush toilets and dual-flush toilets should be considered. Electromechanical toilets use electrically powered mechanical devices such as pumps and compressors to assist the removal of wastewater from toilets and use less than 1.0 gallon (3.8 liters) of water per flush.

Even greater water conservation can be achieved in certain (limited) applications with composting toilets. Because of the size of composting tanks, lack of knowledge about performance, local regulatory restrictions, and higher first costs, composting toilets are rarely an option except in certain unique applications, such as national park facilities. Composting toilets are being used very successfully, for example, at Grand Canyon National Park.

For urinals, water conservation well beyond the standard 1.0 gallon (4.5 liter) per flush can be obtained using waterless urinals. These products, available from, for example, the Waterless Company, Falcon Waterfree Technologies, and the Sloan Valve Company, use a special trap with a lightweight biodegradable oil that allows urine and water to pass through but prevents odors from escaping into the restroom; there are no valves to fail, and clogging does not cause flooding. Water and wastewater savings that can be achieved are truly remarkable. For example, Falcon Waterfree Technologies cites an annual net savings of $12,600 for a 75-unit installation, or about $168 per installed urinal. The payback for this rate of savings is less than 3 years at today's water and wastewater prices, and will be far greater in the future as pressure mounts to optimize the use of increasingly scarce sources of potable water.

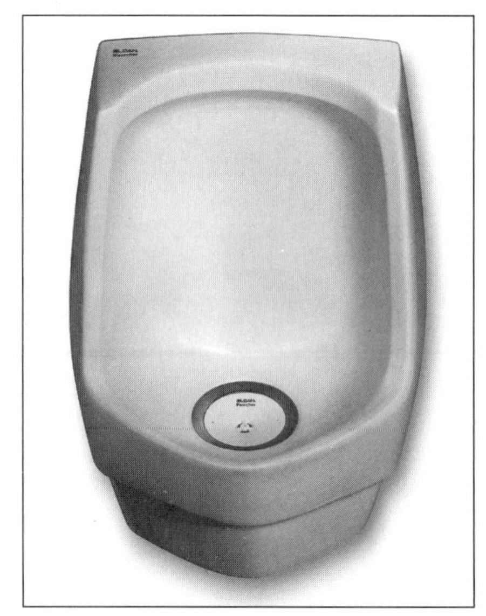

Figure 8.4 Waterless urinals save about 40,000 gallons (151,400 liters) of water per year per fixture. A Sloan Valve Company waterless urinal. (Photograph courtesy of Sloan Valve Company.)

Showers

A conventional showerhead is rated to use 3 to 7 gallons (11 to 27 liters) per minute at normal water pressure, about 80 psi (550 kPa). A 5-minute shower with a conventional showerhead typically consumes 15 to 35 gallons (60 to 130 liters) of water. High-quality replacement showerheads that deliver 1.0 to 2.5 gallons (3.8 to 9.5 liters) per minute can save many gallons per shower when used to replace conventional showerheads. Products vary in price from $3 to $95, and many good models are available for $10 to $20. A variety of spray patterns are also available, ranging from misty to pounding and massaging. These showerheads typically have narrower spray jets and a greater mix of air and water than conventional showerheads, enabling them to provide what feels like a full-volume shower while using far less water.

Flow regulators on the shower controls and temporary cutoff buttons or levers incorporated into the showerhead reduce or stop water flow when the individual is soaping or shampooing, further lowering water use. When the water flow is reactivated, it emerges at the same temperature, eliminating the need to remix the hot and cold water. Flow restrictors are washerlike disks that fit inside showerheads, and they are tempting retrofits. However, flow restrictors provide poor water pressure in most showerheads, and flow-restrictor disks that were given away by many water conservation programs led to poor acceptance of water conservation in general. Permanent water savings are better provided through the installation of well-engineered showerheads.

Faucets

Faucets are generally found in bathrooms, kitchens, and workrooms. Bathroom faucets need no more than 1.5 gallons (5.7 liters) per minute, and residential kitchens rarely need less than 2.5 gallons (9.5 liters) per minute. Institutional bathroom faucets may include automated controls and premixed temperatures. Institutional kitchen faucets may include special features such as swivel heads and foot-activated on/off controls. Older faucets with flow rates of 3 to 5 gallons (11 to 19 liters) per minute waste tremendous quantities of water. Federal guidelines mandate that all lavatory and kitchen faucets and replacement faucet tips (including aerators) manufactured after January 1, 1994, consume no more than 2.5 gallons (9.5 liters) per minute at 80 psi (550 kPa). Metered-valve faucets are restricted to a 0.25-gallon (0.95-liters) per cycle discharge after this date. Metered-valve faucets deliver a preset amount of water and then shut off. For water management purposes, the preset amount of water can be reduced by adjusting the flow valve. The Americans with Disabilities Act (ADA) requires a 10-second minimum on-cycle time. Variations in water pressure can occur in buildings, and pressure-compensating faucets can be used to automatically maintain 2.5 gallons (9.5 liters) per minute at varying water pressures. For kitchens, 8.3- to 9.5-liter (2.2- to 2.5-gallon)-per-minute devices are available. In washrooms, 1.9- to 4.7-liter (0.5- to 1.25-gallon)-per-minute models will often prove adequate for personal washing purposes. Foot controls for kitchen faucets provide both water savings and hands-free convenience. The hot water mix is set, and the foot valve turns the water on and off at the set temperature. Hot water recirculation systems reduce water wasted while users wait for water to warm up as it flows from the faucet. To prevent these water-saving systems from wasting large amounts of energy, hot water pipes should be well insulated.

Drinking Fountains

Self-contained drinking fountains have an internal refrigeration system. Adjusting the exit water temperature to 70°F (21°C) versus the typical 65°F (18°C) will result in substantial energy savings. Insulating the piping, chiller, and storage tank will save energy. If appropriate, adding an automatic timer to shut off the unit during evenings and weekends will add to the savings. Remote chillers or central systems are used in some facilities to supply cold drinking water to multiple locations. To conserve energy, the temperature can be raised from 65° to 70°F (18°C to 21°C); piping should be well insulated, and a timer can be used to turn off the unit when the building is unoccupied. Sensor faucets require either electrical wiring for the connection of AC power or regular replacement of battery power supplies.

ELECTRONIC CONTROLS FOR FIXTURES

Automated controls for faucets, toilets, and urinals can dramatically lower water consumption and potentially eliminate disease transmission via contact with bathroom surfaces and fixtures. These controls are rapidly gaining popularity in all types of commercial and institutional facilities, though the driver is generally hygiene rather than water or energy savings.

Electronic controls can be installed with new plumbing fixtures or retrofitted onto many types of existing fixtures. Though water savings depend greatly on the type of facility and the particular controls used, some facilities report 70 percent water savings. This type of on-demand system can also produce proportional savings in water heating (for faucets) and sewage treatment. Electronic controls for plumbing fixtures usually function by transmitting a continuous beam of infrared (IR) light. With faucet controls, when a user interrupts this IR beam, a solenoid is activated, turning on the water flow. Dual-beam IR sensors or multispectrum sensors are generally recommended because they perform better for a wider range of users. With toilets and urinals, the flush is actuated when the user moves away and the IR beam is no longer blocked. Some brands of no-hands faucets are equipped with timers to defeat attempts to alter their operation or to provide a maximum on cycle—usually 30 seconds. Depending on the faucet, a 10-second handwash typical of an electronic unit will consume as little as 1⅓ cups (0.3 liter) of water. A 10-second cycle is required as a minimum by the ADA.

Electronic controls can also be used for other purposes in restrooms. Sensor-operated hand dryers are hygienic and save energy by automatically shutting off when the user steps away. Soap dispensers can be electronically controlled. Electronic door openers can be employed to further reduce contact with bathroom surfaces. Even showers are now sometimes being controlled with electronic sensors—for example, in prisons and military barracks. Electronic fixtures are particularly useful for handicapped installations and hospitals, greatly reducing the need to manipulate awkward fixture handles and removing the possibility of scalding caused by improper water control. No-touch faucets are available either with (1) the sensor mounted in the wall behind the sink, (2) the sensor integrated into the faucet, or (3) the sensor mounted in an existing hot or cold water handle hole and the faucet body in the center hole. For new installations, the first or second option is usually best; for retrofit installations, the last option may be the only one feasible. At sports facilities where urinals experience heavy use, the entire restroom can be set up and treated as if it were a single fix-

Figure 8.5 The rainwater harvesting system for Rinker Hall at the University of Florida has a cast-in-place cistern (shown here under construction) located under the south stairwell of the building. The rainwater is used for flushing the building's toilets. (Photograph courtesy of Centex-Rooney, Inc.)

ture. Traffic can be detected and the urinals flushed periodically based on traffic rather than per person. This can significantly reduce water use. Computer controls can be used to coordinate water usage to divert water for fire protection when necessary. Thermostatic valves can be used with electronic faucets to deliver water at a preset temperature. Reducing hot water consumption saves a considerable amount of energy. A 24-volt transformer operating off a 120-volt AC power supply is typically used, at least with new installations. The transformer should be Underwriters Laboratory (UL)-listed; and for security reasons, the transformer and the solenoid valve should be remotely located in a chase.

RAINWATER HARVESTING

Rainwater has been considered a crucial source of water for survival for all of human existence. For building applications, rain was typically collected from the roofs of homes and other buildings and conducted into a storage tank or cistern. With the advent of centralized potable water systems, rainwater systems all but disappeared until the emergence of the modern high-performance green building movement. *The Texas Guide to Rainwater Harvesting* cites three factors that are propelling rainwater back into the picture as a viable water source:[6]

1. The escalating environmental and economic costs of providing water by centralized water systems or by well drilling
2. Health concerns regarding the source and treatment of polluted waters
3. The perception that there are cost efficiencies associated with reliance on rainwater

Rainwater systems are appropriate when one or more of the following factors is present:

- Groundwater or aquifer water supplies are limited or fragile. Fragile aquifer systems are those that, when pumped, can threaten ecologically valuable surface waters and springs.
- Groundwater supplies are polluted or significantly mineralized, requiring expensive treatment.
- Stormwater runoff is a major concern.

A rainwater harvesting system generally has the following key components:[7]

- *Catchment area:* With most rainwater harvesting systems, the catchment area is the building's roof. The best roof surface for rainwater harvesting does not support biological growth (e.g., algae, mold, moss), is fairly smooth so that pollutants deposited on the roof are quickly removed by the roof-wash system, and should have a minimal number of overhanging tree branches above it. Galvanized metal is the roofing material most commonly used for rainwater harvesting.
- *Roof-wash system:* This is a system for keeping dust and pollutants that have settled on the roof out of the cistern. It is necessary for systems used as a source of potable water but is also recommended for other systems, as it keeps potential contaminants out of the tank. A roof-wash system is designed to purge the initial water flowing off a roof during rainfall.
- *Prestorage filtration:* To keep large particulates, leaves, and other debris out of the cistern, a domed stainless-steel screen should be secured over each inlet leading to the cistern. Leaf guards over gutters can be added in areas with significant windblown debris or overhanging trees.

- *Rainwater conveyance:* This is the system of gutters, downspouts, and piping used to carry water from the roof to the cistern.
- *Cistern:* This is usually the largest single investment required for a rainwater harvesting system. Typical materials used include galvanized steel, concrete, ferro-cement, fiberglass, polyethylene, and durable wood (e.g., redwood or cypress). Costs and expected lifetimes vary considerably among these options. Tanks may be located in a basement, buried outdoors, or located aboveground outdoors. Light should be kept out to prevent algae growth. Cistern capacity should be sized to meet the expected demand. Particularly for systems designed as the sole water supply, sizing should be modeled on the basis of 30-year precipitation records, with sufficient storage to meet the demand during times of the year having little or no rainfall.
- *Water delivery:* A pump is generally required to deliver water from the cistern to its point of use, though gravity-fed systems are occasionally possible with appropriate placement of system components.
- *Water treatment system:* To protect plumbing and irrigation lines (especially with drip irrigation), water should be filtered through sediment cartridges to remove particulates, preferably down to 5 microns. For systems providing potable water, additional treatment is required to ensure a safe water supply. This can be provided with microfiltration, ultraviolet sterilization, reverse osmosis, or ozonation (or a combination of these methods). With some systems, higher levels of treatment are provided only at a single faucet where potable water is drawn.

Rainwater harvesting systems have immense potential for reducing potable water consumption by introducing a water source that is readily obtainable in many regions of the United States. In spite of this advantage, there are no standard designs or approaches to designing a rainwater harvesting system; hence, currently, each system designed for a building is unique. As a consequence, these systems can be failure-prone and unreliable, resulting in a potential erosion of interest in rainwater as a substitute for potable water. The creation of clear standards, designs, and standard components would go a long way toward resolving this problem and making the implementation of these systems standard practice.

GRAYWATER SYSTEMS

Graywater is generally considered to comprise the nonhuman waste fraction of wastewater. Graywater collection involves separating graywater from blackwater, which, as defined previously, is the human waste–contaminated water from toilets and urinals. Graywater is generally used for landscape irrigation, but it can also be used to flush toilets and urinals.

Buildings with graywater systems must have a dual waste piping system, one for each type of water. Graywater waste lines should run to a central location where a surge tank can collect and hold the water until it drains or is pumped into an irrigation system or for other appropriate end uses. An overflow for the graywater collection system should be provided that feeds directly into the sewer line. If excess graywater fills the system due to a mismatch between supply and outflow, or due to a filter or pumping malfunction, the overflow conducts the excess flow to the sewer system. A controllable valve should also be included so that graywater can be shunted into the sewer line when the area(s) being irrigated become too wet or other reasons preclude the use of graywater.

Graywater should not be stored for extended periods of time before use. Decomposition of the organic material in the water by microorganisms will quickly use up available oxygen and anaerobic bacteria will take over, producing unpleasant odors.

Figure 8.6 A restored wetland at the Lewis Environmental Studies Center at Oberlin College in Oberlin, Ohio, accepts wastewater treated by the building's Living Machine (see Figure 8.7). (Photograph courtesy of Oberlin College.)

Some graywater systems are designed to dose irrigation pipes with a large, sudden flow of water instead of allowing the water to trickle out as soon as it enters the surge tank. For a dosing system, holding the water for some amount of time will be necessary, but this should be limited to no more than a few hours. If a filter is used in the graywater system, it should be one that is easy to clean or self-cleaning. Filter maintenance is a major problem with many graywater systems. For complete protection from pathogens, graywater should flow by gravity or be pumped to a belowground disposed field (subsurface irrigation). Perforated plastic pipe—3 inches (76 millimeters) minimum diameter—is called for in California's graywater regulations, though with filtering, smaller-diameter drip irrigation tubing can also be used. The California standards require that untreated graywater be disposed of at least 9 inches (about 230 millimeters) below the surface of the ground. Some graywater systems discharge into planter beds—sometimes even beds located inside buildings. Some ready-made systems are available by mail order, but these should be modified for specific soil and climate conditions. As a general rule, graywater can be used for subsurface irrigation of lawns, flowers, trees, and shrubs, but it should not be used for vegetable gardens. Drip irrigation systems have not yet proven to be effective for graywater discharge because of clogging or high maintenance costs.

RECLAIMED WATER

The use of reclaimed water for nonpotable purposes can greatly reduce the demand on potable water sources. Municipal wastewater reuse now amounts to about 4.8 billion gallons (18 million cubic meters) per day (about 1 percent of all freshwater withdrawals). Industrial wastewater reuse is far greater— about 865 billion gallons (3.2 billion cubic meters) per day.

In areas of chronic water shortage, the design team should check with the local water utility and inquire whether it has a program to provide reclaimed water to the building's location. Reclaimed water programs are particularly popular in California, Florida, Arizona, Nevada, and Texas.

There are a host of potential applications for reclaimed water: landscaping; golf course or agricultural irrigation; decorative features, such as fountains; cooling tower makeup; boiler feed; once-through cooling; concrete mixing; snowmaking; and fire main water. Making use of reclaimed water is easiest if this is planned for at the outset of building a new facility, but major renovations or changes to a facility's plumbing system provide opportunities as well. For certain uses, such as landscape irrigation, required modifications to the plumbing system may be quite modest. It is important to note, however, that the use of reclaimed water may be restricted by state and local regulations. For locations such as universities or military bases that often have their own wastewater treatment plant (WWTP), there may be an opportunity to modify the plant to provide on-site reclaimed water.

To consider using reclaimed water for a building, one or more of the following situations should be present: (1) high-cost water or a need to extend the drinking water supply; (2) local public policy encouraging or mandating water conservation; (3) availability of high-quality effluent from a WWTP; or (4) recognition by the building owner of environmental or other nontangible benefits of water reuse.

Technologies vary with end uses. A modern WWTP has three stages of treatment—primary, secondary, and tertiary—with each succeeding stage requiring more energy and chemicals than the previous stage. In general, tertiary or advanced secondary treatment is required, either of which usually includes a combination of coagulation, flocculation, sedimentation, and filtration. Virus inactivation is attained by granular carbon adsorption plus chlorination, or by reverse osmosis, ozonation, or ultraviolet exposure. Dual water systems are beginning to appear in some parts of the country where the water supply is limited or where water shortages may constrain development. Buildings may have two water lines coming in, one for potable water and the other for reclaimed water. The former is for all potable uses, the latter for nonpotable uses. Piping and valves used in reclaimed water systems should be color-coded with purple tags or tape. This minimizes piping identification and cross-connection problems when installing systems. Liberal use of warning signs at all meters, valves, and fixtures is also recommended. (Note that potable water mains are usually color-coded blue, while sanitary sewers are green.) Reclaimed water should be maintained at 10 psi (70 kPa) lower pressure than potable water mains to prevent backflow and siphonage in the event of accidental cross-connection. Although it is feasible to use backflow prevention devices for safety, it is imperative never to directly connect reclaimed and potable water piping. One additional precaution is to run reclaimed water mains at least 12 inches (30 centimeters) lower (in elevation) than potable water mains and to separate them from potable or sewer mains by a minimum of 10 feet (3 meters) horizontally.

Although water prices vary greatly throughout the country, reclaimed water costs significantly less than potable water. For example, in Jupiter, Florida, the price of potable water is now \$1.70 per 1,000 gallons (\$0.45 per cubic meter) versus \$0.26 per 1,000 gallons (\$0.07 per cubic meter) for reclaimed water. Similar pricing differences occur wherever reclaimed water is available.

High-Performance Building Wastewater Strategy

Reducing potable water consumption is relatively straightforward compared to the effort needed to change wastewater treatment strategies. Contemporary WWTPs are large, centralized energy- and chemical-intensive operations designed to ensure that public health is protected. However, future high energy costs and increasing public resistance to chemical use are motivating building owners to consider other options for treating wastewater. The fundamental approaches being used today rely on nature, either directly or indirectly, for these alternative approaches. In the direct approach, effluent from buildings is treated by surface or subsurface wetlands. In the indirect approach, nature is brought into the building and enclosed in tanks and vats through which wastewater is passed and cleaned up by plants, light, and bacteria.

CONSTRUCTED WETLANDS

One of the ultimate goals of green building is the application of ecological design to the greatest extent possible, including creating a synergistic relationship among natural systems, buildings, and the humans occupying them. Using nature to perform tasks that would otherwise be accomplished by energy-intense mechanical and electrical systems has four distinct advantages:[8]

1. Nature is self-maintaining, self-regulating, and self-organizing.
2. Nature is powered by solar energy and chemical energy stored in organic materials.
3. Natural systems can degrade and absorb undesirable toxic and metal compounds, converting them into stable compounds.
4. Natural systems are easy to build and operate.

The use of wetlands to treat wastewater from buildings provides precisely this type of opportunity because these ecological systems can break down organic waste, minimizing the need for complex infrastructure and creating nutrients that benefit the species performing these services. Constructed wetlands can be characterized as passive systems for wastewater treatment. They mimic natural wetlands by using the same filtration processes to remove contaminants from wastewater.[9] In addition to removing organic nutrients, constructed wetlands have the ability to remove inorganic substances; thus, they can be used to treat industrial wastewater, landfill leachate, agricultural wastewater, acid mine drainage, and airport runoff. Constructed wetlands also provide the added benefit of environmental amenity and can blend into natural or rural landscapes. Moreover, in addition to treating wastewater, constructed wetlands can provide surge areas for stormwater and treat this often contaminated runoff.

Wetlands remove contaminants by several mechanisms, including nutrient removal and recycling, sedimentation, biological oxygen demand, metals precipitation, pathogen removal, and toxic compound degradation.

A number of site-specific factors must be considered when considering the use of a constructed wetland for wastewater treatment: hydrology (groundwater, surface water, permeability of ground), native plant species, climate, seasonal temperature fluctuations, local soils, site topography, and available area. Constructed wetlands are built for either surface or subsurface flow. Surface flow systems consist of shallow basins with wetlands plants that are able to tolerate saturated soil and aerobic conditions. The wastewater entering the surface system flows slowly through the basins and is released at the end as clean water. Subsurface systems, where the wastewater flows through a substrate such as gravel, have the advantages of higher rates of contaminant removal, compared to surface flow systems, and limited contact for humans and animals. They also work especially well in cold climates due to the Earth's insulating properties.

Cost is an important factor in deciding which approach is best for a particular situation. The good news about constructed wetlands is that both the capital and operating costs are far lower than those for a conventional WWTP.

LIVING MACHINES

In addition to using constructed wetlands to treat wastewater from the built environment, nature can be brought directly into a building in order to break down the materials in the wastewater system. Although there are several approaches, the best known is the Living Machine, created by John Todd, a pioneer in the development of natural wastewater processing systems. A Living Machine differs from a conventional WWTP in four basic respects:[10]

1. The vast majority of a Living Machine's working parts are live organisms, including hundreds of species of bacteria, plants, and vertebrates such as fish and reptiles.
2. The Living Machine has the ability to design its internal ecology in relation to the energy and nutrient streams to which it is exposed.
3. A Living Machine can repair itself when damaged by toxics or when shocked by interruption of energy or nutrient sources.
4. Living Machines can self-replicate through reproduction of the organisms in the system.

The concept of a Living Machine can be applied not only to an alternative WWTP but also to a range of other systems that can generate fuel, grow food, restore

Figure 8.7 The Living Machine built into the Lewis Environmental Studies Center at Oberlin College contains biological organisms that break down wastewater components into nutrients that are then fed into a constructed wetland outside the building. (Photograph courtesy of Oberlin College.)

degraded environments, and even heat and cool buildings. Several successful examples of Living Machines have been integrated into buildings. The most recent of these is the Living Machine in the Lewis Environmental Studies Center at Oberlin College, in Oberlin, Ohio, which processes wastewater from the occupants of this 14,000-square-foot (1,400-square-meter) building.

Landscaping Water Efficiency

Approximately 30 percent of residential water use, or about 32 gallons (121 liters) per person per day, is used for exterior uses; and the bulk of this, as much as 29 gallons (110 liters) per person per day, is used for maintaining landscaping, with wide variations depending on the climatic region. Most of the water used for this purpose is wasted due to overwatering. Water-intensive turf grass creates the major demand for irrigation. In the United States, more than 16,000 golf courses consume 2.7 billion gallons (10.2 billion liters) of water per day.[11]

Several forms of sustainable landscaping are emerging after several decades of evolution. The best known is *xeriscaping,* which emphasizes the use of drought-tolerant native and adapted species of plants and turf grass. (Note: the terms *enviroscaping* and *water-wise landscaping* are sometimes used interchangeably with *xeriscaping.*) Seven principles can be used to ensure a well-designed, water-efficient landscape:

1. Proper planning and design
2. Soil analysis
3. Appropriate plant selection
4. Practical turf grass areas
5. Efficient irrigation
6. Use of mulches
7. Appropriate maintenance

Perhaps an even more sustainable form of landscaping than xeriscaping is *natural landscaping.* Using restorative landscaping principles, natural landscaping supports the use of indigenous plants that, once established, virtually eliminate the need for watering. Even turf grass, the most ubiquitous consumer of water, can be replaced with indigenous species because there are thousands of native species in the United States. The restoration of native landscapes has other benefits as well. Animal species that live in native landscapes are reestablished; natural landscapes filter stormwater effectively; and the natural beauty of the landscape is restored. In 1981, Darrel Morrison, a professor at the University of Georgia and a member of the American Society of Landscape Architects (ASLA), defined three characteristics necessary for natural landscape design:[12]

1. Regional identity (sense of place)
2. Intricacy and detail (biodiversity)
3. Elements of change

Opposition to natural landscaping was initially strong because many people, after having grown accustomed to manicured turf grass lawns, had difficulty accepting landscaping that appeared wild and unconventional. In fact, numerous people were prosecuted for attempting to implement natural landscaping; they were accused of violating weed laws. Fortunately, natural landscaping is now far more widely accepted, and the beauty and aesthetics of this approach are winning over most skeptics. Natural landscaping can include butterfly gardens, native trees and shrubs that attract birds, small ponds, native groundcovers in lieu of turf grass, and gardens composed of native plants. Native plants have several environmental advantages that fit in with the concept of a high-performance green building: they survive without fertilizers or synthetic pesticides and rarely need watering; they provide food and habitat for wildlife; and they contribute to biodiversity.

Connection to LEED-NC

The LEED-NC category covering water and wastewater issues is Water Efficiency (WE). A maximum of 10 total points are available in the WE category, as summarized in Table 8.2.

WE CREDITS 1.1 AND 1.2 (WEc1.1 AND WEc1.2): WATER-EFFICIENT LANDSCAPING (4 POINTS MAXIMUM)

Reducing potable water consumption use for landscaping by 50 percent results in 2 points (WEc1.1); the total elimination of potable water for landscaping results in another 2 points (WEc1.2). Use of graywater or rainwater in place of potable water is allowable for both the 50 percent reduction and elimination credits.

WE CREDIT 2 (WEc2): INNOVATIVE WASTEWATER TECHNOLOGIES (2 POINTS MAXIMUM)

Although the title of this credit indicates that it concerns wastewater, the actual issues addressed are somewhat broader and allow two strategies. The first strategy is to reduce the quantity of wastewater actually being generated that would have to undergo wastewater treatment. Consequently, any strategies that reduce building water consumption also help to reduce wastewater flows. There are two ways to earn the point associated with WEc2: (1) reduce potable water used for sewage conveyance by at least 50 percent or (2) treat at least 50 percent of the water to tertiary standards on-site and then use the water on-site or infiltrate it back into the ground.

TABLE 8.2

Water Efficiency (WE) Credits and Points under LEED-NC 2009

Credit	Name of Credit	Maximum Points
WE Credit 1.1	Water-Efficient Landscaping: 50% Reduction	2
WE Credit 1.2	Water-Efficient Landscaping: No Potable Use or Irrigation	2, in addition to WE 1.1
WE Credit 2	Innovative Wastewater Technologies	2
WE Credit 3	Water Use	2–4
	Total WE Points Available	**10**

Note: The descriptions of the LEED-NC credits and points in this section are abbreviated, and the appropriate LEED Reference Manual should be consulted for detailed information about how to achieve the points.

WE CREDIT 3 (WEc3): WATER USE REDUCTION (4 POINTS MAXIMUM)

The reduction of potable water use in the building by 30 percent results in 2 points; increasing the reduction to 35 percent provides an additional point and reducing by 40 percent earns one more point. In calculating the reduction, potable water use is compared to that of a baseline building meeting the fixture requirements of EPAct of 1992. To earn one or both of these points, the mechanical/electrical/plumbing (MEP) engineer must provide calculations that demonstrate the reduction compared to the baseline building. The use of ULF fixtures, rainwater, and graywater can help to achieve this reduction.

Connection to Green Globes V.1

The high-performance building hydrologic cycle is covered by Green Globes in Part D, which is outlined in Table 8.3.

Summary and Conclusions

Much of the attention to high-performance green building design has focused on superior energy performance because there are demonstrable, easy-to-document savings that can be used to justify investments in energy conservation. But for the building hydrologic system, the savings for water conservation and innovative handling of wastewater are not so easy to document because, in the United States, water has been a heavily subsidized resource, as has been the treatment of wastewater effluent. However, it is water, not energy, that can be the limiting resource for development, as

TABLE 8.3

Outline of Green Globes v.1, Part D (Water)

Section		Description	Points
D.1		**Water**	**40**
	D.1.1	Potable water reduction compared to EPAct	40
D.2		**Water-Conserving Features**	**40**
	D.2.1	Sub-metering of high water use operations	5
	D.2.2	Minimizing cooling tower water consumption	9
	D.2.3	Minimizing potable water for landscape irrigation	10
	D.2.4	Water-efficient landscape irrigation	5
	D.2.5	Water-efficient landscape	5
	D.2.6	Landscaping without lawn	5
	D.2.7	Lawn only for functional purposes (in lieu of D2.6)	1
D.3		**Reduce Off-Site Treatment of Water**	**20**
	D.3.1	Graywater system	10
	D.3.2	Blackwater system or composting toilets	10
		Total Points Available	**100**

demonstrated by several growth moratoriums that have been imposed to limit or stop development and construction activity until the supply or wastewater treatment problem was solved. Though the USGBC LEED-NC building assessment standard includes consideration of water as a significant issue to be addressed by the project team, it allocates only 10 possible points for water issues, compared to 35 possible points for energy and atmosphere measures.

Water is in fact such a critical issue that project teams in many areas of the United States should consider making extraordinary efforts to reduce potable water consumption to exceptionally low levels. Ample experience now shows that the use of rainwater harvesting, reclaimed water, graywater systems, and new waterless fixture technologies is eliminating the need for water use in urinals. ULF fixtures for toilets are also available, requiring only about one-half of the water needed by toilets meeting current plumbing codes. One area where progress still needs to be made is landscape irrigation, where about 50 percent of potable water tends to be used.

If the construction industry does not make significant reductions in the consumptive water profile of most buildings a major goal, growth moratoriums, often instituted because of water or wastewater limitations, will reduce the volume of its business. It is now apparent that finding more appropriate ways of using potable water and treating wastewater will result in a win-win situation for both the public and the construction industry.

Notes

1. Hawken et al. (1999), chapter 11, describes the faulty logic of contemporary building water and wastewater systems and suggests remedies that can ensure the sustainability of the world's potable water supply.
2. From the Letters to the Editor section of the online version (www.elynews.com) of *The Ely Times,* April 7, 2004.
3. From the January 31, 2004, online edition (www.emmitsburg.net) of Emmitsburg.net, a nonprofit Internet source for information about the Emmitsburg area.
4. RMI is a nonprofit organization that provides a wide range of consulting services on energy, water, development, and green building issues. It provides many valuable resources at www.rmi.org.
5. The description of alternative water systems is from Sections 6.1 through 6.6 of *Greening Federal Facilities* (2001).
6. *The Texas Guide to Rainwater Harvesting* (1999), an excellent overview of rainwater harvesting, addresses health issues, materials, and safety concerns in a succinct, informative manner.
7. Information on rainwater harvesting components is from Section 6.7 of *Greening Federal Facilities* (2001).
8. In the early 1970s, the USEPA began investigating alternatives to large, centralized, technically complex WWTPs; part of this effort was the creation of an alternative technology program that encouraged the development of systems that employed ecological systems to break down their own waste. This program and the advantages noted here are discussed by Campbell and Ogden (1999).
9. An excellent summary of the state of the art of constructed wetlands technology can be found in Lorion (2001).
10. Todd (1999) describes the concept of the Living Machine in great detail in chapter 8 of this book.
11. A detailed explanation of landscape water conservation practices can be found in Vickers (2001), chapter 3. Information in this section is from this source, which also contains extensive information about water conservation in general and both technical and policy information on the subject of reducing potable water use.
12. An excellent source of information about natural landscaping is the nonprofit organization Wild Ones: Native Plants, Natural Landscapes. Information and free downloads are available at its website, www.for-wild.org.

References

Ausubel, Kenny. 1997. *The Bioneers: A Declaration of Interdependence.* White River Junction, VT: Chelsea Green.

Campbell, Craig S., and Michael H. Ogden. 1999. *Constructed Wetlands in the Sustainable Landscape.* New York: John Wiley & Sons.

"Graywater Guide." 1995. Sacramento: California Department of Water Resources, Publications Office. Available at www.dwr.water.ca.gov.

Greening Federal Facilities: An Energy, Environmental and Economic Resource Guide for Federal Facility Managers and Designers, 2nd ed. May 2001. Washington, DC: Federal Energy Management Program, Office of Energy Efficiency and Renewable Energy, U.S. Department of Energy, DOE/GO-102001-1165. Available at www.1.eere.energy.gov/femp/pdfs/29267-0.pdf.

Hawken, Paul, Amory Lovins, and L. Hunter Lovins. 1999. *Natural Capitalism: Creating the Next Industrial Revolution.* Boston: Little, Brown.

Kibert, Charles J., ed. 1999. *Reshaping the Built Environment: Ecology, Ethics, and Economics.* Washington, DC: Island Press.

Lorion, Renee. August 2001. "Constructed Wetlands: Passive Systems for Wastewater Treatment," a Technology Status Report for the U.S. Environmental Protection Agency's Technology Innovation Office. Available at www.epa.gov/swertio1/download/remed/constructed_wetlands.pdf.

Ludwig, Art. 1999. *Builder's Greywater Guide.* Santa Barbara, CA: Oasis Design. Available at www.oasisdesign.net.

Texas Guide to Rainwater Harvesting, 2nd ed. 1999. Austin: Texas Water Development Board, in collaboration with the Center for Maximum Potential Building Systems. Available at www.twdb.state.tx.us/publications/reports/RainHarv.pdf or from the Texas Water Development Board.

Todd, John. 1999. *Reshaping the Built Environment,* Charles J. Kibert, Ed. Washington, DC: Island Press.

Vickers, Amy. 2001. *Handbook of Water Use and Conservation.* Amherst, MA: WaterPlow Press. Book details available at www.waterplowpress.com.

Chapter 9

Closing Materials Loops

The selection of building materials and products for a high-performance green building project is by far the most difficult and challenging task facing the project team. In the second edition of *Green Building Materials: A Guide to Product Selection and Specification* (2006), one of the first books about the subject of green building materials, Ross Spiegel and Dru Meadows defined green building materials as "those that use the Earth's resources in an environmentally responsible way."[1] At present, however, there is no clear consensus about the criteria for materials and products that would characterize them as *environmentally preferable, environmentally responsible,* or *green.* As a matter of fact, alternative terminologies are rapidly infiltrating the language of high-performance building green materials and products. For example, *environmentally preferable products* (EPPs) is a commonly used label that can be found in U.S. government specifications for building materials and products. As a result, the question of what is or is not environmentally preferable has not been settled and is open to much controversy. For example, some organizations promote green products based on a narrow range of attributes they specify as being important for this purpose. The Forestry Stewardship Council (FSC), represented in the United States by the SmartWood Program and Scientific Certification Systems, Inc., defines green products as wood products derived from a sustainably managed forest. The Greenguard Environmental Institute instead relies on levels of chemical emissions that affect indoor environmental quality (IEQ) to describe what constitutes a green product.[2]

Clearly, it would be advantageous for green products to carry a certification, or eco-label, to designate them as being preferable on the basis of consensus standards that address each type of building product. In Europe, several eco-labels cover at least some building materials. The Blue Angel eco-label in Germany, the Nordic Swan eco-label of the Nordic countries, and the European Union eco-label all have programs for labeling some types of building products. For example, the Blue Angel Standard, RAL-UZ-38, addresses the requirements for certification of wood panels.[3] Unfortunately, the range of products covered by these labeling programs is very limited; consequently, they provide minimal assistance in identifying those that might be considered green. Thus, the project team must rely on their own best judgment in deciding which materials fit the criteria for environmental friendliness.

On the positive side, several tools are available to assist this process, the most familiar being LCA. LCA provides information about the resources, emissions, and other impacts resulting from the life cycle of materials use, from extraction through disposal, and incorporates a high degree of rigor and science in the evaluation process. Two readily available LCA programs, ATHENA[4] and Building for Environment and Economic Sustainability (BEES),[5] apply to North American projects and can provide the project team with a decision system for materials selection that is based on science.

This chapter addresses the issues of green building materials and products, the criteria for defining environmentally friendly products, and the application of LCA in decision making for materials selection. It also offers information about specific materials and product groups where new technologies and approaches are beginning

Figure 9.1 Partial demolition of the University of Florida Law School Library for a building expansion project. Truly green buildings of the future should be designed for deconstruction to maximize the reuse and recovery of building components and materials. (Photograph: M.R. Moretti.)

to take hold in support of the green building movement. A case study about InterfaceFLOR Corporation's path to becoming a sustainable corporation is included.

Issues in Selecting Green Building Materials and Products

Determining how building materials and products will affect the environment is the central unresolved problem of the green building movement. Even evaluating the relative worth of using recycled versus virgin materials—which should be a relatively simple matter—can result in controversy. One school of thought, here referred to as the *ecological school,* maintains that keeping materials in productive use, as in an ecological system, is of primary importance, and that the energy and other resources needed to feed the recycling system are a secondary matter. Nature, after all, does not use energy *efficiently,* but it does employ it *effectively;* that is, it matches the energy needed to the available energy sources. Another school of thought, here referred to as the *LCA school,* suggests that if the energy and the emissions due to energy production are higher for recycling than for the use of virgin materials, then virgin materials should be used. The LCA school also generally contends that too much attention is given to solid waste and that greater emphasis should be put on global warming.[6]

Nothing, in fact, is obvious when it comes to using renewable resources in construction. Consider wood from old-growth forests: although these forests are certainly a renewable resource, extracting resources from them is generally frowned upon by environmental groups, and the green building movement is in favor of protecting the biodiversity of these beautiful and increasingly rare natural assets. Rather, it is generally agreed, wood should come from plantation forests and, even better, from rapidly renewable species. The USGBC's LEED standard defines a class of materials known as *rapidly renewable resources,* which are species with a growth and harvest cycle of 10 years or less. However, in spite of this strategy to shift extraction from old-growth forests to other resource, plantation forestry, which produces rapidly renewable resources, must be called into question, because it can require large quantities of water, fertilizer, pesticides, and herbicides to support the rapid growth cycle and protect the company's financial investment, not to mention that monoculture forestry runs counter to the notion of biodiversity. The definition of *rapidly renewable* as 10 years or less is itself arbitrary, and any number of other definitions are equally applicable.

Besides determining which materials are environmentally preferable or green, one must decide which products or materials will have low environmental impact. Many building products are selected to help reduce the overall environmental impact of the building, not for their own low environmental impacts. Using an energy recovery ventilator (ERV), for example, a relatively complex device containing desiccants, insulation, wiring, an electric motor, controls, and other materials, contributes to an exceptionally low energy profile for the building, but it cannot be considered inherently green because its constituent materials cannot be readily recycled. Today, one of the greatest challenges in designing a high-performance green building is selecting materials and products that lower the overall impact of the building, including the impact on its site. As time progresses, a hoped-for outcome is the development of more products that both have a low environmental impact and are inherently green—that is, can be disassembled into their recyclable constituent materials.

Distinguishing Between Green Building Products and Green Building Materials

The terms used to refer to the materials and products used in high-performance building can be contradictory and confusing. *Green building products* generally refer to building components that have any of a wide range of attributes that make them preferable to the alternatives. For example, low-emissivity (low-e) glass is a spectrally selective type of glass that allows visible light to pass through but rejects a substantial part of the heat-producing infrared portion of the light spectrum. As a product, it is preferable to ordinary float glass in windows because of its energy performance. *Green building materials* refer to basic materials that may be the components of products or used in a stand-alone manner in a building. Green building materials have low environmental impacts compared to the alternatives. As noted earlier, an example of a classic green building material is wood products certified by the FSC as having been grown using sustainable forestry practices. Wood is a renewable resource, the forest is managed to produce wood at a replenishable rate, and the biodiversity of the local ecosystems is protected. In short, wood meets all the criteria for a green building material as a raw input to the production process. However, the processing of the sustainably harvested wood may produce significant waste, requires large quantities of energy and water, and may contribute to the degradation of the environment. Consequently, although the raw material may be ideal from an environmental point of view, the entire life cycle must be considered to fully assess the environmental performance of a product.

The point is, depending on how they are defined, green building products may not even be made of green building materials. For example, the glass in the low-e window may be difficult or impossible to recycle because of the films utilized to provide spectral selectivity, which are glued to the glass. In contrast, ordinary float glass can be readily recycled; therefore, with respect to materials, it may be considered greener than the low-e product. This example illustrates the complexity of the product and materials selection process for high-performance buildings.

GREEN BUILDING MATERIALS

The basic materials of construction and construction products have changed over time from relatively simple, locally available, natural, minimally processed resources to a combination of synthetic and largely engineered products, especially for commercial and institutional buildings. Vernacular architecture—design rooted in the building's location—evolved to take advantage of local resources such as wood, rock, and a few low-technology products made of metals and glass. Today's buildings are made from a far wider variety of materials, including polymers, composite materials, and metal alloys. A side effect of these evolving building practices and materials technology is that neither buildings nor the products that comprise them can be readily disassembled and recycled. There is some controversy over the relative merits of materials from natural resources versus those of synthetic materials made from a wide variety of materials, some of which do not even exist in nature. Most ecologists would in fact agree that there is nothing fundamentally wrong with synthetic materials. For example, it could be argued that recyclable plastics can be more environmentally friendly than cotton, whose cultivation requires large quantities of energy, water, pesticides, herbicides, and fertilizer. Nonetheless, debate continues in the contemporary green building movement about the efficacy of synthetic materials versus materials derived from nature.

GREEN BUILDING PRODUCTS

A basic philosophical approach to selecting materials for building design is sorely lacking in today's green building movement. Consequently, there are many different schools of thought, many approaches, and abundant controversy. It is not obvious, for example, that building products made from postcommercial, postindustrial, or postagricultural waste are in fact green. Many of the current green building products contain recycled content from these various sources.

To shed light on this topic, this section describes three philosophies or points of view about what constitutes a green building product: The Natural Step, the Cardinal Rules for a Closed-Loop Building Materials Strategy, and a pragmatic approach suggested by *Environmental Building News.*

The Natural Step and Construction Materials

One philosophical approach to designing the built environment is to use the well-known Natural Step, a tool developed to assess sustainability, as guidance for materials, product, and building design. The Natural Step, which is based on four scientifically based "System Conditions," was developed in the 1980s by Dr. Karl Henrik Robèrt, a Swedish oncologist. These conditions are as follows:[7]

1. *In order for a society to be sustainable, nature's functions and diversity are not systematically subjected to increasing concentrations of substances extracted from the Earth's crust.* In a sustainable society, human activities such as the burning of fossil fuels and the mining of metals and minerals will not occur at a rate that causes them to increase systematically in the ecosphere. There are thresholds beyond which living organisms and ecosystems are adversely affected by increases in substances from the Earth's crust. Problems may include an increase in greenhouse gases leading to global climate change, contamination of surface water and groundwater, and metal toxicity, which can cause functional disturbances in animals. In practical terms, the first condition requires society to implement comprehensive metal and mineral recycling programs and decrease economic dependence on fossil fuels.
2. *In order for a society to be sustainable, nature's functions and diversity are not systematically subjected to increasing concentrations of substances produced by society.* In a sustainable society, humans will avoid generating systematic increases in persistent substances such as DDT, PCBs, and freon. Synthetic organic compounds such as DDT and PCBs can remain in the environment for many years, bioaccumulating in the tissue of organisms, causing profound deleterious effects on predators in the upper levels of the food chain. Freon and other

ozone-depleting compounds may increase the risk of cancer due to added ultraviolet radiation in the troposphere. Society needs to find ways to reduce economic dependence on persistent human-made substances.

3. *In order for a society to be sustainable, nature's functions and diversity are not systematically impoverished by overharvesting or other forms of ecosystem manipulation.* In a sustainable society, humans will avoid taking more from the biosphere than can be replenished by natural systems. In addition, they will avoid systematically encroaching upon nature by destroying the habitat of other species. Biodiversity, which includes the great variety of animals and plants found in nature, provides the foundation for ecosystem services that are necessary to sustain life on this planet. Society's health and prosperity depend on the enduring capacity of nature to renew itself and rebuild waste into resources.
4. *In a sustainable society, resources are used fairly and efficiently in order to meet basic human needs globally.* Meeting this System Condition is a way to avoid violating the first three System Conditions for sustainability. Considering the human enterprise as a whole, we need to be efficient with regard to resource use and waste generation in order to be sustainable. If 1 billion people lack adequate nutrition while another 1 billion have more than they need, there is a lack of fairness with regard to meeting basic human needs. Achieving greater fairness is essential for social stability and the cooperation needed for making large-scale changes within the framework laid out by the first three System Conditions. To achieve this fourth System Condition, humanity must strive to improve technical and organizational efficiency around the world and to use fewer resources, especially in affluent areas. System Condition 4 implies an improved means of addressing human population growth. If the total resource throughput of the global human population continues to increase, it will be increasingly difficult to meet basic human needs, as human-driven processes intended to fulfill human needs and wants are systematically degrading the collective capacity of the Earth's ecosystems to meet these demands.

Applying the System Conditions to new building construction, with a particular focus on building materials, produces a matrix, as shown in Table 9.1. The matrix indicates the relationship between the System Conditions and the various major types

TABLE 9.1

Violation of Natural Step System Conditions in the Application of Construction Materials[8]

Item	Violation Examples	1	2	3	4
Durables	Use of less abundant mined metals and minerals (copper, chromium, titanium)	X		X	
	Use of heavy metals (mercury, lead, cadmium)	X			
	Use of persistent synthetic materials (PVC, HCFC, formaldehyde)		X		
	Wood from rainforests and old-growth timber that is harvested unsustainably			X	
Consumables	Use of petroleum-based products (solvents, oils, plastic film)	X	X	X	X
	Excessive packaging and other disposables		X	X	X
Solid Waste	Landfill disposal of construction and demolition waste, including toxic components such as lead and asbestos	X	X	X	X

of materials used or generated in construction: durables, consumables, and solid waste. It also shows which System Conditions are violated when contemporary practices are used.

In practical terms, applying The Natural Step to the employment of building materials would result in the following materials practices:[9]

1. All materials are nonpersistent and nontoxic and procured either from reused, recycled, renewable, or abundant (in nature) sources.
   a. *Reused* means reused or remanufactured in the same form, such as remilled lumber, in a sustainable way.
   b. *Recycled* means that the product is 100 percent recycled and can be recycled again in a closed loop in a sustainable way.
   c. *Renewable* means able to regenerate in the same form at a rate greater than the rate of consumption.
   d. *Abundant* means that human flows are small compared to natural flows—for example, aluminum, silica, and iron.
   e. In addition, the extraction of renewable or abundant materials has been accomplished in a sustainable way, efficiently using renewable energy and protecting the productivity of nature and the diversity of species.
2. Design and use of materials in the building will meet the following criteria in order of priority:
   a. Material selection and design favor deconstruction, reuse, and durability appropriate to the service life of the structure.
   b. Solid waste is eliminated by being as efficient as possible; or,
   c. Where waste does occur, reuses are found for it on-site; or,
   d. For what is left, reuses are found off-site.
   e. Any solid waste that cannot be reused is recycled or composted.

On a systemwide—in this case, planetary—scale, The Natural Step contends that, unless we are willing to severely compromise human health, we ultimately need to eliminate the extraction of ores and fossil fuels mined and extracted to produce energy and materials. Additionally, The Natural Step calls for the ultimate elimination of synthetic materials whose concentration in the biosphere is compromising not only human health, but the very health of the biosphere in which we reside. The Natural Step also cautions against the degradation of the biosphere by human activities because it is the very source of the resources needed to sustain life. And, finally, it addresses the social aspects of sustainability by noting that human needs in all parts of the world must be met. In sum, the message of The Natural Step is to reduce resource extraction, increase reuse and recycling, and minimize emissions that affect both ecosystems and human systems.

Cardinal Rules for a Closed-Loop Building Materials Strategy

A truly green building product should ideally be composed of several different materials that are also green. As pointed out earlier in this chapter, currently there are many green building products that are not themselves inherently green: for example, low-e windows, T-8 lighting fixtures, and ERVs. Although there are many arguments about what constitutes a green building product, perhaps the primary question relates to the ultimate fate of the product and its constituent materials. Presuming that ecology is the ideal model for human systems, and that in nature there is said to be no waste, it follows that the building materials cycle should be closed and as waste-free as the laws of thermodynamics permit. A *closed-loop* building product and materials strategy must address several levels of materials use in its implementation:

the building, the building products, and the materials used in building products and in construction. Ideally, the building materials system should follow the Cardinal Rules for an Ideal Closed-Loop Building Materials Strategy listed in Table 9.2.

The Cardinal Rules provide for the complete dismantling of the building and all of its components so that materials input at the time of the building's construction can be recovered and returned to productive use at the end of the building's useful life. These rules also establish the ideal conditions for materials and products used in building. It is, however, important to point out that very few materials and products today can adhere to these five rules, meaning that the behavior of materials is far from its ideal state. As it stands, devising a system of materials, products, and buildings to support closed-loop behavior is in the distant future. Nonetheless, this thought process can be used as a touchstone for making decisions about the development of new products, materials, and technologies that support the high-performance green building movement.

TABLE 9.2

Cardinal Rules for a Closed-Loop Building Materials Strategy

1. Buildings must be deconstructable.
2. Products must be disassemblable.
3. Materials must be recyclable.
4. Products/materials must be harmless in production and in use.
5. Materials dissipated from recycling must be harmless.

PRAGMATIC VIEW OF GREEN BUILDING MATERIALS

In order to take a pragmatic view of green building materials, it is useful to examine contemporary efforts to wrestle more directly with these issues based on our current understanding, capabilities, and technologies. As noted several times in previous chapters, *Environmental Building News* (EBN) is an excellent source of well-reasoned approaches to most matters concerning high-performance buildings, and the subject of building materials and products is no exception. According to EBN, green building products can be broken down into five major categories:[10]

1. Products made from environmentally attractive materials
   a. Salvaged products
   b. Products with postconsumer recycled content
   c. Products with postindustrial recycled content
   d. Certified wood products
   e. Rapidly renewable products
   f. Products made from agricultural waste material
   g. Minimally processed products
2. Products that are green because of what is not there
   a. Products that reduce material use
   b. Alternatives to ozone-depleting substances
   c. Alternatives to products made from PVC and polycarbonate
   d. Alternatives to conventional preservative-treated wood
   e. Alternatives to other components considered hazardous
3. Products that reduce environmental impacts during construction, renovation, or demolition
   a. Products that reduce the impacts of new construction
   b. Products that reduce the impacts of renovation
   c. Products that reduce the impacts of demolition
4. Products that reduce the environmental impacts of building operation
   a. Building products that reduce heating and cooling loads
   b. Equipment that conserves energy
   c. Renewable energy and fuel cell equipment
   d. Fixtures and equipment that conserve water
   e. Products with exceptional durability or low maintenance requirements
   f. Products that prevent pollution or reduce waste
   g. Products that reduce or eliminate pesticide treatments
5. Products that contribute to a safe, healthy indoor environment
   a. Products that do not release significant pollutants into the building
   b. Products that block the introduction, development, or spread of indoor contaminants
   c. Products that remove indoor pollutants
   d. Products that warn occupants of health hazards in the building
   e. Products that improve light quality

This pragmatic view of building materials and products is a useful starting point because it deals with the contemporary supply chain and with today's technologies and practices. The question then is: how do we evolve closer to the ideal of green building materials and products espoused by The Natural Step and the Cardinal Rules for a Closed-Loop Building Materials Strategy?

Priorities for Selecting Building Materials and Products

There are three priorities in selecting building materials for a project:

1. As with energy and water resources, the primary emphasis should be on reducing the quantity of materials needed for construction.
2. The second priority is to reuse materials and products from existing buildings; this is a relatively new strategy called *deconstruction.* Deconstruction is the whole or partial dismantling of existing buildings for the purpose of recovering components for reuse.
3. The third priority is to use products and materials that contain recycled content and that are themselves recyclable or to use products and materials made from renewable resources.

TECHNICAL AND ORGANIC RECYCLING ROUTES

There are two general routes for recycling: technical and organic. The *technical recycling route* is associated with synthetic materials, that is, materials that do not exist in pure form in nature or are invented by humans. These include metals, plastics, concrete, and nonwood composites, to name a few. As noted earlier, only metals and plastics are fully recyclable; hence, they can potentially retain their engineering properties through numerous cycles of reprocessing. Materials in the technical or synthetic category require major investments of energy, materials, and chemicals for their recycling. Materials recyclable through the *organic recycling route* are described in the previous section under renewable resources. Composting is the best-known organic recycling route. This route is designed to allow nature to recycle building materials and turn them back into nutrients for ecosystems. Although feasible in theory, organic recycling has not been attempted on a large scale in the United States. For the organic route to work, it would have to incorporate products from a wide range of applications, including agricultural waste and landscape clearing debris, as well as organic waste from construction.

GENERAL MATERIALS STRATEGY

Assuming that a building is in fact needed for a given function, minimizing the environmental impacts of building materials suggests the following strategy, in general order of priority:

1. *Reuse existing structures.* By modifying an existing building and reusing as much of its structure and systems as possible, one can minimize the use of new materials, with their accompanying impacts of resource extraction; transportation; and processing energy, waste, and other effects. Clearly, trade-offs must be made when considering a building for reuse. For example, a building that, historically, has been inefficient and would need significant changes to its envelope and mechanical/

electrical systems might incur significant waste, as well as require enormous quantities of new materials, in order for the original structure to be retrofit for its new use.

2. *Reduce materials use.* Using the minimal amount of materials required for a building project also lowers the environmental impact of introducing products produced from virgin resources. In a typical building, however, the opportunities for *dematerialization* are few, and center on the possible elimination of systems that are not absolutely necessary. Rejecting floor finishes in favor of finished concrete is an example of reducing materials use, but probably at the cost of aesthetic appeal. Materials waste caused by handling and conventional construction processes also contributes to unnecessary materials use.

   In general, dematerializing a building is difficult because of building code provisions, the desires of the users, and, sometimes, the need for new systems that are becoming standard in high-performance green buildings. An example of a relatively new system frequently used in green buildings is a rainwater harvesting system that requires cisterns, piping, pumps, power, and controls, which are not present in a conventional building. Fortunately, building performance can often be enhanced by the introduction of more systems and materials that may offset the impacts caused by increasing the mass of materials in the building project. The building materials cycle also can be enhanced by modifying existing building designs so that they incorporate Design for Deconstruction (DfD) as a component of the overall building design strategy and by using materials that will have future value for recycling. DfD is addressed in more detail later in this chapter.

3. *Use materials created from renewable resources.* Materials created from renewable resources offer the opportunity to close materials loops via an organic recycling process. The organic route involves recycling by biodegradation, that is, by composting or aerobic/anaerobic digestion, either by nature itself or by processes that mimic the decomposing action of nature. This approach applies to all products made of wood or other organic materials such as jute, hemp, sisal, wool, cotton, and paper. Recycling of renewable resources or organic products via the organic recycling route can be accomplished with low to zero energy, additional materials, and chemicals. Note, however, that some materials are composites of organic and technical materials, and hence would fall into the technical class for purposes of recycling. Other emerging materials such as polylactic acid (PLA) polymers are hybrid synthetics. PLA is a polymer made from the lactic acid that results from cornstarch fermentation; it is used in plastics that are competitive with and often superior to hydrocarbon-based polymers and is completely renewable. PLA can be engineered to be biodegradable in controlled compost situations, so although it is a synthetic material, it can be recycled through the organic route.

4. *Reuse building components.* Reusing intact building components from deconstructed buildings reduces the environmental impacts of building materials because these components require minimal resources for reprocessing. Progress in the techniques for deconstructing existing buildings, instead of demolishing them, means that used building components are becoming more widely available; likewise, businesses that specialize in the sale of components salvaged from deconstructed buildings are becoming more commonplace. One problem that remains to be solved, however, is how to recertify most used building products. That said, good progress has been made in developing visual regrading standards for some types of dimensional lumber—for example, Western Cedar and Southern Yellow Pine.

5. *Use recyclable and recycled content material.* To close the materials loop in construction, all materials must have the capacity for recycling. Currently, this remains a very ambitious objective simply because few building materials are recyclable and many others can be recycled only into a lower-value application. For example, recycled concrete aggregate can be used as a subbase material but not—at least not

Figure 9.2 A relatively new material emerging to serve the green building market is compressed wheatboard, made from wheat straw, which can be used in millwork or for cabinetry, as shown here in a Rinker Hall laboratory. (Photograph: M.R. Moretti.)

readily, in the United States—as an aggregate in new concrete. Metals and plastics are perhaps the only materials that are fully recyclable without loss of their basic strength and durability properties.

A wide range of recycled content materials are available for the green building market. These generally contain either postindustrial or postconsumer waste. *Postindustrial waste* refers to materials recycled within the manufacturing plant. For example, during the extrusion of plastic lumber made from high-density polyethylene (HDPE), sprools of the HDPE, which peel off during the process, can be recycled back into the plastic being input to the process. *Postconsumer* waste refers to materials that are recycled from home or business use into new products. Plastic lumber made entirely of HDPE from recycled milk bottles would be considered to have 100 percent postconsumer content. Postconsumer waste recycling is far more difficult than postindustrial waste recycling. This fact is reflected in the LEED-NC building assessment system, which weights postconsumer content as double postindustrial content for the purpose of awarding points.

6. *Use locally produced materials.* Examining the resources and emissions associated with transporting materials between the various sites of extraction, materials production, product manufacture, and installation is one of the steps in an LCA evaluation. There is no doubt that minimizing transportation distances by using locally produced materials and locally manufactured products can greatly reduce the overall environmental impacts of materials. Defining what is meant by *local* can, however, be a challenge. The USGBC LEED-NC building assessment standard sets 500 miles (806 kilometers) as the radius within which a product is considered local for the purpose of obtaining points.

   Another difficulty with assigning weight to locally produced products is that improved technologies may be passed over. A classic example is the introduction of Japanese automobiles to the U.S. marketplace in the late 1970s, when the quality and workmanship of these cars compelled a rapid shift away from American products. The subsequent bailout by the U.S. government of the Chrysler Corporation in 1979–1980, coupled with higher energy prices, forced U.S. companies to rethink their products. Ultimately, fundamental changes took place in the design and production of American cars. Today, American cars are almost on a par with Japanese automobiles and exceed many European cars in terms of quality. In short, products not considered local may be far superior, result in lower life-cycle environmental impacts, and encourage improvements in local products.

The materials selection process may be summarized as follows: rely on the three Rs—reduce, reuse, and recycle (with the meaning of *recycle* being extended to address products and materials with recycled content or from renewable resources).

LCA of Building Materials and Products

As stated previously, the most important tool currently being used to determine the impacts of building materials is LCA. LCA can be defined as a methodology for assessing the environmental performance of a service, process, or product, including a building, over its entire life cycle.[11] LCA comprises several steps, which are defined in the ISO 14000 series of standards that address environmental management systems.[12] These steps include inventory analysis, impacts assessment, and interpretation of the impacts.

Put simply, LCA is a methodology for assessing the environmental performance of a product over its full life cycle, often referred to as *cradle-to-grave* or *cradle-to-cradle* analysis. Environmental performance is generally measured in terms of a wide range of potential effects, for example:

- Fossil fuel depletion
- Other nonrenewable resource use
- Water use
- Global warming potential
- Stratospheric ozone depletion
- Ground-level ozone (smog) creation
- Nutrification/eutrophication of water bodies
- Acidification and acid deposition (dry and wet)
- Toxic releases to air, water, and land

Comparing these effects for a building takes careful analysis. For example, the total energy for a building's life cycle is comprised of the embodied energy invested in the extraction, manufacture, transport, and installation of its products and materials, plus the operational energy needed to run the building over its lifetime. For the average building, the operating energy is far greater than the embodied energy, perhaps 5 to 10 times higher. Consequently, the operational stage has far more energy impacts than those up through the construction stage. For other effects, however, the effects of the stages up through construction can be far greater. Toxic releases during resource extraction and the manufacturing process can be far greater than those occurring during building operation. The net result is that the designer using these tools must keep in mind the entire life cycle of the building, not just the stages leading to construction.

ATHENA ENVIRONMENTAL IMPACT ESTIMATOR

The ATHENA Environmental Impact Estimator (EIE) is an LCA tool that focuses on the assessment of whole buildings or building assemblies such as walls, roofs, or floors. It was created and is maintained by the nonprofit ATHENA Institute and is intended to assist project team members make decisions about product selection early in the design stage. The EIE has a regional character, meaning that the user can select the project site from among 12 different North American locations. It accounts for material maintenance and replacement over an assumed building life and distinguishes between owner-occupied and rental facilities, if relevant. If an energy simulation for a design has been completed, it can be entered into the EIE to take account of operating energy impacts and the impacts of generating that energy. The EIE has a database of generic products covering 90 structural and envelope materials. It can simulate more than 1,000 different assembly combinations and can model the structure and envelope systems for over 95 percent of the building stock in North America. The output of the EIE provides cradle-to-grave and region-specific results of a design in terms of detailed flows from and to nature. It also provides summary measures for embodied energy use, global warming potential, solid waste emissions, pollutants to air, pollutants to water, and natural resources use. Graphs and summary tables show energy use by type or form of energy, and emissions by assembly group and life-cycle stage. A comparison dialogue feature allows the side-by-side comparisons of up to five alternative designs. Similar projects with different floor areas can be compared on a unit floor-area basis.[13]

A typical array of information produced by EIE version 3 is shown in Table 9.3. The information in this table is not meaningful unless it is compared to alternative strategies. For example, the building depicted in Table 9.3, an 18-story office building with five levels of underground parking, has a concrete structure and an exterior curtainwall. An alternative would be a steel structure with masonry walls. The purpose of making these comparisons is to determine the building systems that have the lowest life-cycle impacts—assuming, of course that the construction budget is not violated. An LCA program such as the EIE has a very complex array of outputs, an example of which is shown in Figure 9.3.

TABLE 9.3

Example of an LCA Output

Building Components	Embodied Energy (Gj)	Solid Waste (metric tons)	Air Pollution* (index)	Water Pollution* (index)	GWP† (equivalent CO_2 metric tons)	Weighted Resource Use (metric tons)
Structure	52,432	3,273	859.0	147.0	13,701	34,098
Cladding	17,187	281	649.8	24.7	5,727	2,195
Roofing	3,435	145	64.8	5.8	701	1,408
Total	73,054	3,554	1,573.6	177.5	20,129	37,701
Per square meter	2.36	0.11	.05	.006	.65	1.21

*The air and water pollution indices are based on the critical volume measure (method).
†GWP is global warming potential. Energy and emission estimates do not include operating energy.

BUILDING FOR ENVIRONMENTAL AND ECONOMIC SUSTAINABILITY (BEES)

As introduced earlier, BEES is the other prominent North American tool for LCA of building materials and products; it is U.S.-specific. It was developed by the National Institute of Science and Technology (NIST) with support from the USEPA's Environmentally Friendly Purchasing Program. BEES allows side-by-side comparison of building products for the purpose of selecting cost-effective, environmentally preferable products, and includes both LCA and life-cycle costing (LCC) data. The result is that the user obtains both environmental performance and economic comparisons.

In addition to the typical measures of performance, BEES provides data about air pollutants, indoor air quality, ecological toxicity, and human health for each material or product. BEES can compare building elements to determine where the greatest impacts are occurring and which building elements need the most improvement. The user assigns weights to categories, then combines the environmental and economic performance into a single performance score. For example, the user first decides how to weigh environmental versus economic performance, say, 50–50 or

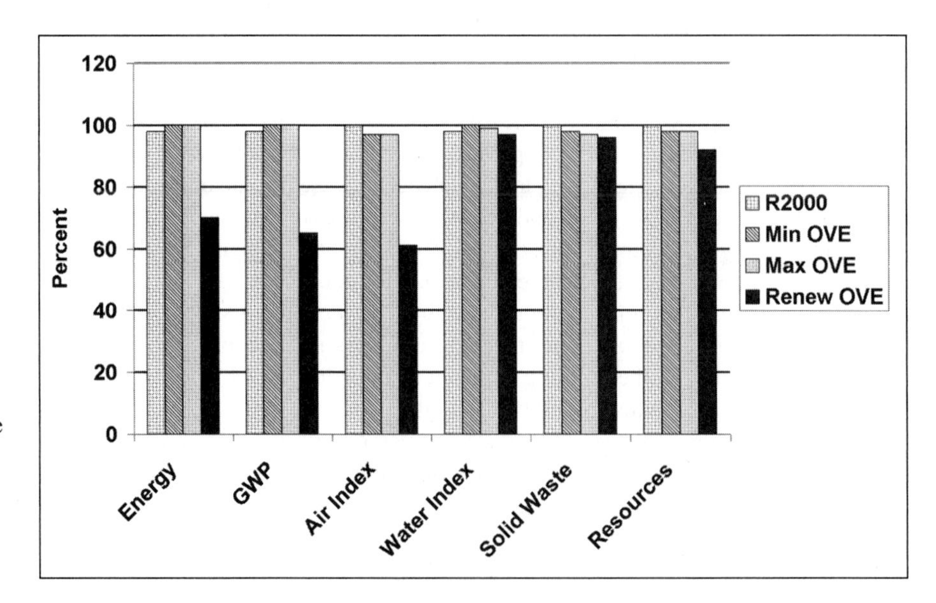

Figure 9.3 Sample output screen from the ATHENA EIE, version 3.0, program showing resource use for a hypothetical steel-framed structure compared to a concrete structure. (Courtesy of the ATHENA Sustainable Materials Institute.)

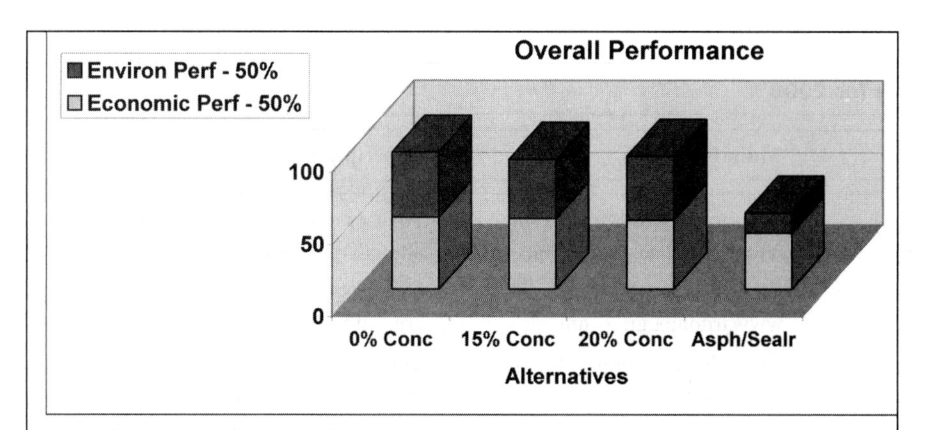

Note: Lower values are better

Category	0% Conc	15% Conc	20% Conc	Asph/Sealr
Economic Perf - 50%	50	49	48	39
Environ Perf - 50%	45	41	44	14
Sum	95	90	92	53

Figure 9.4 Sample output screen from BEES 3.0 showing comparative environmental performance for concrete with varying levels of fly ash and sealed asphalt. BEES is a free LCA program available from the National Institute of Science and Technology.

40–60. The user then selects from among four different weighting schemes for the environmental performance measures. The latest version of BEES, 3.0, includes some 200 building products, including 80 brand-name products. As an example, for floor coverings, there are 18 brand-name products and 17 distinct generic products.[14] A sample output screen for a BEES LCA analysis is shown in Figure 9.4.

Key and Emerging Construction Materials and Products

For many conventional building materials, admirable progress is being made in rethinking their extraction and application in construction. Perhaps the most notable and successful effort has been the inclusion of sustainable forestry as the key criterion for wood products used in construction. LEED-NC, for example, provides a point if a minimum of 50 percent of the building's wood-based materials and products are certified in accordance with the Forestry Stewardship Council's (FSC) Principles and Criteria. Green Globes provides points for wood products certified by FSC, the Sustainable Forestry Initiative, under the Canadian Standards Association Standard for Sustainable Forest Management (CAN/CSA Z809), and the American Tree Farm System. For metal products, the emphasis is on their recycled content, and organizations such as the Steel Recycling Institute ensure that the benefits of their member companies' products are well known.

New technologies are being developed to improve performance or to provide new capabilities of building products. But in part because there is no commonly accepted vision of what constitutes a green building material or product, there are many different approaches to product development. So, one of the most effective ways to track progress is to examine the products emerging to serve the green building marketplace. A second set of recent green building products cited by EBN as the top 10 products of 2006 is shown in Table 9.4.

The following sections address the current issues and status of major classes of construction materials. A comprehensive discussion of the wide range of green building materials and products, both existing and emerging, is beyond the scope of

TABLE 9.4

BuildingGreen's Top Ten Green Products for 2006[15]

Product Type	Product Name	Manufacturer	Description
Polished concrete system	RetroPlate™	Advanced Floor Products, Inc. www.retroplatesystem.com	Polishes old or new concrete slabs into attractive, durable, finished floors
Timber salvaging system	Sawfish Logging Submarine	Triton Logging, Inc. www.tritonlogging.com	Harvests trees submerged in reservoirs created by hydroelectric dams
Electronically tintable glazing	SageGlass®	Sage Electrochromics, Inc. www.sage-ec.com	Tinting of glass is changed using electrochromic control
Water-resistant composite	PaperStone™	Klip Tech Composites www.paperstoneproducts.com	Solid composite material made from postconsumer paper
Interior panels	Varia™	3form, Inc. www.3-form.com	Interior panels for workstations, trim, or toilet partitions, made with 40% preconsumer recycled copolyester
Interior molding	Timbron	Timbron® International, Inc. www.timbron.com	Interior molding profiles made with at least 90% recycled polystyrene
Water-efficient showerhead	H_2O kinetic	Delta Faucet Company www.deltafaucet.com	Showerhead uses just 1.6 gallons of water per minute
Irrigation system controls	WeatherTRAK®	HydroPoint® Data Systems, Inc. www.weathertrak.com	Irrigation control based on local weather data
Evaporative cooler	Coolerado Cooler	Coolerado, LLC www.coolerado.com	Indirect evaporative cooler
Renewable energy credits	REC's	Community Energy, Inc. www.communityenergy.biz	Provides renewable energy credits from its own renewable energy generation capacity

this book. Therefore, the materials discussed here are those considered most important because of the scale of their application in construction: wood and wood products, concrete and concrete products, metals, and polymers.

WOOD AND WOOD PRODUCTS

Wood and products made of wood are very important construction materials, made all the more important because of their renewability. Enormous areas of the United States are considered to be covered with trees, some 747 million acres (302 million hectares), or about one-third of the U.S. landmass. Of this, 504 million acres (204 million hectares) are classified as timberland, that is, productive forest capable of growing at least 20 cubic feet (0.6 cubic meters) of commercial wood per acre per year. Approximately 67 million acres (27 million hectares) are owned by the forest products industry; 291 million (118 million hectares) are held by 10 million individual private landowners; and another 49 million acres (20 million hectares) contained within the National Forest System are available for forest management.[16]

A wide variety of wood products are used in construction, including dimensional lumber, engineered wood products, plywood, oriented strand board, and composite materials with wood fiber content. Wood products used in high-performance green buildings should originate in sustainably managed forests and should bear labels certifying this fact. The major organization governing sustainable forestry internationally is the FSC, whose U.S.-based certifiers are the Smartwood Program and Scientific Certification Systems.[17] A third organization that collaborates with both the FSC and the USGBC to foster the use of certified wood products is the Certified Wood and Paper Association (CWPA).[18] The FSC program is based on a set of 10 principles used as the basis for the criteria to qualify forests for certification (see Table 9.5). And the USGBC LEED-NC building assessment standard provides a point related to certified wood products and for rapidly renewable resources, that is, wood products grown in plantation forests. FSC principle 10 addresses the forestry practices required to earn certification for these types of forests.

The Sustainable Forestry Initiative (SFI) program is a comprehensive system of principles, objectives, and performance measures developed by professional foresters, conservationists, and scientists that combines the perpetual growing and harvesting

TABLE 9.5

FSC Principles for Management of Forests (January 1999 Edition)

Principle 1: Compliance with Laws and FSC Principles. Forest management shall respect all applicable laws of the country in which they occur, and international treaties and agreements to which the country is a signatory, and comply with all FSC Principles and Criteria.

Principle 2: Tenure and Use Rights and Responsibilities. Long-term tenure and use rights to the land and forest resources shall be clearly defined, documented, and legally established.

Principle 3: Indigenous Peoples' Rights. The legal and customary rights of indigenous peoples to own, use, and manage their lands, territories, and resources shall be recognized and respected.

Principle 4: Community Relations and Workers' Rights. Forest management operations shall maintain or enhance the long-term social and economic well-being of forest workers and local communities.

Principle 5: Benefits from the Forest. Forest management operations shall encourage the efficient use of the forest's multiple products and services to ensure economic viability and a wide range of environmental and social benefits.

Principle 6: Environmental Impact. Forest management shall conserve biological diversity and its associated values, water resources, soils, and unique and fragile ecosystems and landscapes, and, by so doing, maintain the ecological functions and the integrity of the forest.

Principle 7: Management Plan. A management plan—appropriate to the scale and intensity of the operations—shall be written, implemented, and kept up to date. The long-term objectives of management, and the means of achieving them, shall be clearly stated.

Principle 8: Monitoring and Assessment. Monitoring shall be conducted—appropriate to the scale and intensity of forest management—to assess the condition of the forest, yields of forest products, chain of custody, management activities, and their social and environmental impacts.

Principle 9: Maintenance of High Conservation Value Forests. Management activities in high conservation value forests shall maintain or enhance the attributes that define such forests. Decisions regarding high conservation value forests shall always be considered in the context of a precautionary approach.

Principle 10: Plantations. Plantations shall be planned and managed in accordance with Principles and Criteria 1 through 9, and Principle 10 and its criteria. While plantations can provide an array of social and economic benefits, and can contribute to satisfying the world's needs for forest products, they should complement the management of, reduce pressures on, and promote the restoration and conservation of natural forests.

of trees with the long-term protection of wildlife, plants, soil, and water quality. On January 1, 2007, the SFI program became a fully independent forest certification program. The multi-stakeholder Board of Directors of the Sustainable Forestry Initiative, Inc., is now the sole governing body over the SFI Standard and all aspects of the program. The diversity of the board members reflects the variety of interests in the forestry community.

The SFI Standard (SFIS) spells out the requirements of compliance with the program. The SFIS is based on nine principles that address economic, environmental, cultural, and legal issues, and a commitment to continuously improve sustainable forest management (see Table 9.6).

Only companies and organizations that have successfully completed an audit by

TABLE 9.6

SFI Standard Principles

1. Sustainable Forestry
To practice sustainable forestry to meet the needs of the present without compromising the ability of future generations to meet their own needs by practicing a land stewardship ethic that integrates reforestation and the managing, growing, nurturing, and harvesting of trees for useful products with the conservation of soil, air and water quality, biological diversity, wildlife and aquatic habitat, recreation, and aesthetics.

2. Responsible Practices
To use and to promote among other forest landowners sustainable forestry practices that are both scientifically credible and economically, environmentally, and socially responsible.

3. Reforestation and Productive Capacity
To provide for regeneration after harvest and maintain the productive capacity of the forestland base.

4. Forest Health and Productivity
To protect forests from uncharacteristic and economically or environmentally undesirable wildfire, pests, diseases, and other damaging agents, and thus maintain and improve long-term forest health and productivity.

5. Long-Term Forest and Soil Productivity
To protect and maintain long-term forest and soil productivity.

6. Protection of Water Resources
To protect water bodies and riparian zones.

7. Protection of Special Sites and Biological Diversity
To manage forests and lands of special significance (biologically, geologically, historically, or culturally important) in a manner that takes into account their unique qualities and to promote a diversity of wildlife habitats, forest types, and ecological or natural community types.

8. Legal Compliance
To comply with applicable federal, provincial, state, and local forestry and related environmental laws, statutes, and regulations.

9. Continual Improvement
To continually improve the practice of forest management and also to monitor, measure, and report performance in achieving the commitment to sustainable forestry.

an independent and accredited certification body can claim certification to the SFI Standard. SFI certification audits are rigorous, on-the-ground assessments, conducted by highly qualified and objective individuals.

Of the leading certification schemes in operation in the United States, only the SFI program has a strict separation between standard setting and accreditation of certifying bodies. Recognized international protocols (ISO) for auditing explicitly require that these functions be separate. To date, over 127 million acres have been independently certified to the SFI Standard. The SFI standard can be found on the Sustainable Forestry Initiative, Inc. website at www.sfiprogram.org.

It should be noted that there are several other third-party certification systems for sustainably harvested wood, including the American Tree Farm System (ATFS) and CSA Sustainable Forest Management (SFM). The Green Globes building assessment system takes all of these into account in awarding points, while the USGBC relies solely on the FSC certification system.

CONCRETE AND CONCRETE PRODUCTS

As one of the mainstays of construction, and one of its oldest and best-known materials, concrete has an enormous and increasing number of roles in construction. Concrete is normally composed of coarse aggregate (rock), fine aggregate (sand), cement, water, and various additives. With respect to high-performance buildings, concrete has many positive qualities: high strength, thermal mass, durability, and high reflectance; is generally locally available; can be used without interior and/or exterior finishes; does not offgas and affect indoor air quality; is readily cleanable; and is impervious to insect damage and fire. Concrete can be designed to be pervious or cast into open-web pavers, thus allowing water to infiltrate directly into the ground to reduce the need for stormwater systems.

The key issue with concrete is the carbon dioxide emitted in the cement manufacturing process. Cement, which comprises 9 to 14 percent of most concrete mixes, is second only to coal-fired utilities in carbon dioxide emissions. For each ton of powder cement produced, up to an equal mass of carbon dioxide is generated. However, during the life cycle of a concrete element, the cement reabsorbs about 20 percent of the carbon dioxide generated in the manufacturing process, at least partially mitigating this effect. Minimizing the quantity of cement in a concrete mix is a strategy that has a number of potential benefits. Fly ash and ground-blast furnace slag, both of which have cementitious properties, can be at least partially substituted for cement and result in increased concrete performance. Fly ash can be readily substituted for over 30 percent of the cement volume, blast furnace slag for more than 35 percent. These substitutions have the advantage of making beneficial use of otherwise industrial waste while simultaneously reducing the quantity of carbon dioxide associated with concrete production. Fly ash and blast furnace slag can also be blended with cement in the cement manufacturing process, resulting in reduced carbon dioxide emissions, reduced energy consumption, and expanded production capacity.

The recycling properties of concrete are generally satisfactory. Crushed concrete can be used as subbase for roads, sidewalks, and parking lots. In the Netherlands, recycled concrete aggregate can substitute for one-third of the virgin aggregate in concrete mixes. In general, recycled concrete aggregate is in high demand and has relatively high value.

METALS: STEEL AND ALUMINUM

Metals in general have high potential for recycling, and most metal products used in typical building applications have significant recycled content. The performance of metal products in building applications can be outstanding, providing high strength and durability with relatively light weight. Additionally, metals are readily recycled, and their dissipation into the environment during the recycling process is benign. Although the LCA and embodied energy impacts associated with metals may appear to be higher than those of alternatives, the inherent recyclability of metals, their durability, and their low maintenance make them competitive for high-performance building applications.

Steel production today incorporates used steel products in the manufacture of new steel in the two production processes still being used. The basic oxygen furnace (BOF) uses 25 to 35 percent scrap steel for products that require drawability—for example, automobile fenders and cans—while the electric arc furnace (EAF) uses almost 100 percent scrap steel for products whose main requirement is strength—for example, structural steel and concrete reinforcement. Steel made from the BOF process generally has a total recycled content of 32 percent, which is composed of 22.6 percent postconsumer content and 8.4 percent postindustrial content. EAF-produced steel generally has a recycled content of about 96 percent, with a postconsumer content of 59 percent and postindustrial content of 37 percent.[20] Recycled steel consumes a fraction of the resources and energy of steel produced from iron ore. Each ton of recycled steel saves 2,500 pounds (1,134 kilograms) of iron ore, 1,400 pounds (635 kilograms) of coal, and 120 pounds (54 kilograms) of limestone. Only one-fifth of the energy needed to produce steel from iron ore is required to recycle scrap steel. Steel recycling systems in the United States are well established, so much so that recycling is dictated less by environmental concerns than by economics.

Aluminum recycling also has marked environmental benefits. Recycled aluminum requires only 5 percent of the energy needed to produce aluminum from bauxite ore, thus eliminating 95 percent of the greenhouse gases that would be generated by manufacturing aluminum from bauxite. Approximately 55 percent of the world's aluminum production is powered by hydropower, which, although controversial because of its environmental impacts, is a renewable resource. Recycling 1 pound (0.45 kilogram) of aluminum saves 8 pounds (3.6 kilograms) of bauxite and 6.4 kWh of electricity. Aluminum recycling in this country is highly successful and well established, with about 65 percent of aluminum being recycled. The recycled content of the average aluminum can is about 40 percent; and improved engineering means that today, 1 pound (0.45 kilogram) of aluminum produces 29 cans, versus 22 cans in 1972. Although there has been controversy over the value of recycling aluminum cans, the industry claims that they can be profitably recycled by individuals and groups. Recycling rates for building applications range from 60 to 90 percent in most countries.[21]

Aluminum panels used in buildings are corrosion-resistant, lightweight, and virtually maintenance-free; aluminum also has high reflectivity, making it extremely useful as a roofing material. Aluminum is also used extensively in electrical wiring applications, as a casing for appliances, and in moldings and extrusions for windows.

PLASTICS

Along with wood and metals, plastics, which are composed of chains of molecules known as *polymers,* are a major constituent of building products, both as virgin materials and as recycled content. Plastics have a high potential for recycling, and the industry has developed a systematic method for designating and labeling the seven major classes of plastics. The Society of Plastics Industry, Inc. (SPI), introduced this system in 1988 to facilitate recycling of the growing quantity of plastics entering the marketplace and the waste stream. Large quantities of postconsumer plastics, particularly high-density polyethylene (HDPE) and polyethylene terephthalate (PET), are being recycled into a range of building products, such as plastic lumber. Construction products are the second highest user of plastics in the United States, exceeded only by packaging.

At present, however, there is little if any recycling of plastic building products into other end uses, which is a serious problem. Closed-loop behavior is, of course, desirable. But there are some success stories in plastic recycling. One is the development of processes that recycle HDPE into high-quality plastic lumber, a product with very high durability that is impervious to rot, insects, and saltwater damage, and with a lifetime measured in hundreds of years. The holy grail of any recycling effort is to develop technologies that can recycle products back into their original use. The United Resource Recovery Corporation (URRC) technology, recently developed in Germany, can recycle PET plastics back into very-high-quality flakes, which can then be used to produce the ubiquitous clear plastic of soft-drink bottles. Recycling rates for HDPE and PET in this country are in the 20 percent range, the highest for the common classes of plastics used in consumer products.

Manufacturers of plastics derived from chlorine or that employ chlorine in their production are under severe pressure from environmental groups such as Greenpeace because of the various impacts associated with their manufacture and disposal. Polyvinyl chloride (PVC), a ubiquitous product in construction (it appears in piping, siding, flooring, and wiring, to name a few), is the main focus of these struggles. To date, recycling rates for PVC are among the lowest for the seven major classes of plastics covered by the SPI—less than 1 percent. And in the United States, PVC is being defended by its industry based on its technical and economic merits, meaning that fundamental changes to the product or its manufacture are not anticipated in the near future. In contrast, the European PVC industry is exploring how to make fundamental changes in the production and disposal of its products, positioning PVC to be regarded as an environmentally responsible product. A Green Paper on PVC was released by the European Commission in 2000, indicating that the major problems with PVC are the use of certain additives (lead, cadmium, and phtalates) and the disposal of PVC waste.[22] According to the Green Paper, only 3 percent of PVC waste is recycled; 17 percent is incinerated and the remaining 80 percent is landfilled, with the total waste stream amounting to 3.6 million tons per year. Risks associated with landfilling PVC, especially the loss of phthalate from soft PVC, were highlighted, along with problems caused by incineration, namely, the generation of dioxins, which are very hazardous chemicals. Unquestionably, PVC recycling must be improved, and reformulation of the basic product must be considered in order to remove the barriers to its recycling. PVC product recycling faces many of the same problems associated with other plastics, namely, the use of additives such as plasticizers, stabilizers, fillers, flame retardants, lubricants, and colorants, which are used to provide specific properties.

A relatively new development in the plastics industry is the production biobased polymers such as PLA. In 2002, Cargill Dow Polymers (CDP) opened a large facility in Blair, Nebraska, to manufacture a plastic product from PLA, the first of its kind, thus marking the introduction of a polymer technology based on a renewable resource, rather than oil, a nonrenewable resource. The product is known as NatureWorks PLA, which CDP says can be produced from other agricultural products such as sugar beets and cassava. Not to be outdone, Dow Chemical introduced a product called BIOBALANCE polymers, which are advanced polyurethane polymers designed to be used as commercial carpet backing. One of the polyurethane components, polyol, is derived from renewable resources. Another Dow Chemical product, WOODSTALK™, is manufactured from formaldehyde-free polyurethane resin and harvested wheat straw fiber, a renewable resource. It is a boardlike material that can be used as an alternative to medium-density fiberboard (MDF) for millwork, cabinetry, and shelving. BioBase 501 is a relatively new low-density, open-cell polyurethane foam insulation partially made from soybeans. The polyol component of BioBase 501 is made of SoyOl©, the soy-based component that is also used in carpet backing. And in Stockholm, Sweden, a new process developed by the Royal Institute of Technology (KTH) uses wood to create polymers known as *hemicellulose-based hydrogels,* as announced in late 2003.

In addition to being produced from renewable resources such as agricultural products and wood, biobased polymers hold the promise of being recyclable via natural processes.

Design for Deconstruction and Disassembly

It is undeniable that the current state of construction is wasteful and will be difficult to change. And as noted at the start of this chapter, closing materials loops in construction remains the most challenging of all green building efforts. More specifically, choosing building materials and products is by far the most daunting challenge.

Criteria for materials and products for the built environment should be similar to those for industrial products in general. Many materials used in buildings, most notably metals, are the same as those used in other industries. But buildings have a distinct character compared to other industrial products. The major factors that make closing materials loops in this segment of the economy particularly difficult are delineated in Table 9.7. The vision of a closed-loop system for the construction industry is, by necessity, one that is integrated with other industries to the maximum extent possible. Many materials—again, metals—can flow back and forth for various uses, whereas others, such as aggregates and gypsum drywall, are unique to construction, so their reuse or recycling would stay within construction. Closing materials loops for the built environment will be much more difficult due to the factors that make its material cycles differ significantly from those of other industries.

To move from wasteful materials practices to closed-loop materials behavior will require that the green building movement embrace the concepts of *deconstruction* and *Design for Disassembly* (DfD). Deconstruction is the whole or partial disassembly of buildings to facilitate component reuse and materials recycling; DfD is the deliberate effort during design to maximize the potential for disassembly, as opposed to demolishing the building totally or partially, to allow the recovery of components for reuse and materials for recycling and to reduce long-term waste generation. To be effective, DfD (a notion that emerged in the early 1990s) must be considered at the design stage.

Experiments in DfD conducted at Robert Gordon University in Aberdeen, Scotland, included a wide range of approaches that can facilitate a greatly improved mate-

TABLE 9.7

Factors That Increase the Difficulty of Closing Materials Loops for the Built Environment

1. Buildings are custom-designed and custom-built by a large group of participants.
2. No single "manufacturer" is associated with the end product.
3. Aggregate, for use in subbase and concrete, brick, clay block, fill, and other products derived from rock and earth, are commonly used in building projects.
4. The connections of building components are defined by building codes to meet specific objectives (e.g., wind load, seismic requirements), not for ease of disassembly.
5. Historically, building products have not been designed for disassembly and recycling.
6. Buildings can have a very long lifetimes exceeding those of other industrial products; consequently, materials have a long "residence" period.
7. Building systems are updated or replaced at intervals during the building's lifetime (e.g., finishes at 5-year intervals; lighting at 10-year intervals; HVAC systems at 20-year intervals).

rials cycle: handling, materials identification, simplicity of construction techniques, exposure of mechanical connections, independence of structure and partitioning, and making short-life-cycle components readily accessible. Research indicates that DfD must be implemented at three levels of the entire materials system in buildings in order to produce sound product design and construction strategies: the systems or building level, product level, and materials level. A number of examples exist to test various DfD ideas. One, a multistory residential housing project in Osaka, Japan, employs a reinforced concrete frame to support independently constructed dwellings that can be replaced on 15-year cycles without removing the supporting frame. Ultimately, closing construction materials loops will necessitate the inclusion of product design and deconstruction together in a process that might be labeled *Design for Deconstruction and Disassembly* (DfDD).

Philip Crowther of Queensland Technical University in Brisbane, Australia, suggests 27 principles for building DfD that are enumerated in Table 9.8.[23] This comprehensive list covers a wide range of thinking about materials selection, product design, and deconstruction.

Crowther's work serves as an excellent starting point in the discussion of a comprehensive approach to developing a seamless framework for closing construction materials loops. Importantly, these principles perhaps generate as many questions as they answer. An example is principle 4, which calls for avoiding composite materials. In the context of materials, *composite* can have many meanings—for

TABLE 9.8

Principles of Design for Disassembly As Applied to Buildings

1. Use recycled and recyclable materials.
2. Minimize the number of types of materials.
3. Avoid toxic and hazardous materials.
4. Avoid composite materials, and make inseparable products from the same material.
5. Avoid secondary finishes to materials.
6. Provide standard and permanent identification of material types.
7. Minimize the number of different types of components.
8. Use mechanical rather than chemical connections.
9. Use an open building system with interchangeable parts.
10. Use modular design.
11. Use assembly technologies compatible with standard building practice.
12. Separate the structure from the cladding.
13. Provide access to all building components.
14. Design components sized to suit handling at all stages.
15. Provide for handling components during assembly and disassembly.
16. Provide adequate tolerance to allow for disassembly.
17. Minimize numbers of fasteners and connectors.
18. Minimize the types of connectors.
19. Design joints and connectors to withstand repeated assembly and disassembly.
20. Allow for parallel disassembly.
21. Provide permanent identification for each component.
22. Use a standard structural grid.
23. Use prefabricated subassemblies.
24. Use lightweight materials and components.
25. Identify the point of disassembly permanently.
26. Provide spare parts and storage for them.
27. Retain information on the building and its assembly process.

Figure 9.5 One of the innovations in the design of Rinker Hall at the University of Florida was to include DfD as a design criterion. Steel connections were bolted and exposed for ready removal. (Photograph: M.R. Moretti.)

example, mixed materials (concrete, steel) or homogeneous layered materials (PVC pipe, laminated wood products). Composites may be very acceptable under certain conditions, where recycling the composite mixture is feasible or where the ability to disassemble the layers readily has been designed into the product. The question is how to develop a systematic approach for determining the acceptability of composites as building materials within the context of attempting to increase reuse and recycling.

In contrast to DfD, deconstruction offers an alternative to demolition that has two positive outcomes: First, it is an improved environmental choice; second, it can serve to create new businesses, to dismantle buildings, transport recovered components and materials, remanufacture or reprocess components, and resell used components and materials. Existing buildings, though not designed to be taken apart, are in fact being disassembled to recover materials. There are distinct benefits to be gained from increasing the recycling rates of materials from buildings from the 20 percent range to in excess of 70 percent, because waste from demolition and renovation activities can comprise up to 50 percent of national waste streams. Economic and noneconomic policy instruments can assist in the shift from demolition to deconstruction by providing financial incentives and aiding in allotting the time needed for deconstruction. In developing countries, building deconstruction practices offer a source of high-quality materials to assist in improving the quality of life and the potential for new businesses, which may provide economic opportunity for their citizens.

Closing Materials Loops in Practice

Of the thousands of companies supplying products for buildings, none has invested more time, resources, and finances in closing their products' materials loops than Interface, Inc., now know as Interface FLORCommercial, Inc. The path to the present has been marked by many experiments, successes, and occasional setbacks.[24] David Hobbs, President of InterfaceFLOR Commercial, describes their journey from 1994 to the present in the following section.

Closing the Loop at Interface

Interface began in 1973 when Ray Anderson, currently chairman of the board, recognized the need for flexible floor coverings for the modern office environment. Anderson established operations in LaGrange, Georgia, and Interface Flooring Systems (now InterfaceFLOR Commercial) began manufacturing and distributing modular carpet tiles in 1974. Over the years, the company has grown to its current status as the world's leading producer of soft-surfaced modular floorcoverings. The original Interface has been augmented by more than 50 acquisitions. The company has diversified into broadloom carpet, interior fabrics, and even a sustainability consulting practice. With the addition (or creation) of each company or brand, the Interface entrepreneurial spirit has steadfastly endured, with each subsidiary company encouraged to maintain and thrive on its unique and/or original brand equity in the marketplace and to sustain its distinctive culture in the workplace.

Something extraordinary happened in our company in 1994. Ray Anderson had his now legendary "spear in the chest" epiphany that transformed the course of his company and an entire industry, as he made what he termed a "mid-course correction" and set out on his quest for sustainability (initially known as the "Journey Up Mt. Sustainability" but now known as "Mission Zero"). Remarkably, Interface associates worldwide enthusiastically joined him. Mr. Anderson uses a simile to describe Sustainability as a mountain to be climbed. His charge to Interface associates led to the creation of an internal task force, which recommended accomplishing environmental objectives through the following "Seven Fronts," or faces of the mountain:

1. *Eliminate Waste:* Eliminating all forms of waste in every area of business
2. *Benign Emissions:* Eliminating toxic substances from products, vehicles, and facilities
3. *Renewable Energy:* Operating facilities with renewable energy sources—solar, wind, landfill gas, biomass, geothermal, tidal, and low-impact/small-scale hydroelectric or non-petroleum-based hydrogen
4. *Closing the Loop:* Redesigning processes and products to close the technical loop using recovered and bio-based materials
5. *Resource-Efficient Transportation:* Transporting people and products efficiently to reduce waste and emissions
6. *Sensitizing Stakeholders:* Creating a culture that integrates sustainability principles and improves people's lives and livelihoods
7. *Redesign Commerce:* Creating a new business model that demonstrates and supports the value of sustainability-based commerce

These fronts would become the building blocks or foundation for achieving what would become the company vision: "To be the first company that, by its deeds, shows the entire industrial world what sustainability is in all its dimensions: people, process, product, place and profits—by 2020—and in doing so we will become restorative through the power of influence."

A NEW CHALLENGE FOR A CHANGING COMPANY

What began as an epiphany by Ray Anderson in 1994 grew into a movement by Interface associates worldwide. Interface associates realized that a new day and a new set of challenges had arrived, requiring the energy, passion, and dedication of all associates worldwide to reduce and eventually eliminate the company's environmental footprint by 2020. It had become painfully clear at Interface that the age-old "take, make, waste" cycle had to be replaced with one emulating nature's closed-loop cycle. Each front of Mission Zero would be used, along with other metrics, to measure progress.

Since embarking on its mission in 1994, Interface's associates have had a significant impact on waste elimination. The overwhelming charge to become responsible for our product's full life cycle not only presented engineering and manufacturing challenges, but also required a complete rethinking of how products were designed, manufactured, and marketed. The resounding success in this area through waste cost avoidance has allowed the company to further accelerate its Mission Zero progress. As we made steps forward in eliminating waste, the other fronts took on a life of their own. It was during this time that the seeds of Front 4: Closing the Loop, began to take shape. Our petroleum-dependent company had to disconnect from the oil head and develop processes to allow it to recycle the materials that had already been mined from the earth. In Front 4, the overall goals for closing the loop in the manufacturing process called for all raw materials used in manufacturing products to be from either recycled or rapidly renewable sources. In addition, the products sold would ultimately be recyclable and reenter the manufacturing process at the end of their normal life cycle.

By the late 1990s, Interface would be exploring ways to design products that were more environmentally friendly and functional for its market segments by using bio-based fibers and other renewable materials. And near the turn of the new century, the introduction of new technology (Cool Blue™) would enable the company to create new backing materials from reclaimed material. This latest development represents the most significant step to date for closing the loop. The new backing technology holds the potential for using a wider range of materials for food sources to produce new products and reduces our reliance on virgin petroleum based raw materials. With each small step along the journey, Interface has embraced change and stepped outside of its comfort zone, even at times learning from its failures and successes, in an attempt to eliminate the environmental footprint of its operations.

GETTING STARTED: ELIMINATING WASTE

With the end in mind, the first step was to embrace the "less is more" concept. The fact was that our product design and manufacturing processes were very wasteful. As a result, initial efforts would focus predominantly on Front 1: Eliminate Waste. Interface defines waste as any cost that does not produce value to our customers. To address the challenge of eliminating waste, Interface engaged all of its associates in a common purpose called QUEST (Quality Utilizing Employee Suggestions and Teamwork). QUEST was launched as our initiative designed to eliminate measurable waste by establishing innovative teams throughout the world to identify, measure, and then eliminate waste streams. Interface honed in on process and material efficiencies in order to minimize materials required to generate product and worked on eliminating gaseous, liquid, and solid waste streams. QUEST teams also formed with the objective of ensuring that all raw materials would either go into products for customers or would become raw materials for a secondary product. This effort would require us to work closely with our vendors and other entities to minimize all waste possible and then utilize what was previously considered waste as raw material "food sources" for closing the loop.

We also had to design products that were less wasteful, with ease of recycling in mind. At the same time, we knew we needed to increase the recycled and renewable raw material contents and manufacture recyclable products to enable us to meet the ultimate objective of closing the loop. Over the coming years, the unforeseen and remarkable success in the area of waste elimination and the resulting cost avoidance would allow Interface to fund activities that would move it along its Mission Zero path.

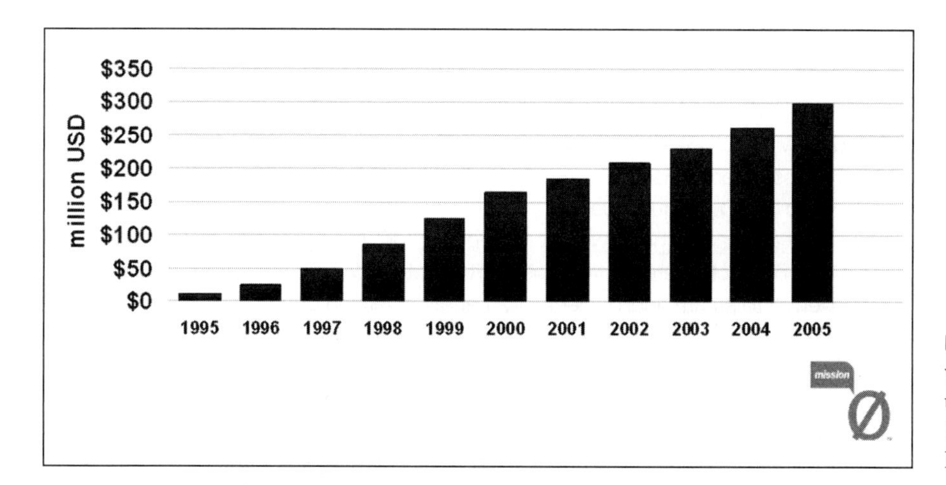

Figure 9.6 Interface's efforts to eliminate waste grew to over $300 million of cumulative avoided costs during the period 1995–2005. (Diagram courtesy of Interface-FLOR Commercial, Inc.)

DIVERTING CARPET FROM LANDFILLS: A BIG STEP TOWARD REDUCING THE ENVIRONMENTAL FOOTPRINT

Part of Interface's strategy was to look for ways to take existing material and either recycle it to make a new product or divert it from the landfill. One of the very first steps in this process included the concept of taking ownership and responsibility for a product at the end of its useful life. In the fall of 1994, Interface began to embrace the idea of *leasing carpet.* This was already being done with automobiles, computers, and the like, but it had never been suggested for carpet. Interface struggled with the concept due to the reluctance of the banking industry to make the necessary loans. They had problems valuing the used carpet.

In 1995, Interface attempted to launch a unique leasing program for carpet tile, called the Evergreen Lease™. The original Evergreen Lease was intended to be an operating lease under which Interface would retain ownership of the carpet at the end of the lease. For a monthly fee charged to the lessee, Interface would guarantee the function and appearance of the floorcovering by installing, maintaining, and, when necessary, replacing carpet tiles. The concept included the idea of a perpetual lease—for example, leasing carpet to a building's owner rather than to the transitory tenants, theoretically for as long as the building remained in use.

But the program ran into a snag in regard to how to account for carpet leasing. Financial accounting standards require the term of an operating lease to be less than 75 percent of the estimated economic life of the product. The standards also require about 15 percent residual value at the end of the lease. The original Evergreen Lease proved to be an unsuccessful initiative mainly because financial institutions questioned whether the carpet tiles met this requirement and they were unwilling to finance the leases. It seemed that the Evergreen Lease program was ahead of its time.

Interface remained dedicated to the concept of the Evergreen Lease and tried many different approaches. It offered a take-back/recycling provision on its standard capital leases. Under this arrangement, the lessee would own the carpet at the end of a 5-year lease. If the carpet had additional years of useful life, it could remain on the floor with the understanding that Interface would take it back and recycle it when it reached the end of its life. This would prevent premature removal of the carpet. This process did not prove successful either; it was not economically competitive with the added costs of take-back and recycling built into the price. We will not give up on Evergreen Leases. One day, a redesign of commerce (see Interface's Front 7) will allow this concept to be successful.

The company's carpet leasing concept opened up new ways of thinking about how to reclaim existing product and divert it from the landfill. In its overall mission to eliminate waste, Interface began to see the tremendous impact it could have in reducing its environmental footprint by diverting carpet from landfills, and it became actively engaged in finding practical ways to do so at the beginning of 1995. DuPont (now Invista), at the time, was already taking used carpet back and finding ways to keep it out of the landfills. During that first year, Interface became the largest user of the DuPont Partners for Carpet Reclamation Program (PCR) by diverting 202,000 square yards.

DuPont had some down-cycling options for broadloom carpet but none for tile. Tile products were for the most part sent to a "Waste to Energy" facility to be converted to steam and/or electricity. Landfill diversion quantities were impacted by the willingness of customers (carpet owners) to pay the cost of diversion, customer interest or lack of interest in recycling, and the many logistical issues.

At about this time, Interface explored the possibility of building a Waste to Energy facility close enough to our plant for supplying our electrical energy needs. This plant facility would also be used for our local municipal waste. After a good deal of study and meetings with community leaders, it was decided that there was insufficient community interest, and the project was abandoned.

Although there was a technology gap in separating face fiber from backing, Interface knew that someday it would find or develop the technologies that would allow it to graduate to a closed-loop process—converting old product into new. With this in mind, Interface began looking for closed-loop recycling solutions for its tile products. For a couple of years and through a partnership with two companies, tiles were cryogenically ground into a powder form so that we could add this recycled powder material to our virgin mix to introduce recycled content. This process was costly and challenging due to the fiber content and resulted in quality problems.

We also pursued "repurpose" opportunities in which carpet still in good condition was offered for free to nonprofit organizations. This is still being done today, but it has its own set of management and cost issues.

Launch of Interface's ReEntry® Program

Between 1996 and 1997, Interface, along with other members of the carpet industry, decided to enter the carpet dealer business by acquiring established businesses. It was then decided that carpet reclamation would fit best within this new dealer network. The Interface carpet reclamation program was launched as ReEntry and was a part-time job within the new Re:Source® dealer business. For several years, responsibility for ReEntry passed through a number of Re:Source managers. Interface and Re:Source tried numerous approaches to encourage ReEntry participation in order to increase diversion of carpet from the landfill. One example was a contest for the sales force and dealers as an incentive for participation. Although this resulted in increased awareness, a larger role in growing the ReEntry program would be played by architecture and design (A&D) and customer education, effective program communication, and increasing support of the program through our sales, marketing, and dealer resources.

In 2001, Interface brought the ReEntry program back into its Troup County, Georgia, manufacturing facility (now InterfaceFLOR Commercial), devoted additional resources to the effort, and agreed to supplement the customer's reclamation cost. These changes resulted in increased growth in participation. Then in 2005, management at InterfaceFLOR Commercial made the decision to supplement the total reclamation cost on only those backed products that we could recycle into new product. The initial challenges in the launch and administration of Interface's ReEntry program were well worth the results our company is able to report in carpet diversion from landfill.

National Carpet Recycling Effort—CARE Organization

In addition to changing processes and creating programs internally to divert carpet from landfills, Interface looked at ways to collaborate with other organizations

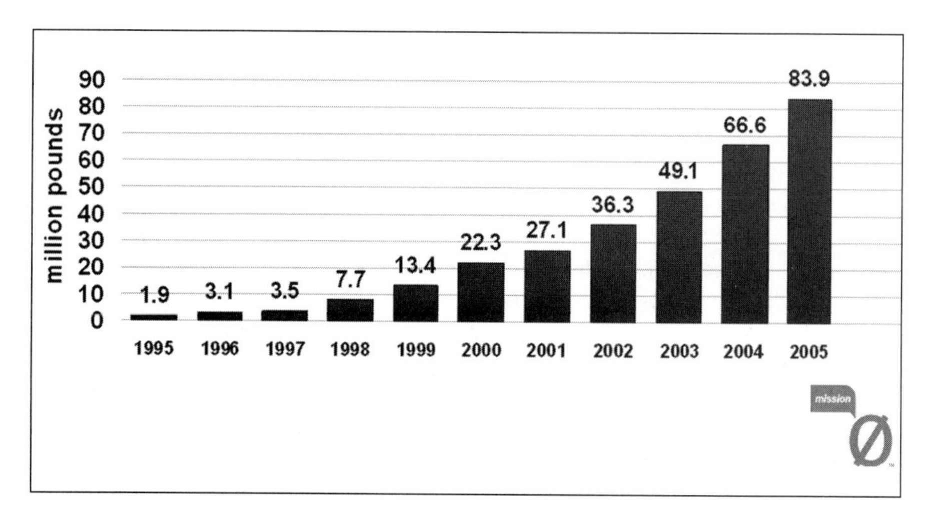

Figure 9.7 Interface created the ReEntry program as its carpet reclamation program. The program prevented 83.9 million pounds (38.1 million kilograms) of carpeting from being landfilled in 2005. (Diagram courtesy of InterfaceFLOR Commercial, Inc.)

to have an even greater impact on reducing the amount of carpet that normally would go to landfills. In the late 1990s, a small group known as The Midwest Work Group was formed, consisting of different groups that were concerned about carpet waste recycling. Made up of representatives from some of the state governments, a few carpet companies that included Interface, a few nongovernmental organizations, and a number of concerned entrepreneurs, the organization's objective is to find market-based solutions to keep carpet out of landfills.

The Midwest Work Group was renamed Carpet America Recovery Effort (CARE), and Interface was one of the charter members to sign the 2002 Memorandum of Understanding (MOU), a pledge to support CARE's recycling objectives. Interface knew that in order to make its closed-loop recycling goal a reality, a national carpet recycling network needed to be in place, as it would to be nearly impossible for any one company to do more than scratch the surface on recycling this massive quantity of carpet waste now going into landfills throughout the country. CARE's work to establish an infrastructure for carpet reclamation is ongoing, with membership and annual meeting attendance continuing to grow. This development of infrastructure and an increased demand for recycled carpet products will help to make Interface's closing-the-loop efforts more financially viable.

INNOVATION: DESIGNING WITH SUSTAINABILITY IN MIND

Interface's history is intertwined with innovation. Designing new products that meet a need in the marketplace and that also take our sustainability mission into account is a crucial part of our strategy and success.

Around 1998, in an attempt to maximize our "less is more" theme, Interface's product development team accepted the challenge of attaching a flat woven face to a tile backing. A project team was formed to explore the possibilities. An initial market analysis found that a market between hard surface and carpet existed and that a cleanable, patternable, flat surface with low-slip characteristics could penetrate the hard surface market.

The design phase subsequently began, with the primary goal to minimize if not eliminate waste. Materials were to be designed at the lowest weight possible. New yarn systems were explored and created. New polymers were studied, extrusion methods deliberated, and construction methods forged. Recycled systems were pondered and piloted in labs. The new Solenium™ product was designed with recycling in mind, with a "zippering" separation of backing from face for easy dismantling.

The excitement of the project led to a rush to bring the product to market, and early deadlines were set for launch dates. Initial research showed that the product's cleanability, appearance retention, rolling resistance, and slip resistance were perfect solution points for the target market.

In 1999 the product was launched with an aggressive performance story, a solid sustainability story, and much fanfare. It was on the market for about a year before failures started to occur. The new product, launched as resilient textile flooring, was very different from carpet. Existing industry standard testing for carpet did not include testing for the failures we experienced with Solenium.

Sustainability has provided Interface with opportunities to venture into unknown territory, in this case the launch of a new category of floor covering. The Solenium launch demonstrated Interface's willingness to take bold steps in its sustainability journey. Although the product line was unsuccessful, it remains a valuable learning experience for our company—from design to launch. The insight gained will be beneficial in the consideration of any complementary or traditional floor products launched by the company as they relate to sustainability and functionality.

Design Success Using Biomimicry: Entropy® and i2™

Interface would continue its efforts to design more environmentally friendly products. After becoming oriented to the concepts of biomimicry, David Oakey, Interface's award-winning designer, let go of the idea that design was only color and pattern. His first product utilizing biomimicry was Entropy, a 50-centimeter tile launched in 2000 that was inspired by the way nature covers its "floors." Leaves that cover the ground have different shapes, colors and sizes—no two are exactly alike. They fall so that there is no set pattern or design, and yet, the effect is beautiful. There exists symmetry without sameness.

Translating that idea into modular floorcovering resulted in the development of one of Interface's most efficient and most successful products, Entropy, which quickly became the top seller. Today, InterfaceFLOR Commercial's products designed using the principles of biomimicry are marketed as i2. Each i2 tile is unique in pattern and color. No two are the same, but they work together as part of a whole design and look great no matter how the tiles are laid on the floor. This product design concept and launch represent a notable success in Interface's sustainability endeavors.

CLOSING THE LOOP

One of the important steps on the journey to sustainability is minimizing and ultimately eliminating Interface's use of virgin petroleum-based raw materials with a move to generating new product using reclaimed materials. Interface also set aggressive goals to increase the recycled and renewable material content of its products, using its EcoMetrics Report (Interface's tracking mechanism of what it takes, makes, and wastes) to measure progress along the way. Interface's Purchasing Department plays an important role in the effort to increase recycled and renewable content. In 1998, the department hosted a "Greening the Supply Chain" meeting with its suppliers. The forum allowed Interface to share its vision and needs with its suppliers, as well as to persuade attendees to join the sustainability journey through a challenge to supply Interface with as much recycled and renewable content possible. Since that kickoff meeting, the process has become our culture and our way of doing business. At least once a year, Interface Purchasing representatives meet with all critical suppliers to review business goals and progress on reducing their own environmental footprint while emphasizing the mutual benefits of integrating this goal into the customer-supplier relationship.

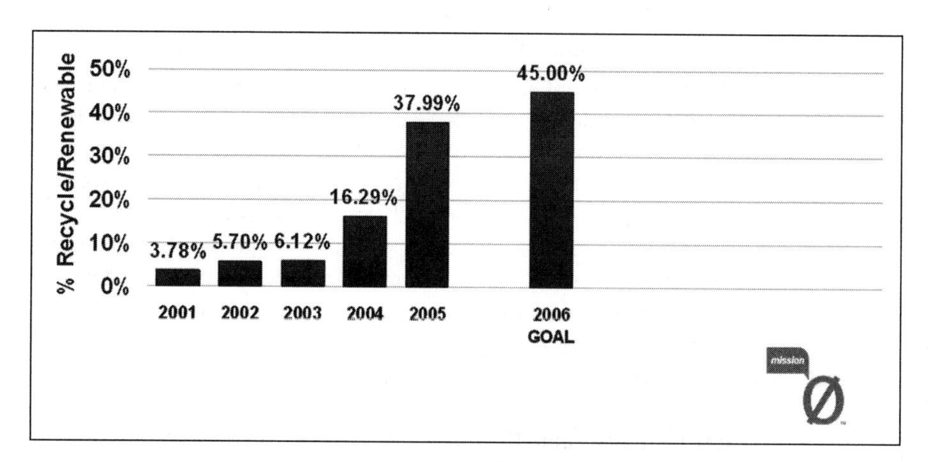

Figure 9.8 Efforts by Interface have resulted in the recycled and renewable content of the product in their Troup County, Georgia, plant rising from nearly zero in 2000 to almost 38 percent in 2005 and perhaps 45 percent in 2006. (Diagram courtesy of InterfaceFLOR Commercial, Inc.)

Design and Development of Renewable Bio-Based Fiber Products: Polylactide Polymer

Conventional carpet is constructed using nylon petroleum-based fibers. In early 2000, Interface began its exploration of bio-polymers and chose to work with bio-based polymers made from polylactic acid (PLA). PLA production uses 20–50 percent less fossil fuel resources than that of traditional hydrocarbon resins, so there are fewer greenhouse gas emissions associated with PLA production. PLA fibers are also completely biodegradable to lactic acid at the end of their service lives. In addition, they are derived from non-food-grade corn and potentially other starch-containing agricultural plant materials and waste products. The corn supplying the PLA market is typically from the overproduction of animal feed, which would be grown with or without the decision to manufacture PLA.

In April 2000, David Oakey Designs (DOD) began designing new Interface products with 5 to 20 percent PLA content. The products were designed by carefully selecting the fiber components necessary in the construction of the products to achieve the desired result. A balance was needed to meet manufacturing process requirements while at the same time maintaining Interface's high standards of product performance. DOD, working with the Interface's Product Development Team, has developed over 25 new products that have an average of 15 to 20 percent PLA content as of this writing. We will continue to develop new products using PLA and other bio-based fibers as we strive to climb to the peak of Mt. Sustainability.

Increasing Recycle Content to Backing: ASG

As carpet tile backing represents the greatest percentage of tile weight, Interface has researched and carefully considered reduction and substitution opportunities in its backing systems. In years past, the use of Aluminosilicate Glass (ASG) was researched as a backing material substitute. ASG is a by-product of coal-fired power plants. The material rapidly accumulates, creating solid waste destined for landfills. By 2004, in alignment with its efforts to reduce dependence on virgin materials and increase the use of recycled materials, Interface replaced calcium carbonate with ASG in all of its GlasBac® carpet tiles manufactured at InterfaceFLOR Commercial in Troup County, Georgia. Utilizing this material not only increases the recycled content in our product, but also reduces the volume of ASG sent to landfill. In addition, the lower specific gravity of ASG delivers enhanced backing construction qualities that result in lower product weights. ASG is also used in many other building products, including concrete, cement, aggregates, wallboard, roofing tiles, and shingles.

The Quest for Recyclable Products: The Face Fiber Challenge

As mentioned, Interface's Front 4: Closing the Loop involves using renewable and recycled materials to manufacture recyclable product. Early on in our sustainability journey, we focused on trying to capture the value of the face fiber nylon. Because the greatest value of the reclaimed carpet is in the value of the nylon, we have investigated and attempted many different techniques to separate face nylon from the carpet backing. Some of those efforts involved shaving the fiber from the carpet surface, mechanical grinding and air separation, chemical separation, and water separation. Interface is still working on a fiber solution for the separated carpet tile components. At the time of this writing, some exciting and promising projects are under way that should move us forward in closing the loop on face fiber.

The Quest for Recyclable Products: Success Via New Carpet Face with Extruded Recycled Back

In the latter part of 1998, Interface became interested in the process of extruding the separated tile backing into a vinyl sheet for lamination to a new carpet substrate—in other words, a new carpet face with an extruded recycled back. After working with several "toll" carpet separation companies, one company with the capability to handle both the separation and extrusion processes expressed interest. In these initial stages, Interface purchased the recycled extruded backing from suppliers and laminated it in-house to the carpet in its backing process using a thin virgin layer of adhesive.

CLOSING THE LOOP: BREAKTHROUGH TECHNOLOGY

In November 1999, Interface discovered a company in Germany that had developed and patented the Thermofix® technology. The equipment provided a full-width remelt of thermoplastic materials into a sheet. This was one of the breakthrough technologies Interface had been searching for that, if successful, would have a tremendous impact on its ability to close the loop. Interface would work through many years of trials, and address legal and patent issues to determine if the technology could be applied to carpet manufacturing.

With successful trial results, Interface moved forward, and in May 2005 the first Thermofix-type machine arrived at InterfaceFLOR Commercial in LaGrange, Georgia. The machine and the process for making carpet with a 100 percent recycled back was named Cool Blue™ as a result of its cooling or climate-neutral effect on the environment. The Cool Blue backing line first allowed InterfaceFLOR Commercial to make its own recycled backing on the new line and then laminate the backing to the carpet during a second pass. This method still required the virgin layer of adhesive. This is a thing of the past; Interface is now able to produce the recycled backing sheet and laminate it to the carpet on the Cool Blue backing line in a single pass without the virgin layer of adhesive. This "cool" approach to producing modular carpet was a huge step toward meeting the goals of Front 4. Cool Blue is a new, more flexible process to produce our GlasBac®RE backing that maintains the standard we have set for our GlasBac products while increasing the recycled content, decreasing reliance on virgin petroleum-based raw materials and fuel sources, reducing our use of waste-to-energy as a last best alternative for waste materials, and providing flexibility for the future. The Cool Blue backing process keeps reclaimed vinyl and potentially other recycled plastics in a closed-loop manufacturing process and out of landfills, where their impact becomes most apparent.

Cool Blue currently utilizes reclaimed vinyl-backed carpet and postindustrial carpet waste (frame scrap) as its primary food sources as we continue to explore other potential food sources. In mid-2005, Interface partnered with Invista through its Partners for Carpet Reclamation program. The logic of this partnership made sense for both companies. Interface needed more vinyl-backed products for its new recycled backing process, and Invista needed more broadloom products for its recycling needs.

The second phase of Cool Blue will eliminate the loss of the face fiber that now goes to a waste-to-energy process once separated from the backing. Instead, the entire product will be used either in the backing process (using supplemental "sweeteners" if required to process the material) or by sending the fiber to be recycled back into yarn.

Cool Blue utilizes a highly flexible technology that can function with a variety of polymers, allowing us to explore new waste streams and renewable materials as future innovations occur. This takes us beyond closed-loop recycling to help create a more dynamic loop. Cool Blue will increase the diversity of raw materials that can be used as industrial food for GlasBacRE far more than would have previously been acceptable due to technological limitations.

Recycled Content in the Face and Backing

Cool Blue increases the total potential recycled content of certain InterfaceFLOR Commercial (IFC) products backed with Cool Blue GlasBacRE to as much as 74 percent (31 percent postindustrial). Entropy®, IFC's best-selling modular carpet

Figure 9.9 Pictured is an installation of InterfaceFLOR Commercial's best-selling product, Entropy in Color Random. All Entropy is now produced using the Blue Chip yarn system and GlasBac RE backing, resulting in the highest total recycled content available at the time of this writing. (Photography courtesy of InterfaceFLOR Commercial, Inc.)

product in the i2 collection, now comes standard with a high-recycled-content solution-dyed nylon 6,6 yarn called Blue Chip™, backed with GlasBacRE. This development enabled IFC to provide the highest recycled-content contract carpet available in the industry at this time.

Renewable Energy Used with Cool Blue

Front 3: Renewable Energy outlines Interface's objective to ensure that by 2020, all fuels and electricity used to operate its facilities will be from renewable sources. Interface has taken an aggressive approach to achieving these goals by supporting markets for renewable energy, testing technologies, installing renewable energy systems at numerous locations, and committing to renewable energy purchasing targets. In line with this effort, Interface pursued a project in Troup County, Georgia, to utilize landfill gas as an energy source. This endeavor, made possible through a landfill gas-to-energy partnership that InterfaceFLOR Commercial initiated with the City of LaGrange, Georgia, would be used in its new Cool Blue Backing process. Through the landfill gas project, the city captures methane gas from a local landfill and pipes it to Interface's LaGrange plant, some 8 miles (13 kilometers) away, to use as process heat. This innovation has decreased the company's reliance on conventional natural gas and has significantly reduced greenhouse gas emissions into the atmosphere. All heat required for Cool Blue's recycled sheet consolidation is provided by landfill gas.

The landfill gas project was initiated in 2001, when InterfaceFLOR approached the City of LaGrange about the potential of tapping the local landfill as a renewable energy source. Today, it is being touted as a renewable energy model for private-public partnerships.

The InterfaceFLOR-City of LaGrange Landfill Partnership won the USEPA's Landfill Methane Outreach Program (LMOP) Energy Partner of the Year Award in 2005.

SUMMARY

Supplying enough recycled raw material for Cool Blue will continue to be a challenge due to increased competition for the recycled materials. The ReEntry program of bringing back used carpet and other recycle materials for Cool Blue is now part of Purchasing and Reverse Supply Chain management responsibility.

Other development work on closing the yarn fiber recycling loop is ongoing. With several promising projects under way, Interface feels confident that a major breakthrough is very close.

Cool Blue is an exciting new technology that will forever change the manufacturing process for Interface, as it has potential for using all sorts of existing materials for food sources to produce new products and reduces our overall reliance on virgin petroleum-based raw materials. It is perhaps the most comprehensive step the company has taken to date to close the loop on the manufacturing process and is just one more step on Interface's journey to accomplish Mission Zero: to eliminate any negative impact the company has on the environment by 2020.

Since the company began its sustainability journey in 1994, Interface has made significant progress in its effort to minimize the environmental impact of its operations. At this writing, several other players in the floorcoverings industry have joined the effort and are working on closing the loop as well. Interface will continue to seek out new technologies and is committed to increasing its knowledge of sustainable business practices and further developing them internally while influencing and teaching others along the way. With the realization that we cannot do it alone, however, Interface will continue to challenge its suppliers and industry as a whole to join the Mission Zero journey. Interface's success is proof that a company can indeed, as Mr. Anderson says, "do well by doing good."[25]

Connection to LEED-NC

The materials provisions of LEED-NC with respect to credits and points in the Materials and Resources (MR) category are listed in Table 9.9. Note that the descriptions of the LEED-NC credits and points in this section are abbreviated and that the appropriate LEED Reference Manual should be consulted for detailed information about how to achieve the points.

MR PREREQUISITE 1 (MRp1): STORAGE AND COLLECTION OF RECYCLABLES

A very interesting and progressive LEED requirement is that an easily accessible area must be set aside for the separation, collection, and storage of materials, which at a minimum include paper, corrugated cardboard, glass, plastics, and metals.

MR CREDIT 1 (MRc1): BUILDING REUSE (4 POINTS MAXIMUM)

Reusing a building to the maximum extent possible is an excellent strategy for reducing the materials impacts of construction. LEED offers a maximum of 4 points for

TABLE 9.9

MR Credits and Points under LEED-NC 2009

Credit	Name of Prerequisite/Credit	Maximum Points
MR Prerequisite 1	Storage and Collection of Recyclables	NA
MR Credit 1.1	Building Reuse: Maintain 55% of Existing Walls, Floors & Roof	1
MR Credit 1.2	Building Reuse—Maintain 75% of Existing Walls, Floors & Roof	1, in addition to MR 1.1
MR Credit 1.3	Building Reuse—Maintain 95% of Existing Walls, Floors & Roof	1, in addition to MR 1.1 and 1.2
MR Credit 1.4	Building Reuse: Maintain 50% of Interior Non-Structural Elements	1, in addition to MR 1.1, 1.2, and 1.3
MR Credit 2.1	Construction Waste Management: Divert 50%	1
MR Credit 2.2	Construction Waste Management: Divert 75%	1, in addition to MR 2.1
MR Credit 3.1	Materials Reuse: Specify 5%	1
MR Credit 3.2	Materials Reuse: Specify 10%	1, in addition to MR 3.1
MR Credit 4.1	Recycled Content: Specify 10% (postconsumer + ½ preconsumer)	1
MR Credit 4.2	Recycled Content: Specify 20% (postconsumer + ½ preconsumer)	1, in addition to MR 4.1
MR Credit 5.1	Regional Materials: 10% Extracted, Processed & Manufactured Regionally	1
MR Credit 5.2	Regional Materials: 20% Extracted, Processed & Manufactured Regionally	1, in addition to MR 5.1
MR Credit 6	Rapidly Renewable Materials	1
MR Credit 7	Certified Wood	1
	Total MR Points Available	**14**

building reuse. MR Credit 1.1 provides one point for reusing 55 percent of the existing walls, floors, and roof; MR Credit 1.2 provides an additional point if 75 percent of these elements are reused. MR Credit 1.4 provides yet another point if 50 percent of the interior nonstructural elements are reused. *Exterior shell/structure* refers to the exterior skin and framing.

MR CREDIT 2 (MRc2): CONSTRUCTION WASTE MANAGEMENT (2 POINTS MAXIMUM)

Reducing construction waste is a critical part of construction operations for the production of green buildings. LEED provides a maximum of 2 points for construction waste diversion: 1 point for a diversion rate of 50 percent and a second point for a diversion rate of 75 percent. For the purposes of this credit, the calculations can be by weight or by volume, but they must be consistent throughout. Construction waste management is covered in more detail in Chapter 11.

MR CREDIT 3 (MRc3): MATERIALS REUSE (2 POINTS MAXIMUM)

Beyond reducing materials use in the building, reusing components of existing buildings has the greatest benefit in lowering overall materials impacts. Reusing building materials and products can result in the project's receiving up to 2 points. One point is achievable if 5 percent of the value of the project's materials are salvaged from a previous building; a second point can be earned if the level of reuse is at least 10 percent. If significant mechanical, electrical, or plumbing (MEP) system components are reused, their value should be included in the computation; otherwise, the calculation would exclude the MEP systems.

MR CREDIT 4 (MRc4): RECYCLED CONTENT (2 POINTS MAXIMUM)

The use of recycled content building materials provides up to 2 points in the LEED building assessment process. One point is achieved if the total recycled content value of the building materials (calculated as the percentage of postconsumer recycled content plus half of the percentage of preconsumer recycled content) is at least 5 percent. A second point is achieved if the total recycled content value is at least 10 percent. The value of the materials is used in these calculations, not their weight. Mechanical and electrical systems are excluded, but plumbing systems may be included.

MR CREDIT 5 (MRc5): REGIONAL MATERIALS (2 POINTS MAXIMUM)

Placing an emphasis on local or regional materials reduces the transportation impacts associated with the LCA of the materials. Two points are achievable for this credit. If at least 10 percent of the value of the materials and products in the project were extracted, harvested, and manufactured within 500 miles of the project site, 1 point is awardable. If 20 percent of the value of these materials were extracted, harvested, and manufactured within the same distance, a second point can be achieved.

MR CREDIT 6 (MRc6): RAPIDLY RENEWABLE MATERIALS (1 POINT MAXIMUM)

For the purposes of LEED, rapidly renewable materials are defined as those that are derived from plants with a total cycle of growth and harvesting that is 10 years or less. To obtain the point associated with this credit, 5 percent of the total materials value used in the project must be from rapidly renewable materials. This credit is under review by the USGBC to address more broadly the issue of biobased materials. The proposal is that MRc6 be changed from a rapidly renewable credit to a biobased credit, under which wood that is not derived from illegal logging would be recognized. With this change, wood certified by the FSC, the SFI, the CSA, and possibly other certification systems could obtain this credit along with credit for all of the rapidly renewable materials currently covered under MRc6.

MR CREDIT 7 (MRc7): CERTIFIED WOOD (1 POINT MAXIMUM)

The use of certified wood has been established as one of the key criteria for green buildings. This credit is awarded if a minimum of 50 percent of the wood-based materials and products in the project are certified and have the seal of an FSC certifier. In the context of LEED-NC, wood-based products include structural and general framing, flowing, finishes, furnishings, and temporary structures for formwork, bracing, and pedestrian barriers that are not rented. This credit, similar to MRc6, is being considered for revision by the USGBC. The proposal is that MRc7 be changed from a wood-only credit to a credit that recognizes certified biobased materials that satisfy as-yet-to-be-developed robust certification criteria, along with waste agricultural products such as straw-based particleboard. Initially, for wood products, it is anticipated that only FSC would satisfy the top-tier certification requirements for this credit at this stage. However, as other certification systems for wood improve and as certification systems for bamboo, cork, or agricultural products emerge that meet the USGBC criteria, they would also be included under MRc7.

Connection to Green Globes v.1

Green Globes v.1 addresses the issues of materials use in Part E, Resources, Building Materials and Solid Waste. The basic structure of Part E is shown in Table 9.10.

TABLE 9.10

Outline of Green Globes v.1, Part E

Section	Description	Points
E.1	**Materials with Low Environmental Impact**	**40**
E.1.1	Use of LCA to select building assemblies	40
E.2	**Minimized Consumption and Depletion of Material Resources**	**30**
E.2.1	Reused building materials	10
E.2.2	Postconsumer recycled content	10
E.2.3	Use of biobased materials	5
E.2.4	Certified wood products	5
E.3	**Reuse of Existing Structures**	**10**
E.3.1	Reuse of existing façade	5
E.3.2	Reuse of existing structure	5
E.4	**Building Durability, Adaptability, and Disassembly**	**10**
E.4.1	Envelope design to control rain penetration	2
E.4.2	Measures to prevent groundwater intrusion	2
E.4.3	Promotion of adaptability in the design	3
E.4.4	Design of building for disassembly	3
E.5	**Reduction, Reuse, and Recycling of Waste**	**10**
E.5.1	Diversion of construction, demolition, and renovation waste from landfill	6
E.5.2	Designated storage space for recyclable waste in building	3
E.5.3	Space for recycling dumpster	1
	Total Points Available	**100**

Summary and Conclusions

Many new materials and products are being developed to serve the high-performance green building movement. But in the face of the rapid changes taking place in this area, there is no clear philosophy that precisely articulates the criteria for this new class of products and materials. One proposal is that LCA should determine what constitutes greenness in the context of building materials and products. But LCA, too, has limitations in that it does not adequately address closed-loop materials behavior, which is how nature behaves. Nor does LCA address whether a product or building can be disassembled and recycled, or the recyclability of products and materials. A material or product could conceivably appear to be very beneficial according to the LCA data, but may not be recyclable and subject to disposal after use. LCA does, however, provide an excellent account of the resources and environmental impacts of a given decision and allows side-by-side comparisons of various approaches—for example, a steel versus a concrete structural system. Combined with other criteria, LCA offers a good way of evaluating the appropriateness of labeling a product or material green.

At this point in the evolution of high-performance green buildings, considering both the production and fate of materials and products should be a high priority. And as pointed out in the Cardinal Rules for a Closed-Loop Building Materials Strategy, the products and materials must be harmless in use and in recycling before they can be considered truly green.

Notes

1. *Green Building Materials: A Guide to Product Selection and Specification,* 2nd ed. (2006) was written by Ross Spiegel and Dru Meadows.
2. See "Navigating the Maze of Environmentally Preferable Products" (November 2003) for a wide-ranging discussion of EPP.
3. The Blue Angel eco-label website is www.blauer-engel.de.
4. The ATHENA Environmental Impact Estimator version 3.0 is available for purchase from The ATHENA Sustainable Materials Institute online at www.athenasmi.ca. ATHENA uses a location-specific database of materials to provide LCA information about whole building systems—for example, wall or roof sections. A demonstration version of ATHENA is available for download from the website.
5. BEES is a product of the Building and Fire Research Laboratory (BFRL) of the National Institute of Standards and Technology (NIST). BEES measures the environmental performance of building products using the LCA approach specified in the ISO 14000 standards. It is available from www.bfrl.nist.gov/oae/software/bees.html.
6. For a fuller discussion on using LCA as the primary tool in making building materials decisions, see Trusty and Horst (2003).
7. The website of The Natural Step's U.S. branch is www.naturalstep.org.
8. Adapted from a working paper by the Oregon Natural Step Construction Industry Group, "Using the Natural Step as a Framework Toward the Construction and Operation of Fully Sustainable Buildings" (2004).
9. Ibid.
10. Excerpted from "Building Materials: What Makes a Product Green?" (January 2000). This article provides a detailed description of each of the various attributes in the five major categories of green building products.
11. Excerpted from Trusty and Horst (2003).
12. An overview of the International Standards Organization (ISO) can be found at its website, www.iso.ch/iso/en/ISOOnline.frontpage. ISO 14000 is one of many standards promulgated by this organization. The ISO 14040 series is the member of the ISO 14000 family of standards that addresses LCA.
13. Excerpted and adapted from Trusty (2003).

14. BEES 3.0 is available as a free download from www.bfrl.nist.gov/oae/software/bees.html.
15. Detailed information on the BuildingGreen Top Ten 2006 Products is available at www.buildinggreen.com.
16. Forestry statistics are from the American Forestry & Paper Association website, www.afandpa.org.
17. The website of the U.S. branch of the FSC is www.fscus.org. The two U.S. certifiers maintain websites at these addresses: the Smartwood Program, www.rainforest-alliance.org/programs/forestry/smartwood, and Scientific Certification Systems, www.scs1.org.
18. The Certified Wood and Paper Association's website is www.scscertified.org.
19. The SFI Standard can be found at the AFPA website, www.afandpa.org.
20. Excerpted from "2002: The Inherent Recycled Content of Today's Steel," available on the Steel Recycling Institute's website, www.recycle-steel.org.
21. Data on aluminum are from the International Aluminum Institute at www.world-aluminum.org.
22. The European Community's "Green Paper: Environmental Issues of PVC" (2000) can be found at http://ec.europa.eu/environment/waste/pvc/en.pdf.
23. Philip Crowther's list of DfD principles was first published in his doctoral dissertation at Queensland University of Technology with the title "Design for Disassembly: An Architectural Strategy for Sustainability" (2002).
24. Detailed information about Interface's sustainability efforts can be found at www.interfacesustainability.com.
25. InterfaceFLOR Commercial's website is www.interfaceflorcommercial.com.

References

"Building Materials: What Makes a Product Green?" January 2000. *Environmental Building News,* 9(11), pp. 1, 10–14.

Crowther, Philip. 2002. "Design for Disassembly: An Architectural Strategy for Sustainability." Doctoral dissertation, School of Design and Built Environment, Queensland University of Technology, Brisbane, Australia.

Demkin, Joseph, ed. 1996. *Environmental Resource Guide.* New York: John Wiley & Sons.

"Navigating the Maze of Environmentally Preferable Products." November 2003. *Environmental Building News,* 12(11), pp. 1–15.

Spiegel, Ross, and Dru Meadows. 2006. *Green Building Materials: A Guide to Product Selection and Specification,* 2nd ed. New York: John Wiley & Sons.

Trusty, Wayne B. 2003. "Understanding the Green Building Toolkit: Picking the Right Tool for the Job." *Proceedings of the U.S. Green Building Council Annual Conference,* Pittsburgh, Pennsylvania. Available at www.athenasmi.ca.

Trusty, Wayne B., and Scot Horst. 2003. "Integrating LCA Tools in Green Building Rating Systems." *Proceedings of the U.S. Green Building Council Annual Conference,* Pittsburgh, Pennsylvania. Available at www.athenasmi.ca.

"Using the Natural Step as a Framework Toward the Construction and Operation of Fully Sustainable Buildings." 2004. The Oregon Natural Step Construction Industry Group. Available at www.ortns.org/documents/THSConstructionPaper.pdf.

Chapter 10

Indoor Environmental Quality

The key emerging benefits of high-performance green buildings appear to be their health and productivity benefits, with paybacks that may be as much as 10 times higher than their energy savings. More and more hard evidence of the effects of good *indoor air quality* (IAQ) is emerging, supporting design and construction efforts that provide excellent building air quality. More recently, the range of health problems connected to buildings has shifted from air quality alone to include a far wider range of human health effects associated with lighting quality, noise, temperature, humidity, odors, and vibration. This broader range of impacts is referred to as *indoor environmental quality* (IEQ) and includes the subject of IAQ.

The impact of buildings on human health is substantial and results from a combination of building design, construction practices, and the activities of the occupants. A study by Fisk and Rosenfeld in 1998,[1] updated in 2002, placed the annual cost of IAQ problems at $100 billion.[2] Table 10.1, which is adapted from this study, shows estimated productivity gains from improvements made to indoor environments. In the United States, people spend about 90 percent of their time indoors, either in their homes, workplaces, schools, shopping malls, fitness centers, or numerous other types of structures. Air quality in some of these buildings is often cited as being far worse than that of the outside air. This poor air quality can be attributed to a number of factors: tight buildings, materials that offgas pollutants into the indoor environment, poor ventilation, and poor moisture control, to name a few. In addition, poor construction practices can contribute to significant indoor environmental quality problems. For example, ductwork that has been stored and handled without being covered and sealed can be contaminated with particulates that are blown into occupied spaces during building operation, potentially affecting the health of the people in the building.

The high-performance green building movement has been highly successful in integrating indoor environmental issues into the criteria for green buildings, in essence taking ownership of IEQ when it comes to new buildings, so it is now expected that a high-performance green building will have excellent IEQ. In particular, the USGBC's LEED suite of standards addresses IEQ and provides points for incorporating at least some of the major IEQ components, generally those concerned with air quality and individual control of temperature and humidity. Green Globes, an emerging competitor to LEED, addresses the same issues as LEED but also includes other important IEQ matters such as acoustic comfort for building occupants and neighbors. Green Globes address noise from air-conditioning systems, plumbing, preventing noise generated in the building from affecting neighbors, protecting building occupants from outside noise, and noise attenuation for the structural system.

The actual savings attributed to a high-quality interior building environment are substantial and are thought to be greater than even the energy savings. A study by Greg Kats of Capital E indicated 20-year life health and productivity savings of $36.89 per square foot (square meter) for LEED-certified Silver buildings and $55.33 per square foot (square meter) for LEED-certified Gold and Platinum buildings.[3] The productivity and health benefits of high-performance green buildings, a

TABLE 10.1

Estimated Potential Productivity Gains from Improvements Made to Indoor Environments

Source of Productivity Gain	Strength of Evidence	U.S. Annual Savings or Productivity Gain
Respiratory disease	Strong	\$6–\$14 billion
Allergies and asthma	Moderate to strong	\$1–\$4 billion
Sick building syndrome	Moderate to strong	\$10–\$100 billion
Worker performance	Moderate to strong	\$20–\$200 billion
Total range		\$37–\$318 billion

result of designing a high-quality indoor environment, dominate the discussion of benefits. For Gold and Platinum buildings, the claim is that the health and productivity benefits are almost 10 times greater than the energy savings, which amount to \$5.79 per square foot (square meter). These results are not only impressive, but startling as well. However, the basis for these claims is rarely scientific; thus, using these results in LCC or in economic analyses should be done only with extreme caution to avoid compromising the justification of an otherwise sound approach.[4]

IEQ Issues

Of the wide variety of issues associated with IEQ, in the recent past two in particular stand out: sick building syndrome (SBS) and building-related illness (BRI). Though both refer to health problems associated with IAQ, there is a very important difference between them. SBS describes an assortment of symptoms experienced by a majority of building occupants for which no specific cause can be identified. Typically, SBS is diagnosed when the affected employees' symptoms disappear almost immediately on leaving the building. In contrast, BRI refers to symptoms of a diagnosable illness that can be attributed directly to a defined IAQ problem.

SBS, also known as *tight building syndrome,* is the "condition in which at least 20 percent of the building occupants display symptoms of illness for more than two weeks, and the source of these illnesses cannot be positively identified."[5] Most of the structures that fall victim to SBS are modern office buildings, the majority of which have been constructed over the past two decades and are tightly sealed, mechanically ventilated, and have few or no operable windows. Symptoms of SBS may include headache; fatigue and drowsiness; irritation of the eyes, nose, and throat; sinus congestion; and dry, itchy skin. These symptoms can occur alone or in combination. The most common complaints include flulike symptoms or respiratory tract infections. Some occupants relate SBS to stresslike headaches, coughs, and the inability to concentrate, while others experience dry skin or rashes.[6]

The economic impacts of SBS can be tremendous, making it a building owner's worst nightmare. The USEPA has estimated that the United States spends over \$140 billion in direct medical costs attributable to IAQ problems.[7] SBS is also believed to be responsible for marked decreases in productivity coupled with increases in absenteeism. Vacant buildings and nonrenewed building leases may also be a direct result of SBS. An example of the high costs associated with SBS is the Polk County Court House in Florida. Located in Lakeland, a community in central Florida, the court house was constructed for \$37 million and opened in the summer of 1987. Due to a severe case of SBS, it was closed in 1992; its occupants, including prison inmates, had to be evacuated and temporarily relocated. It took 3 years and $26 million to literally rebuild the facility to correct the original toxic mold problems that were attributed to design and construction problems.

The wide range of conditions associated with both SBS and BRI, some chemical and some biological—including multiple chemical sensitivity, legionellosis, and allergic reactions—are described in the following subsections.

MULTIPLE CHEMICAL SENSITIVITY

Multiple chemical sensitivity (MCS), a relatively recently identified condition related to IAQ, is marked by sensitivity to a number of chemicals, all at very low concentrations.[8] MCS is characterized by severe reactions to a variety of volatile organic compounds and other organic compounds that are released by building materials and many consumer products. These reactions may occur following one sensitizing exposure or a sequence of exposures. It should be noted, however, that there is currently a great deal of debate over the legitimacy of the condition. Some contend that it is a physical illness, while many others believe the cause to be psychosomatic.

LEGIONELLOSIS

Legionellosis refers to two important bacterial diseases: Legionnaire's disease and Pontiac fever, caused by the virus *Legionella pneumophila.* The diseases are not spread via person-to-person contact, but rather through the soil-air and water-air links both indoors and outdoors. The bacteria can survive in water for up to a year under certain conditions. *Legionella* prefers stagnant water, which is found in drain pans of HVAC units and cooling towers. Fans then can transfer the bacteria, to be inhaled by unsuspecting victims. Sources of *Legionella* in residences and other buildings may also include hot tubs, vaporizers, humidifiers, and contaminated forced-air heating systems. Algae and other aquatic life forms can promote the growth of *Legionella* by providing the bacteria with food.

Pontiac Fever

In July 1968, 95 of 100 people employed in—ironically—a public health building in Pontiac, Michigan, became ill with a flulike ailment. In fact, if the number of cases had not comprised such a high proportion of the employees, the disease probably would have been diagnosed as the flu. The employees all claimed to suffer from headaches, fevers, and muscles aches and pains. Called *Pontiac fever,* the disease was eventually traced back to a faulty HVAC system. However, it was not until the discovery of Legionnaire's disease, nearly 10 years later, that the virus that caused Pontiac fever was finally identified.

Pontiac fever is a mild form of legionellosis. It is characterized by a high attack rate (90 percent) and a short incubation period of 2 to 3 days. The disease lasts for only 3 to 5 days and requires no hospitalization. Symptoms include those exhibited by the employees in 1968, as well as chills, sore throat, coughing, nausea, diarrhea, and chest pain. Many people may never suspect that they have Pontiac fever, as only an estimated 5 to 10 percent of those seeking medical care have lab tests done.

Legionnaire's Disease

Legionnaire's disease is a type of pneumonia caused by *Legionella.* Both the disease and the bacterium were discovered following an outbreak traced to a 1976 American Legion convention in Philadelphia. This disease develops within 2 to 10 days after exposure to *Legionella,* and early symptoms may include loss of energy, headache, nausea, aching muscles, high fever (often exceeding 104°F (40°C)), and chest pains.

Later, many bodily systems, as well as the mind, may become affected. The disease eventually causes death if high fever and antibodies cannot defeat it. Victims who survive may suffer permanent physical or mental impairment. The U.S. Centers for Disease Control and Prevention (CDC) in Atlanta, Georgia, has estimated that the disease infects 10,000 to 15,000 persons annually in the United States; others have estimated as many as 100,000 annual U.S. cases.

Legionnaire's disease is a severe multisystem illness that can affect the lungs, gastrointestinal tract, central nervous system, and kidneys. It is characterized by a low attack rate (2 to 3 percent), a long incubation period (2 to 10 days), and severe pneumonia. Unlike Pontiac fever, hospitalization is required. Most victims are men in their fifties and sixties who are smokers and/or have underlying respiratory problems. Alcohol consumption, diabetes, and recent surgery can also be contributing factors.

ALLERGIC REACTIONS

Allergies are reactions to a form of indoor air pollution that occur when the body responds to nontoxic substances, like pollen, as threats. The body will mimic the effects of a real illness by stimulating the production of white blood cell to combat the allergen. An individual usually does not experience an allergic reaction until after the second exposure to a specific allergen. The first exposure results in the manifestation of the allergy. Allergens that cause an allergic response include viable and nonviable agents. Viable agents include bacteria, fungi, and algae. Common nonviable agents include house dust, insect and arachnid body parts, animal dander, mite fecal pellets, remains of molds and their spores, pollens, and dried animal excretions.

Preventing encounters with offending allergens is easier said than done. They constitute a new variation on the IAQ problem in that the reactions of a building's inhabitants to an allergen can vary more than with other environmental factors. What may send one person gasping to the emergency room may have absolutely no effect on another. Regular cleaning to remove dust, the use of high-efficiency filters, and regular filter changing can help to reduce or eliminate biological contaminants.[9]

Indoor Environmental Factors

The indoor environment of a building has a complex makeup. Table 10.2 provides a list of factors that are thought to affect the indoor environment.[10] These can be classified as chemical, physical, and biological. The sensory systems of the inhabitants interact directly with certain factors, such as the level of sound, light, odor, temperature, humidity, touch, electrostatic charges, and irritants.[11] Hundreds of other substances can also be harmful to inhabitants yet go undetected by the sensory systems. Some of these can actually be more dangerous than those that are detected, as their presence can only be determined through testing. Inhabitants may be exposed to high concentrations of these substances for long periods of time without even knowing it—among them radioactive substances, many toxic substances, carcinogens, and pathogenic microorganisms.

PHYSICAL FACTORS

Physical indoor environmental problems are traceable primarily to the electrical and mechanical infrastructure of a building. They include sound/noise transmission, lighting quality, thermal conditions, and odors. Physical factors are generally nontoxic but are at least a nuisance to building occupants and can lead to health problems after exposure for extended periods.

TABLE 10.2

Building Elements Affecting IEQ

Operation and Maintenance of the Building	Ventilation and performance standards Ventilation system operational routines and schedules Housekeeping and cleaning Equipment maintenance, operator training
Occupants of the Building and Their Activities	Occupant activities: occupational, educational, recreational, domestic Metabolism: activity and body characteristic dependent Personal hygiene: bathing, dental care, toilet use Occupant health status
Building Contents	Equipment: HVAC, elevators Materials: emissions from building products and the materials used to clean, maintain, and resurface them Furnishings Appliances
Outdoor Environment	Climate, moisture Ambient air quality: particles and gases from combustion, industrial processes, plant metabolism (pollen, fungal spores, bacteria), human activities Soil: dust particles, pesticides, bacteria Water: radon, organic chemicals including solvents, pesticides, by-products of treatment process chemical reactions
Building Fabric	Envelope: material emissions, infiltration, water intrusion Structure Floors and partitions

Sound/Noise Transmission

Control of sound and noise transmission in buildings is a major problem, although it is not addressed by the USGBC LEED suite of standards. Green Globes does address issues of acoustic comfort for the building's occupants and its neighbors. Green Globes issues for evaluation can be found in the Green Globes summary at the end of this chapter. Noise from air-handling systems, lights, transformers, and other sources can cause discomfort and even health problems for building occupants. Building designers and engineers are often intimidated by the challenge of dealing with sound and noise transmission because it is a somewhat intangible concept in a world of mostly tangibles: steel, size, color, and so on.

The basic premise in creating an acoustically acceptable indoor environment is to ensure that sound levels in particular areas of a building are at or below an acceptable range for the specific application. For instance, it would be a mistake to locate a helicopter pad just outside of a library. It is clear that sections of a building where low noise levels are required must be separated and insulated from noise-generating areas. When it comes to acoustics, designers can easily prevent obvious problems—for example, taking care not to locate a conference room next to a chiller plant. But more subtle problems may be overlooked, such as neglecting to insulate a wall that separates a restroom from a private office.

A less obvious requirement for ensuring good indoor sound quality is to eliminate as much as possible the subtle background noises that, although not necessarily apparent to building occupants, can be irritating and may, over time, lower morale and decrease productivity. Building systems can generate a wide variety of annoying sounds. Fluorescent light ballasts often buzz when they are not in perfect order, and ventilation systems produce a host of grating yet seemingly untraceable noises. Fan vibrations, too, are a nuisance inherent in ventilating systems that, when isolated, can be dealt with effectively and cheaply. Duct-air noises are more problematic and much more difficult to fix. High-speed air in a duct can create whistling sounds and vibrations that are difficult to eliminate. The solution is to reduce the air velocity. To maintain the same quantity of air at a lower velocity, a duct with a larger cross-sectional area must be used. But this solution itself can pose problems when the ductwork is installed in a tight ceiling space or an HVAC chase. The best answer to this problem is to address it before it happens by including an acoustic specialist in the design of an HVAC system.

As noted, high noise levels in commercial buildings can lead to morale problems and loss of productivity when occupants become irritated and annoyed and thus distracted from their work. The other major noise-related problem for building occupants is caused by exposure to unhealthy noise levels generated by air handlers, transformers, lighting, elevators, machinery, and motors.

Lighting Quality

Problems associated with lighting quality are similar to those associated with noise in that the cause is a poorly understood building support system. As a requirement for a high-quality indoor environment, lighting is probably better understood than sound, but it is nevertheless often overlooked in building design.

It is widely acknowledged that natural sunlight is the best light source for the eye. Unfortunately, these days most people spend an inordinate amount of time indoors and away from natural sunlight. Thus, the ideal healthy indoor light environment is one that allows natural light indoors or whose lighting system replicates natural light as closely as possible. Natural sunlight has an equal spectral distribution of the visible light frequencies combined to appear as white light. In contrast, artificial light sources are bound by the laws of physics, and hence they are limited in the frequencies of visible light that they emit. A list of common artificial light sources and their general color characteristics is shown in Table 10.3.

Incandescent lights, particularly the halogen type, give the best color rendition of natural light. Fluorescent and mercury vapor lights emit white light with a distinct

TABLE 10.3

General Color Characteristics of Typical Building Lighting Systems

Type of Light	Color Characteristics
Incandescent (argon-surrounded filament)	White with yellow tint
Incandescent (halogen-surrounded filament)	White
Fluorescent	White with blue tint
Mercury vapor	White with blue tint
Metal halide	White with blue-green tint
Sodium vapor (high pressure)	Amber white
Sodium vapor (low pressure)	Yellow

preponderance of blue frequencies. Fluorescents can be made to offer more in the warm color range, but the color of fluorescent lighting is not natural and typically tends to produce a too-bright, sterile atmosphere. Sodium-based lights produce a yellowish light and are commonly used for outdoor applications.

In commercial buildings, the primary sources of artificial light are incandescent and fluorescent lighting fixtures. Mercury vapor and metal halide sources are also used in large rooms or high-bay areas. In a typical building, general lighting in office areas is almost entirely fluorescent. Incandescent lights are used for more direct applications where a fluorescent tube is not applicable—for example, accent lighting. Incandescent lights are also used predominantly in dimming applications such as recessed lighting in lecture halls and meeting rooms. Dimming fluorescent fixtures are available, but they have not yet replaced incandescent lights as the choice for dimming applications.

The glow of fluorescent lighting in office settings can often be irritating to occupants. The obvious complaints caused by too much fluorescent light are sore eyes and headache, lowered morale, and decreased productivity. Poor lighting also has more subtle effects on mood. The eye, it is now known, is most comfortable with natural sunlight, which changes in intensity and color throughout the day. Because indoor artificial light is basically unchanging in color and intensity, there may be adverse effects on the health and well-being of those subjected to it. This is an important new field of study in the area of IEQ, so it is not entirely understood.

Flickering lights can also cause irritation and health problems. Ballasted lights—for example, fluorescent, mercury vapor, metal halide, and sodium lights—are subject to flickering when the ballast malfunctions. This can easily lead to sore eyes and headaches, and ultimately lower productivity. Glare is also a problem; however, unlike the others described here, it is not a consequence of artificial light but rather involves the light source and reflector positioning. Windows, desktops, and computer screens, even shiny paper, are all reflectors that can cause uncomfortable glare. Glare, depending on the intensity of the light, can quickly lead to discomfort and headaches, especially when reading, typing, or looking at a computer screen.

Thermal Conditions

The climatic setting in which a person is working has a profound impact on how he or she behaves and how well he or she works. But because everyone is different, what is perfectly comfortable to one person in an office may be profoundly uncomfortable to his or her neighbor. In general, the indoor comfort range is considered to be located in the center of the psychometric chart. Generally accepted ranges for comfort are as follows: in winter, temperatures between 68°F (20°C) and 75°F (24°C) and relative humidity between 30 percent and 60 percent; in summer, temperatures between 72°F (22°C) and 80°F (27°C) and relative humidity between 30 percent and 60 percent. Relative humidity below 30 percent in any season is considered too dry and will lead to discomfort. Typically, lower humidity levels can be tolerated in the winter and higher humidity levels can be tolerated in the summer, but relative humidity levels outside the 30 to 60 percent range are generally uncomfortable in all seasons.

Air velocity, mentioned briefly above, is another variable in the indoor climate that is not a fundamental property of the air. Air velocity varies greatly, depending on where one is in relation to vents, doors, windows, and fans. It is an integral aspect of air conditioning (heating and cooling) that indoor air be circulated; hence, it must have a certain velocity. The goal of HVAC designers is to introduce the highest-velocity air where it has little or no effect on the building occupants, usually along ceilings or walls, so that by the time it comes in contact with people, it has slowed to an undetectable rate. High-velocity air is more likely to cause discomfort in cool indoor climates and, conversely, be welcome in warm indoor climates.

Odors

Odors are one of the most common and annoying indoor environmental problems. Solving these problems is not easy, because the human olfactory system is highly complex and not well understood; moreover, the chemical sources that create many of these odors also are poorly understood. Even simple odors in office settings are complex, consisting of many substances. Typical sources of odors in the indoor environment include tobacco smoke, human body odor, and cleaning and personal grooming products. Offgassing of building materials is another a common source of smells. Complicating this issue is the pronounced difference in individual sensitivity to odors. Visitors to an office are generally far more sensitive to odors than its long-time occupants, for example. Because human reactions to odors are so varied, it is nearly impossible to predict how any one person or group of people will react.

CHEMICALS, RADON, ASBESTOS, AND COMBUSTION BY-PRODUCTS

A wide variety of chemicals can contaminate the indoor environment. Chemicals may be introduced into the indoor environment by painting, installation of carpets, or cleaning products. Chemical factors are classified according to the form they take at room temperature: vapor, gas, liquid, or particulate. Particulates include inorganic fibers; respirable particulates, such as dust and dirt; metals; and a variety of organic materials. Because small particulates can penetrate deep into the lungs, they are a serious concern. The size and density of particulates determine how deeply they can penetrate the respiratory system. Radon, a naturally occurring radioactive gas that has been connected to health problems, is a problem in many regions of the United States; thus, taking measures to mitigate it are important for ensuring a good indoor environment.

Volatile Organic Compounds

Volatile organic compounds (VOCs) are carbon-containing compounds that readily evaporate at room temperature and are found in many housekeeping, maintenance, and building products made with organic (carbon-based) chemicals. Paint, glues, paint strippers, solvents, wood preservatives, aerosol sprays, cleansers and disinfectants, air fresheners, stored fuels, automotive products, and even dry-cleaned clothing and perfume are all sources of VOCs. In any indoor environment, there can be up to 100 different VOCs in varying concentrations. Carbon filters can be used to adsorb VOCs, but they must be replaced regularly, as the odors deplete the carbon.

There are six major classes of VOCs: aldehydes (formaldehyde), alcohols (ethanol, methanol), aliphatic hydrocarbons (propane, butane, hexane), aromatic hydrocarbons (benzene, toluene, xylene), ketones (acetone), and halogenated hydrocarbons (methyl chloroform, methylene chloride).

Formaldehyde is highly reactive and may be found in all three states of matter. It is highly soluble in water and can irritate body surfaces normally containing moisture—for example, the eyes and the upper respiratory tract. Formaldehyde gas is pungent and easily detectable by its odor at concentrations well below 1 part per million (ppm). It is perhaps the most commonly occurring VOC in construction, found in many common products such as paints, wood products, and floor finishes. When combined with other chemicals, it can be used as glues and binders in numerous products. Urea-formaldehyde-foam-insulation, particleboard, interior-grade plywood, wallboard, some paper products, fertilizers, chemicals, glass, and packaging materials contain significant amounts of formaldehyde.

Radon

Radon, a colorless and odorless gas, is the product of the decay of the radium isotope that results from the disintegration of uranium-238. An inert gas, radon itself is fairly harmless, but as it decays, the resulting materials, known as *radon daughters,* are not. Radon's daughters are not chemically inert, and they form compounds that bind to dust particulates in the atmosphere. When inhaled, these particles can lodge in the respiratory system and cause damage due to the alpha particle radiation they emit. The half-life of the daughters is relatively short: they disintegrate in 1 hour or less. Despite this rapid disintegration, radon is a major concern because it may take 10 to 20 years for the first signs of exposure to develop, and it has serious consequences. The inhalation of radon is the second leading cause of lung cancer in America, is suspected in the deaths of 2,000 to 20,000 individuals a year, and is considered one of the most deadly indoor air pollution problems.

Anthony Nero of the Indoor Environment Radon Group at Lawrence Berkeley National Laboratory noted that the average indoor level of radon represents a radiation dose about three times larger than the dose most people get from X-rays and other medical procedures in the course of their lifetime. Hundreds of thousands of Americans living in houses with high radon levels are exposed yearly to as much radiation as people who were living in the vicinity of the Chernobyl nuclear power plant in 1986, when one of its reactors exploded. According to the USEPA, an acceptable maximum level of radon is 4 picocuries per liter (pCi/L) of air.[12] In Canada, the Atomic Energy Control Board (AECB) also set a level of 4 pCi/L for the general public in homes and other nonoccupational settings. If this level is exceeded, action must be taken to reduce it.

In buildings, radon occurs primarily through diffusion from the underlying subsoil into the building structure. Radon gas can enter a building through cracks or openings such as sewer pipe openings, cracks in concrete, wall-floor joints, hollow masonry walls, and other similar pathways. If the foundation of the building is tight, very little or no radon will enter. Because of the ground-up infiltration process of radon, a multistory building will have lower radon concentrations than a single-story building with an identical foundation. Indoor radon concentrations also relate directly to ventilation and fresh air intake of buildings. Due to energy conservation techniques and the resultant tighter buildings, new buildings may actually encourage the infiltration of radon gas by negative pressurization.

Asbestos

Asbestos is another potentially deadly IAQ problem. Unlike radon, however, the health implications of asbestos have been documented in detail, for it has been a major environmental problem for many years. When it was discovered that the threadlike particles in asbestos could lodge in human lungs, its use began to be phased out. Exposure to asbestos has been definitively linked to stomach and lung cancers.

The term *asbestos* refers to a group of silica-based minerals in fibrous bundles. Introduced in the 1930s and widely used in the United States from 1940 to 1973, asbestos comprises a large number of naturally occurring materials that are processed to produce a manageable form for use in construction, insulation, and fire retardation materials. Indoor building materials containing asbestos include thermal insulation on ceilings and walls; insulating materials used on pipes, ducts, boilers, and tanks; and finishing materials such as ceiling and floor tiles and wall boards. The materials that pose the greatest threat are those that can be easily crumbled or powdered by hand pressure.

High-quantity release of asbestos into the airstream usually occurs during maintenance, renovation, and other construction activities, when it becomes dangerous. There is very little danger to human health if the material is left undisturbed; asbestos becomes a health hazard only when its fibers are released into the air. Most experts agree that if asbestos surfaces are not deteriorating or being abraded, thus releasing asbestos fibers, they are best left alone. Removal of asbestos is very costly and can be done safely only by professionals. An unsafe removal process can do more harm than good by releasing more particles into the air, where they can continue to contaminate a building for years.

Combustion By-Products

Combustion by-products are created under conditions of incomplete combustion. The primary sources of combustion by-products that contribute to the contamination of indoor air are gas, wood and coal stoves, unvented kerosene space heaters, fireplaces under downdraft conditions, and tobacco smoke. The major by-products include carbon monoxide carbon dioxide, nitrogen dioxide, sulfur dioxide, and particulates. Their health effects can vary, depending on the type of by-product produced.[13] Each of these will now be described more fully.

Carbon dioxide: Carbon dioxide is a colorless, odorless, and tasteless gas. Although it is a by-product of combustion, it is relatively harmless; it is, after all, also a natural product of respiration. That said, and despite the fact that it is nontoxic, if the concentration of carbon dioxide is too high, the result can be unpleasant and perhaps unhealthy for a building's inhabitants. And since it is a natural product of respiration, it can also be an indicator of the quality of ventilation and IAQ.

Carbon monoxide: Carbon monoxide is another colorless, odorless, and tasteless gas, but it must not be confused with carbon dioxide. The effects of high-level carbon monoxide exposure can range from nausea and vomiting to headaches and dizziness to coma and death. The health effects of low-level carbon monoxide exposure are not clearly defined, but its toxicity is unquestionable. The symptoms of carbon monoxide poisoning, which include nausea, dizziness, confusion, and weakness, may be confused with those of the flu. People with anemia or a history of heart disease can be especially sensitive to carbon monoxide exposure.

Nitrogen dioxide: Concentrated nitrogen dioxide is a dark brown gas with a strong odor. Exposure can cause irritation of the skin and eyes and other mucous membranes. Controlled human exposure studies and epidemiological studies in homes with gas stoves illustrate that, depending on the level of exposure, nitrogen dioxide can alter lung function and cause acute respiratory symptoms. Because of its ability to oxidize, nitrogen dioxide has been shown to damage the lungs directly. Symptoms of exposure may include shortness of breath, chest pains, and a burning sensation or irritation in the chest. People with chronic respiratory illnesses, such as asthma and emphysema, may be especially sensitive to nitrogen dioxide.

Sulfur dioxide: Sulfur dioxide is a colorless gas with a suffocating odor. It is highly soluble in water and is thus readily absorbed by the mucous membranes. Once it is inhaled, sulfur dioxide is dissolved and forms sulfuric acid, sulfurous acid, and bisulfate ions. During normal nasal respiration, sulfur dioxide is absorbed primarily by the nasal tissues; only 1 to 5 percent reaches the lower respiratory tract. However, when a person breathes through the mouth, for example during heavy exercise, significant quantities of sulfur dioxide can penetrate the lower respiratory tract even at low concentrations. The primary physical effect of sulfur dioxide exposure is bronchoconstriction. This begins at considerably lower levels for asthmatics than for healthy individuals. The constriction will develop almost immediately upon exposure, but it will also subside just as quickly when exposure ends. The intensity of the constriction is directly related to the amount of sulfur dioxide per unit of time that reaches the lower respiratory tract, not necessarily the level of the exposure. Also, the effect of sulfur dioxide does not increase with time.

Combustion particulates: Particulates produced by combustion can directly affect lung function. The smaller the particulates, the more deeply they penetrate the lungs and thus the more dangerous they become. The particles can serve as carriers for other contaminants or as mechanical irritants that interact with chemical contaminants.

BIOLOGICAL CONTAMINANTS

Biological contaminants include bacteria, fungi, viruses, algae, insect parts, and dust, which may result in allergenic or pathogenic reactions. There are many sources for these pollutants: pollens from outdoors, viruses and bacteria from humans, and hair and skin flakes from household pets, to name but a few.

Humidity and airflow rates significantly affect the concentrations of biological contaminants. Moisture can act as a breeding ground for molds, bacteria, and mites. Mites are the most prominent cause of house dust allergies. They are found in beds and pillows, especially when humidity levels are high. An indoor moisture level of 30 to 50 percent relative humidity is recommended to maintain good health as well as comfort. Relative humidity is the amount of moisture in the air relative to the amount of moisture the air can hold when it is completely saturated. Biological contaminants may also multiply in standing water, in cooling towers, in water-damaged ceilings, and on surfaces where moisture in the air condenses on cold walls. Additionally, damp organic materials like leather, cotton, furniture stuffing, and carpets can be contaminated with fungi. Airflow rates also have an important effect on the concentrations of airborne biological pollutants. Reduced flow rates tend to provide a favorable medium for molds, dust, and fungi. The HVAC equipment in a building plays a very important role in maintaining proper airflow rates.

HVAC Systems

Proper design of a building's HVAC system is perhaps the most important approach for providing a healthy indoor environment. Conversely, a poorly designed HVAC system can be a harbinger of trouble. The HVAC system provides a means for moving, exchanging, filtering, and conditioning all of the air in a building. Because the HVAC system plays such an important role in IEQ, it is imperative that it be understood and maintained properly. This section describes the advantages offered by an effective HVAC system and the problems caused by a poorly designed system.

SYSTEMS DESIGN

HVAC systems differ greatly from building to building, from simple facilities with a simple forced-air furnace to hospitals with state-of-the-art, computer-controlled, automated systems. In all cases, however, the HVAC system affects IEQ because it moves and conditions air. A typical office building HVAC system is a complex arrangement of equipment, with sources of chilled water and hot water coupled to air handlers. The chilled and hot water can either be generated in the building via chillers and boilers or obtained from a central plant serving a group of buildings. The air handlers are composed of fans, cooling coils, heating coils, filters, and other components arranged in a large container that condition and circulate air through the building. Conditioning means that the air is heated or cooled, cleaned, and humidified if needed to ensure that the desired temperature and humidity conditions in the various building spaces and zones are provided. The total HVAC system consists of one or more air handlers (depending on building size), each of which is responsible for conditioning a specific zone of the building. The HVAC system is responsible for ensuring that the proper quantity of outside ventilation air is provided for the building occupants. The outside ventilation air is probably the most important contribution of the HVAC system to a quality indoor environment. In terms of IAQ, the higher the ventilation rate, the better the air quality in the building. A 2000 study by Wargocki and colleagues showed that a so-called productivity index based on tasks performed by office workers increased as ventilation rates were increased, resulting in a decrease in pollution loads (see Figure 10.1).[14]

CLIMATE CONTROL

Climate control is the general objective of the HVAC system; most likely, it is the reason the system was installed in the first place. Surprisingly though, keeping in mind that the system is designed primarily for climate control, there is often a tremendous amount of dissatisfaction in this area. Before the advent of air conditioning, when opening windows was the only way to help cool a space, building occupants accepted any discomfort as unavoidable.

The state of the air in a space is defined by its psychrometric properties: temperature, relative humidity, enthalpy, and moisture content. Any two properties uniquely define the state of the air. The two most common properties that are controlled by an HVAC system are the temperature and the relative humidity (or moisture content, as the two are different manifestations of the same property). The proper balance of temperature and humidity must be provided by the HVAC system in order to maintain a comfortable indoor environment. Temperatures in the range of 65°F to 78°F (18°C to 26°C) and relative humidity levels between 30 and 60 percent are considered the comfort range for the majority of the population.

The HVAC system must be capable of controlling the supply air in spite of changing conditions in the return of outside sources. For example, if a summer thunderstorm

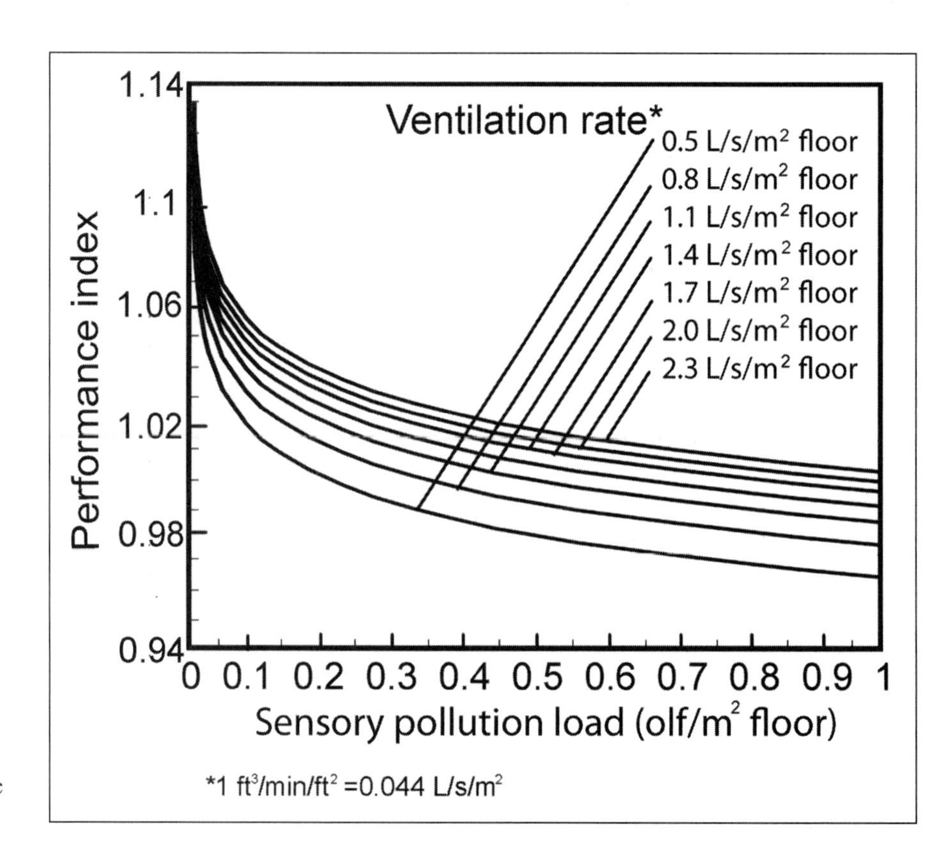

Figure 10.1 The performance of office workers increases (y-axis) as ventilation rates increase (series of curves) for specific pollution loads (x-axis).

saturates the outside air and suddenly increases the humidity level of the air flowing into the HVAC system, the system must be able to adapt and maintain the proper humidity level for the outgoing supply air. Moisture control is a critical yet difficult-to-achieve purpose of the HVAC system. When air is too dry, discomfort is a problem; when air is too moist, discomfort and contaminant generation become problems.

CONTAMINANT GENERATION

The HVAC system can often be the source of several types of airborne contaminants. It is a potential breeding ground for many types of biological contaminants, including molds, spores, and fungi. Certain components of the HVAC can be more easily contaminated than others, particularly porous ductwork linings that are used for insulation and sound control. The HVAC system, because it is responsible for humidification and dehumidification, can also be the cause of uncomfortable humidity levels in the air. High humidity levels help to accelerate the production of biological contaminants. Excess water buildup in the HVAC system, particularly in locations near the evaporator coils or humidifiers, is a major breeding ground for biological contaminants. Water buildup inside components of the HVAC system is sometimes difficult to detect and often expensive to fix. If a well-designed HVAC system is running properly, there should be no excess water buildup.

CONTAMINANT CIRCULATION

When it becomes a circulator of airborne contaminants generated both inside and outside the building, the HVAC system is negatively affecting IAQ. ASHRAE Standard 62.1-2004 defines the requirements for the quantities of outside air ventilation required to remove excess carbon dioxide generated by people. Unfortunately, ventilation sometimes has the effect of introducing new pollutants from the outside. The ASHRAE standard is based on the National Ambient Air Quality Standard (NAAQS) and provides guidelines to follow if the supply of outside air does not meet the standard.

Care must be taken to ensure that the outside air source for a building is not unnecessarily or inadvertently contaminated by an isolated pollutant source. IAQ problems frequently arise when air intakes are positioned near loading docks or other possible sources of pollutants. Air intakes should be located away from exhaust sources, both from automobiles and from other buildings, and they should be high enough above the ground to avoid bringing in ground source contaminants like radon and pesticides.

Interior source contaminant circulation is another major problem in buildings. The most probable cause of unnecessary circulation of internal contaminants is improper zoning. Most buildings have areas designated for specific purposes, and the HVAC system must meet the needs of each area. Some areas, such as laboratories and machine shops, have a greater need for ventilation than others—for example, office space. If the HVAC system is not properly zoned, contaminants from one area may affect the air quality of another area. For example, if an area of a building is turned into a metal shop where welding is performed, but the HVAC system continues to function as if it were an office space, contaminants will be spread to other parts of the building.

Building Materials

All materials have emissions, some more than others, and they all may contribute to deterioration of the air quality. Many health complaints have been linked to new materials installed during the construction or renovation of buildings. "Engineering out" materials known to have an adverse effect on IEQ is perhaps the easiest means of ensuring excellent IAQ. Proper materials selection offers a type of quality control that can save millions of dollars in remediation and lessen legal liability. Table 10.4 lists materials of particular concern that warrant careful selection because of their potential adverse effects on IAQ.[15]

The primary concern with respect to building materials is the types of contaminants they emit. But of additional concern is that some materials act as "sinks" for emissions for other materials or for contaminants that enter the building from other sources. For example, many building materials readily absorb VOCs and rerelease them into the air. In fact, the majority of harmful building material constituents are VOCs, which are typically components of the manufacturing and installation processes. Usually, however, the emission rate will be reduced in proportion to the time the contaminant is exposed to the air.

Due to the increasing awareness of issues related to IAQ, both public agencies and private industry are promoting the use of low-emissions building materials. Communicating IEQ requirements to subcontractors and suppliers is an important step in the process of creating a healthy building, but there is still debate about how to include materials emissions requirements in specifications. The MasterFormat form of specifications developed by the Construction Specifications Institute (CSI) is generally employed to describe the methods and materials of construction. MasterFormat had 16 divisions until 2005, when it was expanded to 50 divisions. Each division covers major aspects or systems of the building; further, each division is divided into sections that cover subsystems. Each section has three parts: Part 1—General, Part 2—Products, and Part 3—Execution. One suggestion for addressing the issue of how to include the required environmental attributes is to expand this three-part for-

TABLE 10.4

Building Material of Particular Concern for IAQ Impacts

Site Preparation and Foundation	
Soil treatment pesticides	
Foundation waterproofing	
Mechanical Systems	
Duct sealants	
External duct insulation	
Internal duct lining	
Building Envelope	
Wood preservatives	Concrete sealers
Curing agents	Caulking and sealants
Glazing compound	Joint fillers
Thermal insulation	Acoustical insulation
Fireproofing materials	
Interior Finishes	
Subfloor or underlayment	Floor or carpet adhesives
Carpet backing or pad	Carpet or resilient flooring
Wall coverings	Adhesives
Paints, stains	Paneling
Partitions	Furnishings
Ceiling tiles	

TABLE 10.5

Chronic RELs for Selected Organic Chemicals Associated with IAQ

Chemical Name	Chemical Abstracts Service (CAS) Number	REL, ppb*	REL μg/cubic meter†
Benzene	71-43-2	20	60
Chloroform	67-66-3	50	300
Ethylene glycol	75-00-3	200	400
Formaldehyde	50-00-0	2	3
Naphthanlene	91-20-3	2	9
Phenol	108-95-2	50	200
Styrene	100-42-5	200	900
Toluene	100-88-3	70	300
Trichloroethylene	79-01-6	100	600
Xylenes	several	200	700

*ppb = parts per billion.
†μg = micrograms.

mat to four parts for each section and to include information on materials emissions requirements and other environmental attributes in the new Part 4—Environmental Attributes.

Another option is to simply introduce an entire section into the general division (Division 1) of the CSI MasterFormat that addresses all the environmental requirements of the project, including materials emissions. This approach is being implemented in California with the creation of Section 01350—Special Environmental Requirements, which includes emissions requirements for materials. Section 01350 covers product selection guidelines, emissions testing protocols, and nontoxic performance standards for cleaning materials.[16] It requires that materials safety data sheets (MSDS) be submitted for each material and that these materials be tested by an acceptable testing laboratory in accordance with American Society for Testing and Materials (ASTM) Standard D5116-97, Guide for Small Scale Environmental Chamber Determination of Organic Emissions from Indoor Materials/Products. Section 01350 also provides information about so-called chemicals of concern, which are carcinogens, reproductive toxicants, and chemicals with an established Chronic Reference Exposure Level (REL). A Chronic REL is an airborne concentration level that would pose no significant health risk for individuals indefinitely exposed to that level. Chronic RELs have been developed for 80 hazardous substances; another 60 chemicals are under review. The modeling of total concentrations of airborne emissions must show that the maximum indoor air concentration of any of the chemicals of concern must not exceed half of the REL level. Table 10.5 lists some of the RELs for common VOCs present in building materials.

ADHESIVES, SEALANTS, AND FINISHES

Adhesives, sealants, caulks, coatings, and finishes are placed in the building when wet and are expected to dry or cure on the premises. The release of VOCs is an inherent part of this process. The solvents used in formulating these materials are the source of most VOCs emitted during drying and later during building occupation.

Adhesives and Sealants

Adhesives are materials or substances that bind one surface to another. They affect a wide range of construction materials; adhesives can be applied with floorings and wall coverings, or they may be a component of a material like plywood, particleboard, movable wall panels, and office workstations. Adhesives are applied in a liquid or viscous state, then cure to a solid or more solid state to achieve bonding. The majority of adhesives release VOCs and pose the greatest threat during their application and curing. When applied, adhesives should be used in areas with increased ventilation (at normal room temperature) for 48 to 72 hours to avoid accumulation of VOCs. The packaging label or other installation information for adhesives should always be consulted for additional product-specific precautions.

One method of characterizing adhesives in terms of their influence on IAQ is to identify the resin used in the base. Resins can be natural or synthetic. Natural resins usually have low emission potential, but in synthetic resins this potential can vary dramatically. Currently, advances are being made in the development of adhesives with no or low emissions.

Sealants are applied to joints, gaps, or cavities to eliminate penetration of liquids, air, and gases. (Note: Though the construction industry differentiates between indoor and outdoor sealants, the former being referred to as *caulks* and the latter as *sealants,* this discussion does not make this distinction.) Sealants are usually selected on the basis of their flexibility and resin base. Like adhesives, sealants can be haz-

TABLE 10.6

Sample VOC Limits on Adhesives South Coast Air Quality Management District Rule 1168*

Architectural Applications	VOC Limit (grams per liter less water)
Indoor carpet adhesives	50
Carpet pad adhesives	50
Wood flooring adhesives	100
Rubber floor adhesives	60
Subfloor adhesives	50
Specialty Applications	**VOC Limit (grams per liter less water)**
PVC welding	510
CPVC welding	490
ABS welding	325
Plastic cement welding	250
Adhesive primer for plastic	550
Substrate-Specific Applications	**VOC Limit (grams per liter less water)**
Metal to metal	30
Plastic foams	50
Porous material (except wood)	50
Wood	30
Fiberglass	80
Sealants	**VOC Limit (grams per liter less water)**
Architectural	250
Nonmembrane roof	300
Roadway	250
Single-ply roof membrane	450

*Through January 7, 2005, amendments.

ardous during installation and curing. Their emission potential is directly related to the percentage of base resins and solids. Sealants, which definitely raise a concern with regard to their VOC emission potential, fortunately are used indoors in small quantities. Alternate water-based sealants manufactured using nontoxic components are now available. One such product for interior use is a vinyl adhesive sealant. An acrylic latex exterior sealant for building joints is also on the market.

The USGBC LEED-NC building assessment standard provides one credit (Materials and Resources Credit 4.1) for the use of low-emission adhesives and sealants. To earn the point associated with this credit, adhesives, sealants, and sealant primers must meet the VOC content limits of the South Coast Air Quality Management District (SCAQMD) Rule 1168 (see Table 10.6).[17] Aerosol adhesives must meet the requirements of the Green Seal Standard for Commercial Adhesives GS-36 requirements as of October 19, 2000.

Finishes

Finishes encompass a wide range of products, including paints, varnishes, stains, and sealers. Finishes are a major component of building materials and furnishings whose primary purpose is to provide protection against corrosion, weathering, and damage. Secondarily, they may also add aesthetic value to building materials. All finishes have similar characteristics. They require resins and oils to form a film and to aid adhesion by promoting penetration into the substrate. All coatings require carriers (water or organic solvents) that provide viscosity for application. Carriers also improve adhesion through evaporation.

Paints and stains require solids, including pigments, to provide various colors. The amount of solids is a good indicator of the VOC emission potential of the finish. Table 10.7 lists the hazardous chemicals associated with particular pigments used in paints.[18] The sanding or burning of finishes generates potential IEQ hazards such as dust from talc, silica, and mica, and especially from lead.

Water-based finishes are typically low-emitting; however, organic solvent-based finishes are more likely to be high-emitting. The current trend is to replace conventional finishes with water-based alternatives, although it is primarily paints that have been targeted in this effort. Very few stains, sealers, and varnishes have been successfully adapted for low VOC emissions, because, to date, alternative finishes generally do not perform as well as their traditional counterparts. The new products often require more applications to achieve results similar to those of traditional products. The color selection of alternative paints is limited as well. And although hypoallergenic, preservative-free paints are available, their shelf life and color selection are limited.

It is also important to point out that water-based products may have low VOCs but contain other hazardous materials. Unlike organic solvent-based paints, water-based paints require preservatives and fungicides such as arsenic disulfide, phenol, copper, and formaldehyde. These additives are considered chemical hazards by the National Institute for Occupational Safety and Health (NIOSH).

TABLE 10.7

Hazardous Chemicals in Pigments

Antimony oxide	Titanium dioxide	Rutile titanium oxide
Cadmium lithopone	Chrome yellow	Molybdate orange
Strontium chromate	Zinc chromate	Phthalocyanine blue
Chrome green	Chromium oxide	Phtalocyanine green
Hydrated chromium oxide	Copper powders	Cuprous oxide

The USGBC LEED-NC building assessment standard provides one credit (Credit 4.2) for the use of low-emission paints and coatings if their VOC emissions do not exceed the VOC and chemical component limits of Green Seal's GS-11 requirements. This standard specifies VOC limits of 150 grams/liter for nonflat interior paints and 50 grams/liter for flat interior paints.[19] There is growing interest in the use of paints with recycled content and in the 2006 Green Seal issued GS-43, Environmental Standard for Recycled Content Latex Paint, setting VOC limits of 250 grams/liter. Although having the environmental attribute of recycled content, paints just meeting this standard would not quality as low-emissions paints under GS-11.[20]

PARTICLEBOARD AND PLYWOOD

Adhesives containing urea formaldehyde (UF) are an integral part of the composition of particleboard and plywood. These materials emit the UF after they have been manufactured and installed in construction. The rate of emission of UF is affected by the temperature and humidity of the installation location.

Particleboard

Particleboard is a composite material made from wood chips or residues, bonded together with adhesives under heat and pressure. Particleboard is relatively inexpensive and is available in 4-foot by 8-foot (1.2 meters by 2 meters), sheets. The major IAQ concern with particleboard is the offgassing of formaldehyde. Most particleboard (about 98 percent) contains UF. The remaining 2 percent contains phenol formaldehyde (PF). Particleboard containing PF emits far less formaldehyde than board made with UF. PF is used in particleboard where a high-moisture environment is anticipated, specifically restrooms and kitchens.

The most common construction application of particleboard is as a core material for doors, cabinets, and a wide variety of furnishings, such as tables and prefabricated wall systems. Particleboard is also used in wood-framed housing, primarily for nonstructural floor underlayment. Usually, a finished floor is installed over particleboard. Particleboard is also used as a backing for paneling. Once the board is covered, the VOC content is inconsequential because the formaldehyde emissions are delayed for as long as it remains covered.

Particleboard, though it can now be manufactured with lower formaldehyde emissions, is still of great concern because of the possibility of large exposed surface areas in relation to the volume of a given space. Emissions of trace amounts of formaldehyde can continue for several months or even years. These emissions do decrease over time, but rates increase as temperature and/or humidity rises. It is estimated that emission rates double with every increase of 12°F (7°C).

Plywood

Plywood is composed of several thin wood layers oriented at alternating 90° angles that are permanently bonded by an adhesive. The exterior plies are referred to as *faces,* and the interior plies are known as the *core.* Plywood is generally classified as hardwood or softwood. Approximately 80 percent of all softwood plywood is used as wall and roof sheathing, siding, concrete framework, roof decking, and subflooring. Hardwood plywood is used for building furniture, cabinets, shelving, and interior paneling.

The type of adhesive used to bond the plies plays a major role in assessing the effects of the plywood on IAQ. The surface area of the plywood in relation to the volume of the space is another determining factor for proper IAQ. Interior-grade plywood is generally bonded with UF resins. The offgassing of UF in plywood can be compounded by finishes or sealants used in conjunction with the plywood. Size, temperature, humidity of the space, surface area, and finish of the plywood all can affect the concentration of formaldehyde emissions.

FLOOR AND WALL COVERINGS

Carpet, resilient flooring, and wall coverings may have VOC-emitting components and may use adhesives that emit VOCs as part of their installation process. New products with zero or low emissions are, fortunately, entering the marketplace to serve the green building industry. As competition and demand increase, the quality of the products will also improve; at the same time, the price will decrease, making these new products very competitive with conventional products.

Carpet

Of all building materials, carpet has generated the most debate, which is ironic considering that emissions from carpet systems are relatively low compared with emissions from other building materials. The majority of the emissions associated with carpeting are actually due to the adhesives used to secure it. Thus, when selecting a carpet, the entire system and the emissions of each constituent must be evaluated. The components of a carpet system are the carpet fiber, carpet backing, adhesive, and a carpet pad (generally used in residential applications only).

- *Carpet backing* is used to hold the fibers in place. Often two backings are used: one keeps the fibers in place, and the other adds strength and stability. The secondary backing is made from fabric, jute, or polypropylene bonded with either styrene-butadiene rubber (SBR) latex or a polymer coating such as synthetic latex. SBR latex contains the chemicals styrene and butadiene, which are known irritants to mucous membranes and skin. SBR latex adhesives are found in primary and secondary backings and emit low but steady amounts of the by-product 4-phenylcyclohexene (4-PC), the chemical that is responsible for the "new carpet" smell and is suspected of being a possible source of building occupant illness complaints.
- *Adhesives* may be used twice in common carpet systems: to glue the backing to the fiber and/or to glue the carpet system to the substrate.
- *Carpet pads* are an optional part of the carpet system. They generally do not contribute to IAQ problems. There are five basic types of pads: bonded urethane, prime polyurethane, sponge rubber, synthetic fiber, and rubberized jute.

There are five basic carpet fiber materials. Wool is the only natural fiber, and it accounts for less than 1 percent of the carpet market. The remaining four—nylon, olefin, polyester, and polyethylene terephthalate—are synthetic fibers. Derived from petrochemicals, synthetic fibers are stronger, more durable, and usually less expensive; they are also less likely than wool to release small fibers into the air.

Both the USGBC LEED-NC and Green Globes building assessment standards provide credit (Indoor Environmental Quality Credit 4.3) for using low-emissions carpeting systems if the system meets or exceeds the requirements of the Carpet and Rug Institute's Green Label Plus Program.[21]

Resilient Flooring

Resilient flooring is a pliable or flexible flooring. Tile and sheet are the two basic forms, both of which are attached to a substrate using adhesives. Resilient flooring can be composed of vinyl, rubber, or linoleum. Vinyl flooring is made primarily of polyvinyl chloride (PVC) resins, with plasticizers, to provide flexibility; fillers; and pigments for color. Rubber flooring comes in two basic forms—smooth surface or molded—and is made from a combination of synthetic rubber (styrene butadiene), nonfading organic pigments, extenders, oil plasticizers, and mineral fillers. Linoleum is a natural, organic, and biodegradable product. Its main components are linseed oil, pine rosin, wood flour, cork powder, pigments, driers, and natural mildew inhibitors. Linoleum tiles are durable, greaseproof, and water- and fire-resistant. They are also easily maintained and long-lasting.

Typically, no individual compound in resilient flooring has high VOC emissions. The plasticizers are the main source of emissions. Using a more rigid, less plastic tile is recommended to avoid potential hazards. Note, however, that low-emitting tiles may be glued with high-emitting adhesives.

Wall Coverings

Wall coverings are a popular alternative to paints. The majority of available coverings pose little or no threat to IAQ. The three basic types of wall coverings are paper, fabric, and vinyl. Paper itself has no impact on IAQ, but the adhesives used to apply it may contain formaldehyde. However, the majority of paper adhesives are purchased as a powder and mixed with water; thus, they emit little or no VOCs.

Fabric wall coverings may contain formaldehyde, which is sometimes used to keep the material from fading and to improve resistance to water. Fabric coverings can also act as a sink by absorbing extraneous VOCs in a building and reemitting them into a space. Two major concerns with vinyl wall coverings are the environmental conditions of the project location and the construction of the walls receiving the finish. In temperate climates, when moisture may not be readily evaporated, vinyl covered walls can become moldy.

INSULATION AND CEILING TILES

Insulation and acoustical ceiling tiles can contribute VOC and particulate contaminants from a variety of sources. Depending on their composition, these materials may incorporate a variety of adhesives and fibrous materials that can combine to complicate the IAQ issue.

Insulation

Most insulation is made of fiberglass, mineral wool, and cellulose (made from recycled wood). Asbestos was also used frequently until the late 1970s. Fiberglass and mineral wool have raised IAQ concerns because of the small fibers that are produced when the material is disturbed. Fiberglass is listed by the International Agency for Research for Cancer as a possible carcinogen. Cellulose insulation is generally spray-applied and is considered a nontoxic material. In this materials category, foam insulation has received most of the attention because of its impacts on the environment rather than on IAQ. That said, VOCs are emitted from synthetic foam during manufacturing or while spray foam is used.

Acoustical Ceiling Tile

The suspended acoustical ceiling is one of the most common structures found in commercial buildings today. Most acoustical ceiling tile (ACT) is made from mineral or wood fibers, which are wetted and compressed to the desired thickness, size, and pattern. They are usually coated with a latex paint at the factory. The primary concern regarding the effects of ACTs on IAQ is the occurrence of microbial growth on either mineral fiber or fiberglass tile exposed to moisture. Another concern is that porous tiles can absorb VOCs and reemit them.

Best Practices for IAQ

Hal Levin, a prominent figure in the green building movement, especially with respect to IAQ issues, suggests 10 points as best practice concepts for creating good IAQ for high-performance green buildings. They are listed in Table 10.8 and discussed more fully in the subsections to follow.[22]

BEST IAQ PRACTICE CONCEPTS 1 TO 3

The first three best IAQ practice concepts are preparatory to the actual consideration of how to handle the sources of pollution. Understanding the role of ventilation in the movement and dilution of air pollution is essential to designing an appropriate building ventilation system. The impact of a specific chemical on human health is a function of its dose or concentration and the duration of the exposure. The higher the concentration or the greater the exposure, the greater the potential damage. In order to provide excellent building IAQ, the entire life cycle of the project, from design through construction and operation, must be factored into the process.

TABLE 10.8

Best IAQ Practice Concepts for High-Performance Green Buildings

1. Relationships between indoor air pollution sources, ventilation, and concentrations
2. Simple dose-response basis for health effects: "the dose makes the poison"
3. Overall design consideration of IAQ: from cradle to grave
4. Source identification
5. Source control options and strategies
6. Ventilation system design and operation
7. Material selection and specification
8. Construction procedures
9. Maintenance and operation
10. Change of use, renovation, adaptive reuse, and demounting

BEST IAQ PRACTICE CONCEPTS 4 TO 6

Potential air pollution sources are numerous and varied: outdoor sources such as water and pesticides; emissions from building materials, especially finishes such as paint and carpeting, but also including adhesives, glues, mastics, and acoustic materials; occupant activities; and HVAC equipment. Identifying these sources is important for deciding on strategies for coping with them. Source control is a function of the origin and type of the pollutant, as its measures include isolation of the source, filtering and cleaning of fresh air, and diluting pollutants with outside fresh air. The design of the building ventilation system has to account for major sources of emissions—for example, from kitchens where exhaust fans and hoods should remove pollutants as they are generated rather than allowing them to disperse into the building. Fresh air ventilation rates need to be designed to address the range of expected emissions in the occupied spaces, especially those generated by the occupants. *Ventilation effectiveness* is a measure of how well the ventilation system succeeds in bringing conditioned supply air to the breathing zone of the occupants.

BEST IAQ PRACTICE CONCEPTS 7 AND 8

The level of construction materials emissions will be a function of the type and quantity of materials that will be used in a project. An estimate of these two factors is important in forecasting the likely level of emissions that will occur. Especially important are materials that will be emitting into interior spaces, especially wet materials such as adhesives, paints, mastics, glues, and others that emit VOCs as part of the drying and curing process. Fortunately, a wide range of products are coming to the marketplace to address the need for low- and zero-emissions materials.

How the construction process is conducted is also important to the success of the IAQ strategy. Materials should be stored to prevent the introduction of moisture or the accumulation of dust, particulates, and other contamination on nonporous surfaces such as ductwork or their absorption by porous materials. Building commissioning plays into the IAQ strategy because a properly functioning and balanced HVAC system is essential to the success of the strategy.

BEST IAQ PRACTICE CONCEPTS 9 AND 10

The operations phase of the building shows whether or not all the planning, design, and construction measures addressing IAQ have been successful. The hoped-for outcome is indoor air that has very low levels of chemical, biological, and radiological pollutants and that creates a healthy, productive environment for the building occupants. Proper maintenance of HVAC systems is necessary to maintain good IAQ; attention must also be paid to how the building is cleaned to prevent the introduction of harmful chemicals into the indoor environment. Alterations and renovations of the building should be carried out with the same level of care as was the construction process to ensure that high-level IAQ is maintained in spite of changes to the building.

Managing IEQ During Construction

Among the most notable positive outcomes of the high-performance green building movement has been the recognition of, and attention to, the possible impacts of construction activities on the building's IAQ. The areas of concern connected to the construction process are (1) storage of materials to prevent moisture and contaminant exposure; (2) protection of HVAC system components prior to installation; and (3) protection of installed HVAC systems from contamination during construction.

BUILDING MATERIALS STORAGE

Materials that have been exposed to moisture and contamination while being stored during construction can negatively affect IAQ. On the construction site are numerous types of construction work–related airborne contaminants (dust, fibers, VOCs, combustion products, and biological agents). These result from handling or disturbing the materials (fibers and particulates from drywall, ceiling tiles, and flooring, to name a few), from offgassing (from carpeting, paint, adhesives, glues, mastics, and varnishes), and from equipment used during construction (combustion by-products, fuel fumes, and particulates from engines, motors, compressors, and welders). In addition, the storage site may be exposed to rain and moisture, which can degrade the materials, especially those that tend to absorb moisture—for example, drywall, insulation, and ceiling tile.

These contaminants and moisture can severely compromise IAQ if measures are not implemented to prevent contamination of building materials during storage. An isolated, protected storage area, with adequate protection for all materials, will not only bring IAQ benefits but will also have a positive effect on the project's bottom line because far fewer materials will be damaged. Finally, materials should be checked for moisture infiltration prior to their installation to verify that they do not contain excessive levels of water that may contribute to mold and mildew in the building.

Figure 10.2 Construction products and materials such as the gypsum board in this project should be kept dry, covered, and off the ground to prevent IAQ problems. (Photograph courtesy of DPR Construction, Inc.)

Protecting HVAC Systems Prior to Installation

It is now considered best practice to ensure that the ductwork and other pieces of equipment that move air through the HVAC system are sealed and covered from the time they arrive on-site to the time they are installed as part of a working system. The ends of ductwork should be sealed with plastic and taped to prevent moisture

(A)

(B)

Figure 10.3 (A) Ductwork should be protected during storage and prior to installation. (B) Openings should be sealed during the installation process to prevent contamination. (Photographs courtesy of DPR Construction, Inc.)

and contamination from entering it. The ductwork and HVAC components (air handlers, registers, grilles, and other products) should be stored in a clean, dry place prior to use. When the HVAC system is started during the final stages of the construction process, filter media should be installed on all return air intakes to prevent contamination caused by dust, paint, and adhesive particles from entering the air distribution system.

Connection to LEED-NC

The LEED-NC category of Indoor Environmental Quality (EQ) has two prerequisites and eight credits that provide a maximum of 15 points. Only the Energy and Atmosphere category carries more weight, with a maximum of 35 points. Table 10.9 gives an overview of the EQ structure of LEED-NC. Note that the descriptions of the LEED-NC credits and points in this section are abbreviated and that the appropriate LEED Reference Manual should be consulted for detailed information about how to achieve these points.

EQ PREREQUISITE 1 (EQp1): MINIMUM IAQ PERFORMANCE

This prerequisite establishes that a minimal level of IAQ performance must be demonstrable by meeting the requirements of ASHRAE Standard 62.1-2007, Ventilation for Acceptable Indoor Air Quality, and mechanical ventilation systems must be designed using either the Ventilation Rate Procedure or local codes, whichever is more stringent.

TABLE 10.9

EQ Credits and Points Under LEED-NC 2009

Prerequisite/Credit	Name of Prerequisite/Credit	Maximum Points
EQ Prerequisite 1	Minimum IAQ Performance	NA
EQ Prerequisite 2	Environmental Tobacco Smoke (ETS) Control	NA
EQ Credit 1	Outdoor Air Delivery Monitoring	1
EQ Credit 2	Increased Ventilation	1
EQ Credit 3.1	Construction IAQ Plan—During Construction	1
EQ Credit 3.2	Construction IAQ Plan—Before Occupancy	1
EQ Credit 4.1	Low-Emitting Materials—Adhesives and Sealants	1
EQ Credit 4.2	Low-Emitting Materials—Paints & Coatings	1
EQ Credit 4.3	Low-Emitting Materials—Flooring Systems	1
EQ Credit 4.4	Low-Emitting Materials—Composite Wood & Agrifiber Products	1
EQ Credit 5	Indoor Chemical and Pollutant Source Control	1
EQ Credit 6.1	Controllability of Systems: Lighting	1
EQ Credit 6.2	Controllability of Systems: Thermal Comfort	1
EQ Credit 7.1	Thermal Comfort: Design	1
EQ Credit 7.2	Thermal Comfort: Verification	1
EQ Credit 8.1	Daylight and Views, Daylight 75% of Spaces	1
EQ Credit 8.2	Daylight and Views, Views for 90% of Spaces	1
	Total EQ Points Available	**15**

EQ PREREQUISITE 2 (EQp2): ENVIRONMENTAL TOBACCO SMOKE (ETS) CONTROL

This prerequisite provides three options for compliance:

Option 1: Prohibit smoking in the building and locate designated external smoking areas at least 25 feet (7.6 meters) from entries, operable windows, and air intakes.

Option 2: Same as Option 1 *and,* if there is a designated smoking room, the design must ensure no ETS infiltration into the rest of the building.

Option 3: Same as Option 1 but applies to residential building. Smoking is to be prohibited in common areas; no transmission of ETS between units must be ensured; and all doors to interior corridors must be weatherstripped to minimize air leakage into corridors.

EQ CREDIT 1 (EQc1): OUTDOOR AIR DELIVERY MONITORING (1 POINT MAXIMUM)

Ventilation rates are positively correlated with good IAQ, and this credit addresses the issue of monitoring the ventilation rate to ensure occupant comfort and well-being. For the point associated with this credit to be earned, permanent monitoring equipment is required to ensure that the design ventilation rate is maintained. For mechanically ventilated spaces, carbon dioxide concentrations for normally occupied spaces must be measured at a level 3 to 6 feet (0.9 to 1.8 meters) above the floor and each mechanical ventilation system serving nondensely occupied spaces (less than 25 people per 1000 square feet (93 square meters)) must be equipped with an airflow measuring device to ensure that ventilation is being provided in accordance with the requirements of ASHRAE Standard 62.1-2007.

EQ CREDIT 2 (EQc2): INCREASED VENTILATION (1 POINT MAXIMUM)

Increasing the ventilation rate can improve IAQ. Increasing the ventilation rate to at least 30 percent above ASHRAE Standard 62.1-2007 for mechanically ventilated spaces is required to earn the point associated with this credit. Natural ventilation systems can also be used to earn this credit, provided that they meet the recommendations in the Carbon Trust Good Practice Guide 237.

EQ CREDIT (EQc3): CONSTRUCTION IAQ MANAGEMENT PLAN (2 POINTS MAXIMUM)

This credit has two distinct phases, each of which can provide 1 point if properly accomplished and documented.

EQ Credit 3.1 (EQc3.1) addresses measures employed during the construction process that enhance air quality for the eventual building occupants. The requirements are that absorptive materials must be protected from moisture damage; filters with a Minimum Efficiency Reporting Value (MERV) of 8 must be used on return air grilles if the HVAC system is used during construction; all filters must be replaced prior to occupancy; and, during installation of the HVAC system, the approaches recommended in the SMACNA IAQ "Guidelines for Occupied Buildings under Construction" (1995) must be followed.

EQ Credit 3.2 (EQc3.2) addresses the period of time between the end of construction and building occupancy and has two options.

Option 1: Prior to occupancy and with all interior finishing installed, the building must be flushed out with 14,000 cubic feet (396 cubic meters) of outside air per square foot of floor area The internal temperature must be at least 60°F, (16°C) with a relative humidity no higher than 60 percent. Alternatively, if occupancy is desired prior to completion of flush-out, the building must be flushed out with 3,500 cubic feet (99 cubic meters) of outdoor air per square foot prior to occupancy and then ventilated at the rate of 0.30 cubic feet (0.01 cubic meters) per minute of outside air, or the design minimum from EAp1, whichever is greater.

Option 2: Conduct baseline IAQ testing after construction ends and prior to occupancy using the USEPA *Compendium of Methods for the Determination of Air Pollutants in Indoor Air.*

EQ CREDIT 4 (EQc4): LOW-EMITTING MATERIALS (4 POINTS MAXIMUM)

Reducing the VOCs emitted by building materials is addressed by this credit, which is subdivided into four parts that address, respectively, adhesives and sealants (EQc4.1); paints and coatings (EQc4.2); flooring systems (EQc4.3); and composite wood and agrifiber products (EQc4.4). Each of these credits carries 1 point if the process is successful and documented.

EQc4.1: Adhesives, sealants, and sealant primers must comply with the VOC content limits of the South Coast Air Quality Management District (SCAQMD) Rule 1168. Aerosol adhesives must comply with the *Green Seal Standard for Commercial Adhesives GS-36* requirements.

EQc4.2: Architectural paints, coatings, and primers used in the interior of the building must show they meet the requirements of Green Seal Standard GS-11 with flat paints having no more than 50 grams per liter of VOCs and nonflat paints not exceeding 150 grams per liter. Anti-corrosive and anti-rust paints applied to interior metal surfaces should not exceed 250 grams per liter of total VOC content as stated in Green Seal Standard GC-03, Anti-Corrosive Paints. Similarly, clear wood finishes should not exceed 350 grams per liter for varnish and 550 grams per liter for lacquer; floor coatings must not exceed 100 grams per liter; stains less than 250 grams per liter; and sealers less than 275 grams per liter; and stains must contain less than 250 grams per liter of total VOCs.

EQc4.3: All carpet systems must meet the requirements of the Carpet and Rug Institute's Green Label Indoor Air Plus program, and carpet adhesives must contain no more than 50 grams per liter of total VOCs.

EQc4.4: Composite wood and agrifiber products must contain no UF resins. This also applies to the laminating adhesives used to fabricate composite wood and agrifiber assemblies.

EQ CREDIT 5 (EQc5): INDOOR CHEMICAL AND POLLUTANT SOURCE CONTROL (1 POINT MAXIMUM)

Designing the building to minimize the entry of pollutants into occupied spaces is the purpose of this credit. First, it calls for the employment of devices such as grilles and grates at building entrances to prevent the entrance of dirt, pesticides, and other materials into the building. Second, when hazardous gases and chemicals may be present or used, a segregated area with a dedicated exhaust system must be provided. Third, in mechanically ventilated buildings, MERV 13 filters or better should be installed prior to occupancy to process both return and outside air.

EQ CREDIT 6 (EQc6): CONTROLLABILITY OF SYSTEMS (2 POINTS MAXIMUM)

The ability of the building occupants to control their lighting conditions and thermal comfort has emerged as an important issue in providing a high-quality indoor environment. This credit comprises two subcredits, each carrying 1 point maximum: EQc6.1 for lighting and EQc6.2 for thermal comfort.

EQc6.1 Lighting: The design must allow at least 90 percent of the occupants to adjust the lighting for their tasks and preferences. For multioccupant spaces, provide lighting system controllability to suit individual tasks and needs.

EQc6.2 Thermal Comfort: The design must allow at least 50 percent of the occupants to adjust the temperature to suit their needs. Operable windows provided within 20 feet (6.1 meters) of occupants or 10 feet (3 meters) to either side can be used in lieu of controls. For shared multioccupant spaces, controls are to be provided that allow conditions to be adjusted for the group's needs. ASHRAE Standard 55-2004, Thermal Environmental Conditions for Human Occupancy, defines conditions for thermal comfort including air temperature, air speed, humidity, and radiant temperature.

EQ CREDIT 7 (EQc7): THERMAL COMFORT (2 POINTS MAXIMUM)

Thermal comfort is an important component of IEQ, and the project can acquire 2 points for demonstrating that the criteria for this measure have been met. EQ Credit 7 has two subcredits as follows, each carrying 1 possible point:

EQc7.1 Design: The project must meet the requirements of ASHRAE Standard 55-2004 with respect to the design of the building envelope and HVAC systems.

EQc7.2 Verification: The owner must agree to perform a thermal comfort survey of the occupants 6 to 18 months after occupancy. A plan of corrective action is required if more than 20 percent of the occupants are dissatisfied with the building's thermal comfort.

EQ CREDIT 8 (EQc8): DAYLIGHT AND VIEWS (2 POINTS MAXIMUM)

The importance of views to the outside is acknowledged by this credit, which is divided into two subcredits, each carrying 1 possible point:

EQc8.1 Daylight for 75 Percent of Spaces: There are three options for this credit.

Option 1: Requires that 75 percent of all spaces regularly occupied for critical visual tasks achieve a minimum Glazing Factor of 2 percent. Glazing Factor is the ratio of external illumination to internal illumination expressed as a percentage.

Option 2: Show through a computer simulation model that a minimum daylight illumination level of 25 footcandles is achieved in at least 75 percent of all regularly occupied areas.

Option 3: Use direct indoor light measurement to show that a minimum daylight level of 25 footcandles is available for 75 percent of all regularly occupied areas.

EQc8.2 View for 90 Percent of Spaces: Requires direct line of sight for occupants to visible glazing in 90 percent of all regularly occupied spaces.

TABLE 10.10

Outline of Green Globes v.1, Part G

Section	Description	Points
G.1	**Effective Ventilation System**	**60**
G.1.1	Features to avoid entraining pollutants in the ventilation air intakes	11
G.1.2	Ventilation meets ASHRAE 62.1-2004 requirements	10
G.1.3	Effective air exchange	15
G.1.4	IAQ monitoring	7
G.1.5	Construction Indoor Air Quality Management Plan	11
G.1.6	Efficient air filtering	6
G.2	**Source Control of Indoor Pollutants**	**45**
G.2.1	Measures to control moisture and prevent biological contaminant growth	4
G.2.2	Access to air handling units for inspection and maintenance	5
G.2.3	Humidification system features to prevent biological contaminant growth	6
G.2.4	Carbon dioxide monitoring in parking garages and areas where there is combustion	5
G.2.5	Separate ventilation system or physical isolation of indoor air pollution at source for areas with printing, copying, and other similar activities	3
G.2.6	Separate ventilation system designed with negative pressure (applies to G.2.5)	2
G.2.7	Measures to minimize microbial contamination from cooling towers	5
G.2.8	Measures to minimize microbial contamination of the domestic water system	5
G.2.9	Low-VOC-emitting materials	10
G.3	**Lighting Design and Integration of Lighting Systems**	**45**
G.3.1	Extent of daylighting	10
G.3.2	View to building exterior or atria	10
G.3.3	Solar shading devices	5
G.3.4	Lighting levels in accordance with IESNA *Lighting Handbook* (2000)	10
G.3.5	Measures to avoid excessive glare	10
G.4	**Thermal Comfort**	**25**
G.4.1	Thermal comfort in accordance with ASHRAE 55-2004 or per an occupant satisfaction survey	20
G.4.2	Size of thermal comfort zones	5
G.5	**Acoustic Comfort**	**25**
G.5.1	Sound levels below 65 dB at the property line	5
G.5.2	Appropriate Sound Transmission Class (STC) levels for the building envelope	5
G.5.3	Noise attenuation in the structure and insulation of primary spaces from impact noise	5
G.5.4	Interior design achieves appropriate ambient noise levels	5
G.5.5	Measures to mitigate mechanical and plumbing system noise	5
	Total Points Available	**200**

Connection to Green Globes v.1

Green Globes v.1 addresses IEQ in Part G, Indoor Environment, which is outlined in Table 10.10.

Summary and Conclusions

IEQ is perhaps the most important human-related issue of green building, as it directly affects the health of the building occupants. Although IEQ covers a wide range of effects, LEED-NC focuses on IAQ, with far less emphasis on noise and lighting quality. Green Globes does address a wider range of IEQ issues such as acoustic comfort and lighting quality, a definite step forward in the evolution of green building rating tools. As a consequence of relatively recent efforts to address building health, a number of new products have emerged, among them paints, carpets, adhesives, furniture, and wood products for millwork and cabinetry, which have zero or low emissions. Furthermore, greater attention is being paid to the proper sizing of HVAC equipment and control of humidity in spaces. The important issue of moisture infiltration and the consequent problems caused by mold and mildew growth are also being addressed by appropriate architectural detailing and the proper design of the building's air distribution system. Daylighting is receiving increased emphasis because of its demonstrated health benefits and its contribution to reductions in energy consumption. Providing exterior views to the building occupants to enhance their well-being is also a component of IEQ, which both the LEED and Green Globes rating systems acknowledge by allocating points for providing exceptional views.

Future versions of LEED should consider increasing the importance of quiet, relatively noise-free building systems as an aspect of an important health issue. And to cover the full array of IEQ issues, lighting quality should receive additional focus and consideration.

Notes

1. From Fisk and Rosenfeld (1998).
2. The updated information is contained in Fisk (May 2002).
3. Cited in Kats (2003), Executive Summary.
4. Portions of this chapter are based on an unpublished manuscript on sustainable construction by Charles Kibert and G. Bradley Guy.
5. As defined in Hays, Gobbell, and Ganick (1995).
6. See Bass (1993).
7. These are estimated productivity losses quoted by Mary Beth Smuts, a toxicologist with the USEPA, in Zabarsky (2002).
8. Excerpted from "Building Air Quality" (1991).
9. See Hays, Gobbell, and Ganick (1995) and Bass (1993).
10. Excerpted from Levin (1999).
11. As described in "IAQ Guidelines for Occupied Buildings Under Construction" (1995).
12. The USEPA's Radon Page is at www.epa.gov/iaq/radon.
13. The combustion by-product information is summarized from Meckler (1991).
14. Excerpted from Wargocki, Wyon, and Fanger (2000).
15. Adapted from Hansen (1991).
16. The latest version of Section 01350 can be found on the California Integrated Waste Management Board (CIWMB) website, www.ciwmb.ca.gov/GreenBuilding/Specs/Section 01350.

17. The latest version of the SCAQMD Rule 1168 can be found at www.arb.ca.gov/DRDB/SC/CURHTML/R1168.PDF.
18. See Hays, Gobbell, and Ganick (1995).
19. The Green Seal GS-11 Standard can be found at the Green Seal website, www.greenseal.org/certification/standards/paints.cfm.
20. The Green Seal Environmental Standard for Recycled Content Latex Paint (August 2006) can be found at www.greenseal.org/newsroom/GS-43_Recycled_Content_Latex_Paint.pdf.
21. The criteria for the Green Label Carpet Testing Program can be found at www.carpet-rug.com/drill_down.cfm?page=5.
22. Hal Levin posted "Ten Basic Concepts for Architects and Other Building Designers: Best Sustainable Indoor Air Quality Practices for Commercial Buildings" on the BuildingGreen, Inc., website, www.buildinggreen.com.

References

Bass, Ed. *Indoor Air Quality in the Building Environment.* 1993. Troy, MI: Business News Publishing.

"Building Air Quality: A Guide for Building Owners and Facility Managers." 1991. Washington, DC: U.S. Environmental Protection Agency, EPA/400/1-91/033.

"Duct Cleanliness for New Construction." 2000. Chantilly, VA: Sheet Metal and Air Conditioning Contractors National Association (SMACNA).

Fisk, W.J. May 2002. "How IEQ Affects Health, Productivity," *ASHRAE Journal,* 44(5), pp. 56, 58–60.

Fisk, W.J., and A.H. Rosenfeld. 1998. "Potential Nationwide Improvements in Productivity and Health from Better Indoor Environments," in *Proceedings of ACEEE Summer Study '98,* 8, pp. 85–97.

Hansen, Shirley. 1991. *Managing Indoor Air Quality.* Liliburn, GA: Fairmont Press.

Hays, S.M., R.V. Gobbell, and N.R. Ganick. 1995. *Indoor Air Quality Solutions and Strategies.* New York: McGraw Hill.

Hennessey, John F., III. July 1992. "How to Solve Indoor Air Quality Problems," *Building Operating Management,* 39(7), pp. 24–28.

IAQ Guidelines for Occupied Buildings Under Construction. 1995. Chantilly, VA: Sheet Metal and Air Conditioning Contractors National Association (SMACNA).

Indoor Air Quality—A Systems Approach. 1998. Chantilly, VA: Sheet Metal and Air Conditioning Contractors National Association (SMACNA).

Kats, Gregory H. October 2003. "The Costs and Financial Benefits of Green Buildings." A report developed for California's Sustainable Building Task Force. Available at the Capital E website, www.cap-e.com.

Levin, Hal. November 1999. "Commercial Building Indoor Air Quality." Prepared for the Northeast Energy Efficiency Partnerships, Inc. Available at www.newbuildings.org/downloads/papers/IAQNEEP_Final.pdf.

Meckler, M., ed. 1991. *Indoor Air Quality Design Guidebook.* Lilburn, GA: Fairmont Press.

Nero, A.V., Jr. May 1988. "Controlling Indoor Air Pollution," *Scientific American,* 258(5), pp. 42–48.

Wargocki, P., D.P. Wyon, and P.O. Fanger. 2000. "Productivity Is Affected by the Air Quality in Offices," in *Healthy Buildings, Volume 1,* pp. 635–640.

Zabarsky, Marsha. August 16, 2002. "Sick-Building Syndrome Gains a Growing Level of National Awareness." *Boston Business Journal.* Available at www.bizjournals.com/boston/stories/2002/08/19/focus9.html.

Part III

Green Building Implementation

Part II provided an overview of the major systems of a green high-performance building: land and landscape, energy, water, materials, and IEQ. Proper design of these systems is the starting point for green building. But without careful execution of the construction phase of the project and thorough commissioning of the finished building, a green building project is incomplete. Part III of this book addresses these two important aspects of a project and how they fit into the overall green building process. In addition, this part covers the economics of green building and offers an overview of the possible life justifications for green buildings, including energy savings, water and wastewater savings, the benefits of commissioning, operations and maintenance savings, and other approaches to addressing the economics of green buildings. This part concludes with an overview of the future of green building and the variety of directions in which this movement may evolve.

Chapter 11, "Construction Operations," and Chapter 12, "Building Commissioning," elaborate on two major aspects of green building that are not covered separately in LEED-NC 2009 or Green Globes but that warrant additional consideration. The construction managers or general contractors who actually execute the design must be made clearly aware of their responsibilities. Therefore, the importance of developing a Site Protection Plan, a Health and Safety Plan, and a Construction and Demolition Waste Management Plan is addressed in Chapter 11. Each plan is an extension, or elaboration, of current building assessment system requirements. The Site Protection Plan includes the Erosion and Sedimentation Control Plan requirements found in the Sustainable Sites category of LEED and the Site category of Green Globes and includes other measures designed to protect the biological and physical integrity of the site. The Health and Safety Plan elaborates on issues during the construction phase and IAQ requirements, and includes additional measures designed to protect the workforce and the building's future occupants. The Construction and Demolition Waste Management Plan is addressed in the Materials Selection category of LEED-NC, which was described in Chapter 9.

Finally, building commissioning, which has emerged as a key step in the third-party certification of high-performance buildings, is thoroughly explored in Chapter 12. The building commissioning process continues to evolve, from its original role of testing and balancing HVAC systems to a more complete check of all building systems, including, for example, building finishes, to ensure that the owner receives the exact building called for in the design. Commissioning is becoming a service that occurs throughout the entire project, from the onset of design, rather than only at the completion of construction. Initial economic analyses of high-performance green buildings indicate that the savings due to building commissioning are truly staggering, even outstripping the financial benefits of energy savings. This is a remarkable outcome, and if future analyses were to confirm this result, these findings would transform a number of fundamental assumptions about buildings. For example, if the savings from commissioning were so marked at the onset of building operation, ongoing commissioning would also have notable benefits.

Economic analysis of green buildings is addressed in Chapter 13. LCC is the key tool for justifying the decisions to create a high-performance building. Initial studies indicate that the added costs for a LEED-NC new building are about 2 percent for a Silver or Gold certification and that total 20-year savings, using conservative financial assumptions, are on the order of $50 to almost $70 per square foot ($500 to $700 per square meter) for an initial additional investment of about $4 per square foot ($40 per square meter) for a $140-per-square-foot ($1,400-per-square-meter) base building construction cost. Some studies report a 1-year simple payback for a green building when all savings are included—energy, water, emissions, and health/productivity benefits.

The future of green building is covered in Chapter 14, the final chapter of this book. LEED, as might be expected, pushes green building in a given direction because the point system for achieving the various levels of certification, although generally performance based, tends to result in a fairly limited range of outcomes. At present, only a few attempts are being made to define the "ultimate" green buildings, those that will emerge in 20 years or more. Thus, the purpose of this chapter is to attempt to remedy this oversight. To that end, three potential future strategies are described: one based on technology, a second on vernacular architecture, and a third on biomimetic models. No one of these is likely to provide the long-range solution; instead, most likely, a synthesis of the key ideas in these three strategies will be the outcome. Future versions of LEED will ideally pave the way for green building and raise the bar for everyone engaged in this movement, from owners to materials suppliers, designers, and builders.

Chapter 11

Construction Operations

The role of the construction team in executing a green building project, in making it a reality, is extremely important and should not be underestimated. A construction management firm or general contracting firm that orients its employees and its subcontractors to the purpose of the project can make an enormous difference in the overall outcome. Several areas are specifically identified in the LEED-NC and Green Globes building assessment standards as being in the realm of the construction team: construction waste management, erosion and sedimentation control, limiting the footprint of construction operations, and construction IAQ. That said, the construction team may make many other important contributions to green building projects that are not specifically covered by building assessment standards. Examples of innovative major contributions include:

- Improving materials handling and storage to reduce construction waste
- Recycling site materials such as topsoil, limerock, asphalt, and concrete into the new building project
- Making provisions for installing products and materials to reduce the potential for IAQ problems
- Extensive training of subcontractors in construction waste management
- Paying attention to moisture control in all aspects of construction to prevent future mold problems
- Ensuring that stringent erosion and sedimentation control measures are instituted
- Minimizing the impact of construction operations, such as compaction and unnecessary destruction of trees, on the site

These are just a few of the many activities that can be undertaken by the construction team to improve the green building project. This chapter will cover some of these innovations in the following general categories of green building projects:

- Site protection planning
- Health and safety planning
- Construction and demolition waste management
- Subcontractor training
- Reducing the footprint of construction operations
- Materials handling and installation
- Protection of IAQ during construction

Site Protection Planning

A Site Protection Plan spells out what is required of the builder to minimize ecological and other damage to the site, as well as how to maintain good community relations during the construction process. Although not specifically required by either LEED-NC or Green Globes, a Site Protection Plan provides in-depth coverage of many site aspects that can help produce a far more environmentally friendly project. Both LEED-NC and Green Globes do, however, require an Erosion and Sedimentation Control Plan (see Chapter 6), which is at least partially addressed in a Site Protection Plan. One entity that does specify a Site Protection Plan is the Department of Design and Construction of the City of New York, whose "High Performance Building Guidelines" (1999) require such a plan for municipal construction projects. It must include:

- A protection plan for vegetation and trees.
- A "tree rescue" plan for those trees and plantings that must be removed (ideally to be given to a park, community garden, nursery, or some other appropriate entity).
- A site access plan, including a designated staging or "lay down" area designed to minimize damage to the environment. This plan must indicate storage areas for salvaged materials, and access and collection areas for recyclable materials, including day-to-day construction waste (packaging, bottles, etc.). It must also designate site-sensitive areas where staging, stockpiling, and soil compaction are prohibited.
- Wastewater runoff and erosion control measures.
- Measures to salvage existing clean topsoil on site for reuse.
- Plans to mitigate dust, smoke, odors, and other impacts.
- Noise control measures, including schedules for particularly disruptive, high-decibel operations, and procedures for compliance with state and local noise regulations.

Figure 11.1 Erosion, a potentially serious problem on construction sites, is exacerbated by construction equipment. Careful planning of roads and construction can eliminate the loss of topsoil and sedimentation of nearby surface waters.

By implementing these measures as part of a Site Protection Plan, the builder will ensure that everything possible is being done to protect the natural systems on the site, minimize erosion off-site into surface waters, maximize infiltration of water into the soil on-site, and generally ensure a more responsible operation.

Health and Safety Planning

A Health and Safety Plan addresses the health of workers on the building site and the health of the building's future occupants. The New York City Department of Design and Construction guidelines cited in the previous section also state that the contract documents require the contractor to produce a Health and Safety Plan. Neither LEED-NC nor Green Globes has this specific requirement, although some of its elements, such as protection of ducts and airways during construction, provide LEED and Green Globes points. The plan should take into account the building's air quality design and provide for:

- Adequate separation and protection of occupied areas from construction areas for building additions.
- Protection of ducts and airways from dust, moisture, particulates, VOCs and microbes resulting from construction/demolition activities.
- Increased ventilation/exhaust air at the construction site.
- Scheduling of construction procedures to minimize exposure of absorbent building materials to VOC emissions. For example, "wet" construction procedures such as painting and sealing should occur *before* storing or installing "dry," absorbent materials such as carpets and ceiling tiles. These porous components act as a sink, retaining contaminants and releasing them during building occupancy.
- A flush-out period, beginning as soon as systems are operable and before or during the furniture, fittings, and equipment installation phase. The process involves flushing the building with 100 percent outside air for a period of not less than 20 days.
- Appropriate steps to control vermin.
- Prevention of pest infestation once the building or renovated portion is occupied, using integrated pest management.

Poor job-site construction practices can undermine even the best building design by allowing moisture and other contaminants to become potential long-term problems. Preventive job-site practices can preclude residual IAQ problems in the completed building and reduce undue health risks for workers.

Table 11.1 shows steps to follow to ensure that provisions are made for good IAQ throughout the entire project.[1] Table 11.2 contains a list of actions the builder should take directly to ensure that IAQ is not compromised by the construction process.[2]

Though the Health and Safety Plan described here does not cover the scope of construction safety planning and execution required by the Occupational Safety and Health Administration (OSHA), it does cover a number of human safety issues not anticipated by OSHA that have become a part of the green building movement.

In Norway, builders help provide excellent IAQ for the building's future occupants by installing a vacuum system that can take in all of the dust generated on the building project each day. This system prevents particulates from ending up in duct-

TABLE 11.1

Steps for Managing IAQ During Construction

1. Incorporate IAQ goals into the bid and construction documents.
2. Ensure that all members of the project team are knowledgeable about IAQ issues and have defined responsibilities for implementation of good IAQ practices.
3. Require the development and use of an IAQ management plan. The purpose of the management plan is to prevent residual problems with IAQ in the completed building and protect workers on the site from undue health risks during construction. The plan should identify specific measures to address:
   a. Problem substances, including construction dust, chemical fumes, offgassing materials, and moisture. The plan will make sure that these problems are not introduced during construction or, if they must be, will eliminate or reduce their impact.
   b. Areas of planning, including product substitutions and materials storage, safe installation, proper sequencing, regular monitoring, and safe, thorough cleanup.
4. Conduct regular inspection and maintenance of IAQ measures, including ventilation system protection and ventilation rate.
5. Conduct safety meetings, develop signage, and establish subcontractor agreements that communicate the goals of the construction IAQ plan. The IAQ construction plan is also a good place to proscribe behaviors unacceptable to the owner that represent a potentially negative impact on long-term IAQ, such as smoking, using chewing tobacco, or wearing contaminated work clothes.
6. Require contractors to provide information on product substitutions sufficient to enable operations and maintenance (O&M) staff to properly maintain and repair low-emitting or otherwise healthy materials in place.

work and eventually in the air of the building's work spaces. The point is, the impacts of construction particulates on future occupant health should not be underestimated, and all possible measures to keep the building clean during construction and to protect ductwork from contamination should be taken.

Construction and Demolition Waste Management

Construction and demolition (C&D) waste management takes advantage of opportunities for source reduction, materials reuse, and waste recycling. Source reduction is most relevant to new construction and large renovation projects, as it involves reduced waste factors in materials ordering, tighter contract language assigning waste management responsibilities to trade contractors, and value engineering of building design and components. During renovation and demolition, building components that still have functional value can be reemployed on the current project, stored for use on a future project, or sold on the ever-growing salvage market. Recycling of building materials can be accomplished whenever sufficient quantities can be collected and markets are readily available.

According to a USEPA study, C&D waste totaled more than 135 million tons (122.5 million metric tons) in the United States in 1998; about 77 million tons (69.9 million metric tons) resulted from commercial work alone. Per-square-foot waste generation ranges from about 4 pounds (19.5 kilograms per square meter) for new construction and renovation to about 155 pounds (757 kilograms per square meter)

TABLE 11.2

Measures for Builders to Implement to Ensure Good IAQ for Building Occupants

- Keep building materials dry. Building materials, especially those like wood, porous insulation, paper, and fabric, should be kept dry to prevent the growth of mold and bacteria.
- Dry water-damaged materials quickly. Water-damaged materials should be dried within 24 hours. Due to the possibility of mold and bacteria growth, materials that are damp or wet for more than 72 hours may need to be discarded.
- Clean spills immediately. If solvents, cleaners, gasoline, or other odorous or potentially toxic liquids are spilled onto the floor, they should be cleaned up immediately.
- Seal unnecessary openings. Seal all unnecessary openings in walls, floors, and ceilings that separate conditioned space (heated or cooled) from unconditioned space.
- Ventilate when needed. Some construction activities can release large amounts of gases into a facility, and if the building is enclosed with walls, windows, and doors, outdoor air can no longer easily flow through the structure and remove the gases. During certain construction activities, temporary ventilation systems should be installed to quickly remove the gases.
- Provide supplemental ventilation. During installation of carpet, paints, furnishings, and other VOC-emitting products, provide supplemental (spot) ventilation for at least 72 hours after work is completed.
- Require VOC-safe masks for workers installing VOC-emitting products (interior and exterior).
- Reduce construction dust. Minimize the amount of dust in the air and on surfaces. Examples include use of vacuum-assisted drywall sanding equipment, and use of vacuums instead of brooms to clean construction dust from floors.
- Use wet sanding for gypsum board assemblies.
- Avoid use of combustion equipment indoors.

for building demolition. On many construction projects, recyclable materials such as wood, concrete and masonry, metals, and drywall make up as much as 75 percent of the total waste stream, presenting opportunities for significant waste diversion. As more C&D landfills reach capacity, new ones become increasingly difficult to site; and as more municipal waste landfills exclude C&D waste, tipping fees will continue to rise. Construction waste—and costs—can be managed just like any other part of the construction process, with positive environmental impacts on land and water resources.

According to the New York City "High Performance Building Guidelines," C&D waste management techniques divert materials from the waste stream, thus preserving valuable resources and landfill space. C&D waste typically includes building components such as doors, lighting fixtures, packaging materials, hazardous materials, and miscellaneous construction waste such as bottles, cans, and paper. For municipal projects in New York City, the contract documents must include a Waste Management Plan written and executed by the building contractor or construction manager. The plan must cover salvaged materials, recycling, packaging, hazardous materials, and other waste protection measures. These measures and requirements are itemized in the following lists.

SALVAGED MATERIALS

- List materials to be salvaged for reuse in the project in the contract documents.

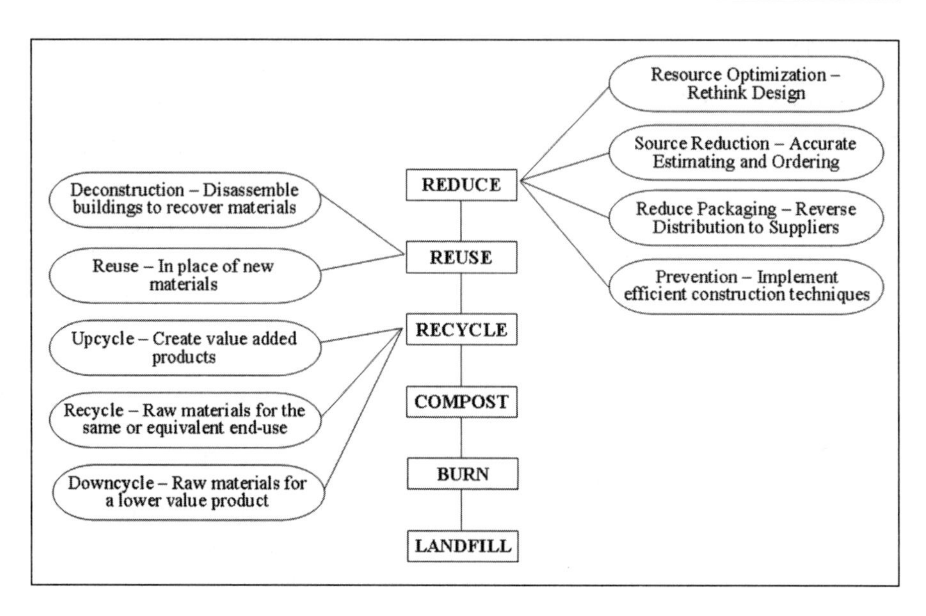

Figure 11.2 The Powell Center for Construction and Environment Waste Management Hierarchy indicates the preferable strategies for dealing with construction waste. Reducing waste (top) has the highest priority; landfilling (bottom) is the least desirable result. Note that recycling can have different outcomes, ranging from upcycling, or increasing the value of materials through recycling, to downcycling, in which the value of the materials is lowered due to their next usage.

- Identify local haulers for salvaged materials and products that will not be reused in the project. List additional materials that are economically feasible for salvaging in the project.

RECYCLING

- Identify licensed haulers of recyclables and document the costs for recycling and frequency of pickups. Confirm with haulers what materials will and will not be accepted. List those materials that are economically feasible for recycling in the project.
- Identify manufacturers and reclaimers that recover construction/demolition scrap of their products for recycling. List materials that are economically feasible for reclamation, along with any special handling requirements for each material. Examples include carpets, ceiling tiles, and gypsum wallboard.
- List procedures to follow to comply with New York City's recycling law. Recyclable materials include bulk metals, corrugated cardboard, bottles, and cans.

PACKAGING

- Identify manufacturers that reclaim their packaging for reuse or recycling.
- Identify manufacturer and distributor options for reduced packaging, where available.

HAZARDOUS MATERIALS

- Develop procedures for separating hazardous waste by-products of construction (e.g., paints, solvents, oils and lubricants) and for disposing of these wastes according to appropriate federal, state, or local regulations.

OTHER WASTE PREVENTION MEASURES

- Educate workers on waste prevention goals and the proper handling and storage of materials.
- Where applicable, reuse salvaged material at the site.
- Coordinate ordering and delivery of materials among all contractors and suppliers to ensure that the correct amount of each material is delivered and

stored at the optimum time and place. This can help prevent material loss, theft, and damage.

Careful training and planning are necessary to significantly reduce the typical volume of construction waste generated during a construction project. A formal plan is needed to ensure that all parties to the construction process understand their roles in reducing and recycling the waste stream.

Subcontractor Training

Perhaps the most important group in a building construction project is the subcontractors. It is generally true in today's construction industry that general contractors are themselves performing less and less of the work involved in the actual erection of the building. Instead, the general contractor or construction manager organizes and orchestrates a diverse group of subcontractors to produce the building. For a green building project to meet its objectives, the subcontractors must be made aware of how the building project differs from a conventional construction project. To that end, job-site training in these key areas will be necessary for all subcontractors:

- *Construction Waste Management Plan:* All subcontractors should have input to the plan to make it effective, and all workers should be informed of the job's waste reduction and waste-handling procedures. The buy-in of the subcontractors is key to successfully minimizing waste.
- *Construction IAQ Plan:* Each subcontractor is at risk of creating hazards that can affect the future IAQ of the building. Protection of ductwork and components of the air handling system is the responsibility of the mechanical contractor. However, due to the extent of the system in many buildings and the possibility that the mechanical subcontractor is not always on the job site, the best practice is to instruct all subcontractors to monitor the implementation of measures to prevent the entry of dust into the air handling system.

Reducing the Footprint of Construction Operations

The very act of constructing a building and the infrastructure that connects it to power, water, communications, sidewalks, and roads causes tremendous changes to the site. The purpose of reducing the footprint of construction on the site is to preserve as much of the original site function and its ecological systems as possible. Among the recommended procedures for reducing the physical footprint of the construction process are the following:[3]

- Document the site's existing natural, historical, and cultural features and make specific plans to preserve them.
- Specify locations for trailers and equipment.
- Specify which areas of the site should be kept free of traffic, equipment, and storage.
- Prohibit clearing of vegetation beyond 40 feet (12.2 meters) from the building perimeter.
- Explain methods for protecting vegetation, such as designating access routes and parking.

- Require methods for clearing and grading the site that are as low-impact as possible.
- Examine how runoff during construction may affect the site. Consider creating stormwater management practices, such as piping systems or retention ponds or tanks, which can be carried over after the building is complete.

Summary and Conclusions

The success of a high-performance green building project is at least in part dependent on the conduct of the construction phase. The construction manager must fulfill several specific responsibilities to ensure that the process embodies the intent of green building, namely, to be environmentally friendly and resource-efficient and to result in a healthy building. By protecting plants and other ecosystem components, keeping the footprint of construction operations as small as possible, and minimizing sedimentation and erosion, the builder can meet the first of these objectives—protecting the environment. The resource efficiency goal can be met by reducing C&D waste and by planning and executing a well-thought-out Construction Waste Management Plan. Finally, the quality of indoor air can be protected for future building occupants by protecting ductwork from fabrication through installation; by properly storing materials to avoid moisture penetration, mold, and mildew; and by appropriately ventilating and flushing out the building prior to occupancy. A thorough training program for subcontractors should be instituted, and requirements peculiar to the construction of a green building should be integrated with other standard training programs, such as construction safety.

Notes

1. Information is from the USEPA IAQ website, www.epa.gov/iaq/schooldesign/construction.html.
2. Ibid.
3. A range of environmental considerations during construction, including how to reduce the footprint of construction operations, can be found at the Department of Energy's Whole Building Design website, www.eere.energy.gov/buildings/info/design/wholebuilding.

References

"High Performance Building Guidelines." 1999. New York City, Department of Design and Construction. Available in pdf format at www.nyc.gov/html/ddc/html/highperf.html.

Chapter 12

Building Commissioning

One of the major contributions of the emerging high-performance green building delivery system is to require building commissioning as a standard practice. This has come about because at least a basic level of commissioning is required for certification under the USGBC LEED-NC building assessment standard and is highly recommended under Green Globes. Building commissioning provides the owner with an unprecedented level of assurance that the building will function as designed, with resultant high reliability and reduced operating costs. The success of building commissioning has culminated in the formation of specialist building commissioning companies and in the development of building commissioning departments in engineering firms whose purpose is to service the green building market.

Initial studies of the impacts of building commissioning indicate that it may reduce building operating costs by a larger margin than energy conservation measures. A report by Greg Kats of Capital E put the 20-year savings in operations and maintenance due to building commissioning at $8.47 per square foot, compared with energy savings of $5.79 per square foot.[1] Unquestionably, then, building commissioning is a powerful tool for ensuring that the design intent—to reduce resource consumption and environmental impacts—is indeed carried out in the construction process. Building commissioning is, however, an additional service, meaning that it adds to the first or construction cost of the building project. Furthermore, and unfortunately, during the cost reduction exercises that are now common practice during design, these additional fees are subject to being cut, regardless of their benefits.

According to the Building Commissioning Association (BCA), "The basic purpose of building commissioning is to provide documented confirmation that building systems function in compliance with criteria set forth in the Project Documents to satisfy the owner's operational needs. Commissioning of existing systems may require the development of new functional criteria in order to address the owner's current systems performance requirements."[2]

Another organization heavily engaged in and committed to building commissioning is the Associated Air Balance Council (AABC), which has a commissioning guideline and a certification program for commissioning agencies.[3] Originally formed to guide the testing and balancing of building HVAC systems, the AABC has developed a more generalized approach that, while performing its traditional role of ensuring that HVAC systems operate as designed, now includes the broader range of roles defined by building commissioning. That means that the AABC Commissioning Guideline can apply to any building system, and the same steps of planning, organizing, systems verification, functional performance testing, and documenting the tasks of the commissioning process that apply to building mechanical systems can also be applied to building electrical systems, control systems, telecommunications systems, and others.

Federal and state governments are increasingly requiring commissioning of their facilities; in fact, several government organizations also publish building commissioning guidelines. For example, the Federal Energy Management Program publishes "The Building Commissioning Guide, Version 2.2"[4] and "The Continuous Commissioning Guidebook for Federal Managers."[5] The latter establishes building commissioning as an ongoing process for use in resolving operating problems in buildings. Another such publication, "New Construction Commissioning Handbook for Facility Managers," was prepared for the Oregon Office of Energy as part of a regional program involving four northwestern states. Its aim is to make building commissioning standard practice.[6]

Essentials of Building Commissioning

According to the BCA, the building commissioning process is controlled and coordinated by a commissioning authority (CA). The following are the essential elements of building commissioning as carried out by the CA:[7]

1. The CA is in charge of commissioning works on behalf of the owner, is an advocate for the owner's interests, and makes recommendations to the owner about the performance of the commissioned systems.
2. The CA must have adequate experience to perform the commissioning tasks and must have recent hands-on experience in building systems commissioning; building systems performance and interaction; operations and maintenance procedures; and building design and construction processes.
3. The scope of commissioning must be clearly defined in the commissioning contract.
4. The roles and scope of all building team members in the commissioning process should be clearly defined in the design and engineering consultants' contracts; [in] the construction contract; [in] the General Conditions of the Specifications; in the divisions of the specifications covering work to be commissioned; and in the specifications for each system or component for which a supplier's support is required.
5. A commissioning plan must be produced to describe how the commissioning process will be carried out, and should identify the systems to be commissioned; the scope of the commissioning process; the roles and lines of communications for each team member; and the estimated commissioning schedule.
6. For new construction, the CA should review systems installation for commissioning issues throughout construction.
7. Commissioning activities and findings are documented exactly as they occur, distributed immediately, and included in the final report.
8. A functional testing program, composed of written, repeatable test procedures, is carried out, indicating expected and actual results.
9. The CA should provide constructive input for the resolution of system deficiencies.
10. A commissioning report is produced that evaluates the operating condition of each system; deficiencies that were discovered and measures taken to correct them; uncorrected operational deficiencies accepted by the owner; functional test procedures and results; documentation of all commissioning activities; and a description and estimated schedule for deferred testing.

Maximizing the Value of Building Commissioning

As noted previously, building commissioning has tremendous potential for generating savings for the building owner. To ensure that the maximum value is obtained for building commissioning, the BCA recommends that the scope of building commissioning include the following:

1. Prior to design, seek assistance in evaluating the owner's requirements such as energy conservation, IEQ, training, operations, and maintenance.
2. During each design phase, review construction documents for compliance with design criteria, commissioning requirements, bidding issues, construction coordination and installation concerns, performance, and facilitation of operations and maintenance.
3. Review equipment submittals for compliance with commissioning issues.
4. Review and verify schedules and procedures for system startup.
5. Ensure that training of operating staff is conducted in accordance with project documents.
6. Ensure that operations and maintenance manuals comply with contract documents.
7. Assist the owner in assessing system performance prior to expiration of the construction contract warranty.

HVAC System Commissioning

Today's process of building commissioning has its roots in the science of Test and Balancing (TAB), traditionally used to verify that the building's HVAC systems are operating as designed. The TAB authority is an independent organization hired to check that the various fans, pumps, dampers, energy recovery systems, hot water heating units, and other components are functioning properly, that the flow rates of hot and chilled water are as designed, and that airflows are properly adjusted so that the quantities of supply air, return air, and ventilation air in each space are also as designed.

AABC is the governing organization for TAB activities in the United States and, with the advent of the high-performance green building movement, it has shifted its activities and nomenclature to cover building commissioning. Commissioning of HVAC systems does, however, remain the dominant commissioning activity, meaning that the migration from TAB to full building commissioning was fairly straightforward. With respect to commissioning HVAC systems, AABC defines the key commissioning activities as those shown in Table 12.1.[8]

Commissioning of Nonmechanical Systems

Although the commissioning of mechanical systems is at the heart of building commissioning, the commissioning process should include all building systems: all electrical components; telecommunications and security systems; plumbing fixtures; rainwater harvesting systems; graywater systems; electronic water controls and items such as finishes, doors, door hardware, windows, millwork, and ceiling tiles; and any other component in the building's drawings and specifications. The commissioning tasks for nonmechanical systems are to:[9]

- Ensure appropriate product selection during design and the design intent review.
- Ensure that product specifications are clear by conducting a specification review.
- Ensure [that] the construction manager or subcontractor selects an acceptable product during the submittal review.
- Ensure [that] the construction manager or subcontractor properly installs the product.

TABLE 12.1

Key HVAC System Commissioning Activities as a Function of Project Phase

Phase	Key Commissioning Activities
Predesign	Establish commissioning as an integral part of the project. Owner selects the commissioning agent. Develop the scope of commissioning. Commissioning agent reviews the design intent.
Design	Review of design to ensure [that] it accommodates commissioning. Write commissioning specifications defining contractor responsibilities. Commissioning agent produces the commissioning plan. The project schedule is established.
Construction	Commissioning agent reviews contractor submittals. Commissioning agent updates commissioning process. Continued coordination of commissioning process. Carry out and document system verification checks. Carry out and document equipment and system startup. Carry out and document TAB activities.
Acceptance	Carry out functional performance tests for all HVAC systems. Train O&M [operations and maintenance] staff for effective ongoing operations and maintenance of all systems. Provide full documentation of HVAC systems.
Postacceptance	Correct any deficiencies and carry out any required testing. Carry out any required "off season" tests. Update documentation as required.

- Ensure [that] adequate operations and maintenance (O&M) documentation is provided so that facility staff can properly maintain the item through an O&M documentation review.
- Ensure [that] facility staff receive adequate training to operate and maintain the item through a training verification.
- Ensure [that] the O&M plan addresses all items through an O&M plan review.
- Ensure [that] the building's indoor environmental quality (IEQ) meets the design objectives.

The more involved the commissioning process, the greater will be the need for a diverse commissioning team whose members can handle the range of systems included in the building commissioning process. The members involved during the design process may be different from those who test the systems when construction is complete. For example, the commissioning team members engaged during design may be experts in multidisciplinary work and have an overall understanding of the process and experience in selecting products and ensuring clear and complete documentation to support all design decisions.

Costs and Benefits of Building Commissioning

Building commissioning provides a wide range of benefits—and, it is important to note, the earlier in the building design and construction process that building commissioning is implemented, the greater will be the benefits. The ideal arrangement is for the CA to be hired at the onset of the project, along with the design team and construction manager. The CA provides another set of inputs to the building team and ensures that throughout design and construction, issues related to commissioning are included in the construction documents. The following are some of the typical benefits that can be expected as a result of including a full scope of building commissioning services from an independent CA:

- Reduced operating costs due to an energy efficiency increase of 5 to 10 percent, attributed to building commissioning
- Increased employee productivity due to improved IEQ resulting from building commissioning
- Improved construction documents resulting from the participation of the CA in the review process during each design phase and the potential for greatly reducing change orders
- Fewer errors in equipment ordering due to the continual review of equipment requirements by the CA
- Fewer equipment installation errors because the CA reviews equipment installation during the construction process
- Fewer equipment failures during building operation due to the testing, calibration, and reporting carried out by the CA
- Complete documentation of systems provided to the owner
- A fully functioning building from the first day of operation

The Oregon Office of Energy provides information on the benefits of building commissioning for energy savings for several building types. These are listed in Table 12.2.[10]

The cost of commissioning is a function of the size of the project, its complexity, and the level of commissioning selected by the owner (see Table 12.3).[11] Buildings with simple conditioning systems, few zones, and simple control systems would be at the lower end of the commissioning cost range shown for various levels of construction cost, whereas buildings with complex conditioning systems and control systems would be at the higher end of the range. Similarly, the benefits of

TABLE 12.2

Energy Savings Attributable to Building Commissioning for Various Building Types

Building Type	Dollar Savings	Energy Savings
110,000 ft^2 office	\$0.11/ft^2/yr (\$12,276/yr)	279,000 kWh/yr
22,000 ft^2 office	\$0.35/ft^2/yr (\$7,630/yr)	130,800 kWh/yr
60,000 ft^2 high-tech manufacturing	\$0.20/ft^2/yr (\$12,000/yr)	336,000 kWh/yr

TABLE 12.3

Cost of Commissioning Services by an Independent Third-Party Service

Construction Cost	Total Commissioning Cost*	Note
<$5 million	2–4%	Costs include moderate
<$10 million	1–3%	travel, but building complexity,
<$50 million	0.8–2.0%	number of site visits, and
>$50 million	0.5–1.0%	other factors may also
Complex projects (labs)	Add 0.25–1%	affect the cost.

*As a percentage of the construction cost.

commissioning for more complex buildings are far greater than those for buildings with relatively simple systems.

Table 12.3, it is important to point out, does not separate the costs of fundamental commissioning, required by the LEED rating system, from those of enhanced commissioning, which is optional under LEED. Fundamental commissioning may be carried out by personnel from the design firms on the building team as long as they are not directly involved in the project, and these costs are sometimes rolled into the design fee to minimize costs.

The Oregon Office of Energy provides another viewpoint of commissioning costs, as shown in Table 12.4.[12]

It should be noted that the commissioning costs in the design phase include the costs for the CA and the architect, with the allocation being approximately 75 percent for the CA and 25 percent for the architect. Similarly, for the construction phase, additional costs are charged for the engineers to attend meetings, create checklists, and participate in testing. These costs are not listed in Table 12.4 and amount to 10 to 25 percent of the CA's fee. There may also be additional costs for the architect's involvement in reviewing the commissioning plan and attending meetings, in the range of 5 to 10 percent of the CA's fee.

Connection to LEED-NC

Building commissioning is an extremely important part of designing and building a high-performance green building. For the purposes of the USGBC LEED-NC build-

TABLE 12.4

Costs of Commissioning During Design and Construction Phases for Typical Systems

Phase	Commissioned System	Total Commissioning Cost
Design	All	0.1% to 0.3%
	HVAC and controls	2.0% to 3.0% of total mechanical cost
Construction	Electrical system	1.0% to 2.0% of total electrical cost
	HVAC, controls, and electrical system	0.5% to 1.5% of total construction cost

ing assessment standard, building commissioning is covered twice, under the Energy & Atmosphere (EA) category, as follows:

- Fundamental Building Commissioning is Prerequisite 1 under EA for LEED certification and is therefore a requirement. A building cannot be certified without carrying out this level of commissioning.
- Enhanced Commissioning (EA Credit 3) provides 2 points for implementing building commissioning in a more comprehensive manner.

Both EA Prerequisite 1 and EA Credit 3 are also described in Chapter 7.

EA PREREQUISITE 1 (EAp1): FUNDAMENTAL BUILDING COMMISSIONING

The purpose of Fundamental Building Commissioning is to ensure that the building operates as intended by the design team. For this to be possible, however, the design team must have adequately carried out its design tasks such that the building's systems have the capability to function as indicated on the plans and in the specifications. The requirements for this level of commissioning are:

1. Designating a qualified person to be the Commissioning Authority (CxA) to lead and oversee the commissioning process. The CxA must have documented commissioning authority experience on at least two previous building projects. The CxA must be independent of the project's design and construction team, but may be an employee of the firms involved in the project. The exception is for projects smaller than 50,000 square feet for which the CxA may include qualified personnel from the project team.
2. The CxA will report directly to the Owner with all results and findings from the commissioning process.
3. The Owner is required to document the Owner's Project Requirements (OPR) and the Basis of Design (BOD) is developed by the design team. The CxA is required to review these documents for consistency and completeness.
4. The CxA shall incorporate commissioning requirements into the construction documents.
5. The CxA shall develop a commissioning plan.
6. The CxA shall verify the installation and performance of the systems to be commissioned.
7. The CxA shall complete a summary commissioning report.

The systems to be commissioned include (1) HVAC&R systems and associated controls; (2) lighting systems and lighting controls; (3) domestic hot water systems; and (4) renewable energy systems.

It is worthwhile noting that this prerequisite does not call for a fully independent commissioning team. The commissioning team may be composed of qualified people from the design and engineering firms engaged in the building projects; however, the members of the commissioning team must not have been directly engaged in the project being commissioned.

EA CREDIT 3 (EAc3): ENHANCED COMMISSIONING

Enhanced Commissioning adds several additional requirements to the Fundamental Building Commissioning category:

1. The CxA must be appointed prior to the construction documents phase and have experience commissioning at least two prior building projects. The CxA must be independent of the design and construction and may not be an employee of the

design or construction firms engaged in the project. The CxA may be a qualified employee of the Owner.

2. The CxA must review the Owner's Project Requirements (OPR), the Basis of Design (BOD), and design documents prior to the mid-point of the construction documents phase.
3. The CxA must review contractor submittals for systems being commissioned for compliance with the OPR and BOD.
4. The CxA must develop a systems manual for the future operating staff so that they will be able to understand and operate the commissioned systems.
5. The CxA must verify that training for operating personnel and building occupants is completed.
6. The CxA must be involved in reviewing building operation within 10 months after substantial completion with operations and maintenance staff and building occupants. A plan for resolving outstanding commissioning-related issues must be included.

Interestingly, in spite of the high return on investment for building commissioning, EA Credit 3 for Enhanced Commissioning is one of the least pursued LEED credits. The reason is that, in general, it is a challenge to convince the owner that the additional costs are warranted to gain the benefit/cost ratio for a full-fledged building commissioning effort. One of the issues is paying for another set of services and associated fees on top of the design and construction management fees that building owners must pay and that they are also intent on minimizing. The additional 1 to 4 percent of the construction cost for building commissioning has not been fully accepted by owners, so most rely on the Fundamental Commissioning called for in EA Prerequisite 1 to meet their essential needs.

Summary and Conclusions

Building commissioning is an important component of the delivery process for high-performance green buildings and has been shown to reap enormous benefits in the form of reducing O&M costs. A diverse range of firms provide building commissioning services in support of the high-performance green building movement. The LEED-NC 2009 building assessment system requires Fundamental Building Commissioning, and a LEED-NC 2009 credit is available for Enhanced Commissioning. The economic returns for building commissioning are very high—greater, according to some accounts, than for energy savings.

For building commissioning to be truly effective, it must occur periodically throughout the building's life cycle because complex systems tend to drift out of specification and even fail. The high-performance green building system has brought the relatively new discipline of building commissioning to the forefront in terms of its value to the building project. The return on investment for building commissioning warrants consideration of an extensive building commissioning program for green building projects.

Notes

1. These savings reflect the total net present worth or the sum of the net present worth of each year's savings over the assumed 20-year life cycle of the building assumed in the study. From Kats (2003).
2. As described on the BCA website, www.bcxa.org.
3. The website for the AABC is www.aabchq.com. Additional information is available in Magee et al. (2002).

4. "The Building Commissioning Guide" was published by Federal Energy Management Program (FEMP) in 1998 and is available at www.eere1.energy.gov/femp/pdfs/29267-9.2.pdf.
5. The "Continuous Commissioning Guidebook for Federal Managers," published in 2002, is available from FEMP at www1.eere.energy.gov/femp/pdfs/ccg03-h7.pdf.
6. "New Construction Commissioning Handbook for Facility Managers" (October 2000).
7. From the Building Commissioning Association website, www.bcxa.org.
8. Excerpted from Magee et al. (2002).
9. Suggestions for nonmechanical systems commissioning are from Portland Energy Conservation, Inc., whose website is www.peci.org/library.htm.
10. From Dasher, Potter, and Strum (2002).
11. Excerpted from Dorgan, Cox, and Dorgan (2002).
12. See Appendix 4, "Estimating Commissioning Costs," in the "New Construction Commissioning Handbook for Facility Managers" (October 2000).

References

"A Practical Guide for Commissioning Existing Buildings," ORNL/TM-1999/34. May 1999. Report prepared by the staff of Portland Energy Conservation, Inc., and Oak Ridge National Laboratory. Available at http://eber.ed.ornl.gov/commercialproducts/retrocx.htm.

Dasher, Carolyn, Amanda Potter, and Karl Strum. 2000. "Commissioning to Meet Green Expectations," published by the Oregon Office of Energy. Available at www.peci.org/library/PECI_CxGreen1_0402.pdf.

Dorgan, Chad E., Robert E. Cox, and Charles Dorgan. 2002. "The Value of Building Commissioning," *The Austin Papers, Best of the 2002 USGBC International Green Building Conference.* Brattleboro, VT: BuildingGreen, Inc.

Kats, Gregory H. 2003. "Green Building Costs and Financial Benefits," published for the Massachusetts Technology Collaborative. Available at the Capital E website, www.cap-e.com.

Magee, James I., Daniel Acri, Joseph Baumgartner, Bret Privitt, John Shelander, Kenneth Sufka, and Cedric Trueman. 2002. "AABC Commissioning Guideline and Certification Program," *The Austin Papers, Best of the 2002 USGBC International Green Building Conference.* Brattleboro, VT: BuildingGreen, Inc.

"New Construction Commissioning Handbook for Facility Managers." October 2000. Prepared for the Oregon Office of Energy by Portland Energy Conservation, Inc. Available at the Oregon Office of Energy website, www.oregon.gov/ENERGY/CONS/BUS/comm/docs/Newcx.pdf.

Chapter 13

Economic Analysis of Green Buildings

Understanding building economics is important for any construction project, but it is especially important for high-performance green buildings because justifying this approach can involve somewhat more complex analysis than for conventional construction. High-performance buildings can produce benefits for their owners in a diverse range of categories: energy, water, wastewater, health and productivity, O&M, maintainability, and emissions, to name a few. To be able to address this scope of benefits, the building team must be able to either quantify the effects of their decisions by using simulation tools or rely on the best available research and evidence gathered from other projects.

This chapter will address the economic and business arguments for high-performance buildings and approaches for quantifying the various benefits achievable by investing in environmentally beneficial buildings.

General Approach

A report to the California Sustainable Building Task Force states that a 2 percent additional investment to produce a high-performance building would produce life-cycle savings that are 10 times greater than the incremental investment.[1] For example, an additional $100,000 investment in a $5 million building should produce at least $1 million in savings for a building with an assumed 20-year life cycle. This is a truly remarkable claim and, if verifiable, makes a virtually unshakable case for high-performance building.

Today, high-performance green buildings are thought to have a higher capital or construction cost than conventional buildings, on the order of 2 percent, or $2 to $5 per square foot.[2] The additional required capital is proportional, at least generally, to the level of the building's LEED-NC rating (see Table 13.1).[3]

An analysis of the financial benefits of high-performance green buildings concluded that significant benefits could be attributed to this type of delivery system, and that there was a correlation between the LEED-NC rating and the financial return. Table 13.2 indicates that, for a typical high-performance building, the total net present value (TNPV) of the energy savings over a 20-year life cycle is $5.79 per square foot, with other notable per square foot savings from reduced emissions ($1.18), water ($0.51), and O&M savings resulting from building commissioning ($8.47).[4] Table 13.2 shows productivity and health savings per square foot of $36.89 for LEED Certified and Silver buildings and $55.33 for LEED Gold and Platinum buildings. Clearly, the productivity and health benefits of high-performance green buildings dominate this discussion; and for Gold and Platinum buildings, the claim is that the savings are almost 10 times greater than the energy savings. It is important to point out, however, that though these claims are generally accepted by high-performance building practitioners, most of those made for productivity and health

TABLE 13.1

Cost Premiums Derived from 33 LEED-NC-Rated Buildings

LEED Rating	Sample Size	Cost Premium
Platinum	1	6.50%
Gold	6	1.82%
Silver	18	2.11%
Certified	8	0.66%
Average	—	1.84%

TABLE 13.2

Value of Various Categories of Savings for Buildings Certified by the USGBC

Category	20-Year Total Net Present Value (TNPV) per Square Foot*
Energy value	$5.79
Emissions value	$1.18
Water value	$0.51
Waste value—construction only, one year	$0.03
Commissioning O&M† value	$8.47
Productivity and health value (Certified and Silver)	$36.89
Productivity and health value (Gold and Platinum)	$55.33
Less green cost premium	($4.00)
Total 20-year NPV (Certified and Silver)	$48.87
Total 20-year NPV (Gold and Platinum)	$67.31

*Net present value (NPV) is the net savings for each year, taking into account the discount rate (time value of money). The 20-year total net present value (TNPV) is the sum of the NPVs for all 20 years and represents the total life-cycle savings.

†O&M commissioning ensures that the building is built and operated according to the design and results in substantially lower O&M costs.

improvements are based on anecdotal information, not scientific research. The total 20-year net present value (NPV) is $48.87 for Certified and Silver buildings and $67.31 for Gold and Silver buildings. The magnitude of these benefits is very impressive when considering that, on average, the incremental construction cost ranges from about $1.50 per square foot for LEED Certified buildings to about $9.50 per square foot for LEED Platinum buildings.

A side-by-side analysis of two prototype buildings by the U.S. Department of Energy's Pacific Northwest National Laboratory (PNNL) and the National Renewable Energy Laboratory (NREL) compared the costs and benefits of investing in high-performance buildings. A base two-story, 20,000-square-foot (1,858 square meters) building with a cost of $2.4 million meeting the requirements of ASHRAE Standard 90.1-1999 was modeled using two energy simulation programs, DOE-2.1e and Energy-10, and compared to a high-performance building that added $47,210 in construction costs, or about 2 percent for its energy-saving features. Table 13.3 summarizes the results of this study.[5] The features listed are those for which an additional investment was made to produce the high-performance version of the NREL prototype building:

- Building commissioning, as noted previously, can produce significant savings by ensuring that the mechanical systems are functioning as designed.
- Natural landscaping and stormwater management produce savings due to the elimination of infrastructure and the use of easily maintainable native plants.
- Raised floors and movable walls produce savings by improving the flexibility of a building, reducing renovation costs.

The results of this comparison are remarkable because they indicate that the annual savings produced by the high-performance version are about equal to the added construction cost, producing a simple payback in just over 1 year. It should be noted, however, that this study did not address this comparison as if the building were to undergo certification through the USGBC's LEED process.

The additional capital costs often associated with high-performance buildings are a function of several factors. First, these buildings often incorporate systems that are

TABLE 13.3

Comparison of Costs and Savings for NREL Prototype Buildings

Feature	Added Cost	Annual Savings
Energy-efficiency measures	$38,000	$4,300
Commissioning	$4,200	$1,300
Natural landscaping, stormwater management	$5,600	$3,600
Raised floors, movable walls	0	$35,000
Waterless urinals	($590)	$330
Total	$47,210	$44,530

not typically present in conventional buildings, such as rainwater harvesting infrastructure, daylight-integrated lighting controls, and energy recovery ventilators. Second, green building certification (fees, compilation of information, preparation of documents, cost of consultants) can add markedly to the costs of a project. And, finally, many green building products cost more than their counterparts, often because they are new to the marketplace and demand is only in the process of developing. In this last category are many nontoxic materials such as paints, adhesives, floor coverings, linoleum, and pressed strawboard used in millwork, to name but a few of the many new green building products appearing to serve the high-performance building market. Conversely, cost reductions for some building systems are achievable in green buildings—for example, in HVAC systems—that can be downsized as a consequence of improved building envelope design. However, additional energy-saving components such as energy recovery ventilators (ERVs), premium high-efficiency motors, variable-frequency drives for variable air volume (VAV) systems, carbon dioxide sensors, and many others all add to the front-end capital cost.

As for every other type of project, understanding the economics of the situation and including them in the decision-making process is of crucial importance. As described earlier in the book, the classical approach used in assessing high-performance building economics is LCC, which includes a consideration of both first cost (sometimes referred to as *construction cost* or *capital cost*) and operating costs (utilities and maintenance). These two major cost factors are combined in a cost model that takes into account the time value of money, the cost of borrowed money, inflation, and other financial factors. They are then combined into a single value, the total net present value (TNPV) of the annual costs, and the selection of alternatives is based on an evaluation of this quantity. In some cases, due to legislated requirements, only the capital cost is considered. For example, the State of Florida allows decisions on building procurement to be made solely on the basis of capital costs, whereas the U.S. government requires that an LCC approach be used. Consequently, producing a high-performance public sector building in Florida can be very challenging; therefore, finding creative mechanisms for investing in higher-quality construction is imperative. One potential mechanism is the creation of a revolving fund from which building owners or users can borrow and that can then be repaid through savings over time.

The Business Case for High-Performance Green Buildings

Making the case for high-performance buildings in the private sector must include a justification of why they make good business sense. In an attempt to address this issue, in 2003 the USGBC produced a brochure, "Making the Business Case for High

Performance Green Buildings," that addresses the advantages to a business of selecting green buildings over conventional facilities.[6]

According to the USGBC, high-performance green buildings:

1. *Recover higher first costs, if there are any.* Using integrated design can reduce first costs, and higher costs for technology and controls reap rapid benefits.
2. *Are designed for cost-effectiveness.* Owners are experiencing significant savings in energy costs, generally in the range of 20 to 50 percent, as well as savings in building maintenance, landscaping, water, and wastewater costs. The integrated design process, which is the hallmark of developing high-performance green buildings, contributes to these lower operational costs.
3. *Boost employee productivity.* Increased daylight, pleasant views, better sound control, and other soft features that improve the workplace can reduce absenteeism, improve health, and boost worker productivity.
4. *Enhance health and well-being.* Improved indoor environments can translate into better results in hiring and retaining employees.
5. *Reduce liability.* Focusing on the elimination of sick buildings and specific problems such as mold can reduce the incidence of claims and litigation.
6. *Create value for tenants.* Improved building performance can reduce employee turnover and maintenance and energy costs, thus contributing to better bottom-line performance. Additionally, the operating costs for building tenants will be substantially lower.
7. *Increase property value.* A key strategy of the LEED-NC building rating system is to differentiate green buildings in the marketplace, with the implicit assumption that lower operating costs and better indoor environmental quality will translate to higher value in the building marketplace. A building carrying a LEED-NC plaque will imply superior operational and health performance; hence buyers will be willing to pay a premium for these features. This would in turn spur demand for more high-performance green buildings.
8. *Take advantage of incentive programs.* Many states, for example, Oregon, New York, Pennsylvania, and Massachusetts, have programs in place that provide financial and regulatory incentives for the development of green buildings. The number of these programs is likely to grow and may include, among other possibilities, shorter project approval times, lower permit fees, and lower property taxes.
9. *Benefit your community.* Green buildings emphasize infill development, recycling, bicycle use, brownfield rehabilitation, and other measures that reduce environmental impacts, improve the local economy, and foster stronger neighborhoods. Businesses opting for high-performance green buildings will be contributing to the overall quality of life in the community and earn a better reputation as a consequence of their efforts.
10. *Achieve more predictable results.* The green building delivery system includes improved decision-making processes, integrated design, computer modeling of energy and lighting, and life-cycle costing, and ensures that the owner will receive a final product that is of a predictable high quality. The best practices beginning to emerge in this era of high-performance buildings will also enable more accurate results forecasting.

In addition to these 10 factors, a number of other benefits can be claimed for high-performance buildings, many of them societal. For example, high-performance buildings can help address other problematic issues, among them:[7]

- High electric power costs
- Worsening power grid problems such as power quality and availability
- Possible water shortages and waste disposal issues
- State and federal pressure to reduce criteria pollutants
- Global warming

- Rising incidence of allergies and asthma, especially in children
- The health and productivity of workers
- The effect of school environments on children's ability to learn
- Increasing O&M costs for state facilities

The Economics of Green Building

There are two schools of thought with respect to the economics of green buildings. One school maintains that the construction cost of these buildings should be the same as or lower than that of conventional buildings. The argument for this line of thinking is that through integrated design and reducing the size of mechanical systems needed to heat and cool an energy-efficient building, the costs of high-performance building construction can be kept in line with those of conventional buildings. The ING Bank building, south of Amsterdam in the Netherlands, completed in 1987, is an example of a high-performance facility that cost about $1,500 per square meter ($150 per square foot), including the land, the building, and its furnishings. At that time, this cost was comparable to or less than that of other bank buildings in the Netherlands.[8] This impressive feat was accomplished for an architecturally complex 50,000-square-meter (500,000-square-foot) building, featuring slanting brick walls and an irregular S-shaped footprint, with gardens and courtyards and a 30,000-square-meter (300,000-square-foot) underground parking lot. It is set in a high-density mixed-use area, with retail, office, and residential buildings surrounding it. If all high-performance buildings could be produced at this high level of architectural quality and at the same or lower cost as conventional buildings, the case for these advanced buildings would be made.

In contrast, the second school of thought is that high-performance green buildings will inevitably have higher capital costs, and that by assessing total building costs on a life-cycle basis, the advantages of high-performance building will be achieved. The additional capital costs occur because high-performance buildings incorporate technologies and systems that are simply not present in conventional buildings, some of them complex and expensive. When attempting to assess the LCC of the many alternatives that can produce a high-performance green building, two distinctly different cost categories can be identified—hard costs and soft costs—defined as follows:

- *Hard costs* are those that are easily documented because the owner receives periodic billing for them—for example, electricity, natural gas, water, wastewater, and solid waste.
- *Soft costs* are those that are less easy to document and for which assumptions must be made for their quantification. Examples of soft costs are maintenance, employee comfort/health/productivity attributable to a building, improved IEQ, and reduced emissions.

An LCC analysis using only hard costs is generally acceptable as justification for alternative strategies that include a trade-off of operational costs versus capital costs. Including soft costs in an LCC analysis is far more difficult to justify because the data cannot be verified with the same degree of rigor as for hard costs. If the results of an analysis of alternatives for a high-performance building are to be subjected to a strict review by financial decision makers, then verifiable hard costs should dominate the analysis. If there is greater latitude in the decision-making process, justifiable soft costs can be employed in the analysis.

The following four key points need to be considered when attempting to develop a case for high-performance buildings based on economic issues:

1. *The primary life-cycle savings for a high-performance building will be a result of superior energy performance.* For some types of buildings, HVAC or mechanical plants may indeed be downsized mechanically as a result of reducing external loads through the employment of superior passive design strategies and the design of a highly thermal-resistant building envelope. A significant reduction in HVAC plant size may also translate to a reduction in the size and cost of the electrical plant. However, for buildings that are dominated by interior loads (people and equipment), the HVAC plant may be unchanged in size compared to a conventional building. A daylit building will certainly require far lower levels of electrically derived light during the day but will still require a full lighting system during the evening. As a result, although it will produce significant operational savings, the daylighting system will not lower the requirements for artificial lighting, and in some cases may actually complicate the design of the conventional lighting system.
2. *Life-cycle savings can also be easily demonstrated for water and wastewater conservation measures because these utilities, like energy, are well known.* As water and wastewater costs rise, especially in water-short areas, their life-cycle savings may in some cases approach the scale of energy savings.
3. *Savings due to good IEQ can potentially exceed all other savings.* For example, for a typical office building, maximum energy savings may be $1 per square foot ($10 per square meter) annually, whereas the worth of a 1 percent improvement in employee productivity translates to $1.40 to $3.00 per square foot ($14 to $30 per square meter). Although these savings are far greater than those of any other category, it is difficult to justify their inclusion in an LCC unless the building owner is especially motivated to include this information in the analysis.
4. *Savings due to materials factors are very difficult to demonstrate.* In many cases, green or environmentally friendly materials may in fact cost more—sometimes far more—than the alternatives. For example, currently, compressed wheatboard used for cabinetry costs as much as 10 times more than the alternative, plywood.

Quantifying Green Building Benefits

An LCC for a green building project can address both hard and soft cost issues, either individually or in a comprehensive LCC that includes all cost factors. The following are general benefits that can be included in the LCC and the range of benefits that can be expected (hard costs) or justified (soft costs).

QUANTIFYING ENERGY SAVINGS

Green buildings use substantially less energy than conventional buildings and generate some of their power on-site from renewable or alternative energy sources. In a Capital E survey of 60 LEED-rated buildings conducted by Gregory Kats in 2003, these buildings consumed an average of 28 percent less energy than their conventional counterparts and generated an average of 2 percent of their energy on-site from photovoltaics, thus reducing total fossil fuel–based energy consumption by about 30 percent. Reducing energy consumption provides a second benefit: a reduction in the emissions of global warming gases, which can also be assigned a cost benefit.

TABLE 13.4

Comparison of Energy Performance for a Building Meeting ASHRAE Standard 90.1-1999 with a High-Performance Green Building

	Base Case Building Annual Energy Cost	High-Performance Building Annual Energy Cost	Percent Reduction
Lighting	$6,100	$3,190	47.7
Cooling	$1,800	$1,310	27.1
Heating	$1,800	$1,280	28.9
Other	$2,130	$1,700	20.1
Total	$11,800	$7,490	36.7

Analyzing the energy advantages of a high-performance green building requires the use of an energy simulation tool such as the aforementioned DOE-2.2 and Energy-10™. A series of alternatives can be tried out and tested to determine the best combination of measures for the particular building and its location. An LCC analysis is also generated at the same time to provide cost and payback information, which is used in tandem with the energy-saving data to optimize energy performance. Using this approach, first costs and operational costs are combined to provide a comprehensive picture of the building's energy performance over an assumed lifetime.

Estimating the energy savings for a particular project relies on using a base case that meets a minimum standard. The case of the two-story NREL prototype buildings was used as an illustration at the beginning of the chapter to discuss the costs and benefits of high-performance green buildings. The two-story, 20,000-square-foot (1,858 square meters) building with a base cost that meets the requirements of ASHRAE Standard 90.1-1999 was modeled using DOE-2.1e and Energy-10 to simulate various measures that would substantially improve its performance. The results of this comparison are shown in Tables 13.4 and 13.5.[9]

TABLE 13.5

Costs, Economic Metrics, and Energy Use: Base Case Compared to High-Performance Green Buildings

	Base Case	High-Performance Case
First cost of building	$2,400,000	$2,440,000
Annual energy cost	$11,800	$7,490
Energy reduction from base case	NA	36.7%
Economic Metrics		
Simple payback (years)	NA	8.65
Life-cycle cost	$2,590,000	$2,570,000
Reduction in life-cycle cost from base case	NA	0.85%
Savings-to-investment ratio	NA	1.47
Energy Consumption, Annual		
Million BTUs	730	477
Reduction from base case	NA	34.6%

TABLE 13.6

Annual Savings Using Waterless Urinals Instead of Flush Urinals

Assumptions	75 Units	100 Units	200 Units
Total facility population	1,500	3,000	5,000
Percent of males	55%	50%	60%
Number of males	825	1,500	3,000
Number of urinals	75	100	200
Uses/day/person	3	3	3
Gallons/flush old urinals	3	3	3
Water cost/1,000 gallons	$2.50	$2.50	$2.50
Sewer cost/1,000 gallons	$2.50	$2.50	$2.50
Operating days/year	260	260	260
Annual Water Savings			
Savings in gallons	1,930,500 gal	3,510,000 gal	7,020,000 gal
Savings in dollars	$4,826	$8,775	$17,550
Annual Sewer Savings			
Savings in gallons	1,930,500 gal	3,510,000 gal	7,020,000 gal
Savings in dollars	$4,826	$8,775	$17,550
Total Water and Sewer Savings	**$9,652**	**$17,550**	**$35,100**
Annual Operating Cost Comparison			
Flush urinal*	$5,625	$7,500	$15,000
Waterless urinal†	$3,217	$5,580	$11,700
Total Operating Cost Savings	**$2,408**	**$1,650**	**$3,300**
Total Annual Savings‡	**$12,060**	**$19,200**	**$38,400**
Annual Savings/Urinal	**$161**	**$192**	**$192**

*Total water savings (3 uses/day × 260 days × number of users × water cost).
†Total sewer savings (3 uses/day × 260 days × number of users × sewer rate).
‡Water/sewer savings plus operating cost savings.

QUANTIFYING WATER AND WASTEWATER SAVINGS

Reductions in water consumption produce significant benefits with respect to water and wastewater. A sample, from Falcon Waterfree Technologies, LLC, of the financial impacts of reducing water consumption through the use of waterless fixtures is shown in Table 13.6. This example indicates that the per-fixture savings for a waterless urinal are on the order of $161 to $192 per year. Although the cost of a waterless urinal, on the order of $300, is much higher than that of a flush urinal, the installation costs are much lower because connection to a source of water for flushing is unnecessary. Consequently, the savings noted in this table are for systems with very similar installation costs. In fact, some studies report that the installation costs for waterless urinals are lower than those for flush urinals.

Another set of examples of waterless urinal savings is shown in Table 13.7 for various occupancies such as an office building, a restaurant, and a school, both existing buildings and new buildings.[10] The basic approach indicated in these examples can be extended to a range of other water alternatives, to include rainwater harvesting, graywater systems, ultra-low-flow fixtures, and composting toilets. That is, reductions in water and potentially wastewater costs can be used to develop an LCC analysis for assessing the financial performance of the alternatives versus conventional practice.

TABLE 13.7

Projected Savings from Using Waterless Urinals Instead of Flush Urinals in Various Occupancies: Existing versus New Buildings

Building Type	No. of Males	No. of Urinals	Uses/ Day	Gal/ Flush	Days/ Year	Water Savings/Gal	Water Savings/Liters
Small office	25	1	3	3.0	260	58,500	220,000
New office	25	1	3	1.0	260	19,500	73,800
Restaurant	150	3	1	3.0	360	54,000	204,000
New restaurant	150	3	1	1.0	360	18,000	68,100
School	300	10	2	3.0	185	33,300	126,000
New school	300	10	2	1.0	185	11,100	42,000

QUANTIFYING HEALTH AND PRODUCTIVITY BENEFITS

Factoring human benefits into LCC analyses must be done cautiously and conservatively. Although there is ample information about the health and productivity benefits of high-performance buildings, rarely has it been compiled scientifically; therefore, it cannot be said to have the same reliability as that for hard costs. Nevertheless, some of the major benefits that have been cited are impressive, for example:

- A paper by William J. Fisk of the Indoor Environment Department at Lawrence Berkeley National Laboratory suggests that enormous savings and productivity gains can be achieved through improved IAQ in the United States. He estimated $6 to $14 billion in savings from reduced respiratory disease, $1 to $4 billion from reduced allergies and asthma, $10 to $30 billion from reduced sick building syndrome (SBS)–related illnesses, and $20 to $160 billion from direct, non-health-related improvements in worker performance.[11]
- Daylighting benefits to human health and performance can potentially provide marked financial returns—if they can be quantified. A study of student performance in daylit schools indicates dramatic improvements in test scores and learning progress. One often-cited study by the Heschong Mahone Group states that students in Orange County, California, schools with daylighting in their classrooms improved their test scores 20 percent faster in math and 26 percent faster in reading than students in schools with the lowest levels of daylighting. The study also looked at students in Seattle, Washington, and Fort Collins, Colorado, where improvements in test scores were 7 to 18 percent.[12]
- Another study by the Heschong Mahone Group compared sales in stores with skylights versus nonskylit stores and found that the skylit stores had 40 percent higher sales.[13]

A reasonable approach to determining how to include productivity and health savings in green buildings was suggested in a report to California's Sustainability Task Force.[14] In this report, the authors recommend assigning a 1 percent productivity and health gain to buildings attaining a USGBC LEED-NC Certified or Silver level and a 1.5 percent gain for buildings achieving a Gold or Platinum level. These gains are derived in a conservative fashion from information about improvements in human performance (see Table 13.8). Savings are the equivalent of $600 to $700 per

TABLE 13.8

Human Performance Improvements Associated with Green Building Attributes

Green Building Attribute	Productivity Benefits
Increased tenant control over ventilation	0.5–34%
Increased tenant control over temperature and lighting	0.5–34%
Control over lighting	7.1%
Ventilation control	1.8%
Thermal control	1.2%

employee per year or about $3 per square foot ($30 per square meter), for a 1 percent gain, and $1,000 per employee per year, or $4 to $5 per square foot ($40 to $50 per square meter), for a 1.5 percent gain.

QUANTIFYING THE BENEFITS OF REDUCING EMISSIONS AND SOLID WASTE

Emissions attributed to the operation of buildings are staggering in scope. High-performance buildings have the potential to dramatically lower these impacts. As a result of energy requirements, buildings in the United States are responsible for the creation of 48 percent of the nation's sulfur dioxide emissions, 20 percent of nitrous oxide, and 36 percent of carbon dioxide. Additionally, buildings produce 25 percent of solid waste, consume 24 percent of potable water, create 20 percent of all waste-water, and cover 15 percent of land area.[15] Construction and demolition waste in the United States amounts to about 150 million tons per year, or about 0.5 tons per capita annually. Converting avoided emissions to benefits attributable to high-performance buildings can be accomplished by calculating the societal costs of emissions. The societal impacts of these emissions can be quantified as follows:

- Sulfur dioxide: $91 to $6,800 per ton ($100 to $7,500 per metric ton)
- Nitrous oxide: $2,090 to $10,000 per ton ($2,300 to $11,000 per metric ton)
- Carbon dioxide: $5.50 to $10 per ton ($6 to $11 per metric ton)

For the NREL prototype building, Tables 13.9 and 13.10 provide a summary of benefits that can be claimed as a result of energy reductions.

Including the maximum emissions reductions benefits has a significant impact on the payback time. The payback time due to energy savings is reduced from 8.7 years to 6.0 years when the societal costs of avoided emissions are included.

Another category of savings that can be included in the life-cycle picture is those due to reduced solid waste generation. For high-performance buildings, solid waste reductions are a result of three factors. First is construction and demolition waste reduction, which is addressed in high-performance building assessment systems such as the USGBC LEED-NC building rating system. For example, LEED-NC awards 1 credit (MR Credit 2.1) for diverting at least 50 percent of construction and demolition waste from landfilling and 2 credits (MR Credit 2.2) for diverting 75 percent or more of this waste stream. Second, high-performance buildings address the generation of solid waste by building occupants by calling for the allocation of building space for the collection and storage of recyclables. In fact, LEED-NC makes this allocation of space a prerequisite for achieving a rating, thereby making it a mandatory requirement (MR Prerequisite 1). Third, high-performance buildings address the

TABLE 13.9

Summary of Energy and Cost Savings for the NREL Prototype Building*

	Base Case	High-Performance
Area (square feet)	20,000	20,000
Total cost	$2,400,000	$2,440,000
Incremental cost	NA	$40,000
Annual energy use (BTUs)	730 million	477 million
Annual energy cost	$11,800	$7,490
Reduction in energy use	NA	34.6%
Reduction in energy cost	NA	36.7%
Simple payback, energy	NA	8.7 years
Simple payback, energy and emissions	NA	6.0 years

*Base case and high-performance case, with simple payback for energy alone and for energy and emissions.

use of recycled content and reuse of building materials, thus creating incentives and demand for closing materials loops and reducing the landfilling of solid waste. LEED-NC provides 1 credit (MR Credit 3.1) for 5 percent resource reuse and 2 credits (MR Credit 3.2) for 10 percent resource reuse. For recycled content, 1 credit (MR Credit 4.1) is provided if 5 percent of materials have postconsumer recycled content or 2 credits (MR Credit 4.2) for 10 percent postconsumer recycled content. Alternatively, LEED-NC provides 1 credit for a 10 percent total of postconsumer plus one-half of postindustrial content and 2 credits for a 20 percent total of postconsumer plus one-half of postindustrial content.

The financial benefits of diverting construction and demolition waste from landfilling can be readily calculated. For a nominal U.S. construction project, waste is generated at the rate of about 7 pounds per square foot (32 kilograms per square meter). The actual savings are a function of the diversion rate. Table 13.11 itemizes the savings from construction waste diversion as a function of diversion rate and tipping fees—that is, the cost of disposal.

QUANTIFYING THE BENEFITS/COSTS OF BUILDING COMMISSIONING

One of the hallmarks of high-performance buildings is that, upon completion of the building, all systems are carefully checked and validated through testing. As a consequence of this movement, building commissioning has become a new profession. Commissioning professionals are engaged in the project from the start, along with members of the design and construction professions. And although commissioning

TABLE 13.10

Avoided Emissions and Annual Benefit for the NREL Prototype Building: High-Performance Case Compared to Base Case

Emission Type	Tons of Emissions Avoided per Year	Annual Benefit
Sulfur dioxide	0.16	$1,090
Nitrous oxide	0.08	$800
Carbon dioxide	10.7	$107
Total	10.94	$1,997

TABLE 13.11

Savings for Diverting Construction Waste from Landfill for the NREL Prototype Building*

Diversion Rate	$50/Ton Tipping Fee	$75/Ton Tipping Fee	$100/Ton Tipping Fee
0%	$0	$0	$0
50%	$1,750	$2,625	$3,500
75%	$2,625	$3,938	$5,250

*Assuming 7 pounds per square foot (32 kilograms per square meter) waste generation for various diversion rates and tipping fees.

does add extra cost to a building, the value of this service is substantial, because it provides assurance that the building will perform as designed. Costs of commissioning for typical buildings are shown in Table 13.12.[16] The benefits of building commissioning are difficult to quantify, but current general practice is to attribute a 10 percent energy savings to commissioning. In the case of the NREL prototype building used as an example in this chapter to quantify energy savings, the payback period for building commissioning is less than 4 years.

QUANTIFYING MAINTENANCE, REPAIR, AND MISCELLANEOUS BENEFITS/COSTS

In attempting to minimize LCC, high-performance buildings are specifically designed to lower maintenance costs, but they can also produce lower costs in other areas. The following are examples of design features that can provide these additional economic benefits for high-performance buildings:[17]

- Durable materials
  - Fluorescent lighting systems with long-life, 10,000-hour lights in place of short-life, 1,000-hour incandescent lights
  - Fly ash and blast furnace slag concrete with higher durability compared to conventional concrete mix design
  - Low-emissions paints with higher durability compared to conventional paints
  - Light-colored roofing materials that have longer life than conventional roofing materials
- Repairability
  - Recycled content carpet tiles that can be replaced in worn areas
  - Mechanical and electrical systems designed for ease of repair and replacement by virtue of space allocation and physical arrangement of equipment, piping, conduit, power and control panels, and other components

TABLE 13.12

Commissioning Costs for Typical New Construction

Scope of Commissioning	Cost
Whole building	0.5–1.5% of construction cost
HVAC and control systems	1.5–2.5% of mechanical system cost
Electrical systems	1.0–1.5% of electrical system cost
Recommissioning existing buildings	$0.17 per square foot ($1.83 per square meter)

- Miscellaneous Costs
  - Designing buildings with areas for recycling that reduce waste disposal costs
  - Sustainable landscape design that reduces the need for irrigation, fertilizer, herbicides, and pesticides
  - Stormwater management using constructed wetlands instead of sewers

Quantifying the financial benefits of improved maintenance and repair must, of course, be accomplished on a case-by-case basis and can be difficult to carry out because a database containing this type of information is not readily available. For the sustainable landscape design and stormwater management entries listed above under "Miscellaneous Costs," an example of how to present the savings is provided in Table 13.13 for the NREL prototype buildings used to illustrate energy savings in this chapter.[18]

These two site-related strategies for the NREL prototype buildings are described in more detail here:

> *Sustainable landscape design.* A mixture of native warm-weather turf and wildflowers is used to create a natural "meadow" area. This strategy is compared with traditional turf landscaping of Kentucky blue grass, which requires substantially more irrigation, maintenance, and chemical application.
>
> *Sustainable stormwater management.* An integrated stormwater management system combines a porous gravel parking area with a rainwater collection system, where rainwater is stored for supplemental irrigation of native landscaping. This porous, gravel-paved parking area is a heavy load-bearing structure filled with porous gravel, allowing stormwater to infiltrate the porous pavement (reducing runoff) and to be moved into an underground rainwater collection system. The water can be used to supplant fresh water from the public supply for uses that do not require potable water. This sustainable system is compared to a conventional asphalt parking area and a standard corrugated pipe stormwater management system without rainwater harvesting.

Although the particular sustainable stormwater system used for the prototype increases the total construction cost by a little over $3,000 (about 0.1 percent of the total building construction cost), it saves over $500 annually in maintenance costs because less labor is required for patching potholes and performing other maintenance

TABLE 13.13

Economic Comparison of Sustainable Stormwater Management and Landscape Practices for the NREL Prototype Buildings

	Incremental First Cost	Incremental First Cost/ 1,000 sq ft (100 square meters)	Total Incremental Cost	Annual Cost Savings/ 1,000 sq ft (100 square meters)	Total Cost Savings	Simple Payback (Years)
Sustainable stormwater management	$3,140	$157 ($169)	$3,140	$28.30 ($30.45)	$566	5.6
Sustainable landscape design	$2,449	$122 ($131)	$2,440	$152.00 ($163.55)	$3,040	0.8

on an asphalt lot. The resulting payback period is less than 6 years. The sustainable landscaping approach shows even more favorable economics: the incremental first cost is nearly $2,500, but this is repaid in less than 1 year with an annual O&M costs savings of $3,045 in avoided maintenance, chemical, and irrigation costs.

Managing First Costs

For many organizations, especially state and local governments, the first or capital cost is the primary factor in making decisions about a project because legislation often dictates the maximum investment in a specific type of building. For example, in Florida, the new school construction cost per student station is limited to approximately $13,500 for elementary schools, $15,500 for middle schools, and $20,500 for high schools. For many other potential green building clients, a similar situation exists, with decision makers heavily constrained by construction cost limitations. Coping with these circumstances requires careful consideration of strategies for producing a high-performance building when LCC may be difficult to bring into the process. The following is a list of recommendations for managing first costs for high-performance building projects:[19]

1. Make sure that senior decision makers support the concept.
2. Set a clear goal early in the process. Ideally, the decision to go green should be made before soliciting design proposals so that contract language reflects the green goal, thus permitting more flexibility in decision making. Certain green measures that can save money (such as site planning) have to be done early.
3. Write contracts and request for proposals (RFPs) that clearly describe your sustainability requirements. For example, specify whether the goal is a LEED Silver rating or the equivalent.
4. Select a team that has experience with sustainable development. Hiring a mechanical/electrical/plumbing (MEP) firm with green experience alone can save 10 percent of the MEP construction costs. Look for team members with a history of creative problem solving.
5. Encourage team members to get further training and develop sources of information on green materials, products, and components and technical/pricing information on advanced systems.
6. Use an integrated design process. Do not make the green components add-ons to the rest of the project. Integrate all the candidate green measures into the base budget. Establishing an integrated design can lead to capital savings. Investing 3 percent of total project costs during design can yield at least 10 percent savings in construction through design simplifications and fewer change orders.
7. Understand commissioning and energy modeling. To minimize up-front costs, use a sampling approach for building commissioning.
8. Look for rebates and incentives from states, counties, cities, and utilities.
9. Educate the decision makers without inundating them with technical information. Stay focused on their objectives. Respect their sense of risk aversion.
10. Manage your time carefully. Select one or two team members to oversee research on green products and systems. Set a specific deadline for research results, and give the discovery manager the power to cut off research.

The following are some design and construction strategies that a team can use to reduce first costs:[20]

- *Optimize site and orientation.* One obvious strategy to reduce first costs is to apply appropriate siting and building orientation techniques to capture solar radiation for lighting and heating in winter, and shade the building using vegetation or other site features to reduce the summer cooling load. Fully exploiting natural heating and cooling techniques can lead to smaller HVAC systems and lower first costs.
- *Reuse/renovate older buildings and use recycled materials.* Reusing buildings, as well as using recycled materials and furnishings, saves virgin materials and reduces the energy required to produce new materials. Reusing buildings may also reduce the time (and therefore money) associated with site planning and permitting.
- *Reduce project size.* A design that is space-efficient yet adequate to meet the building objectives and requirements generally reduces the total costs, although the cost per unit area may be higher. Fully using indoor floor space and even moving certain required spaces to the exterior of the building can reduce first costs considerably.
- *Eliminate unnecessary finishes and features.* One example of eliminating unnecessary items is choosing to eliminate ornamental wall paneling, doors (when privacy is not critical), and dropped ceilings. In some cases, removing unnecessary items can create new opportunities for designers. For example, eliminating dropped ceilings might allow deeper daylight penetration and reduce floor-to-floor height (which can reduce overall building dimensions).
- *Avoid structural overdesign and construction waste.* Optimal value engineering and advanced framing techniques reduce material use without adversely affecting structural performance. Designing to minimize construction debris (e.g., using standard-sized or modular materials to avoid cutting pieces and generating less construction waste) also minimizes labor costs for cutting materials and disposing of waste.
- *Fully explore integrated design, including energy system optimization.* As discussed previously, integrated design often allows HVAC equipment to be downsized. Models such as DOE-2 allow the energy performance of a prospective building to be studied and the sizing of mechanical systems to be optimized. Using daylighting and operable windows for natural ventilation can reduce the need for artificial lighting fixtures and mechanical cooling, thereby lowering first costs. Beyond energy-related systems, integrated design can also reduce construction costs and shorten the schedule. For example, by involving the general contractor in early planning sessions, the design team may identify multiple ways to streamline the construction process.
- *Use construction waste management approaches.* In some locations, waste disposal costs are very high because of declining availability of landfill capacity. For instance, in New York City, waste disposal costs exceed $75 per ton ($82 per metric ton). In such situations, using a firm to recycle construction waste can decrease construction costs because waste is recycled at no cost to the general contractor, thereby saving disposal costs.
- *Decrease site infrastructure.* Costs can be reduced if less ground needs to be disturbed and less infrastructure needs to be built. Site infrastructure can be decreased by carefully planning the site, using natural drainage rather than storm sewers, minimizing impervious concrete sidewalks, reducing the size of roads and parking lots (e.g., by locating near public transportation), using natural landscaping instead of traditional lawns, and reducing other man-made

infrastructure on the site, when possible. For example, land development and infrastructure costs for the environmentally sensitive development on Dewees Island, off the coast of Charleston, South Carolina, were 60 percent below average because impervious roadway surfaces and conventional landscaping were not used.

An excellent study of construction costs for green buildings was conducted by Lisa Fay Matthiessen and Peter Morris of Davis Langdon, a cost consulting company.[21] Their report suggests that there is no statistical difference between high-performance green buildings that used LEED-NC for guidance and conventional buildings, that is, the cost per square foot falls into the same range of costs for both green and conventional buildings of a similar program type. The majority of LEED-NC certified buildings examined by the authors did not require additional funding, and where additional costs were incurred, they were due to certain extraordinary specific features such as photovoltaics. The factors that influence the cost of a green building are:

- *Demographic location:* The location of a project, rural versus urban, creates opportunities and problems in obtaining LEED-NC points. For example, points for transportation and urban development are readily available in urban settings, while stormwater management innovations are more likely in rural areas.
- *Bidding climate and culture:* In some states, such as California, contractors and subcontractors are far more familiar with LEED-NC and are less likely to perceive a project as risky, thus lowering costs.
- *Local and regional design standards, codes, and initiatives:* In states such as Oregon and Pennsylvania, where there has been significant government support of green building efforts, the costs are generally lower because green buildings are more likely to be considered the norm.
- *Intent and values of the project:* A clear statement that the owner is serious about the green building concept will motivate the project team members and ensure that green building features are incorporated from the onset of the project, thus lowering overall costs.
- *Climate:* The paybacks for energy-conserving features vary by location because the costs of energy also vary by geographic region. Additionally, some aspects of passive design may be difficult to achieve in very hot, humid, or very cold climates. As a result, more complex and costly active systems are needed to meet the operational requirements of the owner.
- *Timing and implementation:* Fully incorporating green features from the start of design and ensuring their detailed integration into the project will result in lower costs.
- *Size of the building:* Larger, more complex buildings will typically have higher costs for larger, more complex systems simply due to the scale of the project.
- *Synergies:* Selecting systems that have multiple benefits will produce lower costs. For example, a well-designed landscape can integrate stormwater management and building shading and can be designed to require no irrigation, saving infrastructure and lowering operational costs.

The Davis Langdon study also noted that a well-developed budget methodology could go a long way toward reducing construction cost impacts. The authors recommend that the following measures be followed at every step of design and construction to keep a green building construction within budget:

- Establish team goals, expectations, and expertise.
- Include specific goals in the program.
- Align the budget with the program.
- Stay on track during design and construction.

Integrating green building goals into the project, having appropriate expertise and commitment in the project team, and detailed planning are perhaps the key elements in keeping costs aligned with the budget. In this respect, green building projects are no different from any other well-organized and well-run building project except for the inclusion of team knowledge of the green building concept and requirements. Experience to date is that the learning curve for obtaining the requisite knowledge is not very steep, and that training in and exposure to one green building project provide the foundation for successfully tackling other similar projects.

Tunneling Through the Cost Barrier

The preferred design approach used to create a high-performance green building is sometimes referred to as *integrated design,* which is covered in detail in Chapter 4. The fundamental assumption of integrated design is that by bringing the various disciplines together and forcing them out of their silos, a wide variety of synergies is possible. One of the most commonly cited synergies is in the design of the building energy systems, where architects and mechanical engineers collaborate on the details of the building envelope, resulting in a smaller HVAC plant. The present approach to building design does not promote sustainability because the designers, architects, and engineers each optimize the systems they design, generally resulting in a suboptimal building. Additionally, the fee structures for design professionals are such that maximizing cost and complexity can result in higher fees, clearly the wrong motivation when it comes to creating superior buildings. Consequently, finding the synergies that will produce truly high-performance buildings is a struggle, requiring changes in both attitudes and design contracts.

Amory Lovins of the Rocky Mountain Institute (RMI) describes the effects of producing integrated design synergies as *tunneling through the cost barrier* because the result can be a dramatic reduction in first or capital costs. One example cited by Lovins was the design of an industrial process for the carpet maker Interface for a plant in Shanghai. The initial design for this process called for 95 horsepower of pumping power. When Jan Schilhan of Interface examined the design, he threw out the assumptions engineers normally use for sizing pipes, making the pipes larger in diameter, thus greatly reducing pipe friction because fluid velocity was greatly reduced. Because friction follows with the fifth power of the pipe diameter, doubling the pipe diameter results in a friction reduction of 86 percent, so that pumping power falls by the same amount. Also, contrary to common design practices, Schilhan laid out the pipes with minimal bends and with the pipe lengths as short as possible, because each bend and each foot of pipe causes additional friction losses. This redesign reduced pumping power from the original 92 horsepower to 7 horsepower, a 92 percent or Factor 12 improvement. The result of these changes was not only a significant reduction in energy consumption but also a significant reduction in capital cost due to the far smaller pumps, reduced piping complexity, and a smaller electrical service, far offsetting the slightly higher cost of larger-diameter piping.[22]

Many of the tradition-rooted assumptions used by engineers and architects often result in poor design practices that persist for decades, even generations. Challenging these assumptions is important if superior buildings with lower capital costs are the desired outcome. The key, according to Lovins, is *whole-system engineering,* in which all the benefits of a technology are counted, not just, for example, the energy savings benefits. High-efficiency electric motors have as many as 18 benefits, and superwindows have as many as 10 benefits including better daylighting, radiant comfort, no condensation, and noise blocking, to name but a few. Buildings have ample opportunity for synergies and cost reductions, many of them as yet unexplored. One area ripe for exploration is the integration of buildings into local ecosystems and geological formations. Trees have enormous capacity for stormwater uptake and can selectively allow sunlight to fall on buildings, depending on the time of year, as their leaves can block and absorb solar radiation during the summer and, by dropping off the tree in the fall, allow penetration of the sun during winter days. Living roofs on buildings provide insulation, reduce the heat island effect, store stormwater, and replace the ecological footprint removed by the building. Greenery integrated into buildings contributes to a healthy experience for occupants, as suggested by the Biophilia Hypothesis (see Chapter 2). Coupling the building with the ground and groundwater can help provide heating and cooling while lowering energy consumption. Wetlands and constructed wetlands could also benefit the built environment via wastewater treatment and stormwater storage, leading to reduced capital costs.

Lovins suggests four principles as aiding the attempt to tunnel through the cost barrier.[23]

1. *Capture multiple benefits from single expenditures.* By dematerializing buildings, for example, it may be possible to provide more space at lower cost while proportionately reducing environmental impacts. High-efficiency lighting reduces electrical energy requirements and reduces the heat load to the space and can be coupled with occupancy and daylight sensors.
2. *Start downstream to turn compounding losses into savings.* Rather than focusing on the fan power required to push air through ductwork, more attention on reducing friction losses in ductwork through better layout, reducing the length of duct runs, eliminating unnecessary bends, and increasing the duct cross section results in far lower fan horsepower, smaller and less costly equipment, and quieter operations. Going one step further downstream, designing systems that heat and cool only the bottom 6 feet (1.8 meters) or so of vertical zones, where the occupants actually are, further reduces energy consumption. This is the strategy know as *displacement ventilation* (described in Chapter 7), and by delivering air from an underfloor plenum, it can help reduce floor-to-floor heights. Reducing this dimension results in lower overall building heights and lower material costs.
3. *Get the sequence right.* If the issue is health and productivity, thinking through how people will use the space, have access to daylight, views, and preferably greenery, and maximizing the amount of natural light falling on their workspaces should be the first and foremost matters for consideration. The lighting systems should be designed only after the primary human factors are considered. The result: a better indoor environment and lower energy costs.
4. *Optimize the whole system and not the parts.* This is the crux of whole-system engineering, a collaborative effort among architects and engineers to jointly and creatively design the building and its systems. In Germany, for example, the design disciplines have collaborated to create buildings that have superb passive design, totally eliminating the need for cooling systems, resulting in buildings using one-seventh of the primary energy of conventional U.S. buildings. This represents the essence of integrated design and cost barrier tunneling.

Summary and Conclusions

High-performance buildings have enormous potential benefits: for their owners, for the environment, and for society in general. The ability to express and clearly justify these benefits in an economic analysis is an important factor in determining whether or not the project will be conventional or high-performance in its design and construction. Evidence is beginning to emerge that provides information and tools for the building team to use in developing a model that addresses both hard and soft costs. Hard-cost savings on energy, water, and wastewater are fairly straightforward to quantify and include in an economic analysis. Soft costs, such as human health and productivity savings, as well as savings due to building commissioning, are not so straightforward to justify; hence, care must be exercised when including them in a cost analysis. Hopefully, additional verifiable, peer-reviewed data will emerge in the coming years and the decision to include these data in a green building project analysis will be far easier than it is at present.

Notes

1. Cited in the Executive Summary of "Green Building Costs and Financial Benefits" by Kats (2003).
2. Derived from a survey of 33 green buildings conducted in 2003 by Gregory Kats of Capital-E Analysis for the State of California and the USGBC and reported in Kats (2003).
3. Excerpted from Kats (2003).
4. Kats (2003).
5. From "The Business Case for Sustainable Design in Federal Facilities" (October 2003).
6. Available in the Members section of the USGBC website, www.usgbc.org.
7. Paraphrased from Kats (2003).
8. An excellent description of the ING Bank building can be found in von Weizsäcker, Lovins, and Lovins (1997). This book had great influence on high-performance buildings because it suggested that reducing resource consumption by 75 percent was necessary to achieve sustainability and that, furthermore, the technologies needed to support this reduction already existed. A follow-on concept, Factor 10, suggests that long-term sustainability would require a 90 percent reduction in resource consumption.
9. Both tables are excerpted from "The Business Case for Sustainable Design in Federal Facilities" (October 2003).
10. Excerpted from "Big Savings from Waterless Urinal," *Environmental Building News* (February 1998).
11. From Fisk (November 2000).
12. From "Daylighting in Schools" by the Heschong Mahone Group (August 20, 1999). This company, which specializes in building energy efficiency, has published several landmark reports on the correlation between daylighting and student performance. Recent reports for the California Energy Commission are available from the company website, www.h-m-g.com/.
13. Reported in "Skylighting and Retail Sales" by the Heschong Mahone Group (August 20, 1999).
14. Productivity and health gains are from "The Costs and Financial Benefits of Green Buildings" (Kats, October 2003).
15. Adapted from *2002 Buildings Energy Databook* (2002).
16. Adapted from "What Can Commissioning Do for Your Building?" (1997).
17. Adapted from "The Business Case for Sustainable Design in Federal Facilities" (October 2003).
18. Ibid.
19. Adapted from Syphers, Sowell, Ludwig, and Eichel (November 2003).
20. From "The Business Case for Sustainable Design in Federal Facilities" (October 2003).
21. From Matthiessen and Morris (2004).
22. From Hawken, Lovins, and Lovins (1999), chapter 6.
23. Derived from Lovins (Summer 1997).

References

2002 Buildings Energy Databook. 2002. U.S. Department of Energy. Available at http://buildingsdatabook.eren.doe.gov.

"Big Savings from Waterless Urinal," February 1998. *Environmental Building News,* Vol 7, No. 2.

"The Business Case for Sustainable Design in Federal Facilities." October 2003. Resource Document, U.S. Department of Energy. Available at www1.eere.energy.gov/femp/pdfs/bcsddoc.pdf.

"Daylighting in Schools," August 1999. Heschong Mahone Group. "A report for The Pacific Gas and Electric Company." Available at the Heschong Mahone Group website, www.h-m-g.com/downloads/daylighting/schools.pdf.

Fisk, William J. November 2000. "Health and Productivity Gains from Better Indoor Environments and Their Relationship with Building Energy Efficiency," *Annual Review of Energy and the Environment,* 25, pp. 537–566.

Hawken, Paul, Amory B. Lovins, and L. Hunter Lovins. 1999. *Natural Capitalism.* New York: Little, Brown.

Kats, Gregory H. October 2003. "The Costs and Financial Benefits of Green Buildings." A report developed for California's Sustainable Building Task Force. Available at the Capital-E website, www.cap-e.com.

Kats, Gregory H. 2003. "Green Building Costs and Financial Benefits." Written for the Massachusetts Technology Collaborative. Available at the Capital-E website, www.cap-e.com/.

Lovins, Amory B. Summer 1997. "Tunneling through the Cost Barrier: Why Big Savings Often Cost Less Than Small Ones," *Rocky Mountain Institute Newsletter,* XIII(2), pp. 1–4.

"Making the Business Case for High Performance Green Buildings." 2003. U.S. Green Building Council. Available at www.usgbc.org.

Matthiessen, Lisa Fay, and Peter Morris. 2004. "Costing Green: A Comprehensive Cost Database and Budgeting Methodology." Davis Langdon. Available at www.davislangdon.com/pdf/USA/2004CostingGreen.pdf.

"Skylighting and Retail Sales," August 1999. Heschong Mahone Group, "A report for the Pacific Gas and Electric Company." Available at www.pge.com.

Syphers, Geof, Arnold Sowell, Jr., Ann Ludwig, and Amanda Eichel. "Managing the Cost of Green Buildings." November 2003. Published in the "White Paper on Sustainability," a supplement to *Building Design & Construction.* Available at www.bdcmag.com.

von Weizsäcker, Ernst, Amory B. Lovins, and L. Hunter Lovins. 1997. *Factor Four: Doubling Wealth, Halving Resource Use.* London: Earthscan.

"What Can Commissioning Do for Your Building?" 1997. Portland, OR: Portland Energy Conservation, Inc. (PECI).

Chapter 14

The Cutting Edge and Beyond

The contemporary high-performance green building movement is rapidly gaining momentum in the United States and other countries; it is transforming the entire process of creating the built environment, from design through construction and operation. It is affecting not only new construction but also renovations to existing buildings, building products, design tools, and the education of built environment professionals. In the United States, the USGBC LEED building assessment suite of standards now defines what constitutes a high-performance green building. Although LEED has been an enormous success in the marketplace, two questions remain: What is the ultimate goal of building assessment standards such as LEED and Green Globes, and how will they evolve over time to improve the buildings currently being produced that are using them as guidance? Because of the success of the LEED-NC building assessment standard, the USGBC is focused almost exclusively on the implementation of the existing suite of LEED standards for new construction and for existing buildings, and is working to generate and implement other standards in the suite to cover areas of importance such as commercial interiors and residential construction. Consequently, a long-term vision of what constitutes the high-performance building of the next generation is lacking and, as a result, is hampering progress toward a truly sustainable built environment.

In this final chapter, the cutting edge and future high-performance green buildings are addressed for the purpose of stimulating thinking about the long-range goals of this movement. The first section addresses the emerging issue of *passive survivability*, a new building theme embraced by the green building community in the wake of Hurricane Katrina in 2005. Although not yet being incorporated into new buildings, it is on the cusp of consideration and fits nicely into the general philosophical approach underpinning high-performance green buildings. The second section contains several case studies of newer green buildings to illustrate the best practices being employed today. These buildings, of course, point the way to the future and what may possibly be the norm for the green buildings of the future. The future is uncertain and many different outcomes can be hypothesized, and certainly not all can be covered in detail here. In the third section of this chapter, three main approaches to designing future green buildings are proposed, one based on history, another on technology, and a third on ecology. These represent the main attractors for strategies in this arena, although the likely outcome will be a hybrid of these widely differing but potentially equally successful approaches.

Passive Survivability

Recent severe weather events are causing a shift in thinking that would result in buildings having the capability of assisting human survival in the wake of natural or human-induced disasters. During the Chicago heat wave of 1995, the deaths of more than 700 people in their homes or apartments were attributed to high temperatures. In many apartments, temperatures remained in excess of 90°F (32°C) even at night.

The death toll could have been far higher had Chicago lost power during the heat wave. Ten years after the Chicago heat wave, New Orleans was struck by Hurricane Katrina in September 2005, resulting in thousands of deaths, incredible suffering, enormous dislocation of residents, and severe economic impacts. Temperatures in the New Orleans Superdome rose to 105°F (42°C), creating dangerous conditions inside the very structure where people were sent to survive the immediate aftermath of Katrina. *Passive survivability* is a new term being used to describe how buildings should be designed and built to assist the survival of their human occupants in the wake of disasters. In an editorial in *Environmental Building News* (EBN) in November 2005, passive survivability was defined by Alex Wilson as ". . . the ability of a building to maintain critical life-support conditions if services such as power, heating fuel, or water are lost for an extended period."[1] The term *passive survivability* was first used by the military to describe measures taken to ensure that military vehicles are able to survive attacks. It was included in a set of proposals called *The New Orleans Principles,* resulting from a reconstruction conference held in Atlanta in November 2005.[2] One of these proposals states, "Provide for passive survivability: Homes, schools, public buildings, and neighborhoods should be designed and built or rebuilt to serve as livable refuges in the event of crisis or breakdown of energy, water, and sewer systems."

The fact of climate change, and the probability of higher temperatures and more frequent and more violent hurricanes, should be sufficient to cause a shift to using passive survivability as a design criterion. Backup generators are unlikely to be able to provide the power needed for ventilation and air conditioning for extended periods of time; consequently, buildings need to have several key design features that help ensure passive survivability. Among these key green design features are cooling load avoidance, capability for natural ventilation, a high-efficiency thermal envelope, passive solar gain, and daylighting. In May 2006, EBN published the checklist in Table 14.1 that lays out passive survivability strategies for buildings.

Most of the preliminary efforts at passive survivability have addressed the very real problem faced by regions prone to hurricane activity, thought to be on the increase due to climate change. The same basic principles apply to areas that may be subject to severe winter conditions such as blizzards and ice storms, the emphasis shifting to providing the capability for heating, either through passive solar design or the use of local energy resources such as wood. A 1998 ice storm in eastern Canada left 4 million people without power and forced 600,000 people from their homes, with 28 fatalities, indicating that persons living in colder climates also should consider passive survivability strategies for their built environment. Earthquakes have not yet been addressed in the preliminary literature on passive survivability, although in principle buildings that are designed to survive earthquakes may still have downed utilities and should have the added capability of passive survivability. It is clear, then, that different regions will have different approaches to passive survivability that will depend on the weather and the typical natural hazards in that region.

Passive survivability should also be extended to infrastructure. Cisterns can be located throughout a community and under streets for an emergency water supply and for fire protection. Key control and communications systems such as traffic signals and streetlights could have solar-charged power backups. Sewage infrastructure could also be planned to have normal and passive survivability functions.

Exactly how passive survivability can help mitigate the effects of terrorist attacks remains an open question. Clearly, any area of the country can be subject to the effects of terrorism. Attacks directed at utility infrastructure could be mitigated by a shift to passive survivability as a criterion for building. The impacts of biological or nuclear attacks could also be mitigated, at least for some period of time, by passive survivability, although it is likely that systems that would seal the building and protect the occupants from airborne biological agents or radioactivity would likely not be incorporated into typical construction. The wide variety of potential attacks makes designing buildings for all eventualities impossible. However, for attacks

TABLE 14.1

Checklist for Designing Passive Survivability into Buildings

1. *Create storm resilient buildings.* Design and construct buildings to withstand reasonably expected storm events and flooding.
2. *Limit building height.* Most tall buildings cannot be used during power outages due to their reliance on elevators and air conditioning, and a maximum height of six to eight stories is recommended.
3. *Create a high-performance envelope.* A well-insulated thermal envelope with high-performance glazings will assist in maintaining a reasonable interior temperature.
4. *Minimize cooling loads.* Proper building orientation, overhangs, shading, and high performance glazing can minimize building heat loads.
5. *Provide for natural ventilation.* Provisions for natural ventilation, such as chimney effect air movement, even for buildings that would be normally air conditioned, would provide fresh air for the occupants.
6. *Incorporate passive solar heating.* In climates where heating may be the survivability issue, thermal mass and thermal storage walls can be used to help provide thermal energy for heating.
7. *Provide natural daylighting.* The same daylighting strategies used for green buildings also provide light in a passive survivability mode.
8. *Provide solar water heating.* Solar thermal systems coupled with PV-powered pumps can provide hot water during power outages.
9. *Provide photovoltaic power.* PV can provide electrical energy during outages and, with battery storage, can also provide electricity at night. Note that PV panels need to be mounted and protected from high winds and flying debris.
10. *Configure heating equipment to operate on PV power.* Gas- and oil-fired heating equipment is often dependent on electrical power for operation, and equipment may have to be configured to accept DC power from PV panels or have an inverter to provide AC power.
11. *Where appropriate, consider wood heat.* Especially in rural areas, low-pollution wood-burning stoves, masonry heaters, or pellet stoves can provide heating.
12. *Store water on site; consider using rainwater to maintain a cistern.* Water storage for extended outages can be provided by a cistern. Storing water high in the building, for example, on the roof, can provide pressure with no need for pumps.
13. *Install composting toilets and waterless urinals.* Fixtures that do not rely on water for flushing have a distinct advantage in the aftermath of disasters.
14. *Provide for food production in the site plan.* Land can be set aside for fruit-bearing trees and shrubs as a source of food in passive survival mode.

directed against infrastructure, buildings can certainly be provided with key features that assist the occupants in having a safe place, reasonable temperatures, ventilation, and potable water, the key elements of survival.

The list of measures that can be included in a strategy for passive survivability is remarkably similar to a list of typical green building measures (Table 14.1).[3] Indeed, an argument in support of incorporating passive survivability measures into buildings could also be considered an argument in favor of green building.

Cutting Edge: Case Studies

Of all the high-performance buildings either registered or certified in the United States, several could be considered at the cutting edge of practice, among them the San Francisco Federal Building and the Philadelphia Forensic Science Center. These projects, and the aspects that make them cutting edge high-performance buildings, are described below.

The San Francisco Federal Building

The new 18-story San Francisco Federal Building is referred to by its owner, the General Services Administration (GSA), as "a model of excellence" and rightfully so. Located on a 3-acre site in the South of Market Street neighborhood at the intersection of Seventh and Mission Streets, just a 10-minute walk from downtown, it is a long, slender, translucent tower, 60 feet (18 meters) wide and 234 feet (71 meters) high, providing 600,000 gross square feet (55,742 square meters) of usable space. It is a federal government complex serving the Social Security Administration, the Department of Labor, the Department of Health and Human Services, and the Department of Agriculture. The design was led by Thom Mayne of Morphosis Architects in a major collaboration with the Los Angeles office of Ove Arup for the integrated structural and mechanical design; with Horton Lees Brogden of Culver City, California, for lighting and daylighting design; and with the Building Technology Department of the Lawrence Berkeley National Laboratory for modeling the natural ventilation system. The Smith Group of San Francisco served as Executive Architect and executed all interior space planning for the tenant agencies. The goal of the project was to provide a high-quality government work space within the project budget of $144 million. *High quality* in this context meant that the work space had to be efficient, secure, and flexible to allow change.

The San Francisco Federal Building actually consists of several components, the 18-story tower being the dominant feature. A four-story, broader structure at the southwest base of the tower houses the Social Security Administration, an agency that generates substantial pedestrian traffic and is served by a separate entry for the public. In close collaboration with the ethnically diverse local commu-

Figure 14.1 The San Francisco Federal Building, designed by Morphosis Architects, is a breakthrough structure, with an outstanding passive design strategy coupled with active control systems and all building components optimized for connecting the building to its surrounding environment for cooling, ventilation, and lighting. (Photograph by Petros Raptis)

nity, a rich mix of Filipinos, Mexicans, Vietnamese, and other minority groups, the project team developed the building to provide a landscaped plaza that acts as a bridge to the local community, serving as a local asset, and accommodating the substantial pedestrian activity in the area. The skin of the building unfolds to cover a day-care facility, and a freestanding cafeteria rounds out the facilities on the site. The publicly accessible day-care center and cafeteria are used by both the employees of the Federal Building and the local community, providing an architectural solution with a socially responsible dimension. The design of the building responded to the local residents' desire not to have a massive building that would overshadow the two- and three-story light industrial, commercial, and residential structures (including artists' studios, senior housing, and single-room-occupancy units) that provide the eclectic character of the neighborhood.

EXCELLENCE IN DAYLIGHTING

Lighting for office buildings in the United States is the single largest energy consumer for this building type, accounting for up to 40 percent of the total energy. Consequently, minimizing artificial lighting can have significant economic and environmental benefits. The narrow floor slab—just 65 feet (20 meters) wide—the use of floor-to-ceiling glazing, and a floor-to-floor height of 13 feet (4 meters) provide perfect conditions for substantial, deep-penetrating daylight. In contrast to normal practice, the perimeter of the building has open plan offices, with 52-inch-high partitions that minimize the amount of light being blocked. The interior core contains meeting rooms and enclosed offices, all with clear glass panels to allow natural light to penetrate throughout the space. Fritted glass has been provided for these interior spaces for privacy when needed. The southeast face of the 18-story tower is covered with perforated panels that rotate to control light and provide unobstructed views across the city. The lighting system contains sensors that provide feedback to reduce artificial lighting as daylighting increases during the day and turn off lights when there are no occupants in a space. Task lights at workstations are on only when people are present in the spaces. The net result of the lighting strategies employed in the Federal Building is a 26 percent reduction in lighting energy.

Figure 14.2 The folded, perforated metal skin covering portions of the San Francisco Federal Building assists in the flow of air through the structure and provides an interesting and appealing appearance for the structure, both at ground level and at upper levels of the façade. (Photograph 2007 Jenna Hildebrand)

NATURAL VENTILATION STRATEGY

As was noted in Chapter 7, it is becoming standard practice in Germany to use natural ventilation as the strategy for cooling office buildings even during peak summer days. The result is that state-of-the-art German office buildings use just 100 kWh/square meter (30 kWh/square feet) of annual primary energy, about 20 percent the consumption of code-compliant U.S. buildings. Buildings employing passive cooling strategies based on natural ventilation are rare in the United States, especially large buildings. The San Francisco Federal Building embraces passive cooling and ventilation, taking advantage of the 49°F to 65°F (9°C to 18°C) air currents around the building, exploiting them via the design of building elements that allow and facilitate the deep penetration and circulation of outside air. A combination of computer-controlled air vents at floor level and occupant interaction with windows permits the use of these breezes to provide a comfortable and healthy interior environment. The air currents are admitted through openings on the northwest façade and vented through the southeast wall. The open office spaces are designed so as not to impede airflow across the floor, and even enclosed offices and meeting rooms have walls that stop short of the floor above, providing a pathway for air to cross the building. In the evening the air currents cool the concrete structure, providing a cool sink for the following day. The southeast façade is covered with a perforated metal sunscreen that also helps induce airflow across the face of the building between the sunscreen and the façade, creating a pressure drop that induces warm airflow out of the building. Solid narrow walls on the northeast and southwest sides contain the fire stairs and thus minimize heat gain on those sides of the building. Lower levels of the building require some mechanical cooling, and an innovative underfloor air distribution system combined with conventional heat pumps is used to meet the requirements of these zones. The natural ventilation strategy provides cooling for the building from mid-April through mid-October. November and March are swing months during which the building operates optimally with windows closed and no active heating. During the colder months of December through February a hydronic heating sys-

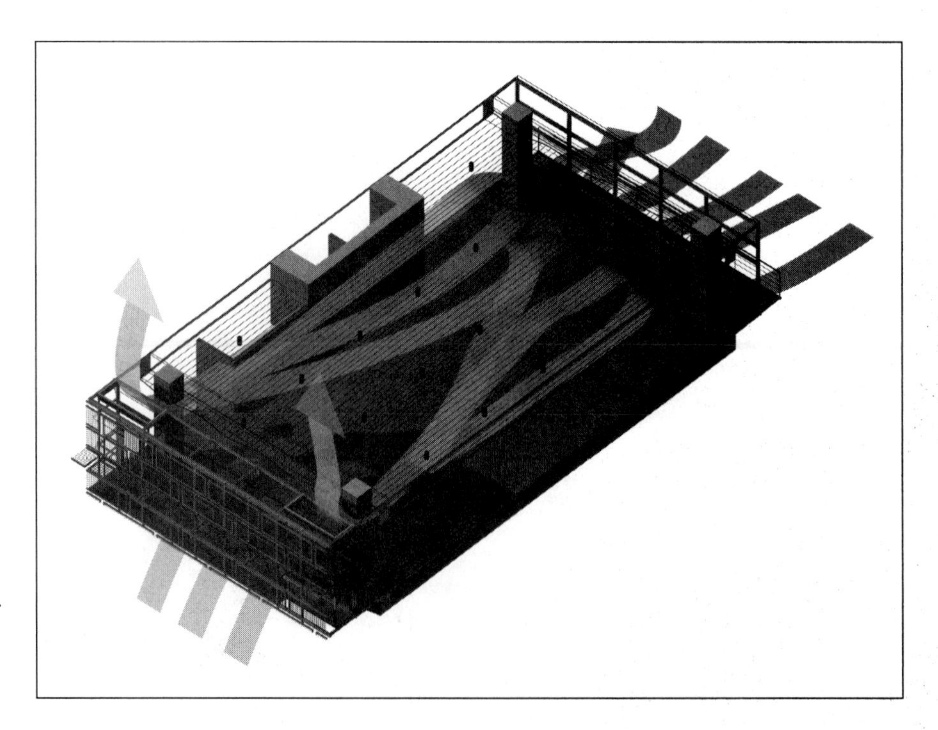

Figure 14.3 The San Francisco Federal Building is cooled and ventilated by controlling openings on either side of the building. (Illustration courtesy of Morphosis Architects.)

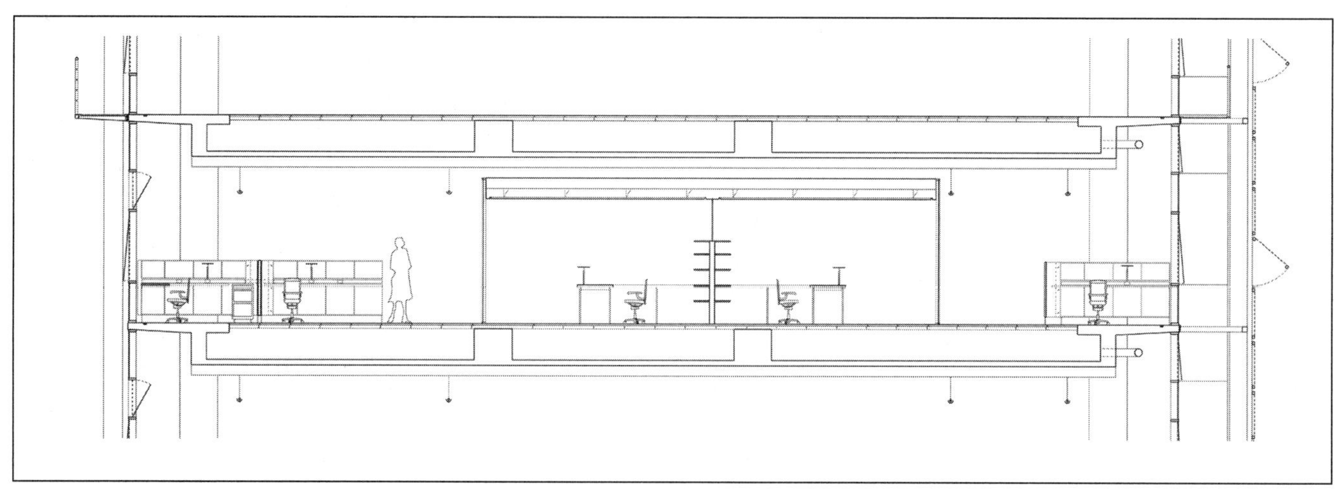

Figure 14.4 This section shows an interior conference room space. Air flows from one side to the other via a pathway over the interior space. (Drawing courtesy of Morphosis Architects.)

tem meets any heating demands; the heat is delivered through a finned-tube convector integrated into the exterior glazing along the entire length of the building. This scheme is estimated to save the federal government a substantial amount of money in annual energy costs, mostly through the reduction in size of mechanical systems. In the true spirit of sustainable construction, the savings realized from downsizing active, energy-consuming mechanical systems were shifted to an investment in intelligent façade design, allowing the employment of passive venti-

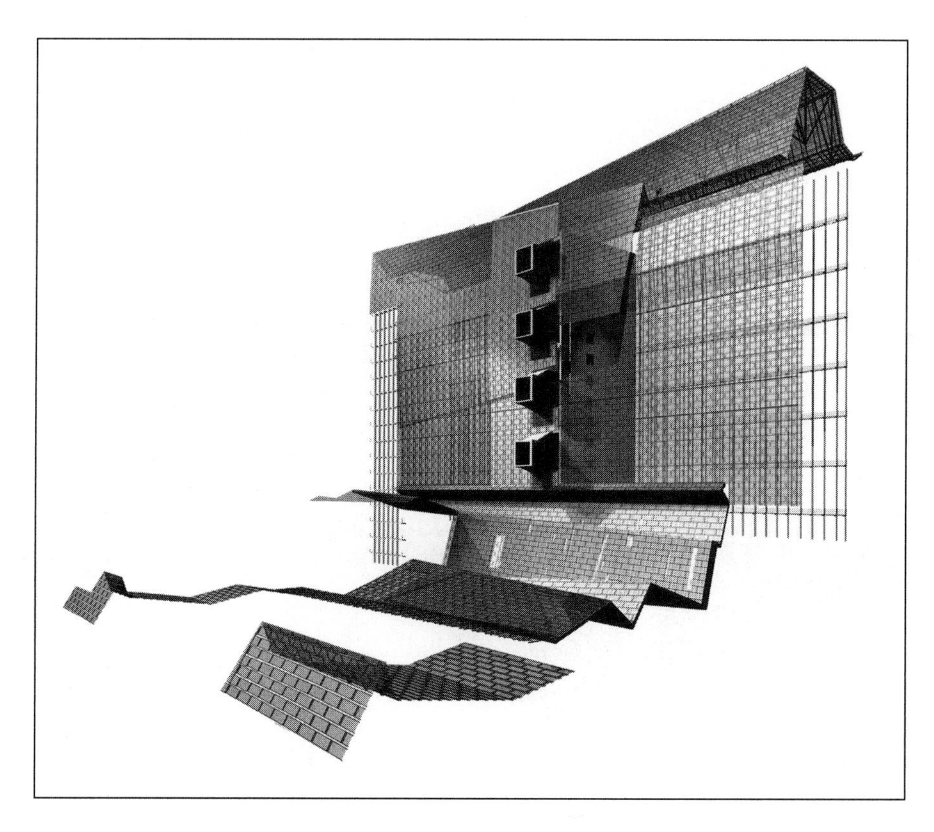

Figure 14.5 The perforated metal skin of the San Francisco Federal Building controls light and airflows through the building. As a result, as noted by architect Thom Mayne, the building "wears" its HVAC system. (Illustration courtesy of Morphosis Architects.)

lation as a cooling strategy. As Thom Mayne of Morphosis described it, "The exterior envelope of the new building is a sophisticated metabolic skin, developed in direct response to light and climate conditions. In lieu of a conventional mechanical plant, the building actually 'wears' the air conditioning like a jacket."

The Federal Building is expected to require only 27,000 BTU per square foot (85 kWh/square meter) annually in comparison to the GSA's national target of 55,000 BTU per square foot (173 kWh/square meter) per year annually and in contrast to a typical consumption of 69,000 BTU per square foot (218 kWh/square meter) annually for GSA buildings.

A FLEXIBLE AND INNOVATIVE INTERIOR STRATEGY

The San Francisco Federal Building also provides highly flexible spaces that can be changed as conditions and tenants change. A raised floor and an easily reconfigurable furniture system allow workstations to be arranged in grids or as single units. Each floor is modular, subdivided by circulation and support areas. The design of the building also promotes collaboration and teamwork through an innovative layout of the vertical transportation system. Starting at the third floor, the elevator stops only at every third floor, where there is a multistory lobby with stairs leading to the floor above and the floor below. A dedicated elevator bank serves the handicapped users of the building as well. The resulting circulation areas and waiting spaces bring people together in unexpected ways, facilitating the exchange of new ideas and information. A three-story interior skygarden starting at the 11th floor, which is landscaped and has a variety of seating, provides a space for reflection and retreat, an inviting place with beautiful vistas. It is a dramatic addition to a dramatic building.

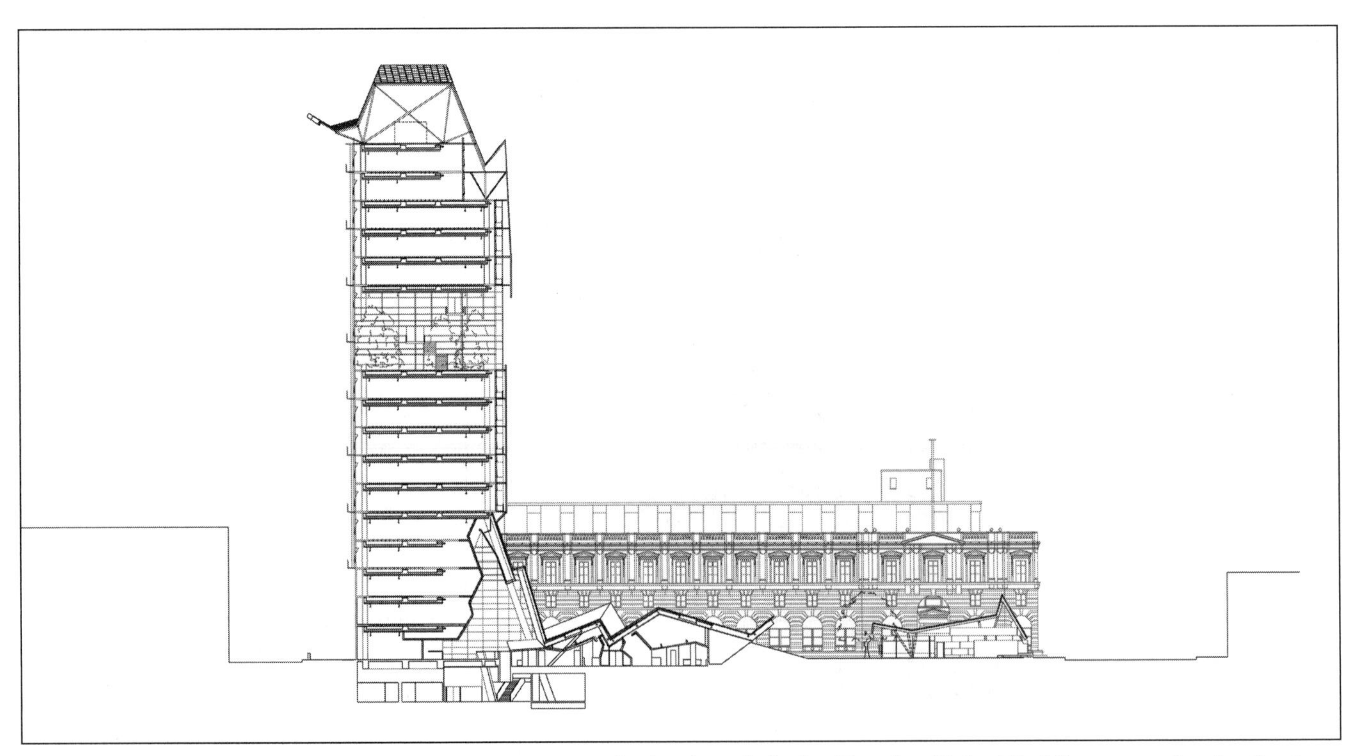

Figure 14.6 Section through the San Francisco Federal Building center, showing the sky lobby that starts at the 11th floor. (Drawing courtesy of Morphosis Architects.)

Figure 14.7 The Philadelphia Forensic Science Center was designed by Croxton Collaborative Architects, P.C. It is a state-of-the-art forensics laboratory, an example of sustainable design, and an example of a well-conceived process for converting a derelict school building into a modern high-performance facility. (Photograph courtesy of the Croxton Collaborative Architects, P.C.)

The Forensic Science Center, Philadelphia, Pennsylvania

The Forensic Science Center for the Philadelphia Police Department was designed by the Croxton Collaborative Architects, P.C., of New York City in a joint venture with Cecil Baker and Associates of Philadelphia. The facility is both a state-of-the-art forensics laboratory facility and a demonstration project for green design. It occupies 58,700 square feet (5,450 square meters) in a four-story building.

The building program includes a firearms unit; a crime-scene unit for gathering evidence; chemistry laboratories for drug analysis; and criminalistics and DNA laboratories for hair, fiber, and blood analysis. The Forensics Science Center handles all crime-scene evidence for the City of Philadelphia, with the exception of evidence from homicides.

The project's many green features include precise mapping and load separation of areas requiring 100 percent outside air to minimize mechanical loads, enve-

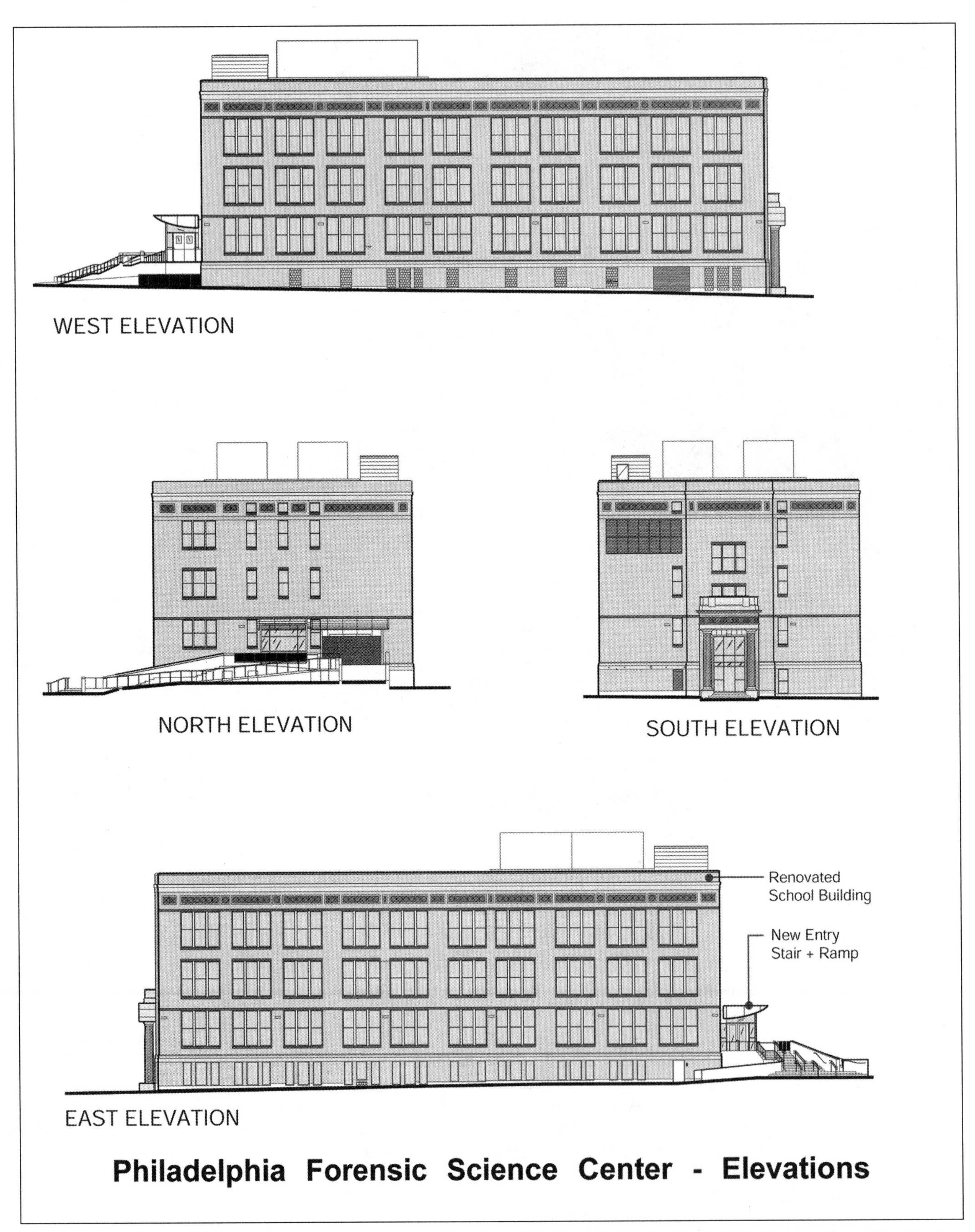

Figure 14.8 Elevations of the Forensic Science Center. (Drawings courtesy of the Croxton Collaborative Architects, P.C.)

Figure 14.9 The laboratory spaces in the Forensic Science Center have excellent daylighting, resulting in a pleasant, healthy, and productive work space. (Photograph courtesy of the Croxton Collaborative Architects, P.C.)

lope upgrades resulting in a superinsulated building, "clean" products and finishes resulting in vastly improved IAQ, deep daylighting achieved by ceiling configurations, and primary access to all mechanical and infrastructure systems outside of lab areas. The project also substantially increases pervious areas of the site, with vegetated swales providing bioremediation of runoff and reduction of input into city sewers. The building is owned and operated by the City of Philadelphia and is typically occupied by 72 people, 50 hours per person per week, with 110 visitors per week, 2 hours per visitor per week.

The City of Philadelphia mandated that the project consume minimal energy, reduce impacts on the region's air and watershed, slow the depletion of natural resources, improve the work environment of 30,000 employees, develop local business opportunities, and save taxpayer dollars. Major challenges included limited financial resources, a multiple prime construction contract, low-bid awards, and difficult communication due to the large number of stakeholders.

From the moment of project initiation, there was a strong feeling that the innovative and green characteristics of the Forensic Science Center would attract funding and that a core group should be established as expert advisors, or advocates, to provide a broader perspective on project potentials and identify sources of funding and support. This group—including representatives from the U.S. Department of Energy, Oak Ridge National Laboratory, and the Philadelphia Municipal Energy Office, among others—participated in two major reviews. The major economic difficulty for the project was a delay in the receipt of significant federal funding, delaying the project for 2 years and ultimately requiring that the project be rebid. The good news was that the additional time allowed the city to

Figure 14.10 Exterior view of the Forensic Science Center showing its context in its northeast Philadelphia neighborhood. (Photograph courtesy of the Croxton Collaborative Architects, P.C.)

apply successfully for a Growing Greener grant from the Pennsylvania Department of Environmental Protection, providing $225,000 for greening the asphalt parking lot. The total project cost, excluding land, was $11,450,000.

SITE AND LANDSCAPING

Located in an underserved neighborhood of north Philadelphia with high crime rates, low income levels, and few services, the Forensic Science Center has helped to breathe new life and a better sense of security into the entire neighborhood. The Center is located in a former K–12 school building, which was distinguished long ago (originally constructed in 1929) but had more recently become derelict and was used only by pigeons at the upper floors. The building sat amid broken glass and other debris on a cracking, completely asphalted lot.

A noticeable upgrade to the entire area has taken place since this building reopened as the Forensic Science Center. Many Philadelphians now see this neighborhood as the next wave of urban improvement. While it is impossible to prove that the Center caused this improvement, it has certainly made a substantial contribution to the improved security and desirability of this zone. An existing train line connecting suburban areas of Philadelphia to Center City runs across the street from the site; active efforts are under way to reopen the stop across the street, which has been closed for many years.

Before this renovation, the site was entirely impervious, contributing to the many annual discharge events carrying stormwater and sewage into the Delaware River rather than for treatment at the Southeast Water Pollution Control Plant. The

Pennsylvania Department of Environmental Protection's Growing Greener grant was a key funding source for sitework, including a system of vegetated swales, "rain gardens," and stone-reinforced water pathways.

WATER CONSERVATION AND USE

The previously impervious site now includes large areas of vegetated swales and buffer vegetation, improving water catchment by roughly 33 percent while still meeting the Center's demanding parking and servicing requirements. Linear vegetated swales paralleling the parking rows filter stormwater and allow it to evaporate or infiltrate the ground before it enters storm drains. Site plantings are drought-resistant, requiring less watering and maintenance than conventional landscaping. All damaged areas were replanted with native vegetation. Waterless urinals reduce water consumption by approximately 176,000 gallons per year (one urinal is included in each of the four bathrooms). Low-flow fixtures were used for all plumbing fixtures.

ENERGY CONSERVATION

Among the greatest assets of the existing school structure were the large windows: 9 feet 6 inches (3 meters) high and 3 feet 2 inches (1 meter) wide and organized in groups of three, they constitute over 30 percent of the exterior walls. Because the building is oriented along a north-south axis, with the long façades facing east and west, these windows receive low-level sun at sunrise and sunset.

Figure 14.11 Detail of a column of the existing school that was converted into the Forensic Science Center, showing the colorful plaster and tile work that was preserved in the design of the new building. (Photograph courtesy of the Croxton Collaborative Architects, P.C.)

Both sides of the building typically receive direct sun for half of the day and ideal and glare-free shade for half of the day. The design team sought to use the design of the windows and shading devices and the placement of circulation areas and workstations to mitigate glare and heat gain. The high-performance glass reduces visible transmittance, and the addition of white thin-line blinds allows for either a self-diffusing light source at the window or a bounce of light toward the sloping ceiling. The circulation corridor was placed along this line of windows, and the workstations were organized perpendicular to the outside wall. The users are delighted with the levels of full-spectrum light and often prefer more direct "slatted" light than was expected.

While laboratory spaces require 100 percent outside air, office spaces do not. Four-pipe fan coil units are used in the office areas to minimize the central plant load. The central plant provides fresh air and ventilation only to offices. The pressurization system requires only minor modifications in order to maintain separation of airflow to offices and labs.

On a room-by-room basis, air systems go into setback mode when there is no occupancy. The entire facility can remain operational even in the event of a failure of one air handler or exhaust (pressure relationships are maintained, but temperature is compromised). Rooftop exhaust-air heat recovery is used to precondition outside supply air. Water-side economizers are used in office fan coil units. Heat exchangers utilize cooling-tower water in lieu of chillers during shoulder seasons. Air-side economizers supply outside air for free cooling. A gas chiller heat exchanger recovers heat to generate hot water for heating and domestic hot water. Domestic hot water is provided through heat-recovery systems and high-efficiency boilers. Fume hoods were designed with limited sash openings in order to reduce air volumes. The setback mode saves energy while maintaining pressure differentials.

Provisions were made for the roof to accept a horizontal roof-tile photovoltaic system of approximately 15 kW, and the roof equipment was configured to accommodate such a system.

The electric lighting system features T-8 lamps and electronic ballasts and separate task and ambient lighting, even in laboratory space. Lighting is controlled by occupancy sensors and daylight-dimming sensors.

CLOSING MATERIALS LOOPS

This project was a restoration of a physically intact but derelict building. The existing linear building had tall windows on both long sides and a center bay with higher structural capacity. This led to a layout that put all occupied spaces along the windowed areas, with heavy equipment, such as mass spectrometers, filing systems, and other support spaces, in the center bay. The existing attic space was used for massive trunk ducts, allowing for the greatest possible separation between intake air and exhaust air.

The existing stairs were reused, and existing ventilation shafts were used for vertical air movement and plumbing infrastructure. The existing subgrade space is used for firearms testing. PVC was avoided for all uses for which there was a reasonable alternative. For example, all piping is stainless steel, glass, cast-iron, or copper; and rubber flooring and base and stainless steel corner guards were used in place of PVC materials. No CFCs or HCFCs are used in any of the equipment in the building, including the water fountains, refrigerators, and mechanical systems. Rapidly renewable products (including linoleum and agrifiber board) and products including recycled content (including cellulose insulation, carpeting, tile, steel, and gypsum board) were used whenever possible. Ductwork is made of galvanized sheet metal, and stainless steel ductwork was used only at the acid fume hood.

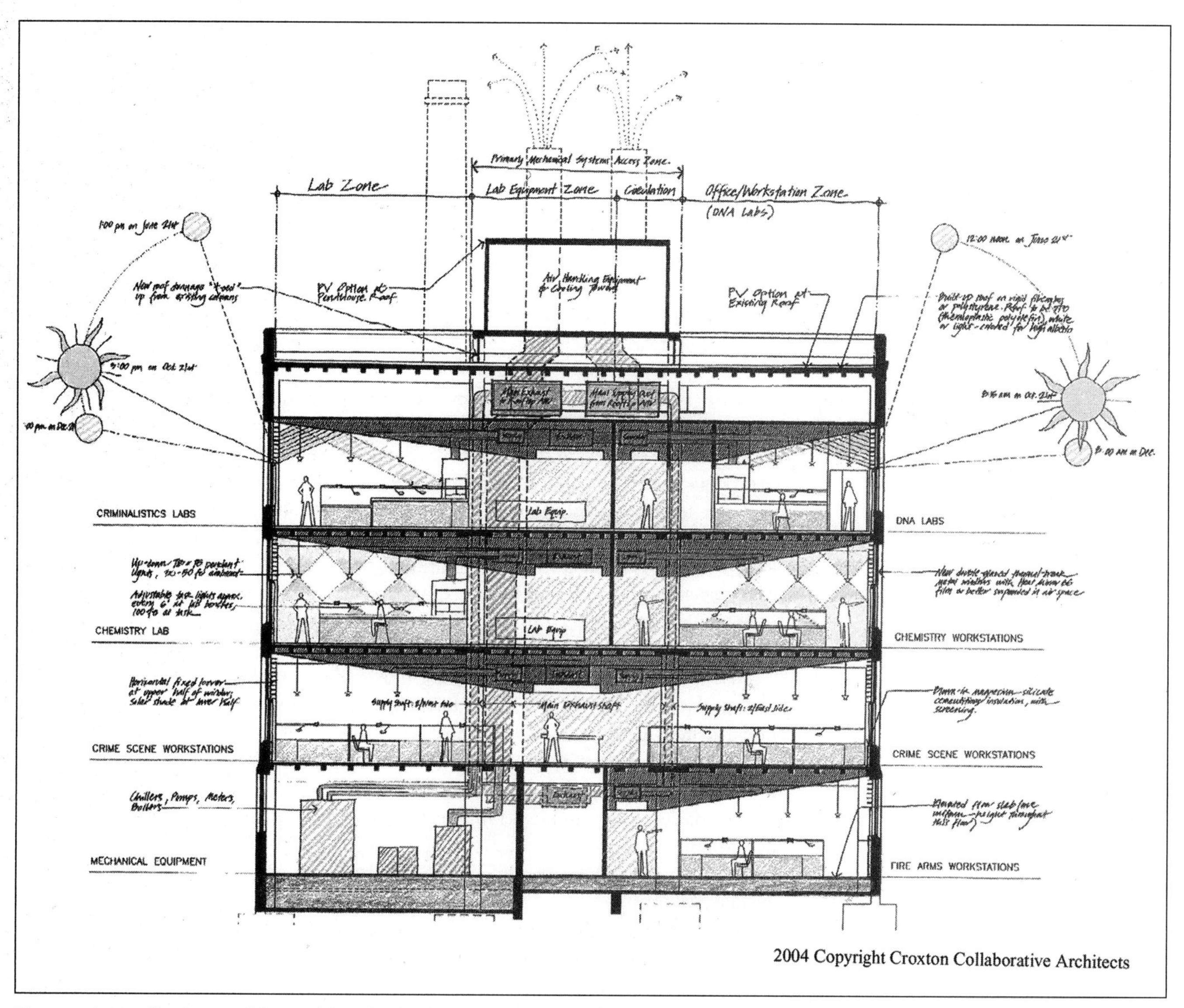

Figure 14.12 Section used for studying the daylighting and physical layout of major ventilation and conditioning ductwork. (Sketch courtesy of the Croxton Collaborative Architects, P.C.)

INDOOR ENVIRONMENTAL QUALITY

Fresh and exhaust airstreams are separated and located at remote points: the exhaust airstream is located at the north end of the roof (and directed straight up), and the supply airstream is located at the south wall (the vertical face of the grill). All duct insulation is external rather than internal. All glues and adhesives were selected for their low emissions of VOCs in order to protect IAQ. The building is extensively daylit, providing virtually all spaces with the health benefits of natural light.

Articulating Performance Goals for Future Green Buildings

One of the major green building issues is to clarify the specific goals of high-performance green buildings. These goals can be expressed in a variety of suitable ways; this section describes four of them. One option is to apply the *Factor 10* approach to buildings and focus on efforts that reduce the consumption of resources in the creation and operation of buildings to one-tenth of their present level, thereby aligning this movement with other sectors and institutions that are striving to behave sustainably.[4] A second option is to express the impact of a building in terms of its *ecological footprint.*[5] The unit of measurement for an ecological footprint is land area, which indicates the impacts by the peoples of different countries based on their lifestyles. The same concept could be applied to buildings, with impacts being stated in hectares or acres per unit area of building. Materials used in building could be measured in part by their *ecological rucksack.*[6] The ecological rucksack is the total mass of materials that must be processed to produce a unit mass of a specific metal or mineral. It is essentially a way to measure impact in terms of transformation of the surface of the planet—a serious matter because humans are now moving twice the amount of materials in natural systems. A fourth way to express the goals is through the routine use of *life-cycle assessment* (LCA), which describes the total inputs and outputs in the production of a given material. Comparisons could be made for different building solutions—for example, wall sections, to determine which approach consumes the least resources and has the fewest emissions.

For the high-performance building movement to make sense, establishing specific and reasonable goals is ultimately necessary to give the various players a direction for their activities. For the most part, the targets set in LEED are based on comparisons to a base building, that is, a building that *just* meets the requirements of the building code.

To project from an ideal future state to the present situation for the purpose of determining the steps that have to be accomplished to create the necessary change, a technique known as *backcasting* is used in the sustainable development arena. This immediately raises questions: What is the ideal future for high-performance green buildings? What do they look like? How do they differ from today's green buildings? Answering these challenging, even daunting, questions is critical if we are to make progress toward a future in which the buildings we construct come far closer to meeting the ultimate standard of high-performance building.

The Challenges

Chrissna du Plessis, a noted research architect and project leader on sustainable development at the Council for Scientific and Industrial Research (CSIR), the national building research institute of South Africa, located in Pretoria, has identified three major challenges we face in defining the future built environment:[7]

- Taking the next technology leap
- Reinventing the construction industry
- Rethinking the products of construction

TECHNOLOGY LEAPS

In the future, technology will undoubtedly play a powerful role in assisting and even accelerating change. In its simplest form, technology is nothing more than applied science, that is, using discoveries of basic science and mathematics for practical purposes, ideally for the benefit of people and natural systems. Technology is clearly a two-edged sword: along with its many benefits typically come a wide variety of impacts. Thus, the challenge is to foster technologies whose benefits are great and whose impacts are low. For the built environment, three general approaches are emerging:

1. Vernacular vision
2. High-technology approach
3. Biomimetic model

Each of these is accompanied by technological approaches. Even the vernacular vision, which focuses on relearning the lessons of history, is also about developing technologies that support today's implementation of those hard-learned lessons.

The Vernacular Vision: Relearning the Past

Vernacular architecture embeds cultural wisdom and an intimate knowledge of place in the built environment. It comprises technology, or applied science, that has evolved by trial and error over many generations all over the planet as people designed and built the best possible habitat with the resources available to them. With respect to designing high-performance buildings, vernacular design comes closest to the ecological design capabilities available today.

Two contrasting examples of vernacular architecture are the traditional styles of the State of Florida and the Southwest. *Cracker architecture* in Florida raises houses and buildings off the ground and creates flow paths for air around and through the structures, opening them to ventilation and conditioning by the prevailing winds. Originating in the early 1800s, the cracker house is well designed for the region's hot, humid climate. It emulates the *chickee* of the Seminole Indians, a covered structure with open sides, in which the floor, an elevated platform 3 feet (0.9 meters) above the often-wet ground, was used for both eating and sleeping. The galvanized metal roof of cracker buildings is durable and reflects Florida's daily intense solar radiation away from the structure. The structure is lightweight and sheds energy; and, rather than absorbing energy, it reflects it, thereby helping to maintain moderate interior temperatures.

Modern cracker architecture buildings, though they retain the appearance of their traditional predecessors, with metal roofs, cupolas, and porches, employ modern technology to meet the needs of contemporary businesses and homes. As is the case with much of today's vernacular architecture, some of the original features, such as the capability for passive ventilation, are for all practical purposes not useful due to year-round reliance on modern HVAC systems. Cracker architecture is generally limited to smaller buildings, as it is difficult to apply to large buildings, because the roof tends to become inordinately large, and for urban office buildings the porches lose their appeal.

Adobe architecture, prevalent in the U.S. Southwest and Mexico, relies on local soils and a relatively massive structure made of adobe clay and straw brick. The large thermal mass of the structure enables the building to take advantage of the great diurnal temperature swings prevalent in high desert areas for heating and cooling. During the day, the thermal mass absorbs solar radiation, storing it for later use, but also provides just enough thermal resistance to keep the interior temperature at a moderate level. As temperatures in the deserts and mountains plunge in the evening, the energy stored in the massive adobe structure is emitted by radiation and convection into the interior spaces.These two historical forms of vernacular architecture, in addition to taking advantage of experience with daily and seasonal weather patterns and the assets of the sites, made use of local materials—long-leaf pinewood in Florida and earth and straw in the Southwest. Incorporating local and

(A)

(C)

(B)

Figure 14.13 Vernacular architecture in north Florida. Early cracker-style houses were lightweight wood-framed structures with wooden siding and metal roofs. The passive aspects of these structures helped them reflect solar radiation and facilitated cross-ventilation; they were raised off the ground for protection from flooding. Modern versions adapt the materials and energy strategies of early cracker architecture to produce hybrid structures that include high-technology windows, composite siding, and energy-efficient air conditioning. (A) The Geiger Residence, Micanopy, Florida (1906). (B) A small cracker vernacular office building in Gainesville, Florida (1996). (C) Interior of Summer House at Kanapaha Botanical Gardens, a larger, 10,000-square-foot (929 square meters) cracker-style building near Gainesville, Florida (1998). (Photographs (A) courtesy of Ron Haase; (B) courtesy of Jay Reeves; and (C) by M.R. Moretti.)

(A)

(B)

(C)

Figure 14.14 Examples of New Mexico adobe architecture. (A) As early as 350 A.D., the Anasazi, the oldest known inhabitants of New Mexico, began to build aboveground masonry structures, the foundations of which are visible here at the base of their cliff dwellings in Bandelier National Park. (B) Communities called *pueblos* flourished around 1250–1300 A.D. and contained intricate arrays of connected flat-roofed, multilevel adobe buildings. (C) A modern office building in Santa Fe, New Mexico, retains the appeal and function of traditional adobe architecture.

regional materials is now a criterion in modern building assessment standards such as LEED. In this way, taking a vernacular approach promises an excellent start to incorporating passive energy design features into a building, because it implies using the site and structural design to assist heating and cooling. Fortunately, there are hundreds of examples of vernacular architecture worldwide that can be used as the basis for designing today's high-performance buildings. The challenge, of course, is to use the wisdom of the past to meet the requirements of modern buildings and current building codes while retaining the positive cultural, environmental, and resource aspects of vernacular design.

The High-Technology Approach

In contrast to the vernacular vision, which uses historical wisdom and cultural knowledge to design buildings, the high-technology approach generally follows the path of current trends in society. Contemporary society, especially in the developed world, has a love affair with technology. The prevalent attitude is that all our problems, including resource shortages and environmental dilemmas, can be solved simply by developing new technologies. For buildings, the high-technology approach centers on devising new energy technologies such as photovoltaics and fuel cells, and on finding technical solutions to the question of how to utilize renewable energy sources more effectively. Typical examples of this approach include windows with spectrally selective coatings and gas-filled panes, control systems and computer systems that respond to optimize energy use based on weather and interior conditions, energy recovery systems that incorporate desiccants to shift both heat and humidity, and materials incorporating postindustrial and postconsumer waste. Contemporary commercial and industrial buildings are equipped with a wide range of telecommunications and computer technologies that would challenge even the most advanced vernacular design approaches simply because of the need to remove the high levels of energy generated by today's workplace tools. Indeed, it could be argued that the technology of the building itself must be carefully matched to the technologies employed by the building occupants.

The high-technology approach to high-performance green building is, in short, an evolution of current practices. Over time, built environment professionals, backed up by experience, research, and the development of better systems and products, will be able to design buildings that are much more resource-efficient than today's green buildings and that will have far lower impacts in their construction and operation. Thus, the key characteristics of the ideal high-performance green building are based on making incremental—as opposed to radical—improvements in existing technology in these areas:

- *Energy:* The ultimate high-performance building consumes just one-tenth of the energy of current buildings and either uses only off-site-generated renewable energy or generates energy from renewable sources on-site for its entire needs. Passive design, assisted by extensive computer modeling, ensure the optimal use of natural ventilation, structural mass, orientation, the building site, building envelope design, landscaping, and daylighting to minimize consumption of electricity and other energy sources so that the building can default to nature if it becomes disconnected from external energy sources.[8] Landscaping is carefully integrated into the project to assist in cooling and heating the structure.
- *Water:* The ideal high-performance building uses only 10 percent of the potable water of contemporary buildings and uses graywater, reclaimed water, or rainwater for nonpotable requirements. Wastewater is recycled for nonpotable building uses or is processed by constructed wetlands or Living Machines for discharge back into nature in as clean a state as it entered.

- *Materials:* All materials employed in the ultimate high-performance building are recyclable; building products can be disassembled and their constituent materials easily separated and recycled; buildings are deconstructable, capable of being disassembled and their components either reused or recycled. The cardinal rule for materials used in construction would be to eliminate those that are not recyclable, that are used in a once-through fashion and become waste after one use. An effective Factor 10 reduction in materials consumption would focus on reducing materials extraction by 90 percent, achievable by dramatically increasing the conservation of materials by deconstruction, materials recovery, and recycling and reuse. Increasing the durability and longevity of the built environment would also help achieve Factor 10 performance. However, this presumes that improvements in design would make buildings so much more valuable to society as cultural artifacts that their removal for economic reasons would be far less likely.
- *Natural systems interface:* The ultimate high-performance building is integrated with natural systems in a synergistic manner such that services and nutrients are exchanged in a mutually beneficial manner. Natural systems provide stormwater uptake and storage, assist cooling and heating, provide amenities, supply food, and break down waste from individual building scale to larger scales, up to the bioregional one. The building is carefully designed to take advantage of the natural assets of the site, the prevailing winds, and the microclimate at the building location.
- *Design:* Ideal high-performance buildings are designed using well-developed principles that are rooted in ecology. A robust version of ecological design is employed to ensure the integration of the building with its site and the natural assets. Architecture, landscape architecture, and engineering are carried out in a seamless, integrated process. The building professionals on the team work in a collaborative fashion, with fees based on the quality of design and construction and the building's performance. These same professionals work to minimize building complexity and maximize adaptability and flexibility.
- *Human health:* All aspects of IEQ in the ultimate high-performance building are carefully addressed, including air quality, noise, lighting quality, and temperature/humidity control. Ventilation rates are optimized to provide exactly the levels of fresh air that support health. Only zero-emissions materials are permitted.

Biomimetic Model

Popularized by Janine Benyus in her book *Biomimicry: Innovation Inspired by Nature,* published in 1997, the idea of using nature's designs and processes as the basis for human goods and services has much appeal when it comes to considering high-performance buildings. A biomimetic strategy, one based on biomimicry, or imitation of nature, is a relatively recent concept, but one that may provide many of the answers to questions relating to the creation of the ultimate high-performance building. Biomimicry is fundamentally about observing nature, then basing materials and energy systems on these observations. Consider, for example, that ceramic-like seashells are produced at ambient water temperatures from materials in the environment, with no waste, the result being elegant products perfectly designed for their function: to protect their inhabitants. Compare ceramics created by human technology, which are produced at temperatures of several thousand degrees, consuming great quantities of energy and producing emissions to air and water and solid waste. Moreover, the materials and resources necessary for the production of the ceramics must often be transported great distances, thereby adding to the energy investment.

Many other examples of biomimicry can be adapted as safe and sound technological approaches: nature's ability to convert sunlight into chemical energy via photosynthesis; the phenomenal information storage and transmission capability of nerves and cells; tremendously strong and lightweight materials; powerful adhesives—to name a few. Using true outside-the-box thinking, Chrissna du Plessis described a fanciful future built environment based on a full-fledged implementation of biomimicry. In it, all components of the building are biologically based and created from proteins, with solar energy collectors embedded in portions of the structure facing the sun. The structure is strong and lightweight and glued together with powerful adhesives based on those used by mussels to attach themselves to rocks in cold, murky water. Temperature and humidity are regulated by membranes that allow energy and moisture to move in and out of the occupied spaces, with embedded nanoprocessors controlling the movement. Like all other components, the membranes are self-repairing, self-regulating, and self-cleaning. Waste from the activities and functions of the building's inhabitants is processed by Living Machines that break down waste into nutrients for use in the food gardens, which are also designed to be self-reproducing and diverse, thereby minimizing pest infestations. At the end of its useful life, the entire building can be "digested," with the organic components cycled for other uses and the mineral and other inorganic materials collected for recycling and reuse.

REINVENTING THE CONSTRUCTION INDUSTRY

The construction industry, referred to in its broadest sense to include design, construction, operation, renovation, and disposal of the built environment, has to change dramatically to meet the future challenges of building. Buildings have become commodities, with little to distinguish one from another in any serious manner, and with little effort to make them—as in the past—cultural artifacts of human existence. Low first cost is the normal order of business, so quality design receives minimal attention; materials and systems are employed that produce minimal performance; the construction process is carried out rapidly and at the lowest possible cost; and the norm is to demolish and landfill buildings at the end of their useful life. Scant attention is paid to the implications of this behavior, both for ecological systems and for human society. Owners focus on buildings that have minimal construction cost and that are designed just to accomplish their functions, with little or no attention given to their aesthetic features. Changing the mindset of this cast of actors is an enormous challenge. To meet that challenge, these changes must take place:

- *Technology:* Technologies that minimize resource consumption and the environmental impact of the built environment need to be developed.
- *Policy:* As a general matter of policy, buildings need to be created based on life-cycle costs as well as first costs.
- *Incentives:* Government needs to develop financial incentives for high-performance construction, such as priority review by building departments, accelerated approval for projects of this type, and reductions in impact fees and/or property taxes for a specified period of time.
- *Education:* All the professionals in the industry need to be educated and trained in the need, process, and approaches for creating high-performance green buildings—owners, architects, engineers, landscape architects, interior designers, construction managers, subcontractors, materials and product manufacturers and suppliers, insurance and bonding companies, real estate agents, building commissioning consultants, and other professionals engaged in the process. This is also necessary for the workforce, the crafts workers, journeymen, and apprentices who work for the broad array of subcontractors that make buildings a physical reality.
- *Performance-based design fees:* Contracts for design and construction services need to be revised to offer incentives to the building team to meet and

exceed project goals with respect to resource consumption and environmental impacts. These goals include targets for energy and water consumption, building health, construction waste, protection of the site's natural assets, and other objectives that contribute to the building's performance.

- *Construction process:* The physical process of construction needs to be changed to ensure that the activities involved in erecting the building have the lowest possible impact. Among these changes are: reduce construction waste and recycle or reuse the residue; understand and implement effective soil and erosion control methods; protect flora and fauna on the site during the construction process; minimize soil compaction during construction; and store materials so that they are protected from wastage and are unlikely to cause IEQ problems.

RETHINKING THE PRODUCTS OF CONSTRUCTION

As this book has pointed out repeatedly, buildings consume enormous quantities of resources and can cause any number of negative impacts on their occupants. In addition to the resources required to build and operate individual buildings, a wide range of additional impacts are the consequence of decisions concerning how to distribute buildings across the landscape. For example, segregating buildings by type (residential, commercial, industrial, government, cultural, etc.) means that people are forced to use their automobiles to get from one type of building to another. The average American makes at least eight automobile trips per day, many of them for no reason other than to socialize. The concepts of *new urbanism* or *traditional neighborhood development* are seeking to reverse this trend by mixing building types and uses and by designing streets and neighborhoods for pedestrian movement. A general goal is that all daily needs must be available within a 10-minute walk from where the individual resides.

Other serious impacts result from the building stock itself. In the United States, buildings and houses are generally very large and consume large quantities of energy, water, and materials to both build and operate them. The extraction of resources to support the construction industry is profound. Some estimates state that 90 percent of all extracted resources in this country are used to create the built environment. Buildings consume two-thirds of all electricity and 35 to 40 percent of primary energy. Three important questions that need to be asked when a new building is proposed are:

- Is the building actually needed or is adequate space already available?
- Can the building be made smaller?
- Can an existing building be renovated for the new purpose?

Revamping Ecological Design

As noted in Chapter 5, contemporary ecological design has only very weak links to ecology. Although virtually any definition of high-performance green building makes reference to ecological design, to date there is little or no evidence of the application of ecology to design. To correct this situation, it is crucial that a new, comprehensive concept of ecological design be developed. In addition to considering ecology in far greater depth in building design, it is imperative to consider the potential for applying *industrial ecology.* Established as a new discipline in 1988, industrial ecology seeks to apply ecological theory to industrial production. Many of the issues and problems faced by an industrial system that builds automobiles and airplanes are also faced by those in the building design and construction professions. Consequently, the experience gained by applying industrial ecology to industrial production will be very useful in creating high-performance green buildings.

In a recent collaboration among architects, ecologists, and industrial ecologists, the possibility of applying current ecological theory to the creation of buildings was explored in great depth to determine which aspects of ecological theory and industrial ecology were applicable to buildings.[9] This collaboration offered a number of insights into how ecology and industrial ecology can better inform building design, construction, and operation. The results of this collaboration are summarized in the following lists:

GENERAL

1. Maximize second-law efficiency (effectiveness) and optimize first-law efficiency for energy and materials.[10]
2. As with natural systems, industry must obey the maximum power principle.[11]
3. Be aware that the ability to predict the effects of human activities on natural systems is limited.
4. Integrate industrial and construction activities with ecosystems functions so as sustain or increase the resilience of society and nature.
5. Interface buildings with nature.
6. Match the intensity of design and materials with the rhythms of nature. In the built environment, move from the "weeds" stage to the "tree" stage for sites that are not frequently disturbed. "Weedy" structure (minimal built structure that is easily and cheaply replaced) may be much more adaptive to sites frequently disturbed by floods, storms, or fires.
7. Consider the life-cycle impacts of materials and buildings on natural systems.
8. Insist that industry take responsibility for the life-cycle effects of its products, to include take-back responsibility.
9. Address the consumption end of the built environment by integrating it with production functions.
10. Increase the diversity and adaptability of user functions in buildings through experiment and education.
11. Explore educational processes beyond academia that instruct through "learning by doing," by involving *all* stakeholders in processes that test different means by which the built environment is produced, sited, deconstructed, and resurrected.
12. Reduce information demands on producers and consumers by testing and improving the means by which materials, designs, and processes are certified as "green." This presupposes the development of a construction ecology based on nature and its laws.
13. Ensure that systems analyses look at system function, processes, and structure from different perspectives and at different scales of analysis.
14. Integrate ecological thinking into all decision-making processes.
15. Follow the precautionary principle to constrain and govern decision making.

MATERIALS

1. Keep materials in productive use, which also implies keeping buildings in productive use.[12]
2. Use only renewable, biodegradable materials or their equivalent, such as recyclable industrial materials.

3. Release materials created by the industrial system only within the assimilative capacity of the natural environment.
4. Eliminate materials that are toxic in use or release toxic components in their extraction, manufacturing, or disposal. Focus first on materials not well addressed by economics—the intermediate consumables (paints, lubricants, detergents, bleaches, acids, solvents) used to create wealth (buildings).
5. Eliminate materials that create "information" pollution—for example, estrogen mimics.
6. Minimize the use and complexity of composites and the numbers of different materials in a building.
7. Realize that not all synthetic materials are harmful and not all natural materials are harmless. Nature has many pollutants that are harmful; for example, natural fibers such as cotton are not necessarily superior to synthetic materials such as nylon.
8. Recognize that the impacts of natural materials extraction can be high—as is the case with agricultural products; or in forestry, in which pesticide use, transportation distances, processing energy, and chemical use are significant factors.
9. Standardize plastics and other synthetic materials based on recycling infrastructure and the potential for recycling and reuse.
10. Rather than for power generation, use fossil fuels to produce synthetic materials, and use renewable energy resources as the primary power source.
11. Acknowledge that it is not possible to rate or compare materials adequately based on a single parameter.

DESIGN

1. Model buildings based on nature.
2. Make structures part of the geological landscape.
3. Design buildings to be deconstructable, using components that are reusable and ultimately recyclable.
4. Design buildings and select materials based on intended use and then measure the outcomes of the design.
5. Incorporate adaptability into buildings by making them flexible for multiple uses.
6. Realize real savings by integrating the production, reuse, and disposal functions.
7. Focus on excellence of design and operation, with greenness as a critical component. Focusing exclusively on greenness trivializes it as a marginal movement.
8. Invest in design that improves building function while minimizing energy use and the number of materials. This will reduce the time and effort required to find and optimize new green materials.
9. Revise designs to take into account major global environmental effects such as global warming and ozone depletion. This is critical at this point in time.
10. Allow for experimentation in green building design to produce structures that, like nature, obey the maximum power principle.
11. Make sure that architects have a strong, fundamental education in ecology.
12. Use performance-based design contracts to develop greener buildings and better architects.

INDUSTRIAL ECOLOGY

1. Make changes needed to create an environmentally responsible industrial ecosystem intelligible to the members of the particular industry.
2. Focus on the clients and key stakeholders of the system. This is necessary due to limits on time, knowledge, and resources. Major stakeholders include the educational system and [the] insurance industry.
3. Make the new paradigm for industry the collaboration of actors versus the possession of technical expertise.
4. Reduce consumption. This is more important than increasing production efficiency as the change agent for industrial ecology.
5. Incorporate ecological engineering into industrial ecology.

CONSTRUCTION ECOLOGY

1. Ensure that construction ecology balances and synchronizes spatial and temporal scales to natural fluxes.
2. Recognize that the corporations leading the way in the production of new, green building materials are a "frontier species" that may be creating a new form of competition, which they are using to their advantage.
3. Be aware that green building probably can be implemented only incrementally because of resistance and potential disruptions from the existing production and regulatory systems.

OTHER ISSUES

1. Better educate government officials and code-writing bodies about ecology.
2. Establish performance standards for buildings and construction to replace existing prescriptive standards. The performance standards need to include provisions for using green building materials.
3. Regard the insurance industry as a major stakeholder in the built environment, as the threat of severe consequences from global warming will drive it to promote green building.
4. Rely on certification only as a starting point; do not rely on it entirely for information on products.

Incorporating High-Performance Green Building Requirements into Building Codes: ASHRAE Standard 189 P

High-performance green buildings have evolved significantly since the appearance of the first version of LEED in 1998 and one of the key challenges has been how to make this approach more readily available to mainstream buildings, not just to the relatively small number of organizations that have been implementing green buildings. According to the USGBC, LEED addresses the top 25 percent of high-performance buildings. By developing a high-performance green building standard using an American National Standards Institute (ANSI) accredited process, the USGBC, in collaboration with ASHRAE and IESNA, is making it possible for green building requirements to be incorporated into building codes, thus addressing the other 75 percent of construction. The full name for this standard, under review as of December 2008, is "Proposed BSR/ASHRAE/USGBC/IESNA Standard 189 P, Standard for the Design of High-

Performance Green Buildings Except Low-Rise Residential Buildings." If this standard were to become incorporated into building codes it would free the USGBC and other green building organizations to set the bar for high-performance buildings even higher. It will also provide a baseline for sustainable design, construction, and building operation in order to drive green building into mainstream construction industry practices.

The standard will apply to new commercial buildings and major renovations. It is being modeled after the LEED building assessment system, including prescriptive measures drawn from all five LEED categories: sustainable sites, water efficiency, energy and atmosphere, materials and resources, and indoor environmental quality.

Beyond the Cutting Edge: Sustainable Geometries

Ecological design is without a doubt the linchpin of sustainable construction and green building. At present, however, it is not well defined, which means that green building design has a shaky foundation. Kim Sorvig, a research professor at the School of Architecture and Planning at the University of New Mexico, Albuquerque, and coauthor with Robert Thompson of *Sustainable Landscape Construction* (2000), has created the notion of fractal architecture as a bridge to ecological design. His reflections on the transition between the built and natural environments are excerpted here.

Processes, Geometries and Principles: Design in a Sustainable Future
by Kim Sorvig
Research Associate Professor
School of Architecture & Planning
University of New Mexico, Albuquerque

Sustainability is about integrating constructed and living systems. To integrate two different processes or entities, each must be clearly understood in its own right. I am convinced that deepening our understanding, not only of ecology, but also of building, is an essential evolution of sustainable design.

Today's understanding of the inherent qualities of constructed systems is detailed and pragmatic, but we often fail to question important principles. Conversely, designers' understanding of ecological systems is generalized, frequently romanticized. In both areas, the relationships between processes (the use of a building, the life-cycle of a watershed) and geometries remain unconsidered, with shapes and patterns designed by habit and without insight. There is a pressing need to understand, factually and concisely, the core qualities of constructed and natural systems, and to use differences and similarities among these systems creatively.

The core qualities that pertain most to sustainable design can be called processes, geometries, and principles. The future of sustainable design may lie less in technical innovation than in whether designers can work out the conflicts between the core qualities of natural systems and those of human development.

Construction does not create its raw materials, but shapes existing substances into units, and assembles those units into structures. This is so obvious that it is often overlooked, but has a critical effect on the processes and geometry of construction.

The essential *processes* of construction are cutting and assembling, plus form-casting. In natural systems, direct parallels to these form-making processes (especially cutting and assembly) are rare. Making bricks and building an arch is categorically different from erosion creating a stone arch, or from a tree-branch growing. Construction is a controlled system, dominated by a selected force (e.g. a saw blade) and excluding extraneous forces (jigs and clamps preventing unplanned movement).

The *geometry* of construction is based on its processes: cutting and assembling are most efficient when using regular, smooth, Euclidean forms. Such forms also lend themselves to easy measurement and calculation.

Form-making in geological and biological systems is markedly different from construction processes. Nature's *processes* are growth, decay, deposition, and erosion, all radically different than the assembly processes of construction. Natural processes are part of an "open" system, with many forces interacting, no one dominating for long.

Mathematical understanding of the geometry of nature is recent, and designers are just beginning to appreciate it. Resulting from growth/decay processes, the forms characteristic of nature are called *fractals.* They result when multiple forces interact repeatedly over time. No matter what scale they are viewed at, and also over time, their forms remain self-similar (like endless variations on a constant theme). Two points are important here:

- Fractals represent long-term dynamic stability among many forces in a system, with no single dominant force (virtually a definition of biodiversity and health).
- Fractals are the optimal geometry for doing what natural systems do—collecting, transporting, and diffusing resources, filtering and recycling wastes, etc.

Construction is ultimately about creating environments from which the forces of climatic and ecological change are excluded (temporarily). Every structure conflicts to some degree with both the processes and the forms of nature. Construction optimizes a few select functions; natural systems appear to optimize for diversity. Construction aims for structural permanence; natural systems self-organize stability through change. Both require resource efficiency, but achieve it differently.

Recognizing core qualities of built and living systems can help generate new sustainability strategies, and evaluate existing ones. One strategy is making built systems more fractal or naturalistic in form: biomimicry (to oversimplify). Cyclical buildings, stability-in-change, and literal integration of landscape with building are other strategies.

Sustainable geometries are, I believe, the next evolution for design. Energy-efficient, materials-efficient structures in the same old shapes and the same old locations are unlikely to be enough. Sustainable design's future is at the edge between buildings and landscapes, where the forms necessary for human structures interface with the forms essential to living systems. Designerly preoccupation with appearance—with buildings that stand out, and that privilege machine-look over naturalism—is clearly not helping. We must apply our visual skills to understand how form and function interact, not just in architecture, not just in nature, but at the borders between the two.

Summary and Conclusions

Describing the qualities of the future high-performance green building is an essential and crucial step toward making real progress in this area. Three possible approaches have been described in this chapter: the vernacular vision, the high-technology approach, and the biomimetic model. Each, or a combination, may be able to answer some of the questions faced today by professionals involved in the high-performance green building movement, which, because it is relatively new, is heavily constrained by a narrow knowledge base, limited availability of appropriate technology, and the absence of a clear vision of the future. A robust theory of ecological design is sorely needed, as, fundamentally, that is what the design of high-performance green buildings is about: developing a human environment that functions in a mutually beneficial relationship with its natural surroundings and that exchanges matter and energy in a symbiotic manner.

Notes

1. The editorial by Alex Wilson in the December 2005 issue of EBN provided this definition for passive survivability. Wilson also noted that the requirements for passive survivability and the sustainable design features of many green buildings were remarkably similar.
2. The New Orleans Principles can be found at http://green_reconstruction.buildinggreen.com/documents.
3. The checklist can be found in Wilson (2006).
4. Factor 10, which is now part of European Union policy, is influencing change to a sustainable system of production and consumption.
5. An *ecological footprint* is the land area in hectares or acres that a person or activity needs to function on a continuing basis. The term can also be applied to the built environment, for which a measurement such as ecological footprint per 1,000 square feet (100 square meters) of building area is a potential metric for comparing building impacts. The term was popularized by Wackernagel and Rees (1996).
6. The *ecological rucksack* of a material is the total mass of materials that must be moved to extract a unit mass of the materials, expressed as a ratio. This term was coined by the Wuppertal Institute in Wuppertal, Germany, to draw attention to mass movements of materials that are changing the surface of the planet. Historically, attention has been paid to the impacts of toxic materials such as DDT and PCB, which are harmful in the microgram range. The ecological rucksack concept looks at the other end of the materials spectrum—the megaton range movements of materials to extract resources. The bottom line is that both micrograms of toxic materials and megatons of less harmful materials should be accounted for with respect to their impacts.
7. Described in du Plessis (2003).
8. "Defaulting to nature" is an expression used by Randy Croxton of the Croxton Collaborative in New York City to describe the ability of a well-thought-out, passively designed building to provide heating, cooling, and lighting for its occupants, thus ensuring its operability in spite of being disconnected from external energy sources—for example, the electric power grid.
9. From "Conclusions" in Kibert (2002).
10. Natural systems match the energy source and its quality to energy use first (effectiveness) and then maximize the system's efficiency. Human-designed systems, in contrast, tend to focus on efficiency alone and neglect energy quality, thus often spending high-quality energy (e.g., electricity) on building needs that could be better served by low-quality energy (e.g., medium-temperature heat). Quality is a measure of the flexibility of applications for a particular energy source. Electricity can be used to drive electric motors and generate power to move vehicles, while moderate-temperature heat, below that of the boiling point of water, has limited application and flexibility. The lower the temperature of the heat source, the lower the quality of the energy. Using electricity for water heating has a very low level of effectiveness because it is using high-quality energy in an application that could use low-quality energy sources. Refer to Kibert, Sendzimir, and Guy (2002), chapter 3.
11. The maximum power principle was hypothesized by the late eminent ecologist H.T. Odum, the founder of a branch of ecology known as *systems ecology.* In its simplest form, the principle states that the dominant natural systems are those that pump the most energy. Refer to Kibert, Sendzimir, and Guy (2002), chapter 2.
12. As is often stated in the green building area, there is no waste in nature; all materials are kept in productive use. Of course, this is very simplified and, strictly speaking, not even true.

References

Benyus, Janine M. 1997. *Biomimicry: Innovation Inspired by Nature.* New York: William Morrow.

du Plessis, Chrissna. May 14, 2003. "Boiling Frogs, Sinking Ships, Bursting Dykes and the End of the World as We Know It," *International Electronic Journal of Construction,* Special Issue on Sustainable Construction. Available at www.bcn.ufl.edu.

Kibert, Charles J., Jan Sendzimir, and G. Bradley Guy, eds. 2002. *Construction Ecology: Nature as the Basis for Green Building.* London: Spon Press.

Sorvig, Kim, and Robert Thompson. 2000. *Sustainable Landscape Construction.* Washington, D.C.: Island Press.

Wackernagel, Mathis, and William Rees. 1996. *Our Ecological Footprint.* Gabriola Island, British Columbia: New Society Publishers.

Wilson, Alex. December 2005. "Passive Survivability," *Environmental Building News,* 14(12), p. 2.

Wilson, Alex. May 2006. "Passive Survivability: A New Design Criterion for Buildings," 15(5), pp. 1, 15–16.

Appendix A

Overview of LEED for New Construction 2009 (LEED-NC 2009)

SUSTAINABLE SITES (SS)		Points
Prerequisite 1	Construction Activity Pollution Prevention	Required
Credit 1	Site Selection	1
Credit 2	Development Density & Community Connectivity	5
Credit 3	Brownfield Redevelopment	1
Credit 4.1	Alternative Transportation: Public Transportation Access	6
Credit 4.2	Alternative Transportation: Bicycle Storage & Changing Rooms	1
Credit 4.3	Alternative Transportation: Low Emitting & Fuel Efficient Vehicles	3
Credit 4.4	Alternative Transportation: Parking Capacity	2
Credit 5.1	Site Development: Protect or Restore Habitat	1
Credit 5.2	Site Development: Maximize Open Space	1
Credit 6.1	Stormwater Design: Quantity Control	1
Credit 6.2	Stormwater Design: Quality Control	1
Credit 7.1	Heat Island Effect: Non- Roof	1
Credit 7.1	Heat Island Effect: Roof	1
Credit 8	Light Pollution Reduction	1
	Possible SS Points	**26**

WATER EFFICIENCY (WE)		Points
Prerequisite 1	Water Use Reduction: 20% Reduction	Required
Credit 1.1	Water Efficient Landscaping: Reduce by 50%	2
Credit 1.2	Water Efficient Landscaping: No Potable Water Use or No Irrigation	2
Credit 2	Innovative Wastewater Technologies	2
Credit 3	Water Use Reduction: 30%-40% Reduction	2-4
	Possible WE Points	**10**

ENERGY & ATMOSPHERE (EA)		Points
Prerequisite 1	Fundamental Commissioning of the Building Energy Systems	Required
Prerequisite 2	Minimum Energy Performance	Required
Prerequisite 3	Fundamental Refrigeration Management	Required

*As of June 2007, the USGBC was balloting a recommendation that all buildings must achieve a minimum of two points under Optimize Energy Performance.

Credit 1	Optimize Energy Performance	1–19
Credit 2	On-Site Renewable Energy	1–7
Credit 3	Enhanced Commissioning	2
Credit 4	Enhanced Refrigeration Management	2
Credit 5	Measurement & Verification	3
Credit 6	Green Power	2
	Possible EA Points	**35**

MATERIALS & RESOURCES (MR)		**Points**
Prerequisite 1	Storage and Collection of Recyclables	Required
Credit 1.1-1.3	Building Reuse: Maintain Existing Walls, Floors and Roof	1-3
Credit 1.4	Building Reuse: Maintain 50% of Interior Non- Structural Elements	1
Credit 2.1	Construction Waste Management: Divert 50% from Disposal	1
Credit 2.2	Construction Waste Management: Divert 75% from Disposal	1
Credit 3.1	Materials Reuse: 5%	1
Credit 3.2	Materials Reuse: 10%	1
Credit 4.1	Recycled Content: 10% (post- consumer + 1/2 pre- consumer)	1
Credit 4.2	Recycled Content: 20% (post- consumer + 1/2 pre- consumer)	1
Credit 5.1	Regional Materials: 10% Extracted, Processed & Manufactured Regionally	1
Credit 5.2	Regional Materials: 20% Extracted, Processed & Manufactured Regionally	1
Credit 6	Rapidly Renewable Resources	1
Credit 7	Certified Wood	1
	Possible MR Points	**14**

INDOOR ENVIRONMENTAL QUALITY (EQ)		**Points**
Prerequisite 1	Minimum IAQ Performance	Required
Prerequisite 2	Environmental Tobacco Smoke (ETS) Control	Required
Credit 1	Outdoor Air Delivery Monitoring	1
Credit 2	Increased Ventilation	1
Credit 3.1	Construction IAQ Management Plan: During Construction	1
Credit 3.2	Construction IAQ Management Plan: Before Occupancy	1
Credit 4.1	Low Emitting Materials: Adhesives and Sealants	1
Credit 4.2	Low Emitting Materials: Paints and Coatings	1
Credit 4.3	Low Emitting Materials: Flooring Systems	1
Credit 4.4	Low Emitting Materials: Composite Wood and Agrifiber Products	1
Credit 5	Indoor Chemical and Pollutant Source Control	1
Credit 6.1	Controllability of Systems: Lighting	1
Credit 6.2	Controllability of Systems: Thermal Comfort	1
Credit 7.1	Thermal Comfort: Design	1
Credit 7.2	Thermal Comfort: Verification	1
Credit 8.1	Daylight and Views: Daylight 75% of Spaces	1
Credit 8.2	Daylight and Views: Views for 90% of Spaces	1
	Possible EQ Points	**15**

INNOVATION IN DESIGN (ID)		**Points**
Credit 1.1	Innovation in Design 1	1
Credit 1.2	Innovation in Design 2	1
Credit 1.3	Innovation in Design 3	1
Credit 1.4	Innovation in Design 4	1

Credit 1.5	Innovation in Design 5	1
Credit 2	LEED Accredited Professional	1
	Possible ID Points	**6**

REGIONAL PRIORITY CREDIT (RPC)		**Points**
Credit 1.1	Regional Priority Credit: Region Defined	1
Credit 1.2	Regional Priority Credit: Region Defined	1
Credit 1.3	Regional Priority Credit: Region Defined	1
Credit 1.4	Regional Priority Credit: Region Defined	1
	Possible RPC Points	**4**

Summary for LEED-NC 2009

For LEED-NC 2009, there are 100 possible base points plus 6 points for Innovation in Design and 4 points for Regional Priority Credit, a total of 110 points available.

Certification levels for LEED-NC 2009 are as follows:

Certified: 40–49 points
Silver: 50–59 points
Gold: 60–79 points
Platinum: 80–110 points

Appendix B

Overview of LEED for Existing Buildings: Operations and Maintenance 2009 (LEED-EB O&M 2009)

SUSTAINABLE SITES (SS)		Points
Credit 1	LEED Certified Design and Construction	4
Credit 2	Building Exterior and Hardscape Management Plan	1
Credit 3	Integrated Pest Management, Erosion Control, and Landscape Management Plan	1
Credit 4.1	Alternative Commuting Transportation: 10% Reduction	3
Credit 4.2	Alternative Commuting Transportation: 25% Reduction	7
Credit 4.3	Alternative Commuting Transportation: 50% Reduction	11
Credit 4.4	Alternative Commuting Transportation: 75% Reduction	15
Credit 5	Site Development: Protect or Restore Open Habitat	1
Credit 6	Stormwater Design: Quantity Control	1
Credit 7.1	Heat Island Effect: Non- Roof	1
Credit 7.1	Heat Island Effect: Roof	1
Credit 8	Light Pollution Reduction	1
	Possible SS Points	**26**

WATER EFFICIENCY (WE)		Points
Prerequisite 1	Minimum Indoor Plumbing Fixture and Fitting Efficiency	Required
Credit 1.1	Water Performance Measurement: Whole Building Metering	1
Credit 1.2	Water Performance Measurementt: Submetering	1
Credit 2	Additional Indoor Plumbing Fixture and Fitting Efficiency	1–5
Credit 3	Water Efficient Landscaping	1–5
Credit 4.1	Cooling Tower Water Management: Chemical Management	1
Credit 4.2	Non-Potable Water Source Use	1
	Possible WE Points	**14**

ENERGY & ATMOSPHERE (EA)		Points
Prerequisite 1	Energy Efficient Best Management Practices: Planning, Documentation, and Opportunity Assessment	Required
Prerequisite 2	Minimum Energy Efficiency Performance: Energy Star Rating: 69	Required
Prerequisite 3	Refrigeration Management: Ozone Protection	Required
Credit 1	Optimize Energy Efficiency Performance	1–18
Credit 2.1	Existing Building Commissioning: Investigation and Analysis	2

Credit 2.2	Existing Building Commissioning: Implementation	2
Credit 2.3	Existing Building Commissioning: Ongoing Commissioning	2
Credit 3.1	Performance Measurement: Building Automation System	1
Credit 3.2	Performance Measurement: System-Level Metering	1–2
Credit 4	Renewable Energy	1–6
Credit 5	Refrigeration Management	1
Credit 6	Emissions Reduction Reporting	1
	Possible EA Points	**35**

MATERIALS & RESOURCES (MR)		**Points**
Prerequisite 1	Sustainable Purchasing Policy	Required
Prerequisite 2	Solid Waste Management Policy	Required
Credit 1	Sustainable Purchasing: Ongoing Consumables 40% of Purchases	1
Credit 2.1	Sustainable Purchasing: Durable Goods, Electric	1
Credit 2.2	Sustainable Purchasing: Durable Goods, Furniture	1
Credit 3	Sustainable Purchasing: Facility Alterations and Additions	1
Credit 4	Sustainable Purchasing: Reduced Mercury in Lamps 90 pg/lum-hr	1
Credit 5	Sustainable Purchasing: Food	1
Credit 6	Solid Waste Management: Waste Stream Audit	1
Credit 7	Solid Waste Management: Ongoing Consumables 50% Waste Diversion	1
Credit 8	Solid Waste Management: Durable Goods	1
Credit 9	Solid Waste Management: Facility Alterations and Additions	1
	Possible MR Points	**10**

INDOOR ENVIRONMENTAL QUALITY (EQ)		**Points**
Prerequisite 1	Minimum IAQ Performance	Required
Prerequisite 2	Environmental Tobacco Smoke (ETS) Control	Required
Prerequisite 3	Green Cleaning Policy	Required
Credit 1.1	IAQ Best Management Practices: IAQ Management Program	1
Credit 1.2	IAQ Best Management Practices: Outdoor Air Delivery Monitoring	1
Credit 1.3	IAQ Best Management Practices: Increased Ventilation	1
Credit 1.4	IAQ Best Management Practices: Reduce Particulates in Air Distribution	1
Credit 1.5	IAQ Management Plan: During Construction	1
Credit 2.1	Occupant Comfort: Occupant Survey	1
Credit 2.2	Controllability of Systems: Lighting	1
Credit 2.3	Occupant Comfort: Thermal Comfort Monitoring	1
Credit 2.4	Occupant Comfort: Daylight and Views: 50% Daylight / 45% Views	1
Credit 3.1	Green Cleaning: High Performance Cleaning Program	1
Credit 3.2	Green Cleaning: Custodial Effectiveness Assessment–Score of < 3	1
Credit 3.3	Green Cleaning: Sustainable Cleaning Products and Materials Purchases	1
Credit 3.4	Green Cleaning: Sustainable Cleaning Equipment	1
Credit 3.5	Green Cleaning: Indoor Chemical and Pollutant Source Control	1
Credit 3.6	Green Cleaning: Indoor Integrated Pest Management	1
	Possible EQ Points	**15**

INNOVATION IN OPERATION UPGRADES & MAINTENANCE (ID)		Points
Credit 1.1	Innovation in Design 1	1
Credit 1.2	Innovation in Design 2	1
Credit 1.3	Innovation in Design 3	1
Credit 1.4	Innovation in Design 4	1
Credit 2	LEED Accredited Professional	1
Credit 3	Documenting Sustainable Building Cost Impacts	1
	Possible ID Points	**6**

REGIONAL PRIORITY CREDIT (RPC)		Points
Credit 1.1	Regional Priority Credit: Region Defined	1
Credit 1.2	Regional Priority Credit: Region Defined	1
Credit 1.3	Regional Priority Credit: Region Defined	1
Credit 1.4	Regional Priority Credit: Region Defined	1
	Possible RPC Points	**4**

Summary for LEED-EB O&M 2009

For LEED-EB O&M 2009, there are 100 possible base points plus 6 points for Innovation in Design and 4 points for Regional Priority Credit, a total of 110 points available.

Certification levels for LEED-EB O&M 2009 are as follows:

Certified:	40–49 points
Silver:	50–59 points
Gold:	60–79 points
Platinum:	80–110 points

Appendix C

Overview of LEED for Core and Shell 2009 (LEED-CS 2009)

SUSTAINABLE SITES (SS)		Points
Prerequisite 1	Construction Activity Pollution Prevention	Required
Credit 1	Site Selection	1
Credit 2	Development Density & Community Connectivity	5
Credit 3	Brownfield Redevelopment	1
Credit 4.1	Alternative Transportation: Public Transportation Access	6
Credit 4.2	Alternative Transportation: Bicycle Storage & Changing Rooms	2
Credit 4.3	Alternative Transportation: Low Emitting & Fuel Efficient Vehicles	3
Credit 4.4	Alternative Transportation: Parking Capacity	2
Credit 5.1	Site Development: Protect or Restore Habitat	1
Credit 5.2	Site Development: Maximize Open Space	1
Credit 6.1	Stormwater Design: Quantity Control	1
Credit 6.2	Stormwater Design: Quality Control	1
Credit 7.1	Heat Island Effect: Non- Roof	1
Credit 7.1	Heat Island Effect: Roof	1
Credit 8	Light Pollution Reduction	1
Credit 9	Tenant Design & Construction Guidelines	1
	Possible SS Points	**28**

WATER EFFICIENCY (WE)		Points
Prerequisite 1	Water Use Reduction: 20% Reduction	Required
Credit 1.1	Water Efficient Landscaping: Reduce by 50%	2
Credit 1.2	Water Efficient Landscaping: No Potable Water Use or No Irrigation	2
Credit 2	Innovative Wastewater Technologies	2
Credit 3	Water Use Reduction: 30%-40% Reduction	2–4
	Possible WE Points	**10**

ENERGY & ATMOSPHERE (EA)		Points
Prerequisite 1	Fundamental Commissioning of the Building Energy Systems	Required
Prerequisite 2	Minimum Energy Performance: 10% New Bldgs or 5% for Existing Bldgs	Required
Prerequisite 3	Fundamental Refrigeration Management	Required
Credit 1	Optimize Energy Performance	3–21
Credit 2	On-Site Renewable Energy	4
Credit 3	Enhanced Commissioning	2
Credit 4	Enhanced Refrigeration Management	2
Credit 5.1	Measurement & Verification: Base Building	3

Credit 5.2	Measurement & Verification: Tenant Submetering	3
Credit 6	Green Power	2
	Possible EA Points	**37**

MATERIALS & RESOURCES (MR)		**Points**
Prerequisite 1	Storage and Collection of Recyclables	Required
Credit 1.1–1.5	Building Reuse: Maintain Existing Walls, Floors and Roof	1–5
Credit 2.1	Construction Waste Management: Divert 50% from Disposal	1
Credit 2.2	Construction Waste Management: Divert 75% from Disposal	1
Credit 3.1	Materials Reuse: 5%	1
Credit 4.1	Recycled Content: 10% (post- consumer (½ pre- consumer)	1
Credit 4.2	Recycled Content: 20% (post- consumer (½ pre- consumer)	1
Credit 5.1	Regional Materials: 10% Extracted, Processed & Manufactured Regionally	1
Credit 5.2	Regional Materials: 20% Extracted, Processed & Manufactured Regionally	1
Credit 7	Certified Wood	1
	Possible MR Points	**13**

INDOOR ENVIRONMENTAL QUALITY (EQ)		**Points**
Prerequisite 1	Minimum IAQ Performance	Required
Prerequisite 2	Environmental Tobacco Smoke (ETS) Control	Required
Credit 1	Outdoor Air Delivery Monitoring	1
Credit 2	Increased Ventilation	1
Credit 3	Construction IAQ Management Plan: During Construction	1
Credit 4.1	Low Emitting Materials: Adhesives and Sealants	1
Credit 4.2	Low Emitting Materials: Paints and Coatings	1
Credit 4.3	Low Emitting Materials: Flooring Systems	1
Credit 4.4	Low Emitting Materials: Composite Wood and Agrifiber Products	1
Credit 5	Indoor Chemical and Pollutant Source Control	1
Credit 6.1	Controllability of Systems: Thermal Comfort	1
Credit 7	Thermal Comfort: Design	1
Credit 8.1	Daylight and Views: Daylight 75% of Spaces	1
Credit 8.2	Daylight and Views: Views for 90% of Spaces	1
	Possible EQ Points	**12**

INNOVATION IN DESIGN (ID)		**Points**
Credit 1.1	Innovation in Design 1	1
Credit 1.2	Innovation in Design 2	1
Credit 1.3	Innovation in Design 3	1
Credit 1.4	Innovation in Design 4	1
Credit 1.5	Innovation in Design 5	1
Credit 2	LEED Accredited Professional	1
	Possible ID Points	**6**

REGIONAL PRIORITY CREDIT (RPC)		**Points**
Credit 1.1	Regional Priority Credit: Region Defined	1
Credit 1.2	Regional Priority Credit: Region Defined	1
Credit 1.3	Regional Priority Credit: Region Defined	1
Credit 1.4	Regional Priority Credit: Region Defined	1
	Possible RPC Points	**4**

Summary for LEED- CS 2009

For LEED-CS 2009, there are 100 possible base points plus 6 points for Innovation in Design and 4 points for Regional Priority Credit, a total of 110 points available.

Certification levels for LEED-CS 2009 are as follows:

Certified: 40–49 points
Silver: 50–59 points
Gold: 60–79 points
Platinum: 80–110 points

Appendix D

Overview of LEED for Commercial Interiors 2009 (LEED-CI 2009)

SUSTAINABLE SITES (SS)		Points
Credit 1	Site Selection: Select a LEED Certified Building	5
OR Locate the tenant space in a building with certain characteristics		1–5
Credit 2	Development Density and Community Connectivity	6
Credit 3.1	Alternative Transportation: 10% Public Transportation Access	6
Credit 3.2	Alternative Transportation: Bicycle Storage and Changing Rooms	2
Credit 3.3	Alternative Transportation: Parking Availability	2
	Possible SS Points	**21**

WATER EFFICIENCY (WE)		Points
Prerequisite 1	Water Use Reduction, 20% Reduction	Required
Credit 1	Water Use Reduction, 30 to 40%	6 to 11
	Possible WE Points	**11**

ENERGY & ATMOSPHERE (EA)		Points
Prerequisite 1	Fundamental Commissioning of the Building Energy Systems	Required
Prerequisite 2	Minimum Energy Performance	Required
Prerequisite 3	Fundamental Refrigeration Management	Required
Credit 1	Optimize Energy Performance: Lighting Power	1–5
Credit 1.2	Optimize Energy Performance: Lighting Controls	1–3
Credit 1.3	Optimize Energy Performance: HVAC	5–10
Credit 1.4	Optimize Energy Performance: Equipment and Appliances	1–4
Credit 3	Enhanced Commissioning	5
Credit 5	Measurement & Verification	2–5
Credit 6	Green Power	5
	Possible EA Points	**37**

MATERIALS & RESOURCES (MR)		Points
Prerequisite 1	Storage and Collection of Recyclables	Required
Credit 1.1–1.3	Building Reuse: Maintain Interior Non- Structural Components	1–3
Credit 2.1	Construction Waste Management: Divert 50% from Disposal	1
Credit 2.2	Construction Waste Management: Divert 75% from Disposal	1
Credit 3.1	Materials Reuse: 5%	1
Credit 3.2	Materials Reuse: 10%	1
Credit 3.3	Materials Reuse: 30% Furniture and Furnishings	1
Credit 4.1	Recycled Content: 10% (post- consumer (½ pre- consumer)	1
Credit 4.2	Recycled Content: 20% (post- consumer (½ pre- consumer)	1

Credit 5.1	Regional Materials: 10% Extracted, Processed & Manufactured Regionally	1
Credit 5.2	Regional Materials: 20% Extracted, Processed & Manufactured Regionally	1
Credit 6	Rapidly Renewable Resources	1
Credit 7	Certified Wood	1
	Possible MR Points	**14**

INDOOR ENVIRONMENTAL QUALITY (EQ)		**Points**
Prerequisite 1	Minimum IAQ Performance	Required
Prerequisite 2	Environmental Tobacco Smoke (ETS) Control	Required
Credit 1	Outdoor Air Delivery Monitoring	1
Credit 2	Increased Ventilation	1
Credit 3.1	Construction IAQ Management Plan: During Construction	1
Credit 3.2	Construction IAQ Management Plan: Before Occupancy	1
Credit 4.1	Low Emitting Materials: Adhesives and Sealants	1
Credit 4.2	Low Emitting Materials: Paints and Coatings	1
Credit 4.3	Low Emitting Materials: Flooring Systems	1
Credit 4.4	Low Emitting Materials: Composite Wood and Agrifiber Products	
Credit 4.5	Low Emitting Materials: Systems Furniture and Seating	1
Credit 5	Indoor Chemical and Pollutant Source Control	1
Credit 6.1	Controllability of Systems: Lighting	1
Credit 6.2	Controllability of Systems: Thermal Comfort	1
Credit 7.1	Thermal Comfort: Design	1
Credit 7.2	Thermal Comfort: Verification	1
Credit 8.1	Daylight and Views: Daylight 75% of Spaces	1
Credit 8.2	Daylight and Views: Views for 90% of Spaces	1
Credit 8.3	Daylight and Views: Views for 90% of Seated Spaces	1
	Possible EQ Points	**17**

INNOVATION IN DESIGN (ID)		**Points**
Credit 1.1	Innovation in Design 1	1
Credit 1.2	Innovation in Design 2	1
Credit 1.3	Innovation in Design 3	1
Credit 1.4	Innovation in Design 4	1
Credit 1.5	Innovation in Design 5	1
Credit 2	LEED Accredited Professional	1
	Possible ID Points	**6**

REGIONAL PRIORITY CREDIT (RPC)		**Points**
Credit 1.1	Regional Priority Credit: Region Defined	1
Credit 1.2	Regional Priority Credit: Region Defined	1
Credit 1.3	Regional Priority Credit: Region Defined	1
Credit 1.4	Regional Priority Credit: Region Defined	1
	Possible RPC Points	**4**

Summary for LEED-CI 2009

For LEED-CI 2009, there are 100 possible base points plus 6 points for Innovation in Design and 4 points for Regional Priority Credit, a total of 110 points available.

Certification levels for LEED-CI 2009 are as follows:

Certified:	40–49 points
Silver:	50–59 points
Gold:	60–79 points
Platinum:	80–110 points

Appendix E

Overview of LEED for Schools 2009 (LEED-S 2009)

SUSTAINABLE SITES (SS)		Points
Prerequisite 1	Construction Activity Pollution Prevention	Required
Prerequisite 2	Environmental Site Assessment	Required
Credit 1	Site Selection	1
Credit 2	Development Density & Community Connectivity	4
Credit 3	Brownfield Redevelopment	1
Credit 4.1	Alternative Transportation: Public Transportation Access	4
Credit 4.2	Alternative Transportation: Bicycle Storage & Changing Rooms	1
Credit 4.3	Alternative Transportation: Low Emitting & Fuel Efficient Vehicles	2
Credit 4.4	Alternative Transportation: Parking Capacity	2
Credit 5.1	Site Development: Protect or Restore Habitat	1
Credit 5.2	Site Development: Maximize Open Space	1
Credit 6.1	Stormwater Design: Quantity Control	1
Credit 6.2	Stormwater Design: Quality Control	1
Credit 7.1	Heat Island Effect: Non- Roof	1
Credit 7.1	Heat Island Effect: Roof	1
Credit 8	Light Pollution Reduction	1
Credit 9	Site Master Plan	1
Credit 10	Joint Use of Facilities	1
	Possible SS Points	**24**

WATER EFFICIENCY (WE)		Points
Prerequisite 1	Water Use Reduction: 20% Reduction	Required
Credit 1.1	Water Efficient Landscaping: Reduce by 50%	2
Credit 1.2	Water Efficient Landscaping: No Potable Water Use or No Irrigation	2
Credit 2	Innovative Wastewater Technologies	2
Credit 3	Water Use Reduction: 30%-40% Reduction	2–4
Credit 4	Process Water Use Reduction, 20% Reduction	1
	Possible WE Points	**11**

ENERGY & ATMOSPHERE (EA)		Points
Prerequisite 1	Fundamental Commissioning of the Building Energy Systems	Required
Prerequisite 2	Minimum Energy Performance	Required
Prerequisite 3	Fundamental Refrigeration Management	Required
Credit 1	Optimize Energy Performance	1–19
Credit 2	On-Site Renewable Energy	1–7
Credit 3	Enhanced Commissioning	2

Credit 4	Enhanced Refrigeration Management	1
Credit 5	Measurement & Verification	2
Credit 6	Green Power	2
	Possible EA Points	**33**

MATERIALS & RESOURCES (MR)		**Points**
Prerequisite 1	Storage and Collection of Recyclables	Required
Credit 1.1–1.2	Building Reuse: Maintain Existing Walls, Floors and Roof	1–2
Credit 1.3	Building Reuse: Maintain 50% of Interior Non- Structural Elements	1
Credit 2.1	Construction Waste Management: Divert 50% from Disposal	1
Credit 2.2	Construction Waste Management: Divert 75% from Disposal	1
Credit 3.1	Materials Reuse: 5%	1
Credit 3.2	Materials Reuse: 10%	1
Credit 4.1	Recycled Content: 10% (post- consumer (½ pre- consumer)	1
Credit 4.2	Recycled Content: 20% (post- consumer (½ pre- consumer)	1
Credit 5.1	Regional Materials: 10% Extracted, Processed & Manufactured Regionally	1
Credit 5.2	Regional Materials: 20% Extracted, Processed & Manufactured Regionally	1
Credit 6	Rapidly Renewable Resources	1
Credit 7	Certified Wood	1
	Possible MR Points	**13**

INDOOR ENVIRONMENTAL QUALITY (EQ)		**Points**
Prerequisite 1	Minimum IAQ Performance	Required
Prerequisite 2	Environmental Tobacco Smoke (ETS) Control	Required
Prerequisite 3	Minimum Acoustical Performance	Required
Credit 1	Outdoor Air Delivery Monitoring	1
Credit 2	Increased Ventilation	1
Credit 3.1	Construction IAQ Management Plan: During Construction	1
Credit 3.2	Construction IAQ Management Plan: Before Occupancy	1
Credit 4	Low Emitting Materials:	1–4
Credit 4.1	Adhesives and Sealants	
Credit 4.2	Paints and Coatings	
Credit 4.3	Flooring Systems	
Credit 4.4	Composite Wood and Agrifiber Products	
Credit 4.5	Furniture	
Credit 5	Indoor Chemical and Pollutant Source Control	1
Credit 6.1	Controllability of Systems: Lighting	1
Credit 6.2	Controllability of Systems: Thermal Comfort	1
Credit 7.1	Thermal Comfort: Design	1
Credit 7.2	Thermal Comfort: Verification	1
Credit 8.1	Daylight and Views: Daylight	1–3
Credit 8.2	Daylight and Views: Views for 90% of Spaces	1
Credit 9	Enhanced Acoustical Performance	1
Credit 10	Mold Prevention	1
	Possible EQ Points	**19**

INNOVATION IN DESIGN (ID)		**Points**
Credit 1.1	Innovation in Design 1	1
Credit 1.2	Innovation in Design 2	1
Credit 1.3	Innovation in Design 3	1
Credit 1.4	Innovation in Design 4	1

Credit 1.5	Innovation in Design 5	1
Credit 2	LEED Accredited Professional	1
	Possible ID Points	**6**

REGIONAL PRIORITY CREDIT (RPC)		**Points**
Credit 1.1	Regional Priority Credit: Region Defined	1
Credit 1.2	Regional Priority Credit: Region Defined	1
Credit 1.3	Regional Priority Credit: Region Defined	1
Credit 1.4	Regional Priority Credit: Region Defined	1
	Possible RPC Points	4

Summary for LEED-S 2009

For LEED-S 2009, there are 100 possible base points plus 6 points for Innovation in Design and 4 points for Regional Priority Credit, a total of 110 points available.

Certification levels for LEED-S 2009 are as follows:

Certified:	40–49 points
Silver:	50–59 points
Gold:	60–79 points
Platinum:	80–110 points

Appendix F

Green Globes v.1

The Green Globes v.1 Assessment Protocol is organized into seven sections:

A. Project Management—Policies and Practices (50 points)
B. Site (115 points)
C. Energy (360 points)
D. Water (100 points)
E. Resources, Building Materials, and Solid Waste (100 points)
F. Emissions and Effluents (75 points)
G. Indoor Environmental Quality (200 points)

A total of 1000 points is available, and an award of one to four Green Globes can be made if the project undergoes third-party certification by a Verifier who visits the project, interacts with the project team, and audits the points claimed by the project team. For Green Globes, unlike LEED-NC 2.2, the number of points actually achievable is variable, depending on the project conditions. One Green Globe is awarded for attaining at least 35 percent of the total achievable points; two Green Globes, for 55 percent; three Green Globes, for 70 percent; and four Green Globes, for 85 percent. The seven sections of Green Globes v.1 are outlined below, together with the points structure for each section.

SECTION A

Project Management—Policies and Procedures (50 points)

Section	Description	Points
A.1	**Integrated Design**	**20**
A.1.1	Green Design Coordinator designated	3
A.1.2	Project Initiation Stage collaboration	3
A.1.3	Sustainability performance goals set during the Project Initiation Stage	5
A.1.4	At least two collaboration sessions held prior to preparation of contract documentation	5
A.1.5	Records of decisions and to-do lists distributed	2
A.1.6	Green Design Coordinator provided information to the client	2
A.2	**Environmental Purchasing**	**5**
A.2.1	Third-party certified environmentally preferable products specified	5

SECTION A continued

Project Management—Policies and Procedures (50 points)

Section	Description	Points
A.3	**Commissioning—Documentation**	**20**
A.3.1	Best practice commissioning procedures implemented	20
A.4	**Emergency Response Plan (ERP)**	**5**
A.4.1	Environmental goals and procedures for ERP for site preparation phase of construction	5
	Total Points Available	**50**

SECTION B

Site (115 points)

Section	Description	Points
B.1	**Site Development Area**	**45**
B.1.1	Site location	30
B.1.2	Avoiding sensitive locations	15
B.2	**Reduce Ecological Impacts**	**40**
B.2.1	Leaving slopes greater than 15 percent undisturbed	2
B.2.2	Limiting construction site activities	2
B.2.3	Protection of trees	1
B.2.4	Erosion control Best Management Practices	8
B.2.5	Heat island mitigation, site hardscape	10
B.2.6	Heat island mitigation, roof	10
B.2.7	Light trespass and sky glow minimized	7
B.3	**Enhancement of Watershed Features**	**15**
B.3.1	Controlling stormwater from damaging the project or waterways	10
B.3.2	Roof runoff control	5
B.4	**Site Ecology Improvement**	**15**
B.4.1	Native landscape vegetation	6
B.4.2	Minimizing turf	5
B.4.3	Design to avoid bird collisions	4
	Total Points Available	**115**

There are two paths for compliance for Section C. Path A is for buildings over 20,000 square feet for which sections C.1, C.2, C.4. and C.5 must be completed. Path B is for buildings less than or equal to 20,000 square feet for which sections C.1, C.2, C.4, and C.5 OR sections C.2, C.3, C.4, and C.5 must be completed.

SECTION C

Energy (360 points)

Section	Description	Points
C.1	**Energy Consumption**	**110**
C.1.1	Building energy consumption vs. baseline	100
C.1.2	Energy modeling	10
C.2	**Energy Demand Minimization**	**135**
	Response to Microclimate and Topography	30
C.2.1	Optimization for microclimatic conditions	13
C.2.2	Wind mitigating measures	5
C.2.3	Natural ventilation system	12
	Daylighting	30
C.2.4	Daylighting strategy	20
C.2.5	Glazing strategy for daylighting	10
	Building Envelope	40
C.2.6	Thermal resistance of envelope—walls	5
C.2.7	Thermal resistance of envelope—roof	5
C.2.8	Thermal resistance of envelope—windows	7
C.2.9	Window solar heat gain coefficient (SHGC)	6
C.2.10	Building air barrier	9
C.2.11	Vapor retarder best practices	8
	Building Controls and Energy Metering	35
C.2.12	Energy submetering	7
C.2.13	Daylighting controls	4
C.2.14	Lighting control zones	4
C.2.15	Room occupancy lighting controls	4
C.2.16	Building automation system	5
C.2.17	Natural ventilation controls	3
C.2.18	Outside air damper controls	6
C.2.19	Vertical transport energy conservation	2
C.3	**Right Sized Energy Systems (Path B only)**	**110**
C.3.1	Lighting power density	15
C.3.2	Energy efficient lighting system	10
C.3.3	High efficiency cooling equipment	25
C.3.4	Part load conditioning strategy	10
C.3.5	High efficiency heating equipment	25
C.3.6	High efficiency heat pump	5
C.3.7	Thermal zoning	5
C.3.8	Efficient air handling system	15
C.4	**Renewable Energy**	**45**
C.4.1	Percent renewable energy	45
C.5	**Energy-Efficient Transportation**	**70**
C.5.1	Accessible public transit	50
C.5.2	Car/van pool parking	6
C.5.3	Covered secure bicycle storage	10
C.5.4	Showers and changing facilities	4
	Total Points Available	**360**

SECTION D

Water (100 points)

Section	Description	Points
D.1 Water	**40**	
D.1.1	Potable water reduction compared to EPAct of 1992	40
D.2 Water-Conserving Features		**40**
D.2.1	Sub-metering of high water use operations	5
D.2.2	Minimizing cooling tower water consumption	5
D.2.3	Minimizing potable water for landscape irrigation	10
D.2.4	Water-efficient landscape irrigation	3
D.2.5	Water-efficient landscape	5
D.2.6	Landscaping without lawn	5
D.2.7	Lawn only for functional purposes (in lieu of D2.6)	1
D.3 Reduce Off-Site Treatment of Water		**20**
D.3.1	Graywater system	10
D.3.2	Blackwater system or composting toilets	10
	Total Points Available	**100**

SECTION E

Resources, Building Materials, and Solid Waste (100 points)

Section	Description	Points
E.1 Materials with Low Environmental Impact		**40**
E.1.1	Use of LCA to select building assemblies	40
E.2 Minimize Consumption and Depletion of Material Resources		**30**
E.2.1	Reused building materials	10
E.2.2	Post-consumer recycled content	10
E.2.3	Use of bio-based materials	5
E.2.4	Certified wood products	5
E.3 Reuse of Existing Structures		**10**
E.3.1	Reuse of existing facade	5
E.3.2	Reuse of existing structure	5
E.4 Building Durability, Adaptability, and Disassembly		**10**
E.4.1	Envelope design to control rain penetration	2
E.4.2	Measures to prevent groundwater intrusion	2
E.4.3	Promotion of adaptability in the design	3
E.4.4	Design of building for disassembly	3
E.5 Reduction, Reuse, and Recycling of Waste		**10**
E.5.1	Diversion of construction, demolition, and renovation waste	6
E.5.2	Designated storage space for recyclable waste in building	3
E.5.3	Space for recycling dumpster	1
	Total Available Points	**100**

SECTION F

Emissions and Other Impacts (200 points)

Section	Description	
F.1	**Air Emissions**	**15**
F.1.1	Low nitrous oxide, low carbon monoxide boilers	15
F.2	**Ozone Depletion and Global Warming**	**30**
F.2.1	Project completely avoids use of ozone depleting and global warming refrigerants OR F2.2, F.2.3., and F2.4	30
F.2.2	Relative ozone depletion and global warming effects of HFC and HCFC refrigerants	15
F.2.3	Global warming potential (GWP) of refrigerant is less than 150	5
F.2.4	Leak detection for refrigerants	5
F.3	**Contamination of Sewers or Waterways**	**12**
F.3.1	Measures to prevent contamination from entering sewers or waterways	12
F.4	**Water and Land Pollution**	**9**
F.4.1	For retrofits, storage tanks are safe	3
F.4.2	For retrofits, PCBs meet regulatory requirements	1
F.4.3	For retrofits, asbestos has been removed or abated	2
F.4.4	Measure to prevent radon and methane entry	3
F.5	**Integrated Pest Management**	**4**
F.5.1	Measure to avoid pest infestations	4
F.6	**Storage of Hazardous Materials**	**5**
F.6.1	Storage area designed to prevent hazardous materials from entering occupied spaces	3
F.6.2	Fire-rated storage for flammable materials	2
	Total Points Available	**75**

SECTION G

Indoor Environmental Quality (200 points)

Section	Description	Points
G.1	**Effective Ventilation System**	**60**
G.1.1	Features to avoid entraining pollutants in the ventilation air intakes	11
G.1.2	Ventilation meets ASHRAE 62.1-2004 requirements	10
G.1.3	Effective air exchange	15
G.1.4	Indoor air quality monitoring	7
G.1.5	Indoor Air Quality Management Plan	11
G.1.6	Efficient air filtering	6
G.2	**Source Control of Indoor Pollutants**	**45**
G.2.1	Measures to control moisture and prevent biological contaminant growth	4
G.2.2	Access to air handling units for inspection and maintenance	5
G.2.3	Humidification system features to prevent biological contaminant growth	6

SECTION G continued

Indoor Environmental Quality (200 points)

Section	Description	Points
G.2.4	Carbon monoxide monitoring in parking garages and areas where there is combustion	5
G.2.5	Separate ventilation system or physical isolation of indoor air pollution at source for areas with printing, copying, and other similar activities	3
G.2.6	Separate ventilation system designed with negative pressure	2
G.2.7	Measures to minimize microbial contamination from cooling towers	5
G.2.8	Measures to minimize microbial contamination of the domestic ater system	5
G.2.9	Low-VOC emitting materials	10
G.3	**Lighting Design and Integration of Lighting Systems**	**45**
G.3.1	Extent of daylighting	10
G.3.2	View to building exterior or atria	10
G.3.3	Solar shading devices	5
G.3.4	Lighting levels in accordance with **IESNA Lighting Handbook 2000**	10
G.3.5	Measures to avoid excessive glare	10
G.4	**Thermal Comfort**	**25**
G.4.1	Thermal comfort in accordance with ASHRAE 55-2004 or per occupant satisfaction survey	20
G.4.2	Size of thermal comfort zones	5
G.5	**Acoustic Comfort**	**25**
G.5.1	Sound levels below 65 dB at property line	5
G.5.2	Appropriate Sound Transmission Class (STC) levels for the building envelope	5
G.5.3	Noise attenuation in structure and insulation of primary spaces from impact noise	5
G.5.4	Interior design meets appropriate ambient noise levels	5
G.5.5	Measures to mitigate mechanical and plumbing system noise	5
	Total Points Available	**200**

INDEX

For these and other Wiley books on sustainable design, visit www.wiley.com/go/sustainabledesign

Alternative Construction: Contemporary Natural Building Methods
by Lynne Elizabeth and Cassandra Adams

Cities People Planet: Liveable Cities for a Sustainable World
by Herbert Girardet

Design with Nature
by Ian L. McHarg

Ecodesign: A Manual for Ecological Design
by Ken Yeang

Green Building Materials: A Guide to Product Selection and Specification, Second Edition
by Ross Spiegel and Dru Meadows

Green Development: Integrating Ecology and Real Estate
by Rocky Mountain Institute

The HOK Guidebook to Sustainable Design, Second Edition
by Sandra Mendler, William O'Dell, and Mary Ann Lazarus

Land and Natural Development (Land) Code
by Diana Balmori and Gaboury Benoit

Sustainable Construction: Green Building Design and Delivery
by Charles J. Kibert

Sustainable Commercial Interiors
by Penny Bonda and Katie Sosnowchik

Sustainable Design: Ecology, Architecture, and Planning
by Daniel Williams

Sustainable Healthcare Architecture
by Robin Guenther and Gail Vittori

Sustainable Residential Interiors
by Associates III

Sustainable Urbanism: Urban Design with Nature
by Douglas Farr